Electrical Wiring

RESIDENTIAL

Based on the 1999
NATIONAL ELECTRICAL CODE®

Electrical Wiring

RESIDENTIAL

Thirteenth Edition

RAY C. MULLIN

Delmar Publishers

an International Thomson Publishing company I(T)P®

Albany • Bonn • Boston • Cincinnati • Detroit • London • Madrid
Melbourne • Mexico City • New York • Pacific Grove • Paris • San Francisco
Singapore • Tokyo • Toronto • Washington

NOTICE TO THE READER

Cover Design: Courtesy of Brucie Rosch

Delmar Staff:
Publisher: Alar Elken
Acquisitions Editor: Mark Huth
Developmental Editor: Jeanne Mesick
Project Editor: Megeen Mulholland
Production Coordinator: Toni Bolognino
Art and Design Coordinator: Cheri Plasse
Editorial Assistant: Dawn Daugherty

COPYRIGHT © 1999
By Delmar Publishers
an International Thomson Publishing company IP®

The ITP logo is a trademark under license
Printed in the United States of America

For more information, contact:

Delmar Publishers
3 Columbia Circle, Box 15015
Albany, New York 12212-5015

International Thomson Publishing Europe
Berkshire House 168-173
168-173 High Holborn
London, WC1V 7AA
United Kingdom

Nelson ITP, Australia
102 Dodds Street
South Melbourne,
Victoria, 3205 Australia

Nelson Canada
1120 Birchmont Road
Scarborough, Ontario
M1K 5G4, Canada

International Thomson Publishing France
Tour Maine-Montparnasse
33 Avenue du Maine
75755 Paris Cedex 15, France

International Thomson Editores
Seneca 53
Colonia Polanco
11560 Mexico D. F. Mexico

International Thomson Publishing GmbH
Königswinterer Strasse 418
53227 Bonn
Germany

International Thomson Publishing Asia
60 Albert Street
#15-01 Albert Complex
Singapore 189969

International Thomson Publishing Japan
Hirakawa-cho Kyowa Building, 3F
2-2-1 Hirakawa-cho, Chiyoda-ku,
Tokyo 102, Japan

ITE Spain/Parninfo
Calle Magallanes, 25
28015-Madrid, Espana

1 2 3 4 5 6 7 8 9 10 XXX 03 02 01 00 99 98

Library of Congress Cataloging-in-Publication Data

Mullin, Ray C.
Electrical wiring, residential / Ray C. Mullin. — 13th ed.
    p. cm.
"Based on the 1999 national electrical code."
ISBN 0-8273-8610-9 (HC: alk. paper) ISBN 0-8273-8607-9 (PB: alk. paper)
1. Electric wiring, Interior. 2. Dwelling—Electric equipment.
I. Title.
TK3285.M84 1998
621.319'24—dc21   98-4135
                                                           CIP

CONTENTS

Plans for Single-Family Dwelling

Sheet 1 of 10 Basement Plan
Sheet 2 of 10 Floor Plan
Sheet 3 of 10 South Elevation; Window Schedule, Door Schedule
Sheet 4 of 10 East Elevation; Interior Elevations
Sheet 5 of 10 Plot Plan; North Elevation
Sheet 6 of 10 West Elevation; Room Finish Schedule
Sheet 7 of 10 Section A-A
Sheet 8 of 10 Electrical Basement Plan
Sheet 9 of 10 Electrical Floor Plan
Sheet 10 of 10 Code Requirements for Swimming Pool Wiring
Blank Floor Plan First Floor
Blank Floor Plan Basement

THE ROLE OF THE ELECTRICAL INSPECTOR AND THE IMPORTANCE OF PROPER TRAINING

When one considers how different groups within the electrical industry affect the end product used by consumers, it is easy to see the importance of those groups working together in a coordinated manner. From the time electrical products are designed to the time when electrical installations are completed or when electrical equipment is purchased by the consumer, many steps must be taken to make it all possible. During this process, close attention to safety standards must be given to the design, installation, and use of electrical products. Product safety standards include testing guidelines that can be used to help ensure that products perform in a safe manner. Products so evaluated by appropriate third-party testing laboratories may be labeled with authorized markings as a means of verifying that the product complies with the minimum requirements of the safety standard.

The *National Electrical Code*® is an electrical safety standard containing rules affecting products, installation procedures, and use of electrical equipment. It is the most widely recognized safety standard in this country. The *NEC*® is usually adopted through a legislative process of a governmental agency and becomes enforceable as a law within that jurisdiction. The *NEC*® is amended every three years in response to such things as the development of new products, the recognition of different installation and use procedures, and the discovery of potential hazards to persons and property. Electrical inspectors have the responsibility of enforcing the safety requirements of the *National Electrical Code*®.

Electrical inspectors are an important part of the safety program. They are involved in the development of safety rules as well as in the application of those rules to electrical installations. The responsibility borne by electrical inspectors is significant. To be responsible for knowing safety rules, making proper interpretation of those rules, and properly applying them to a job site is a major endeavor. In order for the electrical inspector to properly do his or her job, knowledge and skill must be developed. This is achieved through adequate education and experience. It is vital that electrical inspectors be knowledgeable of the rules in the *National Electrical Code*® and know how to apply those rules.

The need for adequate training and experience is not limited to the electrical inspector. All who are a part of the electrical construction industry can benefit by it. In order for one to maintain an acceptable level of proficiency in a profession or trade, one must stay current with the ever-changing electrical field. That proficiency is achieved through continuing education and training. There is no substitute for such training and experience. To emphasize this point, I refer to a philosophy taught by Ray Mullin: "The cost of education is small when compared to the price paid for ignorance."

Philip H. Cox
Executive Director
International Association
of Electrical Inspectors

INTRODUCTION

The thirteenth edition of *Electrical Wiring—Residential* is based on the 1999 *National Electrical Code* (*NEC*).® The *NEC*® is used as the basic standard for the layout and construction of residential electrical systems. In this text, thorough explanations are provided of Code requirements as they relate to residential wiring. To gain the greatest benefit from this text, the student must use the *National Electrical Code*® on a continuing basis. Please note that throughout this text, the *National Electrical Code*® is represented by the abbreviation *NEC*.®

Why Is This Text Different from Other House Wiring Books?

It is extremely difficult to learn the requirements of the Code by merely reading the *NEC*.® Try reading a dictionary from beginning to end. Do you remember what you read? Probably not. But, look up the definition of a word at the time you need to know about it, and you will remember its definition. In much the same manner, *Electrical Wiring—Residential* does not cover the *NEC*® in a boring manner. Instead, it stresses specific Code requirements at the appropriate time. For example, the Code requirements for electric clothes dryers are discussed when studying the electric clothes dryer hook-up.

This text is unique in that it includes a full set of actual house plans. The format is for the student to simultaneously use the text, the plans, and the *NEC*,® taking the learner through a typical house, room by room, circuit by circuit. Also included are many recommendations that are above and beyond the basic *NEC*® requirements.

The *NEC*® (NFPA 70) becomes mandatory only after it has been adopted by a city, county, state, or other governing body. Until officially adopted, the *NEC*® is merely advisory in nature. State and local electrical codes may contain modifications of the *NEC*® to meet local requirements. In most cases, local codes will adopt certain more stringent regulations than those found in the *NEC*.® For example, the *NEC*® recognizes nonmetallic-sheathed cable as an acceptable wiring method for house wiring. Yet, the city of Chicago and surrounding counties do not permit nonmetallic-sheathed cable for house wiring. In these areas, all house wiring is done with electrical metallic tubing (thinwall).

There are also instances where a governing body has legislated action that waives specific *NEC*® requirements, feeling that the *NEC*® was too restrictive on that particular issue. Such instances are very rare. The Instructor is encouraged to furnish students with any local variations from the *NEC*® that would affect this residential installation in a specific locality.

Electrical wiring is a skilled trade. Wiring should not be done by anyone not familiar with the hazards involved. It is a highly technical skill that requires much training. This text provides all of the electrical codes and standards information needed to approach house wiring in a safe manner. That is why the *NEC*® defines a *qualified person* as "one familiar with the construction and operation of the equipment and the hazards involved."

Do not work on live circuits! Always de-energize the system before working on it! There is no compromise when it comes to safety! Many injuries and deaths have occurred when individuals worked on live equipment. The question is always: "Would the injury or death have occurred had the power been shut off?" The answer is "No!"

National Electrical Code® and *NEC*® are registered trademarks of the National Fire Protection Association, Inc., Quincy, MA 02269.

All mandatory safety-related work practices are found in the Federal Regulation Occupational Safety and Health Act (OSHA), Title 29, Subpart S—Electrical, Sections 1910.331 through 1910.360.

Electrical Wiring—Residential is the most comprehensive text available anywhere. It covers residential wiring topics from the most advanced *NEC®* standpoint, to the fundamental requirements necessary for someone wanting to learn the basics. Every code rule is covered by text, illustrations, examples, and wiring diagrams.

Most electrical inspectors across the country are members of the International Association of Electrical Inspectors. This organization publishes one of the finest technical bimonthly magazines, devoted entirely to the *NEC®* and related topics. This organization is open to individuals that are not electrical inspectors. Electrical instructors, vo-tech students, apprentices, electricians, consulting engineers, contractors, and distributors are encouraged to join the IAEI so they can stay up to date on all *NEC®* issues, changes, and interpretations. An application form that explains the benefits of membership in the IAEI can be found in the Appendix of this text.

THE ELECTRICAL TRADE

Electricity is safe when it is handled properly. If electrical installations are made in an unsafe manner, great potential exists for damage to property as a result of fire. But more importantly, electrical installations that do not meet the minimum requirements of applicable codes and standards can and do result in personal injury and/or death.

Most communities, cities, counties, and states adopt regulations by reference to the *NEC®*. These regulations also specify any exceptions to the Code to cover local conditions.

Most building codes and standards contain definitions for the various levels of competency of workers in the electrical industry.

Apprentice shall mean a person who is required to be registered as such under *Section XYZ*, who is in compliance with the provisions of this article, and who is working at the trade in the employment of a registered electrical contractor and is under the direct supervision of a licensed master electrician, journeyman electrician, or residential wireman.

Residential Wireman shall mean a person having the necessary qualifications, training, experience, and technical knowledge to wire for and install electrical apparatus and equipment for wiring one-, two-, three-, and four-family dwellings. A residential wireman is sometimes referred to as a *Class B Electrician*.

Journeyman Electrician shall mean a person having the necessary qualifications, training, experience, and technical knowledge to wire for, install, and repair electrical apparatus and equipment for light, heat, power and other purposes, in accordance with standard rules and regulations governing such work.

Master Electrician means a person having the necessary qualifications, training, experience, and technical knowledge to properly plan, lay out, and supervise the installation and repair of wiring apparatus and equipment for electric light, heat, power and other purposes, in accordance with standard codes and regulations governing such work, such as the *NEC®*.

Electrical Contractor means any person, firm, copartnership, corporation, association, or combination thereof who undertakes or offers to undertake for another the planning, laying out, supervising and installing, or the making of additions, alterations, and repairs in the installation of wiring apparatus and equipment for electrical light, heat, and power.

THE THIRTEENTH EDITION

Continuing in the tradition of previous editions, this text thoroughly explains how Code changes affect wiring installations. New and revised full-color illustrations supplement the explanations to ensure that electricians understand the new Code requirements. New photos reflect the latest wiring materials and components available on the market. Revised review questions test student understanding of the new content. New tables that summarize Code requirements offer a quick reference tool for students. Another reference aid is the tables reprinted directly from the 1999 edition of the *NEC.* The extensive revisions for the thirteenth edition make *Electrical Wiring—Residential* the most up-to-date and well-organized guide to house wiring. Coverage of the *NEC* has been expanded to well over 950 Code references.

A modern residence blueprint serves as the basis for the wiring schematics, cable layouts, and discussions provided in the text. Additional plans may be obtained by contacting Delmar Publishers. Each unit dealing with a specific type of wiring is referenced to the appropriate plan sheet. All wiring systems are described in detail—lighting, appliance, heating, service entrance, and so on.

The house selected for this edition is scaled for current construction practices and costs. Note, however, that the wiring, lighting fixtures, appliances, number of outlets, number of circuits, and track lighting are not all commonly found in a home of this size. The wiring may incorporate more features than are absolutely necessary. This was done to present as many features and Code issues as possible to give the student more experience in wiring a residence.

This text does *not* focus on such basic skills as drilling a hole, splicing two wires together, stripping insulations from wire, or fishing a wire through walls. These skills are certainly necessary for good workmanship, but it is assumed that the student (electrical apprentice or journeyman electrician) has mastered the mechanical skills on the job.

This text *does* focus on the technical skills required to perform electrical installations. It covers such topics as calculating conductor sizes, calculating voltage drop, determining appliance circuit requirements, sizing service, connecting electric appliances, grounding service and equipment, installing recessed fixtures, and much more. These are critical skills that can make the difference between an installation that "meets Code" and one that does not. The electrician must understand the reasons for following Code regulations to achieve an installation that is essentially free from hazard to life and property.

CHANGES FOR THIS EDITION

The following list highlights changes and additions made in this thirteenth edition. Many of these changes and additions are the result of suggestions sent to the author by users of the text.

1. The 13th edition of *Electrical Wiring—Residential* is now in full color.

2. All Code references have been updated for conformance to the 1999 *National Electrical Code.*

3. A new Key Word (Glossary) has been added.

4. All measurements are now shown in both English and Metrics.

5. The Internet Web Site addresses of major manufacturers of electrical equipment have been added.

6. Stringent requirements of the Uniform Building Code relative to drilling and notching of wood joists have been added since these rules are not found in the *National Electrical Code.*

7. More text has been added to cover nonmetallic raceways.

8. More text has been added on service-entrance cables.

9. Added material on carbon monoxide detectors—*NFPA 720.*

10. A new table showing the recommended heights of switches, receptacles, and lighting outlets has been added.

11. The Instructor's Guide has been greatly expanded to provide more detailed explanations for the answers—A boon to the electrical instructor.

12. New wiring diagrams have been included for electric water heaters to show the latest in color coding of conductors and connections for thermostats and high temperature limit controls.

13. Cautions to be followed when drilling and notching factory-manufactured I-joists and beams have been added.

14. The Home Automation chapter has been totally revised to reflect the latest residential systems available today.

NEW CODE REQUIREMENTS FOR THE 1999 *NATIONAL ELECTRICAL CODE*®

1. New material has been added on *arc fault circuit interrupters*.

2. New material on applying *Correction Factors* to the special ampacity values for service-entrance conductors.

3. New requirements for identifying grounded conductors now required by the 1999 *NEC*®

4. New requirements for the installation of underground Type UF cables where solid rock is encountered.

5. New text on the correct way to install recessed lighting fixtures.

6. Complete rearrangement of *Article 250—Grounding*.

7. All *NEC*® references (over 950) have been checked and updated for conformance to the *1999 National Electrical Code.*® The *1999 NEC*® contains a considerable amount of renumbering and rearranging of Code sections and tables.

8. New permitted use for nonmetallic-sheathed cable in multistory homes.

9. New permission to connect other equipment to the separate 20-ampere branch-circuit required for bathroom receptacles.

10. Tougher requirements for the resistance values when supplemental ground rod(s) are installed.

11. New, simplified method for making the optional service-entrance size calculations.

12. New, tough requirements for marking the white conductors in cables whenever they are used as "hot" conductors.

13. New requirements for the number of receptacles needed when a basement is partially finished.

14. A new, detailed diagram shows a new, permitted way to run nonmetallic-sheathed cables into the top of a panel. It is now OK to run a number of cables through a short, single raceway into the top of the panel. Some restrictions apply, and the text points out these restrictions.

15. Hundreds of other *NEC*® requirements "fine tuned" in the text to meet the intent of the 1999 edition of the *NEC*.®

SUPPLEMENTS

An Instructor's Guide is available and consists of the following information: selected text references, answers to unit-end reviews, a final examination covering the content of the entire text, blank service-entrance calculation form, and answers to the final examination.

ABOUT THE AUTHOR

This text was written by Ray C. Mullin, former electrical instructor for the Wisconsin Schools of Vocational, Technical, and Adult Education. He is a former member of the International Brotherhood of Electrical Workers. He is a member of the International Association of Electrical Inspectors, the Institute of Electrical and Electronic Engineers, and the National Fire Protection Association, Electrical Section, and has served on Code Making Panel 4 of the *National Electrical Code.*

Mr. Mullin completed his apprenticeship training and worked as a journeyman and supervisor for residential, commercial, and industrial installations. He has taught both day and night electrical apprentice and journeyman courses, has conducted engineering seminars, and has conducted many technical Code workshops and seminars at International Association of Electrical Inspectors Chapter and Section meetings, and has served on their Code panels.

He has written many technical articles that have appeared in electrical trade publications. He has served as a consultant to electrical equipment manufacturers regarding conformance of their products to industry standards, and on legal issues relative to personal injury lawsuits resulting from the misuse of electricity and electrical equipment. He has served as an expert witness.

Mr. Mullin presents his knowledge and experience in this text in a clear-cut manner that is easy to understand. This presentation will help students to fully understand the essentials required to pass the residential licensing examinations and to perform residential wiring that "meets Code."

Mr. Mullin is the author of *House Wiring with the NEC*—a text that focuses entirely on the *National Electrical Code* requirements for house wiring. He is co-author of *Electrical Wiring —Commercial, Illustrated Electrical Calculations,* and *The Smart House.* He has contributed technical material for Delmar's *Electrical Grounding,* and to the International Association of Electrical Inspectors texts *Soares Book On Grounding* and *Ferm's Fast Finder.*

He served on the Executive Board of the Western Section of the International Association of Electrical Inspectors, on their Code Clearing Committee, and, in the past, served as Secretary/Treasurer of the Indiana Chapter of the IAEI.

Mr. Mullin is past Director, Technical Liaison for a major electrical manufacturer. In this position, he was deeply involved in electrical codes and standards as well as contributing and developing technical training material for use by this company's field engineering personnel.

Mr. Mullin attended the University of Wisconsin, Colorado State University, and the Milwaukee School of Engineering.

DELMAR PUBLISHERS IS YOUR ELECTRICAL BOOK SOURCE!

Whether you're a beginning student or a master electrician, Delmar Publishers has the right book for you. Our complete selection of proven best-sellers and all-new titles is designed to bring you the most up-to-date, technically accurate information available.

NATIONAL ELECTRICAL CODE

National Electrical Code® 1999/NFPA
Revised every three years, the *National Electrical Code®* is the basis of all U.S. electrical codes.

Order # 0-8776-5432-8
Loose-leaf version in binder
Order # 0-8776-5433-6

National Electrical Code® Handbook 1999/NFPA
This essential resource pulls together all the extra facts, figures, and explanations you need to interpret the 1999 *NEC®*. It includes the entire text of the Code, plus expert commentary, real-world examples, diagrams, and illustrations that clarify requirements.

Order # 0-8776-5437-9

Illustrated Changes in the 1999 National Electrical Code®/O'Riley
This book provides an abundantly illustrated and easy-to-understand analysis of the changes made to the 1999 *NEC®*.

Order # 0-7668-0763-0

Understanding the National Electrical Code,® 3E/Holt
This book gives users at every level the ability to understand what the *NEC®* requires, and simplifies this sometimes intimidating and confusing code.

Order # 0-7668-0350-3

Illustrated Guide to the National Electrical Code®/Miller
Highly detailed illustrations offer insight into Code requirements, and are further enhanced through clearly written, concise blocks of text that can be read very quickly and understood with ease. Organized by classes of occupancy.

Order # 0-7668-0529-8

Interpreting the National Electrical Code,® 5E/Surbrook
This updated resource provides a process for understanding and applying the *National Electrical Code®* to electrical contracting, plan development, and review.

Order # 0-7668-0187-X

Electrical Grounding, 5E/O'Riley
Electrical Grounding is a highly illustrated, systematic approach for understanding grounding principles and their application to the 1999 *NEC®*.

Order # 0-7668-0486-0

ELECTRICAL WIRING

Electrical Raceways and Other Wiring Methods, 3E/Loyd
The most authoritative resource on metallic and non-metallic raceways, provides users with a concise, easy-to-understand guide to the specific design criteria and wiring methods and materials required by the 1999 *NEC®*.

Order # 0-7668-0266-3

Electrical Wiring—Residential, 13E/Mullin
Now in full color! Users can learn all aspects of residential wiring and how to apply them to the wiring of a typical house from this, the most widely used residential wiring book in the country.

Softcover Order #0-8273-8607-9
Hardcover Order # 0-8273-8610-9

House Wiring with the NEC®/Mullin
The focus of this new book is the applications of the *NEC®* to house wiring.

Order # 0-8273-8350-9

Electrical Wiring—Commercial, 10E/Mullin and Smith
Users can learn commercial wiring in accordance with the *NEC®* from this comprehensive guide to applying the newly revised 1999 *NEC®*.

Order # 0-7668-0179-9

Electrical Wiring—Industrial, 10E/Smith
This practical resource has users work their way through an entire industrial building—wiring the branch-circuits, feeders, service entrances, and many of the electrical appliances and subsystems found in commercial buildings.

Order # 0-7668-0193-4

Cables and Wiring, 2E/AVO
This concise, easy-to-use book is your single-source guide to electrical cables—it's a "must-have" reference for journeyman electricians, contractors, inspectors, and designers.
Order # 0-7668-0270-1

ELECTRICAL MACHINES AND CONTROLS

Industrial Motor Control, 4E/Herman and Alerich
This newly revised and expanded book, now in full color, provides easy-to-follow instructions and essential information for controlling industrial motors. Also available are a new lab manual and an interactive CD-ROM.
Order # 0-8273-8640-0

Electric Motor Control, 6E/Alerich and Herman
Fully updated in this new sixth edition, this book has been a long-standing leader in the area of electric motor controls.
Order # 0-8273-8456-4

Introduction to Programmable Logic Controllers/Dunning
This book offers an introduction to Programmable Logic Controllers.
Order # 0-8273-7866-1

Technician's Guide to Programmable Controllers, 3E/Cox
Uses a plain, easy-to-understand approach and covers the basics of programmable controllers.
Order # 0-8273-6238-2

Programmable Controller Circuits/Bertrand
This book is a project manual designed to provide practical laboratory experience for one studying industrial controls.
Order # 0-8273-7066-0

Electronic Variable Speed Drives/Brumbach
Aimed squarely at maintenance and troubleshooting, *Electronic Variable Speed Drives* is the only book devoted exclusively to this topic.
Order # 0-8273-6937-9

Electrical Controls for Machines, 5E/Rexford
State-of-the-art process and machine control devices, circuits, and systems for all types of industries are explained in detail in this comprehensive resource.
Order # 0-8273-7644-8

Electrical Transformers and Rotating Machines/Herman
This new book is an excellent resource for electrical students and professionals in the electrical trade.
Order # 0-7668-0579-4

Delmar's Standard Guide to Transformers/Herman
Delmar's Standard Guide to Transformers was developed from the best-seller *Standard Textbook of Electricity* with expanded transformer coverage not found in any other book.
Order # 0-8273-7209-4

DATA AND VOICE COMMUNICATION CABLING AND FIBER OPTICS

Complete Guide to Fiber Optic Cable System Installation/Pearson
This book offers comprehensive, unbiased, state-of-the-art information and procedures for installing fiber optic cable systems.
Order # 0-8273-7318-X

Fiber Optics Technician's Manual/Hayes
Here's an indispensable tool for all technicians and electricians who need to learn about optimal fiber optic design and installation as well as the latest troubleshooting tips and techniques.
Order # 0-8273-7426-7

A Guide for Telecommunications Cable Splicing/Highhouse
A "how-to" guide for splicing all types of telecommunications cables.
Order # 0-8273-8066-6

Premises Cabling/Sterling
This reference is ideal for electricians, electrical contractors, and inspectors needing specific information on the principles of structured wiring systems.
Order # 0-8273-7244-2

ELECTRICAL THEORY

Delmar's Standard Textbook of Electricity, 2E/Herman
This exciting full-color book is the most comprehensive book on DC/AC circuits and machines for those learning the electrical trades.
Order # 0-8273-8550-1

Industrial Electricity, 6E/Nadon, Gelmine and Brumbach
This revised, illustrated book offers broad coverage of the basics of electrical theory and is perfect for those who wish to be industrial maintenance technicians.
Order # 0-7668-0101-2

EXAM PREPARATION

Journeyman Electrician's Exam Preparation, 2E/Holt
This comprehensive exam prep guide includes all of the topics on the journeyman electrician competency exams.
Order # 0-7668-0375-9

Master Electrician's Exam Preparation, 2E/Holt
This comprehensive exam prep guide includes all of the topics on the master electrician's competency exams.
Order # 0-7668-03767

REFERENCE
ELECTRICAL REFERENCE SERIES

This series of technical reference books is written by experts and designed to provide the electrician, electrical contractor, industrial maintenance technician, and other electrical workers with a source of reference information about virtually all of the electrical topics that they encounter.

Electrician's Technical Reference—Motor Controls/Carpenter
Electrical Reference—Motor Controls is a source of comprehensive information on understanding the controls that start, stop, and regulate the speed of motors.
Order # 0-8273-8514-5

*Electrician's Technical Reference—Motors/*Carpenter
Electrician's Technical Reference—Motors builds an understanding of the operation, theory, and applications of motors.
Order # 0-8273-8513-7

Electrician's Technical Reference—Theory and Calculations/Herman
Electrician's Technical Reference—Theory and Calculations provides detailed examples of problem-solving for different kinds of DC and AC circuits.
Order # 0-8273-7885-8

*Electrician's Technical Reference Transformers/*Herman
Electrician's Technical Reference—Transformers focuses on the theoretical and practical aspects of single-phase and 3-phase transformers and transformer connections.
Order # 0-8273-8496-3

Electrician's Technical Reference—Hazardous Locations/Loyd
Electrician's Technical Reference—Hazardous Locations covers electrical wiring methods and basic electrical design considerations for hazardous locations.
Order # 0-8273-8380-0

Electrician's Technical Reference—Wiring Methods/Loyd
Electrician's Technical Reference—Wiring Methods covers electrical wiring methods and basic electrical design considerations for all locations, and shows how to provide efficient, safe, and economical applications of various types of available wiring methods.
Order # 0-8273-8379-7

Electrician's Technical Reference—Industrial Electronics/Herman
Electrician's Technical Reference—Industrial Electronics covers components most used in heavy industry, such as silicon control rectifiers, triacs, and more. It also includes examples of common rectifiers and phase-shifting circuits.
Order # 0-7668-0347-3

RELATED TITLES

Common Sense Conduit Bending and Cable Tray Techniques/Simpson
Now geared especially for students, this manual remains the only complete treatment of the topic in the electrical field.
Order # 0-8273-7110-1

Practical Problems in Mathematics for Electricians, 5E/Herman
This book details the mathematics principles needed by electricians.
Order # 0-8273-6708-2

Electrical Estimating/Holt

This book provides a comprehensive look at how to estimate electrical wiring for residential and commercial buildings with extensive discussion of manual versus computer-assisted estimating.

Order # 0-8273-8100-X

Electrical Studies for Trades/Herman

Based on Delmar's *Standard Textbook of Electricity*, this new book provides non-electrical trades students with the basic information they need to understand electrical systems.

Order # 0-8273-7845-9

ACKNOWLEDGMENTS

I would like to thank my wife, Helen J. Mullin, for her help, encouragement, understanding, and patience while I devoted unlimited time to revising this full-color edition of *Electrical Wiring—Residential.*

Special thanks go to the team at Delmar that worked so hard at getting this full-color edition out on a timely basis to coincide with the *1999 National Electrical Code.*® Adding new text and revising text to conform to the 1999 *NEC,*® redrawing every diagram and illustration in color, and using full-color photographs made this edition a challenge having a magnitude never before experienced. The Delmar Team did an outstanding job! Their dedication brought this edition to its highest state of excellence.

I would also like to express sincere appreciation to my good friend Ron O'Riley, one of the country's outstanding Code experts, for his suggestions and detailed review of this edition. Special thanks to Robert Boiko for his technical input relating to water heaters and heating system controls, and to Brooke Stauffer, Director, Codes and Standards, National Electrical Contractors Association for his contribution to the chapter relating to home automation.

The author and Delmar Publishers would like to thank the following reviewers:

Max Schroeder
Barton County Community College
Hoisington, KS

Robert Coy
Kentucky Technical College
Elizabethtown, KY

David Gehlauf
Tri-County Vocational School
Glouster, OH

Larry Catron
Scott County Vocational School
Gate City, VA

Ed Dwyer
Triangle Technical College
Pittsburgh, PA

Warren DeJardin
Northeast Wisconsin Technical College
Oneida, WI

Charles Trout
Electrical Code Consultant
Boca Raton, FL

The author and Delmar Publishers wish to thank the following companies for their contributions of data, illustrations, art work, photos, and technical information:

AFC Cable Systems

American Home Lighting Institute

BRK Electronics (First Alert), A
    Division of Sunbeam Corporation

Bussmann Division, Cooper Industries

Carlon, A Lamson & Sessions Company

Certified Ballast Manufacturers

Chromalox

Cooper Lighting (Halo), Cooper
    Industries

Electri-Flex Co.

General Electric Company

Heyco Molded Products, Inc.

Honeywell

Hubbell Incorporated, Wiring Devices
    Division

International Association of Electrical
    Inspectors

Juno Lighting, Inc.

The Kohler Co.

Leviton Manufacturing Co., Inc.

Lutron Electronics Co., Inc.

Midwest Electric Products, Inc.

Milbank Manufacturing Co.

Motorola

NuTone Inc.

Paragon Electric Company, Inc.

Progress Lighting

RACO, A Division of Hubbell
Incorporated

Reiker Enterprises, Inc.

Rheem Manufacturing Co.

Seatek Co., Inc.

Square D, Groupe Schneider

Superior Electric Co.

THERM-O-DISC

Thomas & Betts Corporation

Wiremold Co.

General Information for Electrical Installations

OBJECTIVES

After studying this Unit, the student will be able to

- understand the basic safety rules for working on electrical systems.

- explain how electrical wiring information is conveyed to the electrician at the construction or installation site.

- demonstrate how the specifications are used in estimating cost and in making electrical installations.

- explain why symbols and notations are used on electrical drawings.

- list the Nationally Recognized Testing Laboratories (NRTL) that are responsible for establishing electrical standards and ensuring that materials meet the standards.

- develop a better understanding of the metric system of measurement.

- begin to refer to the latest edition of the *NEC.*®

- have a better understanding of the new *Residential Electrical Maintenance Code for One and Two-Family Dwellings*.

SAFETY IN THE WORKPLACE

Before venturing into the study of residential wiring, let's discuss safety. The electrician might be involved in new construction where the only electrical power is temporary power, or in remodel work where the home already has electrical power.

Serious injury can result if the power has not been properly turned off before beginning to work on an electrical problem. Electricity is dangerous. The voltage levels in a home are 120 volts between one "hot" conductor and the "neutral" conductor or a grounded surface. Between the two "hot" conductors, the voltage is 240 volts. These are the *line-to-ground* or *line-to-line* voltages. From basic electrical theory, *line voltage appears across an open in a series circuit*. Getting caught "in series" with a 120-volt circuit will give you a 120-volt shock. For example, open-circuit voltage across the two wires of a single-pole switch on a lighting circuit is 120 volts, when the switch is in the OFF position and the lamp(s) are in place.

Likewise, getting caught "in series" with a 240-volt circuit will give you a 240-volt shock. Working on electrical equipment with the power turned on can result in death or serious injury either as a direct result of electricity (electrocution or burns), or from an indirect secondary reaction such as falling off a ladder or jerking away from the "hot" conductor into moving parts of equipment, such as into the turning blades of a furnace fan. Dropping a metal tool onto live parts, allowing metal shavings from a drilling operation to fall onto live parts of electrical equipment, or cutting into a "live" conductor and a "neutral" conductor at the same time, or touching the "live" wire and the "neutral" wire or a grounded surface at the same time can cause injury. A short-circuit or ground fault can result in an *arc blast* that can cause serious injury or death. The heat of an electrical arc has been determined to be hotter than the sun. Tiny hot "balls" of copper can fly into your eye or onto your skin. Working on switches, receptacles, fixtures, or appliances with the power turned on is dangerous.

Dirt, debris, and moisture can also set the stage for equipment failures and personal injury. Neatness and cleanliness in the workplace are a must.

It's the Law!

Not only is it a good idea to use proper safety measures as you work on and around electrical systems, it is *required* by law. As an electrician or electrical contractor, you need to be aware of these regulations.

The Federal Regulations in the **Occupational Safety and Health Act (OSHA)** Number 29, Subpart S, in Part 1910.332 discuss the training needed for those who face a risk of electrical injury. Proper training means being trained in and familiar with the safety-related work practices required by paragraphs 1910.331 through 1910.335. Numerous texts are available that cover the OSHA requirements in greater detail.

The *NEC®* defines a *qualified person* as "one familiar with the construction and operation of the equipment and the hazards involved." Merely telling someone or being told "be careful" does not meet the definition of proper training, and does not make the person qualified.

Only **qualified persons** are permitted to work on or near exposed energized equipment. To become qualified, a person must have the skill and techniques necessary to distinguish exposed live parts from the other parts of electrical equipment, must be able to determine the voltage of exposed live parts, and must be trained in the use of special precautionary techniques such as personal protective equipment, insulations, shielding material, and insulated tools.

Subpart S, 1910.333 requires that safety related work practices shall be employed to prevent electric shock or other injuries resulting from either direct or indirect electrical contact. Live parts to which an employee may be exposed shall be de-energized before the employee works on or near them, unless the employer can demonstrate that de-energizing introduces additional or increased hazards.

Working on equipment "live" is acceptable only if there would be a greater hazard if the system were de-energized. Examples of this would be hospital life support systems, some alarm systems, certain ventilation systems in hazardous locations, and the power for critical illumination circuits. Working on energized equipment requires proper insulated tools, proper nonflammable clothing, rubber gloves, protective shields and goggles, and in some cases, rubber blankets. None of these reasons to work on electrical equipment "hot" are found in residential wiring.

OSHA regulations allow only qualified personnel to work on or near electrical circuits or equipment that have not been de-energized. The OSHA regulations provide rules regarding "lockout and tagging" to make sure that the electrical equipment being worked on will not inadvertently be turned on while someone is working on the supposedly "dead" equipment. As the OSHA regulations state, "a lock and a tag shall be placed on each disconnecting means used to de-energize circuits and equipment."

Some electricians' contractual agreements require that as a safety measure, two or more qualified electricians must work together when working on energized circuits. They do not allow untrained apprentices to work on "live" equipment but do allow apprentices to stand back and observe.

The **National Fire Protection Association** publications *Safety Related Work Practices NFPA 70E* and *Electrical Equipment Maintenance NFPA 70B* present much of the same material regarding electrical safety as does the OSHA regulation.

Safety cannot be compromised! The rule is *turn off and lock off the power, then properly tag the disconnect with a description of exactly what that particular disconnect controls.* With safety utmost in our minds, let us begin our course on the wiring of a typical residence.

Temporary Wiring

There is an ever-present electrical shock hazard on construction sites. The *NEC®* addresses this in *Article 305*. This is covered in Unit 6.

PLANS

The architect uses a set of working drawings or **plans** to make the necessary instructions available to the skilled craftspersons that are to build the structure shown in the plans. The sizes, quantities, and locations of the materials required and the construction features of the structural members are shown at a glance. These details of construction must be studied and interpreted by each skilled construction craftsperson—masons, carpenters, electricians, and others—before the actual work is started.

The electrician must be able to: (1) convert the two-dimensional plans into an actual electrical installation, and (2) visualize the many different views of the plans and coordinate them into a three-dimensional picture, as shown in Figure 1-1.

The ability to visualize an accurate three-dimensional picture requires a thorough knowledge of blueprint reading. Because all of the skilled trades use a common set of plans, the electrician must be able to interpret the lines and symbols that refer to the electrical installation and also those used by the other construction trades. The electrician must know the structural makeup of the building and the construction materials to be used.

SPECIFICATIONS

Working drawings are usually complex because of the amount of information that must be included. To prevent confusing detail, it is standard practice to include with each set of plans a set of detailed written **specifications** prepared by the architect.

These specifications provide general information to be used by all trades involved in the construction. In addition, specialized information is given for the individual trades. The specifications

Figure 1-1   Three-dimensional view of house wiring.

include information on the sizes, the type, and the desired quality of the standard parts to be used in the structure.

Typical specifications include a section on "General Clauses and Conditions," which is applicable to all trades involved in the construction. This section is followed by detailed requirements for the various trades—excavating, masonry, carpentry, plumbing, heating, electrical work, painting, and others.

The plan drawings for the residence used as an example are included with this text. The specifications for the electrical work indicated on the plans are given in the Appendix of this text.

In the electrical specifications, the listing of standard electrical parts and supplies frequently includes the manufacturers' names and the catalog numbers of the specified items. Such information ensures that these items will be of the correct size, type, and electrical rating, and that the quality will meet a certain standard. To allow for the possibility that the contractor will not always be able to obtain the specified item, the phrase "or equivalent" is usually added after the manufacturer's name and catalog number.

The specifications are also useful to the electrical contractor in that all of the items needed for a specific job are grouped together and the type or size of each item is indicated. This information allows the contractor to prepare an accurate cost estimate without having to find all of the data on the plans.

If there is a difference between the plans and specifications, the specifications will take preference. The electrical contractor should discuss the matter with the homeowner, architect, and engineer. The cost of the installation might vary considerably because of the difference(s), so obtain any changes to the plans and/or specifications in writing.

SYMBOLS AND NOTATIONS

The architect uses **symbols** and **notations** to simplify the drawing and presentation of information concerning electrical devices, appliances, and equipment. For example, an electric range outlet looks like this:

Most symbols have a standard interpretation throughout the country as adopted by the American National Standards Institute (ANSI). Symbols are described in detail in Unit 2.

A notation will generally be found on the plans (blueprints) next to a specific symbol calling attention to a variation, type, size, quantity, or other necessary information. In reality, a symbol might be considered to be a notation because symbols represent words, phrases, numbers, quantities, etc.

Another method of using notations to avoid cluttering up a blueprint is to provide a system of symbols that refer to a specific table. For example, the written sentences on plans could be included in a table referred to by a notation. Figure 2-9 is an example of how this could be done. The special symbols that refer to the table would have been shown on the actual plan.

NATIONAL ELECTRICAL CODE® (NEC®)

The *National Electrical Code®* is published by the National Fire Protection Association, and is referred to as NFPA 70. Because of the ever-present danger of fire or shock hazard through some failure of the electrical system, the electrician and the electrical contractor must use listed materials and must perform all work in accordance with recognized standards. The *NEC®* is the basic standard that governs electrical work. The purpose of the Code is to provide information considered necessary for the safeguarding of people and property against electrical hazards. It states that the installation must be essentially hazard-free, but that such an installation is not necessarily efficient, convenient, or adequate, *Section 90-1(b)*. (Note: All *National Electrical Code®* section references used throughout this text are printed in italics.) It is the electrician's responsibility to ensure that the installation meets these criteria. In addition to the *NEC®*, the electrician must also consider local and state codes. The purpose and scope of the *NEC®* are discussed in *Article 90* of the Code book and should be studied by the student at this time.

Code Arrangement

It is important to understand the arrangement of the *NEC®*.

- Introduction   Explains the purpose of the *NEC®*, what is covered, what is not covered, who enforces the *NEC®*.

- Chapters 1, 2, 3, and 4 — These chapters contain general requirements that cover most situations.

- Chapters 5, 6, and 7 — These chapters pertain to special occupancies, special equipment, and special conditions. The requirements in these chapters are in addition to, or are changes to (modify or amend) the general rules found in Chapters 1, 2, 3, and 4.

- Chapter 8 — Covers communication, telephone, fire and burglar alarm, radio, television, cable systems, etc.

- Chapter 9 — This chapter contains examples and tables.

- Appendix A — Lists text that has been extracted from other NFPA codes. These are identified throughout the *NEC*® in particular code sections with the superscript letter "x."

- Appendix B — Contains useful information not actually enforceable as a *NEC*® rule.

- Appendix C — Contains 24 tables showing the number of conductors permitted in various types of raceways.

- Appendix D — Examples of load calculations.

- Appendix E — A cross-reference to *Article 250* showing the new numbering system compared to the old numbering system.

- Index — The alphabetical index for the *National Electrical Code*®

Chapters are divided into *Articles*. *Articles* are divided into *Sections*. *Sections* are divided into *subsections* and *exceptions*.

The first *NEC*® was published in 1897. It is revised and updated every three years by changing, editing, adding technical material, adding sections, adding articles, adding tables, and so on, so as to be as up-to-date as possible relative to electrical installations. For example, the Code added requirements pertaining to circuits supplying computers and similar electronic equipment.

As material, equipment, and technologies change, the members of the Code-making panel solicit comments and proposals from individuals in the electrical industry or others interested in electrical safety. The panel members then meet to act upon the many proposals for changes to the Code. Final action (voting) on these proposals is taken at the National Fire Protection Association's annual meeting.

The Code-making panel consists of individuals representing over 65 organizations such as the International Association of Electrical Inspectors, the Institute of Electrical and Electronic Engineers, the National Electrical Contractors Association, the Edison Electric Institute, the Independent Electrical Contractors Association, Inc., Underwriters Laboratories, Inc., Consumer Product Safety Commission, International Brotherhood of Electrical Workers, the National Electrical Manufacturers Association, and many others.

The *NEC*® does not become law until adopted by official action of the legislative body of a city, county, or state.

What Edition of the *NEC*® to Use?

This text is based on the 1999 edition of the *NEC.*® However, be aware that some municipalities, counties, and states have not yet adopted the 1999 edition, and might still be using the 1996 or earlier editions. Always check on this with the local authority having jurisdiction before starting a wiring project to make sure your installation complies with the proper electrical code.

Copies of the latest edition of the *National Electrical Code*® *NFPA 70* may be ordered from:

Delmar Publishers
P.O. Box 15015
Albany, New York 12212-5015
1-800-347-7707

Residential Electrical Maintenance Code for One- and Two-Family Dwellings

This code is published by the National Fire Protection Association, and is referred to as *NFPA 73*. It is a rather brief seven-page code that provides requirements for evaluating installed electrical systems within and associated with *existing* one- and two-family dwellings to identify safety, fire, and shock hazards—such as improper installations, overheating, physical deterioration, and abuse. This code lists most of the electrical things to inspect in an existing dwelling that could result in a fire or shock hazard if not corrected. It only points out things to look for that are visible. It does not get into examining concealed wiring that would require removal of permanent parts of the structure. It also does not get into calculations, location requirements, and complex topics as does the *National Electrical Code® NFPA 70.*

This code uses simple statements such as: *The service shall be adequate to serve the connected load; overcurrent protective devices shall be properly rated for conductor ampacities; conductors shall be properly terminated and supported at panelboards, boxes, and devices; raceways shall be securely fastened in place; cables and cable assemblies shall be properly secured and supported; fixture canopies shall be in place and properly secured; covers shall be in place and properly secured;* and so forth.

This code can be an extremely useful guide for electricians doing remodel work, and for electrical inspectors wanting to bring an existing dwelling to a reasonably safe condition. Many localities require that when a home changes ownership, the wiring must be brought up to some minimum standard, but not necessarily as extensively as would be the case for new construction. For example, a local code might require that a minimum 100-ampere, 120/240-volt service be installed, that the service ground be brought up to the requirements of the latest edition of the *NEC,®* that 20-ampere appliance circuits be installed in the kitchen and dining room, that **ground fault circuit interrupter** (GFCI) receptacles be installed in bathrooms and other specified areas, and that a separate 20-ampere branch-circuit be installed for the laundry equipment (clothes washer).

Homes for Physically Challenged People

The *NEC®* does not address the wiring of homes for the physically or mentally challenged, or disabilities associated with the elderly.

Each installation for the physically challenged must be based upon the specific need(s) of the individual(s) who will occupy the home. Some physically challenged people are bedridden, require the mobility of a wheelchair, have trouble reaching, have trouble bending, etc. There are no hard and fast rules that *must* be followed—only many suggestions to consider.

Some of these are:

- Install more ceiling lighting fixtures instead of switching receptacles. Cords and lamps are obstacles.

- Install lighting fixtures having more than one bulb.

- Go "overboard" in the amount of lighting for all rooms, entrances, stairways, stairwells, closets, pantries, bathrooms, etc.

- Use higher wattage bulbs—not to exceed the wattage permitted in the specific fixture.

- Consider installing lighting fixtures in certain areas (such as bathrooms and hallways) to be controlled by motion detectors.

- Consider installing exhaust fans in certain areas (such as laundries, showers, and bathrooms) that turn on automatically when the humidity reaches a predetermined value.

- Consider height of switches and thermostats. Usually lower (42" or 1.07 m) instead of the standard 46–52" (1.27 to 1.32 m).

- Consider rocker-type switches instead of toggle type.

- Install pilot light switches.

- Consider "jumbo" switches.

- Be sure stairways and stairwells are well lit.

- Consider stair tread lighting.

- Position lighting switches so as not to be over stairways or ramps.

- Locate switches and receptacles to be readily and easily accessible—not behind doors or other hard-to-reach places.

- Consider installing wall receptacle outlets higher (24–27" or 610 mm to 686 mm) than normal 12" (305 mm) height.

- Install lighted doorbell buttons.

- Chimes: consider adding a strategically located "dedicated" lamp(s) that will turn on when door bell buttons are pushed. The wiring diagram for this is found in Figure 25-25.

- Telephones: consider adding visible light(s) strategically located that will flash at the same time the telephone is ringing.

- Consider installing receptacle outlets and switches on the face of kitchen cabinets. Wall outlets and switches can be impossible for the physically challenged person to reach.

- Consider fire, smoke, and security systems that are directly connected to a central office for fast response to emergencies that do not depend upon the disabled to initiate the call.

- Consider installing the fuse box or breaker panel on the first floor instead of in the basement.

- Consider the advanced home systems concept of remote control of lighting, receptacles, appliances, television, telephones, etc. The control features can make life much easier for a disabled person.

The American National Standards Institute (ANSI) publication *ANSI A117.1–1998 Accessible and Usable Buildings and Facilities* contains many suggestions and considerations for buildings and facilities for the physically challenged. The address is:

American National Standards Institute
11 West 42nd Street
New York, NY 10036
212-642-4900

Building Codes

The majority of the building departments across the country have for the most part adopted the *NEC®* rather than attempting to develop their own electrical codes. As you study this text, you will note numerous references to the *NEC®,* the electrical inspector, or the authority having jurisdiction.

An authority's level of knowledge varies. The electrical inspector may be full-time or part-time and may also have responsibility for other trades, such as plumbing or heating. The heads of the building department in many communities are typically called the Building Commissioners or Directors of Development. Regardless of title, they are responsible for ensuring that the building codes in their communities are followed.

Communities, rather than writing their own codes, belong to one of the major building officials organizations, and have adopted the building codes of that particular organization. These organizations are:

- **BOCA**   Building Official & Code Administrators International, Inc.
4051 W. Flossmoor Road
Country Club Hills, Illinois 60478-5794
Phone: 708-799-2300

- **ICBO**   International Conference of Building Officials
5360 Workman Mill Road
Whittier, California 90601-2298
Phone: 562-699-0541

- **SBCCI**   Southern Building Code Congress International, Inc.
900 Montclair Road
Birmingham, Alabama 35212-1206
Phone: 205-591-1853

- **IAPMO**   International Association of Plumbing and Mechanical Officials
5032 Alhambra Avenue
Los Angeles, CA 90032-3490
Phone: 213-223-1471

- **ICC**   International Code Council
5203 Leesburg Pike, Suite 708
Falls Church, VA 22041
Phone: 703-931-4533

ICC is an organization that provides technical, educational, and informational products and services in support of international codes. BOCA, ICBO, and SBCCI are members of this organization. Chapters 39 through 46 of the ICC *One- and Two-Family Dwelling Code* contain the basic electrical requirements for one- and two-family dwellings based on the *NEC®.* The remaining chapters cover plumbing, heating, and construction requirements.

Because local electrical codes may differ from the *NEC®,* you should check with the local inspection authority to determine which edition of the *NEC®* is enforced, and what, if any, local requirements take precedence over the *NEC®.*

AMERICAN NATIONAL STANDARDS INSTITUTE

The **American National Standards Institute** is an organization that coordinates the efforts and results of the various standards-developing organizations, such as those mentioned in previous paragraphs. Through this process, ANSI approves standards that then become recognized as American national standards. One will find much similarity between the technical information found in ANSI standards, the Underwriters standards, the International Electronic and Electrical Engineers standards, and the *NEC.*®

Code Definitions

The electrical industry uses many words (terms) that are unique to the electrical trade. These terms need clear definitions to enable the electrician to understand completely the meaning intended by the Code.

Article 100 of the *NEC*® is a "dictionary" of these terms. *Article 90* also provides further clarification of terms used in the *NEC.*® A few of the terms defined in the Code follow.

Ampacity: The current in amperes a conductor can carry continuously under the conditions of use without exceeding its temperature rating.

Approved: Acceptable to the authority having jurisdiction.

Authority Having Jurisdiction: An organization, office, or individual responsible for "approving" equipment, an installation, or a procedure, *Section 90-4.*

Dwelling Unit: One or more rooms for the use of one or more persons as a housekeeping unit with space for eating, living, and sleeping, and permanent provisions for cooking and sanitation. (Note: The terms *dwelling* and *residence* are used interchangeably throughout this text.)

Fine Print Notes (FPN): Fine Print Notes (FPNs) are found throughout the Code. FPNs are "explanatory" in nature, in that they refer to other Sections of the Code. FPNs also define things where further description is necessary. See *Section 90-5.*

Identified (as applied to equipment): Recognizable as suitable for a specific purpose, function, use, environment, or application, where described in a particular Code requirement. Suitability of use, marked on or provided with the equipment, may include labeling or listing.

Identified (as applied to conductors and terminals): The grounded circuit, feeder or service-entrance conductor is "identified" by its white or natural gray color. Terminals on wiring devices, panels, and disconnect switches have their terminals for the grounded circuit conductor "identified" with silver colored screws. In the electrical industry, you will hear the terms *grounded circuit conductor*, *neutral conductor*, or *the white wire*. For all practical purposes, these terms all mean the same thing. They refer to the grounded circuit conductor. An exception to this is that a neutral conductor is always a grounded conductor, but a grounded circuit conductor is not always a neutral as in the case of industrial wiring where a three-phase, grounded "B" phase system might be used. See *Section 250-20.*

Labeled: Equipment or materials to which has been attached a label, symbol, or other identifying mark of an organization acceptable to the authority having jurisdiction and concerned with product evaluation, that maintains periodic inspection of production of labeled equipment or materials and by whose labeling the manufacturer indicates compliance with appropriate standards or performance in a specified manner.

Listed: Equipment, materials, or services included in a list published by an organization that is acceptable to the authority having jurisdiction and concerned with evaluation of products or services, that maintains periodic inspection of production of listed equipment or materials or periodic evaluation of services, and whose listing states that either the equipment, material, or services meets identified standards or has been tested and found suitable for a specified purpose.

Shall: Indicates a mandatory rule, *Section 90-5.* As you study the *NEC,*® think of the word "shall" as meaning "must." Some examples found in the *NEC*® where the word "shall" is used in combination with other words are: *shall be, shall have, shall not, shall be permitted, shall not be permitted, and shall not be required.*

One of the most far-reaching *NEC*® rules is *Section 110-3(b).* This section states that the installation and use of listed or labeled equipment must

PREFIX	SYMBOL	MULTIPLIER	SCIENTIFIC NOTATION (POWERS OF TEN)	VALUE
tera	T	1 000 000 000 000	10^{12}	one trillion (1 000 000 000 000/1)
giga	G	1 000 000 000	10^{9}	one billion (1 000 000 000/1)
mega	M	1 000 000	10^{6}	one million (1 000 000/1)
kilo	k	1 000	10^{3}	one thousand (1 000/1)
hecto	h	100	10^{2}	one hundred (100/1)
deka	da	10	10^{1}	ten (10/1)
unit		1	—	one (1)
deci	d	0.1	10^{-1}	one tenth (1/10)
centi	c	0.01	10^{-2}	one hundredth (1/100)
milli	m	0.001	10^{-3}	one thousandth (1/1 000)
micro	μ	0.000 001	10^{-6}	one millionth (1/1 000 000)
nano	n	0.000 000 001	10^{-9}	one billionth (1/1 000 000 000)
pico	p	0.000 000 000 001	10^{-12}	one trillionth (1/1 000 000 000 000)

Figure 1-2   Metric prefixes, symbols, multipliers, powers, and values.

conform to any instructions included in the listing or labeling. This means that an entire electrical system and all of the system's electrical equipment must be *installed* and *used* in accordance with the *NEC®* and the numerous standards to which the electrical equipment has been tested.

CODE USE OF METRIC (SI) MEASUREMENTS

The metric system is a "base-10" or "decimal" system in that values can be easily multiplied or divided by "ten" or "powers of ten." The metric system as we know it today is known as the International System of Units (SI) derived from the French term "le Système International d'Unités." The metric system is used by all industrial nations in the world but only to a limited extent in the United States.

As each edition of the *NEC®* comes out, more and more metric measurements appear. The electrician must become familiar with the metric system. The 1981 *NEC®* introduced the use of metric measurements in addition to English measurements. Metric measurements appear in the Code as follows:

- in the Code paragraphs, the approximate metric measurement appears in parentheses following the English measurement.
- in the Code tables, a footnote shows the SI conversion factors.

A metric measurement is not shown for conduit size, box size, wire size, horsepower designation for motors, and other "trade sizes" that do not reflect actual measurements.

Guide to Metric Usage

In the metric system, the units increase or decrease in multiples of 10, 100, 1,000, and so on. For instance, one megawatt (1,000,000 watts) is 1,000 times greater than one kilowatt (1,000 watts).

By assigning a name to a measurement, such as a *watt*, the name becomes the unit. Adding a prefix to the unit, such as *kilo*, forms the new name *kilowatt*, meaning 1,000 watts. Refer to Figure 1-2 for prefixes used in the metric system.

The prefixes used most commonly are *centi*, *kilo*, and *milli*. Consider that the basic unit is a meter (one). Therefore, a centimeter is 0.01 meter, a kilometer is 1,000 meters, and a millimeter is 0.001 meter.

Some common measurements of length and equivalents are shown in Figure 1-3.

one inch	=	2.54	centimeters
	=	25.4	millimeters
	=	0.025 4	meter
one foot	=	12	inches
	=	0.304 8	meter
	=	30.48	centimeters
	=	304.8	millimeters
one yard	=	3	feet
	=	36	inches
	=	0.914 4	meter
	=	914.4	millimeters
one meter	=	100	centimeters
	=	1 000	millimeters
	=	1.093	yards
	=	3.281	feet
	=	39.370	inches

Figure 1-3   Some common measurements of length and their equivalents.

inches (in) × 0.025 4	=	meters (m)
inches (in) × 0.254	=	decimeters (dm)
inches (in) × 2.54	=	centimeters (cm)
centimeters (cm) × 0.393 7	=	inches (in)
inches (in) × 25.4	=	millimeters (mm)
millimeters (mm) × 0.039 37	=	inches (in)
feet (ft) × 0.304 8	=	meters (m)
meters (m) × 3.280 8	=	feet (ft)
square inches (in²) × 6.452	=	square centimeters (cm²)
square centimeters (cm²) × 0.155	=	square inches (in²)
square feet (ft²) × 0.093	=	square meters (m²)
square meters (m²) × 10.764	=	square feet (ft²)
square yards (yd²) × 0.836 1	=	square meters (m²)
square meters (m²) × 1.196	=	square yards (yd²)
kilometers (km) × 1 000	=	meters (m)
kilometers (km) × 0.621	=	miles (mi)
miles (mi) × 1.609	=	kilometers (km)

Figure 1-4 Useful conversions (English/SI-SI/English) and their abbreviations.

Electricians will find it useful to refer to the conversion factors and their abbreviations shown in Figure 1-4.

Refer to the Appendix of this text for a comprehensive metric conversion table. The table includes information on how to "round off" numbers for practical use on the job.

NATIONALLY RECOGNIZED TESTING LABORATORIES

How does one know if a product is safe to use? Manufacturers, consumers, regulatory authorities, and others recognize the importance of independent, "third-party" testing of products in an effort to reduce safety risks. Unless you have all of the necessary test equipment and knowledge of how to properly test a product for safety, the surest way is to accept the findings of a third-party testing agency. In other words, "look for the label." **Nationally recognized testing laboratories (NRTL)** have the knowledge and wherewithal to test and evaluate products for safety. Make sure the product has a listing marking on it from a nationally recognized testing laboratory. If the product is too small to have a listing mark on the product itself, then look for the marking on the carton the product came in. Of course, a listed product must also be used and installed properly to assure safety. The basic rule is found in *Section 110-3(b)*, which states that

▶ *"listed or labeled equipment shall be installed and used in accordance with any instructions included in the listing or labeling."* ◀

There are a number of nationally recognized laboratories. The following laboratories do a considerable amount of testing and listing of electrical equipment.

Underwriters Laboratories Inc. (U.L.)

Underwriters Laboratories Inc. (U.L.), founded in 1894, is a highly qualified, nationally recognized testing laboratory. U.L. develops standards, and performs tests to these standards. Most reputable manufacturers of electrical equipment submit their products to U.L. where the equipment is subjected to numerous tests. These tests determine if the product can perform safely under normal and abnormal conditions to meet published standards. After U.L. tests, evaluates, and determines that a product complies with the specific standard, the manufacturer is then permitted to *label* its product with the U.L. Mark. The products are then *listed* in a U.L. Directory.

It should be noted that U.L. *does not approve* a product. Rather, U.L. *lists* those products that conform to a specific safety standard. A U.L. Listing Mark on a product means that representative samples of the product have been tested and evaluated to nationally recognized safety standards with regard to fire, electric shock, and related safety hazards.

Useful U.L. publications are:

Electrical Construction Equipment Directory (Green Book)

Electrical Appliance and Utilization Equipment Directory (Orange Book)

General Information for Electrical Equipment Directory (White Book)

It is extremely useful for an electrician, electrical contractor, and/or electrical inspector to refer to these directories when looking for specific requirements, permitted uses, limitations, and so on, for a certain product. These directories can be purchased from Underwriters Laboratories Inc. You may not be able to purchase all of the previous directories because of cost restraints. In this case, choose what is probably the best companion to the *NEC,*® which is the White Book.

Many times the answer to a product-related question cannot be found in the *NEC.*® It is quite possible that the answer might be found in the U.L. Directories.

The Green and Orange Directories provide technical information regarding a particular product, plus they list the names and addresses of manufacturers and the manufacturers' identification numbers. The White Book provides the technical information by which a product is tested, but does not show manufacturers' names and addresses.

The previous directories can be obtained by writing or calling:

Underwriters Laboratories Inc.
333 Pfingsten Road
Northbrook, Illinois 60062-2096
847-272-8800

Underwriters Laboratories Inc. and the Canadian Standards Association have worked out an agreement whereby either agency can test, evaluate, and list equipment for the other agency. For example, U.L. might test and list an air-conditioner unit to the requirements of U.L. Standard 1995 (Heating and Cooling Equipment) because the Canadian Standard C22.2 No. 236-M90 is a mirror image of U.L. Standard 1995. One by one, the U.L. and CSA Standards are becoming similar.

When U.L. tests and lists products that comply to the requirements of a particular CSA standard, the U.L. logo shown here will appear with a "C" outside of and to the left of the circle.

This means the product has been tested and evaluated for compliance *only* with Canadian requirements.

This harmonization of Codes and Standards is going on in the world today as a result of the North American Free Trade Act (NAFTA). Discussions are also going on with Mexico. When all of this is finalized, electrical equipment standards may be the same in the United States, Canada, and Mexico.

Canadian Standards Association (CSA)

The Canadian Standards Association is the Canadian counterpart of Underwriters Laboratories Inc. in the United States. CSA is the source of the *Canadian Electrical Code (CEC)* and of the Canadian Standards for the testing, evaluation, and listing of electrical equipment in Canada. The *Canadian Electrical Code* is quite different from the *NEC.*® A Canadian version of *Electrical Wiring— Residential* is available in Canada.

The *Canadian Electrical Code* and CSA standards can be obtained by contacting:

Canadian Standards Association
178 Rexdale Blvd.
Rexdale, Ontario, Canada
M9W 1R3

Intertek Testing Services (ITS)

Formerly known as Electrical Testing Laboratories, ITS is a nationally recognized testing laboratory. They provide testing, evaluation, labeling, listing, and follow-up service for the safety testing of electrical products. This is done in conformance to nationally recognized safety standards, or to specifically designated requirements of jurisdictional authorities.

Information can be obtained by writing to:

Intertek Testing Services
3933 U.S. Route 11
Cortland, NY 13045-0950
Phone: 1-800-345-3851

National Electrical Manufacturers Association (NEMA)

The National Electrical Manufacturers Association represents nearly 600 manufacturers of electrical products. NEMA develops electrical equipment standards, which, in many instances, are very similar to U.L. and other consensus standards. NEMA has representatives on the Code Making Panels for the *NEC.*® Additional information can be obtained by contacting:

National Electrical Manufacturers Association
11300 North 17th Street, Suite 1847
Rosslyn, VA 22209
Phone: 703-841-3200

REVIEW

Note: Refer to the *NEC®* or the plans in the back of this text where necessary.

1. What is the purpose of specifications? _____

_____

2. In what additional way are the specifications particularly useful to the electrical contractor? _____

_____

3. What is done to prevent a plan from becoming confusing because of too much detail?

_____

_____

4. Name three requirements contained in the specifications regarding material.

a. _____   c. _____

b. _____

5. The specifications state that all work shall be done _____

_____

6. What phrase is used when a substitution is permitted for a specific item? _____

_____

7. What is the purpose of an electrical symbol? _____

_____

8. What is a notation? _____

_____

9. Where are notations found? _____

10. List at least twelve electrical notations found on the plans for this residence. Refer to the plans at the back of the text. _____

_____

_____

_____

_____

_____

_____

11. What three parties must be satisfied with the completed electrical installation?

a. _____   b. _____   c. _____

12. What Code sets standards for electrical installation work? _____

_____

13. What authority enforces the standards set by the Code? _____

_____

14. Does the Code provide minimum or maximum standards? _____

15. What do the letters *U.L.* signify? _____

    _____

16. What section of the Code states that all listed or labeled equipment shall be installed or used in accordance with any instructions included in the listing or labeling?

    _____

17. When the words "shall be" appear in a Code reference, they mean that it (must)(may) be done. (Underline the correct word.)

18. What is the purpose of the *NEC®*?

    _____

    _____

    _____

    _____

    _____

19. Does compliance with the Code always result in an electrical installation that is adequate, safe, and efficient? Why? _____

    _____

    _____

    _____

    _____

20. Name two nationally recognized testing laboratories. _____

    _____

    _____

21. a. Do Underwriters Laboratories and the other recognized testing laboratories "approve" products? _____

    b. What do these testing laboratories do? _____

    _____

    _____

    _____

22. a. Has the *NEC®* been officially adopted by the community in which you live? _____

    b. By the state in which you live? _____

    c. If your answer is YES to a. or b., are there amendments to the *NEC®*? _____

    d. If your answer is YES to c., list some of the more important amendments.

    _____

    _____

    _____

    _____

23. Does the *NEC®* make suggestions about how to wire a house that will be occupied by handicapped persons? _____

24. A junction box on a piece of European equipment is marked 200 cubic centimeters. Convert this to cubic inches. _____

    _____

25. Convert 4,500 watts to Btu/hour. _____

    _____

26. Residential lighting loads are based upon 3 volt-amperes per square foot. Determine the minimum volt amperes (watts) required for an area of 186 square meters. _____

    _____

Electrical Symbols and Outlets

OBJECTIVES

After studying this Unit, the student will be able to

- identify and explain the electrical outlet symbols used in the plans of the single-family dwelling.
- discuss the types of outlets, boxes, fixtures, and switches used in the residence.
- explain the methods of mounting the various electrical devices used in the residence.
- understand the meaning of the terms *receptacle outlet* and *lighting outlet*.
- understand the preferred way to position receptacles in wall boxes.
- understand issues involved in remodel work.
- know how to position wall boxes in relation to finished wall surfaces.
- know how to make surface extensions from concealed wiring methods.
- understand how to determine the number of wires permitted in a given size box.
- discuss interchangeable wiring devices.
- understand the concept of fire resistance rating of walls and ceilings.

ELECTRICAL SYMBOLS

Electrical symbols used on an architectural plan show the location and type of electrical device required. A typical electrical installation as taken from a plan is shown in Figure 2-1.

The *NEC®* describes an *outlet* as "a point on a wiring system where current is taken to supply utilization equipment."

A **receptacle outlet** is "an outlet where one or more receptacles are installed," Figure 2-2.

A **lighting outlet** is "an outlet intended for the direct connection of a lampholder, a lighting fixture, or a pendant cord terminating in a lampholder," Figure 2-3.

The *NEC®* defines a *device* as "a unit of an electrical system which is intended to carry but not utilize electric energy."

Figure 2-1 Use of electrical symbols and notations on a floor plan.

►ALLOW NOT LESS THAN 6 INCHES (152 mm) OF CONDUCTOR MEASURED FROM WHERE THE CONDUCTORS EMERGE FROM CABLE OR RACEWAY TO THE END OF THE CONDUCTOR ◄

►ALLOW NOT LESS THAN 3 INCHES (76 mm) OF CONDUCTOR MEASURED FROM THE FRONT EDGE OF OPENING TO THE END OF THE CONDUCTOR WHERE THE OPENING IS LESS THAN 8 INCHES (203 mm) IN ANY DIMENSION. ◄

►THESE DIMENSIONAL MINIMUMS DO NOT APPLY TO CONDUCTORS THAT ARE NOT SPLICED OR TERMINATED IN THE BOX. ◄

Figure 2-2 Where a receptacle is connected to the branch-circuit wires, the outlet is called a *receptacle outlet*. *Section 300-14* states the minimum length of conductors at an outlet, junction, or switch point for splices or the connection of fixtures or devices. Leaving the conductors too long can cause crowding of the wires in the box, leading to possible short-circuits and/or ground faults.

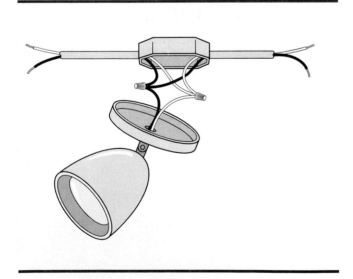

Figure 2-3 When a lighting fixture is connected to the branch-circuit wires, the outlet is called a *lighting outlet*.

By definition, toggle switches and receptacles are devices because they carry current, but do not consume power.

The term *opening* is widely used by electricians and electrical contractors when estimating the cost of an installation. The term *opening* covers all lighting outlets, receptacle outlets, junction boxes, switches, etc. The electrician and/or electrical contractor will estimate a job at "X dollars per lighting outlet," "X dollars per switch," "X dollars per receptacle outlet," and so on. These estimates will include the time and material needed to complete the job. Each type of electrical *opening* is represented on the electrical plans as a symbol. In Figure 2-1, the electrical openings are shown by the symbols:

Receptacle outlet:

Three-way switch: S_3

Ceiling lighting outlet:

Standard electrical symbols commonly found on architectural and electrical plans are detailed in Figures 2-4 through 2-9.

OUTLETS	CEILING	WALL
INCANDESCENT		
LAMPHOLDER WITH PULL SWITCH		
RECESSED INCANDESCENT		
SURFACE FLUORESCENT		
RECESSED FLUORESCENT		
SURFACE OR PENDANT CONTINUOUS ROW FLUORESCENT		
RECESSED CONTINUOUS ROW FLUORESCENT		
BARE LAMP FLUORESCENT STRIP		
SURFACE OR PENDANT EXIT		
RECESSED CEILING EXIT		
BLANKED OUTLET		
OUTLET CONTROLLED BY LOW-VOLTAGE SWITCHING WHEN RELAY IS INSTALLED IN OUTLET BOX		
JUNCTION BOX		

Figure 2-4   Lighting outlet symbols.

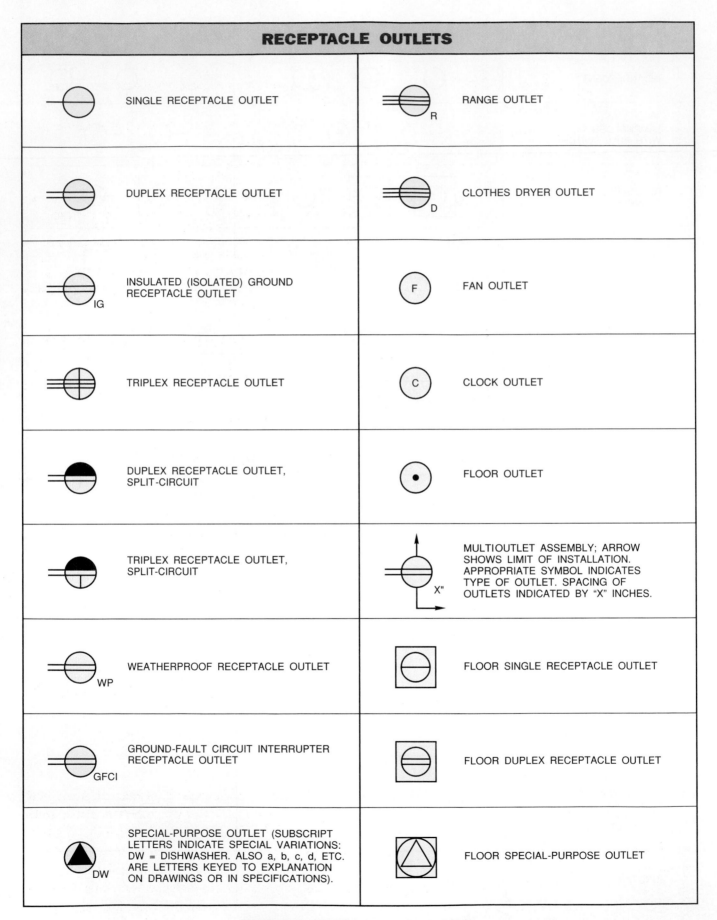

Figure 2-5 Receptacle outlet symbols.

SWITCH SYMBOLS

S	SINGLE-POLE SWITCH
S_2	DOUBLE-POLE SWITCH
S_3	THREE-WAY SWITCH
S_4	FOUR-WAY SWITCH
S_D	DOOR SWITCH
S_{DS}	DIMMER SWITCH
S_K	KEY SWITCH
S_L	LOW-VOLTAGE SWITCH
S_{LM}	LOW-VOLTAGE MASTER SWITCH
S_P	SWITCH WITH PILOT LAMP
S_R	VARIABLE-SPEED SWITCH
S_T	TIME SWITCH
S_{WP}	WEATHERPROOF SWITCH

Figure 2-6 Switch symbols.

The dash lines in Figure 2-1 run from the outlet to the switch or switches that control the outlet. These lines are usually curved so that they cannot be mistaken for invisible edge lines. Outlets shown on the plan without curved dash lines are independent outlets and have no switch control.

A study of the plans for the single-family dwelling shows that many different electrical symbols are used to represent the electrical devices and equipment used in the building.

Figures 2-4 through 2-9 list the standard, approved electrical symbols and their meanings. Many of these symbols can be found on the accompanying plans of the residence. Note that in these figures several symbols have the same shape. However, differences in the interior presentation indicate that the meanings of the symbols are different. For example, different meanings are shown in Figure 2-10 for the outlet symbol. A good practice to follow in studying symbols is to learn the basic forms first and then add the supplemental information to obtain different meanings.

In drawing electrical plans, most architects, designers, and electrical engineers use symbols approved by the American National Standards Insti-

Figure 2-7 Circuiting symbols.

tute (ANSI) wherever possible. However, plans may contain symbols that are not found in these standards. When such unlisted (nonstandard) symbols are used, the electrician must refer to a legend that interprets these symbols. The legend may be included on the plans or in the specifications. In many instances, a notation on the plan will clarify the meaning of the symbol.

Figure 2-8 Miscellaneous symbols.

SYMBOL	NOTATION
1	PLUGMOLD ENTIRE LENGTH OF WORKBENCH. OUTLETS 18" (457 mm) O.C. INSTALL 48" (1.2 m) TO CENTER FROM FLOOR. GFCI PROTECTED.
2	TRACK LIGHTING. PROVIDE 5 LAMPHOLDERS.
3	TWO 40-WATT RAPID START FLUORESCENT LAMPS IN VALANCE. CONTROL WITH DIMMER SWITCH.

Figure 2-9   Example of how certain notations might be added to a symbol when a symbol itself does not fully explain its meaning. The architect or engineer has a choice of explaining fully the meaning directly on the plan if there is sufficient room; if insufficient room, then a notation could be used.

FIXTURES AND OUTLETS

The location of lighting outlets is determined by the amount and type of illumination required to provide the desired lighting effects. It is not the intent of this text to describe how proper and adequate lighting is determined. Rather, the text covers the proper methods of installing the circuits for such lighting.

Lighting fixture manufacturers publish catalogs that provide a tremendous amount of information regarding recommendations for residential lighting. These publications are available at electrical distributors, lighting showrooms, and home centers. The Instructor's Guide lists some excellent publications relating to residential lighting that can be obtained by requesting same.

Architects often include in the specifications a certain amount of money for the purchase of lighting fixtures. The electrical contractor includes this amount in the bid, and the choice of fixtures is then left to the homeowner. If the owner selects fixtures whose total cost exceeds the fixture allowance, the owner is expected to pay the difference between the actual cost and the specification allowance. If

Figure 2-10   Variations in significance of outlet symbols.

the fixtures are not selected before the roughing-in stage of wiring the house, the electrician usually installs outlet boxes having standard fixture mounting studs.

Most modern surface-mount lighting fixtures can be fastened to a fixture stud in the box (Figure 2-11, top row, picture A) or an outlet box or plaster ring using appropriate #8-32 metal screws and mounting strap furnished with the fixture (Figure 2-11, third row, raised plaster cover; outlet boxes).

Symbols Used to Represent Lighting Outlets and Switches

Some of the standard symbols for various types of commonly used switches are shown in Figure 2-12. Wiring diagrams are also shown. Typical lighting outlet symbols are shown in Figure 2-11.

OUTLET, DEVICE, AND JUNCTION BOXES

Article 370 contains much detail about outlet boxes, device (switch) boxes, pull boxes, junction boxes, and conduit bodies (LB, LR, LL, C, etc.) all of which are used to: provide the necessary space for making electrical connections, contain wiring, and support lighting fixtures, etc. *Article 370* is quite lengthy. For the moment, we will focus on some of the basic Code requirements for outlet boxes, device boxes, and junction boxes. Additional Code requirements are covered on an "as needed" basis in this chapter and in later chapters.

Some Basic Code Requirements

- *Sections 300-11* and *370-23:* Boxes must be securely mounted and fastened in place.

SYMBOL	TYPE OF FIXTURE OR OUTLET	OUTLET BOXES THAT MAY BE USED
CEILING LIGHTING OUTLET	SURFACE-MOUNTED LIGHTING FIXTURE	(A) ADJUSTABLE BAR HANGER WITH FIXTURE STUD (B) OCTAGON BOX ON BRACKET (C) OCTAGON BOX AND OFFSET BAR HANGER BOXES ARE AVAILABLE WITH EITHER INTERNAL OR EXTERNAL CABLE CLAMPS.
RECESSED CEILING LIGHTING OUTLET	RECESSED LIGHTING FIXTURE	CABLE TERMINATES IN BOX THAT IS FURNISHED WITH FIXTURE. SPECIAL CODE RULES GOVERNING THE INSTALLATION ARE FOUND IN *SECTIONS 410-64* THROUGH *410-72* OF THE *NATIONAL ELECTRICAL CODE.* SEE UNIT 7.
WALL LIGHTING OUTLET	WALL-MOUNTED LIGHTING FIXTURE	4"-SQUARE BOX WITH SIDE BRACKET AND PROPER RAISED PLASTER COVER. A 4"-OCTAGON BOX COULD ALSO BE USED.
DUPLEX RECEPTACLE OUTLET		SECTIONAL SWITCH DEVICE BOXES

Figure 2-11   Types of fixtures and outlets. *Photo courtesy* of **Progress Lighting.**

SYMBOL	FLUSH TOGGLE SWITCH	OPERATION	CONNECTIONS
S SINGLE-POLE		ON OFF	
S₂ DOUBLE-POLE		ON OFF	
S₃ THREE-WAY		POSITION 1 POSITION 2	
S₄ FOUR-WAY		1 2	
Sₚ SWITCH AND PILOT LIGHT		FOR CONTROLLING LIGHTS FROM ONE POINT WITH PILOT LIGHT INDICATION	ALSO AVAILABLE IN THREE-WAY TYPE OF CONTROLLING LIGHT FROM TWO POINTS WITH PILOT LIGHT INDICATION

Figure 2-12 Standard switches and symbols.

Figure 2-13   A box (or fitting) must be installed wherever there are splices, outlets, switches, or other junction points. Refer to the points marked X. A Code violation is shown at point XX, *Section 300-15(a)*.

- *Section 300-15(a):* Where conduit, electrical-metallic tubing, nonmetallic-sheathed cable, Type AC cable, or other cables are installed, a box or conduit body must be installed at each conductor splice connection point, outlet, switch point, junction point, or pull point. This requirement is illustrated in Figure 2-13.

- *Section 300-15(c):* Where cables enter or exit from conduit or tubing that is used to provide cable support or protection against physical damage, a fitting shall be provided on the end(s) of the conduit or tubing to protect the cable from abrasion. Figure 2-14 shows a transition from a cable to a raceway.

- *Section 370-16:* Boxes must be large enough for all of the enclosed conductors and wiring devices. Installing boxes too small for the number of conductors, wiring devices, fixture studs, and splices is a real problem! This topic is covered in detail later.

- *Section 370-17(c):* The exception allows multiple cables to be run through a single knockout opening in a nonmetallic box. This permission is not given for metallic boxes. See Figure 2-15.

Figure 2-14   *Section 300-15(c)* permits a transition to be made from one wiring method to another wiring method. In this case, the armor of the Type AC cable is removed, allowing sufficient length of the conductors to be run through the conduit. A proper fitting must be used at the transition point, and the fitting must be accessible after installation.

Figure 2-15 Multiple cables may run through a single knockout opening in a nonmetallic box, *Section 370-17(c), Exception.*

- *Section 370-20:* In walls or ceilings constructed of wood or other combustible material, install boxes so they are flush with the finished wall surface. In walls or ceilings constructed of concrete, tile, or other noncombustible material, it is acceptable to have the front edge of the box set back not more than ¼ inch (6.35 mm) from the finished surface. See Figure 2-20.

- *Section 370-22:* A surface extension from a box of a concealed wiring system is made by mounting and mechanically securing a box or extension ring over the concealed box. An exception to this is that a surface extension is permitted to be made from the cover of a concealed box where the cover is designed so it is unlikely to fall off, or be removed if its securing means becomes loose. The wiring method must be flexible and arranged so that any required grounding continuity is independent of the connection between the box and cover. This is shown in Figure 2-16.

- *Section 370-25:* In completed installations, all boxes shall have a cover, faceplate, or fixture canopy. Do not leave any electrical boxes uncovered.

- *Section 370-27(a):* Boxes used at lighting outlets must be designed or installed so that a lighting fixture may be attached to it.

- *Section 370-27(a), Exception:* ▶A wall-mounted lighting fixture not over 6 lb. (2.72 kg) and not more than 16 inches (406 mm) in any dimension may be supported to a box by not less than two No. 6/32 or larger screws.◀ This new Code rule permits small wall-mounted lighting fixtures to be supported by device boxes or plaster rings that are tapped for No. 6/32 screws.

- *Section 370-27(c):* ▶Standard outlet boxes must not be used as the sole support for ceiling-suspended (paddle) fans unless they are specifically "listed" for this purpose.◀ See Unit 9 for detailed discussion of this issue.

- *Section 370-29:* Conduit bodies, junction, pull, and outlet boxes must be installed so the wiring contained in them is accessible without removing any part of the building. Never install outlet boxes, junction boxes, pull boxes, or conduit bodies where they will be inaccessible behind or above permanent finished walls or ceilings. Should anything ever go wrong, it would be a nightmare, if not impossible, to troubleshoot and locate hidden splices. Wiring located above "lay-in" ceilings is considered accessible because it is easy to remove a panel(s) to gain access to the space above the ceiling. The *NEC®* definition of *accessible* is *"Capable of being removed or exposed without damaging the building structure or finish, or not permanently closed in by the structure or finish of the building."*

  The same accessibility requirement is true for underground wiring. Boxes must be accessible without excavating sidewalks, paving, earth, or other material used to establish the finished grade.

- *Section 370-71:* Pull and junction boxes must provide adequate space and dimensions for the installation of conductors, and they shall comply with the specific requirements of this section.

- *Section 370-72:* A portion of this Code section is similar to *Section 370-29* regarding accessibility to outlet boxes, and to using "suitable covers" for closing outlet boxes.

CEILING

OUTLET BOX CONCEALED
IN CEILING

EXPOSED EXTENSION–CONDUIT

A

VIOLATION. IT IS AGAINST THE CODE
TO MAKE A RIGID EXTENSION FROM A COVER THAT IS ATTACHED TO
A CONCEALED OUTLET BOX, *SECTION 370-22.* IT WOULD BE DIFFICULT TO
GAIN ACCESS TO THE CONNECTIONS INSIDE THE BOX.

CEILING

OUTLET BOX CONCEALED
IN CEILING

EXTENSION RING

EXPOSED EXTENSION–CONDUIT

B

THIS MEETS CODE. A BOX OR EXTENSION
RING MUST BE MOUNTED OVER AND MECHANICALLY SECURED TO THE
CONCEALED OUTLET BOX, *SECTION 370-22.*

SEPARATE EQUIPMENT
GROUNDING CONDUCTOR

CEILING

OUTLET BOX CONCEALED
IN CEILING

FLEXIBLE EXTENSION

EXTENSION MUST PROVIDE
A SEPARATE EQUIPMENT
GROUNDING CONDUCTOR.

C

THIS MEETS CODE. IT IS PERMITTED TO
MAKE A SURFACE EXTENSION FROM A COVER FASTENED TO A CONCEALED
BOX WHERE THE EXTENSION WIRING METHOD IS FLEXIBLE AND WHERE A
SEPARATE GROUNDING CONDUCTOR IS PROVIDED SO THAT THE GROUNDING
PATH IS NOT DEPENDENT UPON THE SCREWS THAT ARE USED TO FASTEN
THE COVER TO THE BOX.

IT WOULD BE A CODE VIOLATION TO MAKE THE EXTENSION IF A
FLEXIBLE WIRING METHOD WAS USED, BUT HAD NO PROVISIONS FOR
A SEPARATE GROUNDING CONDUCTOR.

Figure 2-16   How to make an extension from a flush-mounted box.

- *Section 410-12:* This is another repeat of the requirement that boxes must be covered. The "cover" could be a blank cover, a fixture canopy, a lampholder, a switch or receptacle, or similar device with faceplate.

- *Section 410-16(a):* "Listed" outlet boxes are suitable for hanging lighting fixtures that weigh 50 pounds (22.7 kg) or less. ▶Lighting fixtures that weigh more than 50 pounds (22.7 kg) must be supported independently of the outlet box, or be "listed" for supporting more than 50 pounds (22.7 kg).◀

Boxes Where Fixtures Will Be Installed.   Be careful when roughing in boxes for lighting fixtures. U.L. states in its Green and White Books that it has not investigated supporting lighting fixtures from nonmetallic device boxes, nonmetallic device plaster rings, metallic device boxes, or metallic device plaster rings. Therefore, unless the box, device plaster ring, or carton is marked to indicate that it has been "listed" for the support of a lighting fixture, do not use where lighting fixtures are to be hung. See Figure 2-17.

What to Do Where Recessed Lighting Fixtures Will Be Installed.   This topic is covered in detail in Unit 7.

METALLIC SWITCH
(DEVICE) BOX
OR DEVICE
PLASTER RING

ANY NONMETALLIC BOX

Figure 2-17   ▶Wall-mounted lighting fixtures are permitted to be secured to these boxes and plaster rings by No. 6/32 or larger screws provided that they do not weigh more than 6 pounds (2.72 kg) and do not exceed 16 inches (406 mm) in any dimension, *Section 370-27(a), Exception.*◀

NONMETALLIC OUTLET AND DEVICE BOXES

Section 370-3 permits nonmetallic outlet and device boxes to be installed where the wiring method is nonmetallic sheathed cable or nonmetallic raceway. *Section 370-3 (Exceptions)* allows metal raceways and metal-jacketed cables (BX) to be used with non-metallic boxes provided all metal raceways or cables entering the box are bonded together to maintain the integrity of the grounding path to other equipment in the installation.

The house wiring system usually is formed by a number of specific circuits. Each circuit consists of runs of cable from outlet to outlet or from box to box. The residence plans show many branch circuits for general lighting, **appliances**, electric heating, and other requirements. The specific Code rules for each of these circuits are covered in later Units.

GANGED SWITCH (DEVICE) BOXES

A flush switch or receptacle for residential use fits into a standard 2" × 3" sectional switch box (some-times called a *device box*). When two or more switches or receptacles are located at the same point, the switch boxes are ganged or fastened together to provide the required mounts, Figure 2-16.

Three switch boxes can be ganged together by removing and discarding one side from both the first and third switch boxes and both sides from the second (center) switch box. The boxes are then joined together as shown in Figure 2-18. After the switches are installed, the gang is trimmed with a gang plate having the required number of switch handle or receptacle openings. These plates are called two-gang wall plates, three-gang wall plates, and so on, depending upon the number of openings.

The dimensions of a standard sectional switch box (2" × 3") are the dimensions of the opening of the box. The depth of the box may vary from 1½ inches to 3½ inches, depending upon the requirements of the building construction and the number of conductors and devices to be installed. See Figure 2-19 for a complete listing of box dimensions.

Figure 2-18 Standard flush switches installed in ganged sectional device boxes.

QUIK-CHEK BOX SELECTION GUIDE
FOR METAL BOXES GENERALLY USED FOR RESIDENTIAL WIRING

DEVICE BOXES

WIRE SIZE	3x2x1½ (7.5 in³)	3x2x2 (10 in³)	3x2x2¼ (10.5 in³)	3x2x2½ (12.5 in³)	3x2x2¾ (14 in³)	3x2x3 (16 in³)	3x2x3½ (18 in³)
#14	3	5	5	6	7	8	9
#12	3	4	4	5	6	7	8

SQUARE BOXES

WIRE SIZE	4x4x1½ (21 in³)	4x4x2⅛ (30.3 in³)
#14	10	15
#12	9	13

OCTAGON BOXES

WIRE SIZE	4x1½ (15.5 in³)	4x2⅛ (21.5 in³)
#14	7	10
#12	6	9

HANDY BOXES

WIRE SIZE	4x2⅛x1½ (10.3 in³)	4x2⅛x1⅞ (13 in³)	4x2x2⅛ (14.5 in³)
#14	5	6	7
#12	4	5	6

RAISED COVERS

WHERE RAISED COVERS ARE MARKED WITH THEIR VOLUME IN CUBIC INCHES, THAT VOLUME MAY BE ADDED TO THE BOX VOLUME TO DETERMINE MAXIMUM NUMBER OF CONDUCTORS IN THE COMBINED BOX AND RAISED COVER.

NOTE: BE SURE TO MAKE DEDUCTIONS FROM THE ABOVE MAXIMUM NUMBER OF CONDUCTORS PERMITTED FOR WIRING DEVICES, CABLE CLAMPS, FIXTURE STUDS, AND GROUNDING CONDUCTORS. THE CUBIC INCH (IN³) VOLUME IS TAKEN DIRECTLY FROM *TABLE 370-16(a)* OF THE *NEC.*® NONMETALLIC BOXES ARE MARKED WITH THEIR CUBIC INCH CAPACITY.

Figure 2-19   Quik-chek box selection guide.

BOX MOUNTING

Section 370-20 states that boxes must be mounted so that they will be set back not more than ¼ inch (6.35 mm) when the boxes are mounted in non-combustible walls or ceilings made of concrete, tile, or similar materials. When the wall or ceiling construction is of combustible material (wood), the box must be set flush with the surface, Figure 2-20.

These requirements are meant to prevent the spread of fire if a short circuit occurs within the box.

Ganged sectional switch (device) boxes can be installed using a pair of metal mounting strips. These strips are also used to install a switch box between wall studs, Figure 2-21. The use of a bracket box in such an installation may result in an off-center receptacle outlet. When an outlet box is to

CONCRETE, TILE, OR OTHER NONCOMBUSTIBLE MATERIAL

BOX SET BACK NOT MORE THAN 1/4 INCH (6.35 mm)

A

WOOD OR OTHER COMBUSTIBLE MATERIAL

BOX SET FLUSH WITH FINISHED SURFACE

REPAIR THESE GAPS SO GAPS OR OPENINGS ARE NO GREATER THAN 1/8 INCH

B

C

Figure 2-20 **Box position in walls and ceilings constructed of various materials. Where the box for some reason is not flush with the finished wall, then install a switch box extender available in both metal and plastic of the type shown in (C). See *Sections 370-20* and *370-21*.**

Figure 2-21 Switch (device) boxes installed between studs using metal mounting strips, often referred to as "Kruse strips." If wood strips are used, they must have a cross-sectional dimension of not less than 1 inch × 2 inches, *Section 370-23(b)(2)*.

be mounted at a specific location between joists, as for ceiling-mounted fixtures, an offset bar hanger is used, Figure 2-11.

The Code states that when a switch box or outlet box is mounted to a stud or ceiling joist by nailing through the box, the nails must be not more than ¼ inch (6.35 mm) from the back or ends of the box, Figure 2-22. This requirement ensures that when the nail passes through the box, it does not interfere with the wiring devices in the box.

Most residences are constructed with wood framing. However, metal studs and joists are being used more and more. Figure 2-23 shows types of boxes used for fast, easy attachment to metal framing members. These boxes "snap" onto the metal stud, and do not require drilling the metal studs and fastening the box with nuts and bolts. The first one is a 4-inch square box with ½-inch knockouts. The second shows a 4-inch square box with nonmetallic sheathed cable (Romex) clamps. The third shows a 4-inch square box with armored cable (BX) clamps. See Unit 4 for details about running cables through metal framing members.

Another type of device used for fastening electrical boxes to steel framing members is shown in Figure 2-24. The bracket is screwed to the steel framing members, then the box or boxes are attached to the bracket.

Remodel (Old Work)

When installing switches, receptacles, and lighting outlets in remodel work where the walls and ceilings (paneling, drywall, sheetrock and plaster, or lath and plaster) are already in place, first make sure

Figure 2-22 Using nails to install a sectional switch box, *Section 370-23(b)(1)*.

Figure 2-23   Types of boxes that can be used with steel framing construction. Note the side clamps that "snap" directly onto the steel framing members.

you will be able to run cables to where the switches, receptacles, and lighting outlets are to be installed. After making sure you can get the cables through the concealed spaces in the walls and ceilings, you can then proceed to cut openings the size of the specific type of box to be installed at each outlet location. Cables are then "fished" through the stud and joist spaces behind or above the finished walls and ceil-

ings to where you have cut the openings. Then, boxes having plaster ears and snap-in brackets, Figure 2-25, can be inserted from the front, through the holes already cut at the locations where the wall or ceiling boxes are to be installed. After the boxes are snapped into place through the hole from the front, they become securely locked into place.

Another popular method of fastening wall and

Figure 2-24   A bracket for use on steel framing members. The bracket spans the space between two studs. Note that more than one box can be mounted on this bracket. *Courtesy* Erico,® Inc.

Figure 2-25 Nonmetallic boxes are used for remodel work so that when they are inserted into an opening cut into a wall or ceiling, they will "snap" into place, thus conforming to the *NEC*® requirements for the adequate support of boxes.

ceiling boxes in existing walls and ceilings is to use boxes that have plaster ears. Insert the box through the hole from the front. Then position a metal support into the hole on each side of the box. Be careful not to let the metal support fall into the hole. Next, bend the metal support over and into the box, Figure 2-26. Two metal supports are used. Over the years, these metal supports have become known as "Madison Holdits."

Also available for old work are boxes that have a screw-type support on each side of the box, Figure 2-27. This type of box has plaster ears. It is inserted through the hole cut in the wall or ceiling, then the screws are tightened. This pulls up a metal support

Figure 2-26 Metal supports referred to as "Madison Holdits."

Figure 2-27 These boxes are commonly used in remodel work. The box is installed from the front, then the screws are tightened. The box is adequately supported.

tightly behind the wall or ceiling, firmly holding the box in place. The action is very similar to that of Molly screw anchors.

Because these types of remodel boxes are supported by the drywall or plaster, as opposed to new work where the supporting of the boxes is to the framing members, be careful not to hang heavy lighting fixtures from them. Be sure that there is proper support if a ceiling fan is to be installed. This is covered in Unit 9.

Spread of Fire

To protect lives and property, fire must be contained and not be allowed to spread.

This text covers one-family dwelling units. In walls, partitions, and ceilings that are *fire resistance rated*, special consideration must be given when installing wall and ceiling electrical boxes where you might want to mount the boxes "back-to-back" or in the same stud or joist space of common walls or ceilings, as might be found in multifamily buildings. Building codes such as the International Code Council (ICC) *One- and Two-Family Dwelling Code* require fire-resistance-rated walls and ceilings between occupancies in two-family dwellings and in townhouses. These building codes also require fire resistance rating between habitable areas and a garage, even in single-family dwellings. Ratings of fire-resistant materials are expressed in hours. Terms such as 1-hour fire resistance rating, 2-hour fire resistance rating, and so on, are used. Electrical wall boxes installed back-to-back or installed in the same stud or joist space defeat the fire resistance rating of the wall or ceiling.

U.L. Standard 263 *Fire Tests of Building Construction and Materials* and NFPA 251 cover fire resistance ratings. In U.L. Standard 263, and in all building codes, you will find requirements such as: "the surface area of individual metallic outlet or switch boxes shall not exceed 16 square inches" (i.e., a 4-inch square box is $4 \times 4 = 16$ square inches), "the aggregate surface area of the boxes shall not exceed 100 square inches per 100 square feet of wall surface," "boxes located on opposite sides of walls or partitions shall be separated by a minimum horizontal distance of 24 inches," "the metallic outlet or switch boxes shall be securely fastened to the studs and the opening in the wallboard

facing shall be cut so that the clearance between the box and the wallboard does not exceed ⅛ inch," and "the boxes shall be installed in compliance with the *National Electrical Code.®*"

The U.L. *Fire Resistance Directory* covers fire-resistant materials and assemblies, and contains a rather detailed listing of *Outlet Boxes and Fittings Classified for Fire Resistance (CEYY)*, showing all manufacturers' product part numbers for *nonmetallic boxes* to be installed in walls, partitions, and ceilings that meet the fire-resistance standard. This category also covers special-purpose boxes for installation in floors.

When using nonmetallic boxes in fire-resistance-rated walls, the restrictions are more stringent than for metallic boxes. For example, the U.L. *Fire Resistance Directory* shows that one manufacturer of nonmetallic boxes limits the product to be installed so that no opening exceeds 10 square inches (64.5 cm²). Another manufacturer limits the product to be installed so that no opening exceeds 25 square inches (161 cm²).

For both metallic and nonmetallic electric outlet boxes, the *minimum* distance between boxes on opposite sides of a fire-rated wall or partition is 24 inches (610 mm). There is an exception, which is discussed later.

For both metallic and nonmetallic electric outlet boxes, the maximum "gap" between the box and the wall material is ⅛ inch (3.2 mm). This holds true for fire-resistance-rated walls and non-fire-resistance-rated walls.

Of interest to an electrician is an item listed in the U.L. *Fire Resistance Directory* that is an *intumescent* (expands when heated) fire-resistant material that comes in "pads." This moldable putty can be used to wrap electrical boxes that are installed in the same wall cavity. This moldable putty inhibits heat transfer from the fire-resistance-rated side of the wall to the non-fire-resistance rated side of the wall. It is pressed onto the metallic electrical outlet box, around electrical raceways that enter the box, and is pressed into the interface where the box, stud, and gypsum wallboard meet. Electricians commonly use the 6" × 7" pad, ⅛ inch thick. The ⅛ inch (3.2 mm) thickness provides a 1-hour fire resistance rating. A 2-hour rating is obtained with ¼ inch (6.4 mm) thickness. When this material is properly installed, the 24-inch (610 mm) separation between boxes is not required. However, the electrical outlet boxes *must not* be installed back-to-back. Using this material meets the requirements of *Section 300-21* of the *NEC*® and other building codes.

Rule 402.7 in the International Code Council (ICC) *One- and Two-Family Dwelling Code* states that "firestopping be provided to cut off all concealed draft openings (both vertical and horizontal) and to form an effective fire barrier between stories, and between a top story and the roof space." This includes openings around vents, pipes, ducts, chimneys, and fireplaces at ceiling and floor level. Oxygen supports fire. Firestopping cuts off or significantly reduces the flow of oxygen. Most communities that have adopted the ICC code require firestopping around electrical conduits. In residential work, this is generally done by the insulation installer, before the dry wall installer closes up the walls.

The previous discussion makes it clear that the electrician must become familiar with certain building codes, in addition to the *NEC*.® Knowing the *NEC*® is not enough!

It is extremely important to check with the electrical inspector and/or building official for complete clarification on the matter of fire-resistance requirements before you find yourself cited with serious violations and lawsuits regarding noncompliance with fire codes. Changes can be costly, and will delay completion of the project. Improper wiring is hazardous. The *NEC*® addresses this issue in *Section 300-21*.

Figures 2-28, 2-29, 2-30, and 2-31 illustrate the basics for installing electrical outlet and switch boxes in fire-resistance-rated partitions. This is an extremely important issue. If in doubt, check with your local electrical inspector and/or building official.

Figure 2-28 Building codes do not permit electrical wall boxes to be mounted back-to-back. Electrical wall boxes must be kept at least 24 inches apart. This is to maintain the integrity of the fire-resistance rating of the wall.

Figure 2-29 Building codes permit boxes to be mounted back-to-back or in the same wall cavity when the fire-resistance rating is maintained. In this detail, gypsum board has been installed in the wall cavity between the boxes, creating a new fire resistance barrier. Verify if this is acceptable to the building authority in your area.

THIS SIDE OF WALL IS NON-FIRE-RESISTANCE-RATED.

THIS SIDE OF WALL HAS A 1-HOUR FIRE-RESISTANCE RATING.

Figure 2-30   Building codes permit electrical outlet boxes to be mounted in the same stud space when the fire-resistance rating is maintained. In this detail, "Wall Opening Protective Material" is packed around the electrical outlet box, thereby maintaining the fire-resistance rating of the wall. Electrical outlet boxes are not permitted to be installed back-to-back.

Figure 2-31   This illustration shows an electrical outlet box that has been covered with fire-resistance "putty" to retain the wall's fire-resistance rating per building codes.

BOXES FOR CONDUIT WIRING

Some local electrical ordinances require conduit rather than cable wiring. Conduit wiring is discussed in Unit 18. Examples of conduit fill (how many wires are permitted in a given size conduit) are presented.

When conduit is installed in a residence, it is quite common to use 4-inch square boxes trimmed with suitable plaster covers. There are sufficient knockouts in the top, bottom, sides, and back of the box to permit a number of conduits to run to the box. Plenty of room is available for the conductors and wiring devices. Note how easily these 4-inch square outlet boxes can be mounted back-to-back by installing a small fitting between the boxes. This is illustrated in Figures 2-32(B) and (C).

Four-inch square outlet boxes can be trimmed with one-gang or two-gang plaster rings where wiring devices will be installed. Where lighting fixtures will be installed, a plaster ring having a round opening should be installed.

Any unused knockouts and other openings in electrical boxes, cabinets, panels, and load centers must be closed per *Sections 110-12(a)* and *370-18*, Figure 2-33.

Close the Gap Around Boxes!

Section 370-21 requires that openings around electrical boxes shall be repaired so that "there will be no gaps or open space greater than ⅛ inch (3.18 mm) at the edge of the box or fitting." This is to minimize the spread of fire. The U.L. *Fire Resistance Directory* mirrors the *NEC*® rule by stating that "the outlet or switch boxes shall be securely fastened to the studs and the opening in the wallboard facing shall be cut so that the clearance between the box and the wallboard does not exceed ⅛ inch." See Figure 2-20(B).

Figure 2-32   (A) shows electrical metallic tubing run to a 4-inch square box. (B) and (C) show 4-inch square boxes attached together for "back-to-back" installations. When mounting boxes back-to-back, be sure to consider possible transfer of sound between the rooms. Also consider the spread of fire and follow fire-resistance rating codes if applicable.

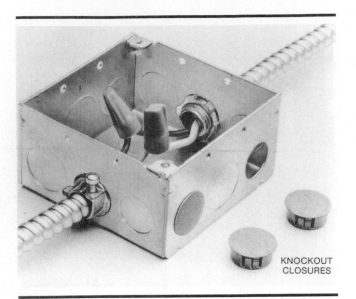

KNOCKOUT
CLOSURES

Figure 2-33 Unused openings in boxes must be closed according to *Sections 110-12(a)* and *370-18*. This is done to contain electrical short-circuit problems inside the box or panel and to keep rodents out. Note that the conductors *do not* meet the 3-inch and 6-inch (152 mm) conductor length requirement of *Section 300-14*. This is a good example of a Code violation. See Figure 2-2 for proper installation regarding conductor length at boxes.

This is easier said than done. The electrician installs the wiring (rough-in), followed by an inspection by the electrical inspector. Next comes the dry wall/plaster/panel installer, who many times cuts out the box openings much larger than the outlet or switch box. In most cases, the walls and ceilings are painted or wall papered before the electrician returns to install the receptacles, switches, and lighting - fixtures. For gaps or openings greater than ⅛ inch (3.18 mm) around the electrical boxes, should the electrician repair the gap with patching plaster, possibly damaging or marring the finished wall? Whose responsibility is it? Is it the electrician's? Is it the dry wall/plaster/panel installer's?

The electrician should check with the dry wall/plaster/panel installer to clarify who is to be responsible for seeing to it that gaps or openings around electrical outlet and switch boxes do not exceed ⅛ inch (3.18 mm). *Section 370-21* puts the responsibility on the electrician.

YOKE

We use the term "yoke." The *National Electrical Code®* uses the term "yoke." Yet, the *NEC®* does not define a "yoke." What is a "yoke"? It is the metal or polycarbonate mounting strap on which a wiring device or devices are attached. The yoke in turn is attached to a device box or device cover with No. 6-32 screws. One yoke may have one, two, or three wiring devices attached to it.

Visit an electrical distributor or home center. You will find many combination wiring devices on a single yoke or strap. Selecting one of these combination devices may eliminate the need to use combinations of interchangeable wiring devices.

SPECIAL-PURPOSE OUTLETS

Special-purpose outlets are usually indicated on the plans. These outlets are described by a notation and are usually detailed in the specifications. The plans indicate special-purpose outlets by a triangle inside a circle with subscript letters. In some cases, a subscript number is added to the letter.

When a special-purpose outlet is indicated on the plans or in the specifications, the electrician must check for special requirements. Such a requirement may be a separate circuit, a special 240-volt circuit, a special grounding or polarized receptacle, or other preparation.

A set of Plans and Specifications for this residence is found at the end of this text. The Specifications include a *Schedule of Special Purpose Outlets*.

NUMBER OF CONDUCTORS IN BOX

Section 370-16 dictates that outlet boxes, switch boxes, and device boxes should be large enough to provide ample room for the wires in that box, without having to jam or crowd the wires into the box. Jamming the wires into the box can not only damage the insulation on the wires, but can result in heat buildup in the box that can further damage the insulation on the wires. The Code specifies the maximum number of conductors allowed in standard metal outlet, device, and junction boxes, see *Tables 370-16(a)* and *(b)*. Nonmetallic boxes are marked by the manufacturer with their cubic-inch capacity.

Table 370-16(a). Metal Boxes

Box Dimension in Inches, Trade Size, or Type	Minimum Capacity (in.^{3})	Maximum Number of Conductors*						
		No. 18	No. 16	No. 14	No. 12	No. 10	No. 8	No. 6
4 × 1¼ round or octagonal	12.5	8	7	6	5	5	4	2
4 × 1½ round or octagonal	15.5	10	8	7	6	6	5	3
4 × 2⅛ round or octagonal	21.5	14	12	10	9	8	7	4
4 × 1¼ square	18.0	12	10	9	8	7	6	3
4 × 1½ square	21.0	14	12	10	9	8	7	4
4 × 2⅛ square	30.3	20	17	15	13	12	10	6
4¹¹/₁₆ × 1¼ square	25.5	17	14	12	11	10	8	5
4¹¹/₁₆ × 1½ square	29.5	19	16	14	13	11	9	5
4¹¹/₁₆ × 2⅛ square	42.0	28	24	21	18	16	14	8
3 × 2 × 1½ device	7.5	5	4	3	3	3	2	1
3 × 2 × 2 device	10.0	6	5	5	4	4	3	2
3 × 2 × 2¼ device	10.5	7	6	5	4	4	3	2
3 × 2 × 2½ device	12.5	8	7	6	5	5	4	2
3 × 2 × 2¾ device	14.0	9	8	7	6	5	4	2
3 × 2 × 3½ device	18.0	12	10	9	8	7	6	3
4 × 2⅛ × 1½ device	10.3	6	5	5	4	4	3	2
4 × 2⅛ × 1⅞ device	13.0	8	7	6	5	5	4	2
4 × 2⅛ × 2⅛ device	14.5	9	8	7	6	5	4	2
3¾ × 2 × 2½ masonry box/gang	14.0	9	8	7	6	5	4	2
3¾ × 2 × 3½ masonry box/gang	21.0	14	12	10	9	8	7	4
FS — Minimum internal depth 1¾ single cover/gang	13.5	9	7	6	6	5	4	2
FD — Minimum internal depth 2⅜ single cover/gang	18.0	12	10	9	8	7	6	3
FS — Minimum internal depth 1¾ multiple cover/gang	18.0	12	10	9	8	7	6	3
FD — Minimum internal depth 2⅜ multiple cover/gang	24.0	16	13	12	10	9	8	4

Note: For SI units, 1 in.^{3} = 16.4 cm^3.
*Where no volume allowances are required by Sections 370-16(b)(2) through 370-16(b)(5).

Table 370-16(b). Volume Allowance Required per Conductor

Size of Conductor (AWG)	Free Space Within Box for Each Conductor (in.^{3})
18	1.50
16	1.75
14	2.00
12	2.25
10	2.50
8	3.00
6	5.00

Note: For SI units, 1 in.^{3} = 16.4 cm^3.

Reprinted with permission from NFPA 70-1999.

When conductors are the same size, the proper metal box size can be selected by referring to *Table 370-16(a)*. When conductors are of different sizes and nonmetallic boxes are installed, refer to *Table 370-16(b)* and use the cubic-inch volume for the particular size of wire being used.

Tables 370-16(a) and *(b)* do not take into consideration the space taken by fixture studs, cable clamps, hickeys, switches, pilot lights, or receptacles that may be in the box. These require additional space. Figure 2-34 shows the additional allowances that must be considered for different situations.

- If box contains no fittings, devices, fixture studs, cable clamps, hickeys, switches, receptacles, or equipment grounding conductors . . .
  - refer directly to *Tables 370-16(a)* or *(b)*.

- **Clamps.** If box contains one or more internal cable clamps . . .
  - add a single-volume based on the largest conductor in the box.

- **Support Fittings.** If box contains one or more fixture studs or hickeys . . .
  - add a single-volume for each type based on the largest conductor in the box.

- **Device or Equipment.** If box contains one or more wiring devices on a yoke . . .
  - add a double-volume for each yoke based on the largest conductor connected to a device on that yoke.

- **Equipment Grounding Conductors.** If a box contains one or more equipment grounding conductors . . .
  - add a single-volume based on the largest equipment grounding conductor in the box.

- **Isolated Equipment Grounding Conductor.** If a box contains one or more additional "isolated" (insulated) equipment grounding conductors as permitted by *Section 250-146(d)* for "noise" reduction . . .
  - add a single-volume based on the largest equipment grounding conductor in the box.

- For conductors running through the box without being spliced . . .
  - add a single-volume for each conductor that runs through the box.

- For conductors that originated outside of the box and terminate inside the box . . .
  - add a single-volume for each conductor that originates outside the box and terminates inside the box.

- If no part of the conductor leaves the box—for example, a "jumper" wire used to connect three wiring devices on one yoke, or pigtails as illustrated in Figure 8-9 . . .
  - don't count this (these). No additional volume required.

- For small equipment grounding conductors or not more than four fixture wires smaller than No. 14 that originate from a fixture canopy or similar canopy (like a fan) and terminate in the box . . .
  - don't count this (these). No additional volume required.

- For small fittings, such as lock-nuts and bushings . . .
  - don't count this (these). No additional volume required.

Figure 2-34 Quick checklist for possibilities to be considered when determining proper size boxes. See *Section 370-16(b)(1 through 5).*

SELECTING THE CORRECT SIZE BOX

When using a nonmetallic box where all conductors are the same size. A nonmetallic box is marked as having a volume of 22.8 cubic inches. The box contains no fixture stud or cable clamps. How many No. 14 AWG conductors are permitted in this box?

From *Table 370-16(b)*, the volume requirement for a No. 14 AWG conductor is 2.00 cubic inches. Therefore, the maximum number of No. 14 AWG conductors permitted in this box is:

$$\frac{22.8}{2} = 11.4 \text{ (Round down to 11 conductors)}$$

When using a metal box where all conductors are the same size. A box contains one fixture stud and two internal cable clamps. Four No. 12 conductors enter the box.

four No. 12 conductors	4
one fixture stud	1
two cable clamps (only count one)	1
Total	6

Referring to *Tables 370-16(a)* and *(b)*, a 4" × 1½" octagon box is suitable for this example.

When using a metal or nonmetallic box where the conductors are different sizes. What is the minimum cubic-inch volume required for a box that will contain one internal cable clamp, one switch, and one receptacle, all mounted on one yoke? Two No. 14 and two No. 12 wires enter the box. The wiring method is armored cable.

two No. 14 wires @ 2 cubic inches per wire	=	4.00 cubic inches
two No. 12 wires @ 2.25 cubic inches per wire	=	4.50 cubic inches
one cable clamp @ 2.25 cubic inches	=	2.25 cubic inches
one switch and one receptacle on one yoke @ 2.25 cubic inches × 2	=	4.50 cubic inches
Total		15.25 cubic inches

Select a box having a minimum volume of 15.25 cubic inches of space. The cubic-inch volume may be marked on the box, otherwise refer to the second column of *Table 370-16(a)* entitled "Minimum Capacity."

The Code in *Section 370-16(a)(2)* requires that all boxes *other* than those listed in *Table 370-16(a)* be durably and legibly marked by the manufacturer with their cubic-inch capacity. Nonmetallic boxes have their cubic-inch volume marked on the box. When sectional boxes are ganged together, the calculated volume is the total cubic-inch volume of the assembled boxes.

When installing a box that will have a raised cover attached to it. When marked with their cubic-inch volume, the *additional* space provided by plaster rings, domed covers, raised covers, and extension rings is permitted to be used when determining the overall volume. This is illustrated in Figure 2-35.

How many No. 12 conductors are permitted in this box and raised plaster ring? Refer to *Section 370-16* and *Tables 370-16(a)* and *(b)*.

4" x 1 1/2"
SQUARE BOX

4" SQUARE, 3/4" DEEP
RAISED PLASTER
RING (RAISED SECTION
MEASURES 2" x 3" x 3/4")

25.5 CUBIC
INCHES
TOTAL SPACE

21 CUBIC INCHES

4 1/2 CUBIC INCHES
MARKED ON COVER

$$\frac{25.5 \text{ cubic inches of total space}}{2.25 \text{ cubic inches per No. 12 conductor}} = 11 \text{ No. 12 conductors}$$

This calculation actually resulted in 11.33. The conductor fill numbers in *Table 370-16(a)* were created by dropping any excess above the whole number after the calculations were made. Following the same procedure, this conductor fill calculation dropped off the excess.

Electricians and electrical inspectors have become very aware of the fact that GFCI receptacles, dimmers, and timers take up a lot more space than regular wiring devices. Therefore, it is a good practice to install boxes that will provide plenty of room for the wires, instead of pushing, jamming, and crowding the wires into the box.

Figure 2-36 shows that transformer leads No. 18 or larger must be counted when selecting a proper size box.

The minimum required size of equipment grounding conductors is shown in *Table 250-122*. The equipment grounding conductors are the same size as the circuit conductors in cables having No. 14, No. 12, and No. 10 circuit conductors. Thus, for normal house wiring, box sizes can be calculated using *Table 370-16(a)*. Refer to Figure 2-37.

Figure 2-19, a quik-chek box selection guide, shows some of the most popular types of boxes used in residential wiring. This guide also shows the box's cubic-inch volume.

3/4" RAISED
COVER

ADDITIONAL WIRING SPACE PROVIDED BY THE
RAISED COVER IS 2" x 3" x 3/4" = 4 1/2 CUBIC INCHES

Figure 2-35 Raised cover. Raised covers are sometimes called plaster rings.

Figure 2-36 Transformer leads No. 18 AWG or larger must be counted when selecting the proper size box, *Section 370-16(b)(1)*. In the example shown, the box is considered to contain four conductors.

FOR CODE PURPOSES,
THIS BOX IS CONSIDERED
TO CONTAIN FOUR
CONDUCTORS.

Figure 2-37 Four or less fixture wires, smaller than No. 14 AWG, and/or small equipment grounding conductors that originate in the fixture and terminate in the box, need not be counted when calculating box fill, *Section 370-16(b)(1), Exception*.

Ⓐ BY LOOPING THE CONDUCTORS THROUGH THE BOX, THE CONDUCTOR COUNT IN THIS ILLUSTRATION IS FOUR.

Ⓑ WHEN USING CABLE, THE CONDUCTOR COUNT IN THIS ILLUSTRATION IS FIVE. THE TWO EQUIPMENT GROUNDING CONDUCTORS ARE COUNTED AS ONE.

Ⓒ WHEN USING CABLE AND A BOX THAT HAS INTERNAL CABLE CLAMPS, THE CONDUCTOR COUNT IN THIS ILLUSTRATION IS SIX. THE TWO EQUIPMENT GROUNDING CONDUCTORS ARE COUNTED AS ONE. THE TWO CABLE CLAMPS ARE COUNTED AS ONE.

Figure 2-38 **Example of conductor count for both metal conduit and cable installations.**

Figure 2-38 illustrates typical wiring using cable and conduit. Note in this figure how the conductor count is determined.

Using electrical metallic tubing (EMT) makes it possible to "loop" conductors through the box, which counts as one conductor only. Looping conductors through a box is not possible when using cable as the wiring method.

The basic rules for box fill are summarized in Figure 2-34. Boxes that are intended to support ceiling fans are discussed in detail in Unit 9.

Means for Connecting Equipment Grounding Conductors in a Box

Section 370-40(d) requires that "a means shall be provided in each metal box for the connection of **equipment grounding conductors**." Generally, these are tapped $10/32$ holes in the box that are marked GR, GRN, GRND, or similar identification. The screws used for terminating the equipment grounding conductor must be hexagon shaped, and must be green. Electrical distributors sell green, hexagon shaped $10/32$ screws for this purpose.

HEIGHT OF RECEPTACLE OUTLETS

There are no hard-and-fast rules for locating most outlets. A number of conditions determine the proper height for a switch box. For example, the height of the kitchen counter backsplash determines where the switches and receptacle outlets are located between the kitchen countertop and the cabinets.

The residence featured in this text is electrically heated and is discussed in Unit 23. The type of electric heat could be:

- electric furnace (as in this text).

- electric resistance heating buried in ceiling plaster or "sandwiched" between two layers of drywall material.

- electric baseboard heaters.

Let us consider the electric baseboard heaters. In most cases, the height of these electric baseboard units from the top of the unit to the finished floor seldom exceeds 6 inches (152 mm). The important issue here is that the manufacturer's receptacle accessories may have to be used to conform to the receptacle spacing requirements, as covered in *Section 210-52*.

Electrical receptacle outlets are not permitted to be located above an electric baseboard heating unit. Refer to the section "Location of Electric Baseboard Heaters in Relation to Receptacle Outlets," in Unit 23.

It is common practice among electricians to consult the plans and specifications to determine the proper heights and clearances for the installation of electrical devices. The electrician then has these dimensions verified by the architect, electrical engineer, designer, or homeowner. This practice avoids unnecessary and costly changes in the locations of outlets and switches as the building progresses.

POSITIONING OF RECEPTACLES

Although no actual Code rules exist on positioning receptacles, there is a popular concept in the electrical industry that there is always the possibility of a metal wall plate coming loose and falling downward onto the blades of an attachment plug cap that is loosely plugged into the receptacle, thereby creating a potential shock and fire hazard. Here are the options:

Recommended

Grounding hole to the top. A loose metal plate could fall onto the grounding blade of the attachment plug cap, but no sparks would fly.

Not Recommended

Ground*ing* hole to the bottom. A loose metal plate could fall onto both the grounded neutral and "hot" blades. Sparks would fly.

Recommended

Ground*ed* neutral blades on top. A loose metal plate could fall onto these grounded neutral blades. No sparks would fly.

Not Recommended

"Hot" terminal on top. A loose metal plate could fall onto these live blades. If this were a split-circuit receptacle fed by a 3-wire, 120/240-volt circuit, the short would be across the 240-volt line. Sparks would fly.

To ensure uniform installation and safety, and in accordance with long-established custom, standard electrical outlets are located as shown in Figure 2-39.

SWITCHES	
Regular	46" (1.17 m). Some homeowners want wall switches mounted 30–36" so as to be easy for children to reach and easy for adults to reach when carrying something. The choice is yours!
Between counter and kitchen cabinets. Depends on backsplash.	44"–46" (1.12 m–1.17 m)

RECEPTACLE OUTLETS	
Regular	12" (305 mm)
Between counter and kitchen cabinets. Depends on backsplash.	44"–46" (1.12 m–1.17m)
In garages	46" (1.17 m)
In unfinished basements	46" (1.17 m)
In finished basements	12" (305 mm)
Outdoors (above grade or deck)	18" (457 mm)

WALL LIGHTING FIXTURE OUTLETS	
Outside entrances. Depends on fixture.	66" (1.68 m). If fixture is "upward" from box on fixture, mount wall box lower. If fixture is "downward" from box on fixture, mount wall box higher.
Inside wall brackets	60" (1.52 m)
Side of medicine cabinet	60" (1.52 m)
Above medicine cabinet	You need to know the measurement of the medicine cabinet. Check the rough-in opening. Mount box approx. 6" to center of box above rough-in opening.

Note: All dimensions are from finished floor to center of the electrical box. If possible, try to mount wall boxes for lighting fixtures based upon the type of fixture to be installed. Verify all dimensions before "roughing in." If wiring for physically handicapped, the above heights may need to be lowered in the case of switches, and raised in the case of receptacles.

Figure 2-39   Typical heights for switches, receptacles, and wall lighting outlets. Local height preferences prevail.

These dimensions usually are satisfactory. However, the electrician must check the blueprints, specifications, and details for measurements that may affect the location of a particular outlet or switch. The cabinet spacing, available space between the countertop and the cabinet, and the tile height may influence the location of the outlet or switch. For example, if the top of wall tile is exactly 48 inches (1.22 m) from the finished floor line, a wall switch should not be mounted 48 inches (1.22 m) to center. This is considered poor workmanship. The switch should be located entirely within the tile area or entirely out of the tile area, Figure 2-40. This situation requires the full cooperation of all craftspersons involved in the construction job.

Faceplates

Faceplates for switches and receptacles shall be installed so as to completely cover the wall opening and seat against the wall surface, *Sections 380-9* and *410-56(e)*.

Faceplates should be level! No matter how good the wiring inside the wall is, no matter how good the connections and splices are, no matter how proper the equipment grounding is, the only thing the homeowner will see is the wiring devices and the faceplates. That establishes the impression the homeowner has of the overall wiring installation. Because of the elongated slots in the yoke of a wiring device, it is rather easy to level a wiring device and faceplate on a single-gang box even if the box is not level. This is not true for multi-gang boxes. It is extremely important to make sure multi-gang boxes are level. If a multi-gang box is not level, it will be virtually impossible to make the wiring devices and faceplate level.

Another nice detail that can improve the appearance is to align the slots in all of the faceplate mounting screws in the same direction, Figure 2-41.

WHERE A WALL IS PARTIALLY TILED, A SWITCH OR CONVENIENCE OUTLET MUST BE LOCATED ENTIRELY OUT OF THE TILE AREA (A) OR ENTIRELY WITHIN THE TILE AREA (B), (C).

THE FACEPLATE IN (D) DOES NOT "HUG" THE WALL PROPERLY. THIS INSTALLATION IS CONSIDERED UNACCEPTABLE BY MOST ELECTRICIANS, AND IS A VIOLATION OF THE LAST SENTENCE OF *SECTION 380-9* OF THE *NEC.*®

Figure 2-40   Locating an outlet on a tiled wall.

ALIGN SCREW–SLOTS
LIKE THIS...

NOT LIKE THIS.

Figure 2-41 Aligning the slots of the faceplate. Mounting screws in the same direction makes the installation look neater. Make sure the faceplate is level.

REVIEW

Note: Refer to the *NEC*® or the plans where necessary.

PART 1—ELECTRICAL FEATURES

1. What does a plan show about electrical outlets? _____
   _____
   _____
   _____

2. What is an outlet? _____
   _____
   _____

3. Match the following switch types with the proper symbol.
   a. single-pole   **S$_p$**
   b. three-way   **S$_4$**
   c. four-way   **S**
   d. single-pole with pilot light   **S$_3$**

4. The plans show dash lines running between switches and various outlets. What do these dash lines indicate? _____
   _____

5. Why are dash lines usually curved? _____

6. a. What are junction boxes used for? _____

   b. Are junction boxes normally used in wiring the first floor? Explain. _____
      _____
      _____

   c. Are junction boxes normally used to wire exposed portions of the basement? Explain. _____
      _____

7. How are standard sectional switch (device) boxes mounted? _____
   _____
   _____
   _____

8. a. What is an offset bar hanger? _____
      _____
      _____

   b. What types of boxes may be used with offset bar hangers? _____
      _____
      _____

9. What methods may be used to mount lighting fixtures to an outlet box fastened to an offset bar hanger? _____

_____

_____

_____

10. What advantage does a 4-inch octagon box have over a 3¼-inch octagon box?

_____

_____

11. What is the size of the opening of a switch (device) box for a single device? _____

_____

_____

12. The space between a door casing and a window casing is 3½ inches (88.9 mm). Two switches are to be installed at this location. What type of switches will be used?

_____

_____

13. Three switches are mounted in a three-gang switch (device) box. The wall plate for this assembly is called a _____ plate.

14. The mounting holes in a device (switch) box are tapped for ⁶⁄₃₂ screws. The mounting holes in an outlet box are tapped for ⁸⁄₃₂ screws. The mounting holes in metal boxes for attaching equipment grounding conductors are tapped for _____ screws.

15. a. How high above the finished floor are the switches located in the garage of this dwelling? _____

b. In the living room of this dwelling? _____

16. How high above the finished floor are the receptacle outlets in the garage located?

_____ In the living room? _____

17. Outdoor receptacle outlets in this dwelling are located _____

_____ inches above grade.

18. In the spaces provided, draw the correct symbol for each of the descriptions listed in (a) through (r).

a. _____ Lighting panel

b. _____ Clock outlet

c. _____ Duplex outlet

d. _____ Outside telephone

e. _____ Single-pole switch

f. _____ Motor

g. _____ Duplex outlet, split-circuit

h. _____ Lampholder with pull switch

i. _____ Weatherproof outlet

j. _____ Special-purpose outlet

k. _____ Fan outlet

l. _____ Range outlet

m. _____ Power panel

n. _____ Three-way switch

o. _____ Push button

p. _____ Thermostat

q. _____ Electric door opener

r. _____ Multioutlet assembly

19. The front edge of a box installed in a combustible wall must be _____ with the finished surface.

20. List the maximum number of No. 12 AWG conductors permitted in a

    a. 4" × 1½" octagon box. _____

    b. 4¹¹⁄₁₆" × 1½" square box. _____

    c. 3" × 2" × 3½" device box. _____

21. When a switch (device) box is nailed to a stud, and the nail runs through the box, the nail must not interfere with the wiring space. To accomplish this, keep the nail _____ (Select a, b, or c.)

    a. halfway between the front and rear of the box.

    b. a maximum of ¼ inch (6.35 mm) from the front edge of the box.

    c. a maximum of ¼ inch (6.35 mm) from the rear of the box.

22. Hanging a ceiling fixture directly from a plastic outlet box is permitted only if

    _____

    _____

    _____

    _____

23. It is necessary to count fixture wires when counting the permitted number of conductors in a box according to *Section 370-16*.

    (True)   (False)   Underline or circle the correct answer.

24. *Table 370-16(a)* allows a maximum of ten wires in a certain box. However, the box will have two cable clamps and one fixture stud in it. What is the maximum number of wires allowed in this box? _____

    _____

25. When laying out a job, the electrician will usually make a layout of the circuit, taking into consideration the best way to run the cables and/or conduits and how to make up the electrical connections. Doing this ahead of time, the electrician determines exactly how many conductors will be fed into each box. With experience, the electrician will probably select two or three sizes and types of boxes that will provide adequate space that will "meet Code." *Table 370-16(a)* of the Code shows the maximum number of conductors permitted in a given size box. In addition to counting the number of conductors that will be in the box, what is the additional volume that must be provided for the following items? Enter *single* or *double* volume allowance in the blank provided.

    a. one or more internal cable clamps: _____ volume allowance.

    b. for a fixture stud: _____ volume allowance.

    c. for one or more wiring devices on one yoke: _____ volume allowance.

    d. for one or more equipment grounding conductors: _____ volume allowance.

26. Is it permissible to install a receptacle outlet above an electric baseboard heater?

    _____

    _____

27. What is the maximum weight of a lighting fixture permitted to be hung directly to an outlet box in a ceiling? _____ pounds.

28. Two No. 12/2 and two No. 14/2 nonmetallic-sheathed cables enter a box. Each cable has an equipment grounding conductor. The No. 12 conductors are connected to a receptacle. Two of the No. 14 conductors are connected to a toggle switch. The other two No. 14 conductors are spliced together since they serve as a switch loop. The box contains two cable clamps. Calculate the minimum cubic-inch volume required for the box.

29. Using the same number and size of conductors as in question 28, but using electrical metallic tubing, calculate the minimum cubic inch volume required for the box. There will be no separate equipment grounding conductors, nor will there be any clamps in the box.

30. ►To allow for adequate conductor length at electrical outlet and device boxes to make up connections, *Section 300-14* requires that not less than (3"), (6"), (9") of conductor length be provided. This length is measured from where the conductor emerges from the cable or raceway to the end of the conductor. For box openings having any dimension less than 8 inches, the minimum length of conductor measured from the box opening in the wall to the end of the conductor is (3"), (6"), (9"). Circle the correct answers. ◄

31. When wiring a residence, what must be considered when installing wall boxes on both sides of a common partition that separates the garage and a habitable room? _____
_____
_____

32. Does the *NEC*® allow metal raceways to be used with nonmetallic boxes?

Yes _____ No _____ , Section _____

PART 2—STRUCTURAL FEATURES

1. To what scale is the basement plan drawn? _____

2. What is the size of the footing for the steel Lally columns in the basement? _____
_____

3. To what kind of material will the front porch lighting bracket fixture be attached?
_____

4. Give the size, spacing, and direction of the ceiling joists in the workshop. _____

_____

5. What is the size of the lot on which this residence is located? _____

_____

6. The front of the house is facing in what compass direction? _____

7. How far is the front garage wall from the curb? _____

8. How far is the side garage wall from the property lot line? _____

9. How many steel Lally columns are in the basement, and what size are they? _____

_____

10. What is the purpose of the I-beams that rest on top of the steel Lally columns? _____

_____

11. The garage walls are finished with _____.

12. Where is access to the attic provided? _____

13. Give the thickness of the outer basement walls. _____

14. What material is indicated for the foundation walls? _____

15. Where are the smoke detectors located in the basement? _____

_____

16. What is the ceiling height in the basement workshop from bottom of joists to floor?

_____

17. Give the size and type of the front door. _____

_____

18. What is the stud size for the partitions between the bathrooms in the bedroom area
where substantial plumbing is to be installed? _____

19. Who is to furnish the range hood? _____

_____

20. Who is to install the range hood? _____

Determining the Required Number and Location of Lighting and Small Appliance Circuits

OBJECTIVES

After studying this Unit, the student will be able to

- understand the fundamental Code requirements for calculating branch-circuit sizing and loading.
- understand "volt-amperes per square foot" for arriving at total calculated general lighting loading.
- calculate the occupied floor area of a residence.
- determine the minimum number of lighting branch-circuits required.
- determine the minimum number of small appliance branch-circuits required.
- know where receptacle outlets must be installed in homes.
- know where GFCI receptacles must be installed in homes.
- know where switched and nonswitched lighting outlets must be installed in homes.

It is standard practice in the design and planning of dwelling units to permit the electrician to plan the circuits. Thus, the residence plans do not include layouts for the various branch-circuits. The electrician may follow the general guidelines established by the architect. However, any wiring systems designed and installed by the electrician must conform to the *NEC,*® as well as local and state code requirements.

This Unit focuses on lighting branch-circuits and small appliance branch-circuits. The circuits supplying the electric range, oven, clothes dryer, and other specific circuitry not considered to be a lighting branch-circuit or a small appliance branch-circuit are covered later in this text. Refer to the index for the specific circuit being examined.

BASICS OF WIRE SIZING AND LOADING

The *NEC*® establishes some very important fundamentals that weave their way through the decision-making process for an electrical installation. They are presented here in brief form, and are covered in detail as required throughout this text.

The *NEC*® defines a **branch-circuit** as "the circuit conductors between the final overcurrent device protecting the circuit and the outlet(s)." See

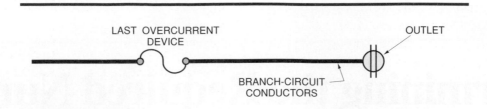

Figure 3-1 The branch-circuit is that part of the wiring that runs from the final overcurrent device to the outlet. The rating of the overcurrent device, not the conductor size, determines the rating of the branch-circuit.

Figure 3-1. In the residence discussed in this text, the wiring to wall outlets, the dryer, the range, etc., are all examples of a branch-circuit.

▶ The *NEC*® defines a **feeder** as "All circuit conductors between the service equipment, the source of a separately derived system, or other power supply source and the final branch-circuit overcurrent device."◀ In the residence discussed in this text, the wiring between Main Panel A and Subpanel B is a feeder.

The **ampacity** (current-carrying capacity) of a conductor must not be less than the rating of the **overcurrent device** protecting that conductor, *Section 210-19, 210-20,* and *240-3.* A common exception to this is a motor branch-circuit, where it is quite common to have overcurrent devices (fuses or breakers) sized larger than the ampacity of the conductor. Motors and motor circuits are covered specifically in *Article 430.* The ampere rating of the branch-circuit overcurrent protective device (fuse or circuit breaker) determines the rating of the branch-circuit. For example, if a 20-ampere conductor is protected by a 15-ampere fuse, the circuit is considered to be a 15-ampere branch-circuit, *Section 210-3.*

Standard branch-circuits that serve more than one receptacle outlet or more than one lighting outlet are rated 15, 20, 30, 40, and 50 amperes. A branch-circuit that supplies an individual load can be of any ampere rating, *Sections 210-3* and *210-23.*

Where the ampacity of the conductor does not match up with a standard rating of a fuse or breaker, the next higher standard size overcurrent device may be used, provided the overcurrent device does not exceed 800 amperes, *Section 240-3(b).* This deviation is not permitted, however, when the circuit supplies receptacles where "plug-connected" appliances, etc.,

could be used, because too many "plug-in" loads could result in an overload condition, *Section 210-19(b).* ▶ You may go to the next standard size overcurrent device only when the circuit supplies other than receptacles for cord and plug-connected portable loads.◀

For instance, when a conductor having an allowable ampacity of 25 amperes *(see Table 310-16)* No. 12 Type TW is derated to 70%:

$$25 \times 0.70 = 17.5 \text{ amperes}$$

It is permitted to use a 20-ampere overcurrent device if the circuit supplies *only* fixed lighting outlets or other fixed loads.

If the previous example were to supply receptacle outlets, then the rating of the overcurrent device would have to be dropped to 15 amperes; otherwise it is possible to overload the conductors by "plugging-in" more load than the conductors can safely carry.

The allowable ampacity of conductors commonly used in residential occupancies is found in *Table 310-16.* The ampacities in *Table 310-16* are subject to **correction factors** that must be applied where high ambient temperatures are encountered—for example, in attics. See bottom of *Table 310-16.*

Conductor ampacities are also **derated** when more than three current-carrying conductors are installed in a single raceway or cable. See *Table 310-15(b)(2).*

For loads that are *continuous,* the load shall not exceed 80% of the rating of the branch-circuit. The *NEC*® defines *continuous* as "a load where the maximum current is expected to continue for three hours or more." The lighting loads in residences are not considered to be "continuous" loads as would commercial lighting loads, such as store lighting. The

NEC® permits 100% loading on an overcurrent device only if that overcurrent device is *listed* for 100% loading. At the time of this writing, Underwriters Laboratories (U.L.) has *no* listing of circuit breakers of the molded-case type used in residential installations that are suitable for 100% loading.

This is another judgment call on the part of the electrician and/or the electrical inspector. Will a branch-circuit supplying a heating cable in a driveway likely be on for 3 hours or more? The answer is "probably." Good design practice and experience tell us "never load a circuit conductor or overcurrent protective device to more than 80% of its rating."

The minimum lighting load per square foot of floor area for dwellings is 3 volt-amperes, *Section 220-3(a)* and *Table 220-3(a)*.

Most receptacle outlets in a residence are considered by the Code to be part of "general illumination," and additional load calculations are *not* necessary. See *Section 220-3(b)(10)*. Appliance receptacle outlets in the kitchen, dining room, laundry, and workshop are *not* to be considered part of general illumination. Additional loads are figured in for these receptacle outlets and are discussed in detail later in this text.

VOLTAGE

All calculations throughout this text use **voltages** of 120 volts and 240 volts. This is in conformance to *Section 220-2(a)*, which states that "unless other voltages are specified, for purposes of computing branch-circuits and feeder loads, nominal system voltages of 120, 120/240, 208Y/120, 240, 480Y/277, 480, and 600 volts shall be used." This is repeated in the second paragraph of *Appendix D* of the *NEC*® which states that "for uniform application of Article 210, 215, and 220, a nominal voltage of 120, 120/240, 240, and 208Y/120 shall be used in computing the ampere load on the conductor."

The word *nominal* means "in name only, not a fact." For example, in our calculations, we use 120 volts even though the actual voltage might be 110, 115, or 117 volts. This provides us with a uniformity that, without it, would lead to misleading and confusing results in our computations.

CALCULATING FLOOR AREA
First Floor Area

To estimate the total load for a dwelling, the occupied floor area of the dwelling must be calculated. Note in the residence plans that the first floor area has an irregular shape. In this case, the simplest method of calculating the occupied floor area is to determine the total floor area using the outside dimensions of the dwelling. Then, the areas of the following spaces are subtracted from the total area: open porches, garages, or other unfinished or unused spaces if they are not adaptable for future use, *Section 220-3(a)*.

Many open porches, terraces, patios, and similar areas are commonly used as recreation and entertainment areas. Therefore, adequate lighting and receptacle outlets should be provided.

For practicality, the author has chosen to round up dimensions for the determining of total square footage and to round down dimensions for those areas (garage, porch, and portions of the inset at the front of the house) not to be included in the computation of the general lighting load. This produces a slightly larger result as opposed to being on the conservative side.

Figure 3-2 shows the procedure for calculating the total square footage of this residence.

Basement Area

The basement area of a home is not generally included in the calculating of general lighting loads because it could be classified as an "unfinished space not adaptable for future use," *Section 220-3(a)*. Yet in this residence, more than half of the total basement area is finished off as a recreation room, which certainly is considered living area. The workshop area also is intended to be used.

To simplify the calculation for this residence, we will consider the entire basement as useable space. The basement square footage area is the same as the area of the first floor.

The combined occupied area of the dwelling is found by adding the first floor and basement areas together:

First Floor	1616 ft² (149 m²)
Basement	1616 ft² (149 m²)
Total	3232 ft² (297 m²)

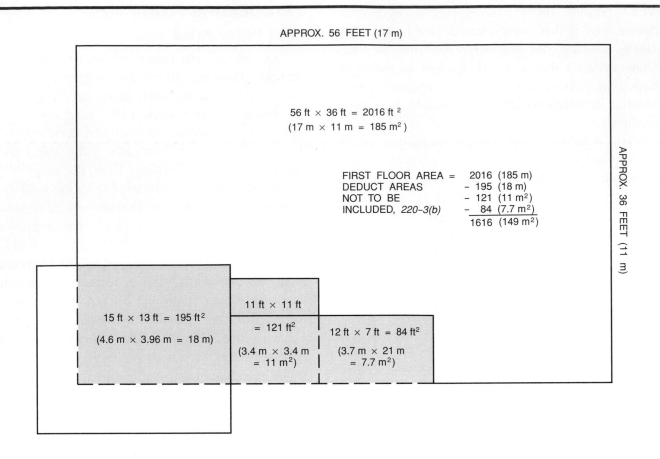

APPROX. 56 FEET (17 m)

56 ft × 36 ft = 2016 ft ^{2}
(17 m × 11 m = 185 m^2)

FIRST FLOOR AREA =   2016 (185 m)
DEDUCT AREAS   − 195 (18 m)
NOT TO BE   − 121 (11 m^2)
INCLUDED, *220–3(b)*   − 84 (7.7 m^2)
                       1616 (149 m^2)

APPROX. 36 FEET (11 m)

15 ft × 13 ft = 195 ft^2

(4.6 m × 3.96 m = 18 m)

11 ft × 11 ft

= 121 ft^2

(3.4 m × 3.4 m
= 11 m^2)

12 ft × 7 ft = 84 ft^2

(3.7 m × 21 m
= 7.7 m^2)

Figure 3-2   Determining the first floor square footage area. The basement area is the same as the first floor. See text for explanation of how the dimensions of the residence were rounded off to make the calculations much simpler and much more practical for the electrician and electrical inspector to compute.

DETERMINING THE MINIMUM NUMBER OF LIGHTING BRANCH-CIRCUITS

To determine the minimum number of 15-ampere lighting branch-circuits required in a residence:

Step 1: $\dfrac{3 \text{ volt-amperes} \times \text{square feet}}{120 \text{ volts}} = \textbf{amperes}$

Step 2: $\dfrac{\text{amperes}}{15} = $ minimum number of 15-ampere circuits required

This equates to a *minimum* of:

- one 15-ampere branch-circuit for every 600 square feet (55.8 m^2).

Although 20-ampere lighting branch-circuits are rarely used in residential installations, using the same procedure as indicated in Steps 1 and 2 would result in:

- one 20-ampere lighting branch-circuit for every 800 square feet (74.4 m^2).

Load calculations for electric water heaters, ranges, heat pumps, air-conditioning units, dryers, motors, electric heat, small appliance branch-circuits, and other specific loads are in addition to the general lighting load. These are discussed throughout this text as they appear.

Table 220-3(a) shows that the minimum load required for dwelling units is 3 **volt-amperes** (VA) per square foot (0.093 m^2) of occupied area. The calculated load for the total occupied area, 3232 ft^2 (297m^2), is:

$$1616 + 1616 = 3232 \text{ ft}^2$$

$$3232 \text{ (ft}^2\text{)} \times 3 \text{ (VA/ft}^2\text{)} = 9696 \text{ volt-amperes}$$

The total required amperage is determined as follows:

$$\text{Amperes} = \frac{\text{volt-amperes}}{\text{volts}}$$

$$= \frac{9696}{120} = 80.8 \text{ amperes}$$

The minimum number of 15-ampere lighting circuits required is equal to the total amperage divided by the maximum amperes of each circuit, *Section 220-4*.

$$\frac{80.8 \text{ amperes}}{15 \text{ amperes}} = 5.39 \text{ circuits}$$

Therefore, a minimum of six lighting branch-circuits is required.

The *NEC®* in *Appendix D* provides a number of examples that show calculations for services, feeders, neutrals, motor circuits, and other important issues.

In *Appendix D*, we also find that "except where the computations result in a major fraction of an ampere (0.5 or larger) such fractions may be dropped."

Table 3-1 is the schedule of the 15-ampere lighting and 20-ampere small appliance branch-circuits in this residence.

**SUMMARY OF 15-AMPERE LIGHTING
AND 20-AMPERE RECEPTACLE CIRCUITS**

15-Ampere Circuits		20-Ampere Circuits	
A14	Bathrooms, Hall Ltg.	A18	Workbench Receptacles
A15	Front Entry/Porch	A20	Workshop Receptacles, Window Wall
A16	Front Bedroom Ltg. & Outdoor Receptacle	A22	Master Bath Receptacle
A17	Workshop Ltg.	A23	Hall Bath Receptacle
A19	Master Bedroom Ltg. & Outdoor Receptacle		
A21	Study/Bedroom Ltg.		
B7	Kitchen/Nook Ltg.	B13	Kitchen/Nook North & East Wall Receptacles
B9	Wet Bar Ltg. & Receptacles	B15	Kitchen Receptacles, West Countertop
B10	Laundry, Rear Entry, Powder Room, Attic Ltg.	B16	Kitchen Receptacles, South Countertop, Refrigerator, & Island
B11	Recreation Room Receptacles	B18	Washer Receptacle
B12	Recreation Room Ltg.	B20	Laundry Room Receptacles & Outdoor Receptacle
B14	Garage, Post Light	B21	Powder Room Receptacle
B17	Living Room & Outdoor Receptacle		

Note: Panel "A" is the main panel located in the workshop. Panel "B" is the subpanel located in the corner of the recreation room.

Table 3-1

Section 240-3(d) tells us that the maximum overcurrent protection is:

15 amperes for No. 14 AWG copper conductors

20 amperes for No. 12 AWG copper conductors

30 amperes for No. 10 AWG copper conductors

In conformance to *Section 210-3*, a branch-circuit is rated according to the rating or setting of the overcurrent device. For example, a lighting branch circuit is rated 15 amperes even if No. 12 copper conductors are installed.

TRACK LIGHTING LOADS

Track lighting loads are considered to be part of the general lighting load for residential installations based upon the 3 volt-amperes per square foot as previously discussed. There is no need to add more wattage (volt-amperes) to the load calculations for track lighting in homes.

▶As required by *Section 410-101(b)*, the connected load on a lighting track shall not exceed the rating of the track. Also, be sure that a lighting track is supplied by a branch-circuit having a rating not more than that of the track.◀

DETERMINING THE NUMBER OF SMALL APPLIANCE BRANCH-CIRCUITS

Sections 220-16(a) and *(b)* state that an additional load of not less than 1500 volt-amperes shall be included for each two-wire small appliance circuit and each laundry circuit.

The receptacle for the refrigerator in the kitchen is permitted to be supplied by a separate 15- or 20-ampere branch-circuit rather than connecting it to one of the two required 20-ampere small appliance circuits. This permission is found in *Section 210-52(b)(1), Exception No. 2*. Because this separate branch-circuit diverts the refrigerator load from the receptacles that serve the countertop areas, it is not considered to be a small appliance branch-circuit. An additional 1500 volt-amperes *does not* have to be included in the load calculations, *Section 220-16(a), Exception*. There is nothing in the Code that prohibits adding another 1500 volt-amperes to the load calculations for this additional branch-circuit, as this would result in more capacity in the electrical system.

Section 210-11(c)(1) requires that two or more 20-ampere small appliance circuits must be provided for supplying receptacle outlets as specified in *Section 210-52(b)*.

Section 210-52(b)(1) states that the 20-ampere appliance circuits shall feed only the receptacle outlets in the kitchen, pantry, dining room, breakfast room, or similar rooms of dwelling units. *Section 210-52(b)(2)* also requires that these 20-ampere small appliance circuits shall not supply other outlets.

Exceptions in *Section 210-52(b)(2)* allow clock receptacles, and receptacles that are installed to plug in the cords of gas-fired ranges, ovens, or counter-mounted cooking units in order to power up their timers and ignition systems, Figure 3-3.

Section 210-52(b)(3) requires that not less than two 20-ampere small appliance circuits must supply the receptacle outlets serving the countertop areas in kitchens.

Food waste disposers and/or dishwashers shall not be connected to these required 20-ampere small appliance circuits. They must be supplied by other circuits, as is discussed later. No. 12 AWG wire is used in small appliance branch-circuits. The use of this larger wire, rather than No. 14 AWG wire, helps improve appliance performance and lessens the danger of overloading the circuits.

Automatic washers draw a large amount of current during certain portions of their operating cycles. Thus, *Section 210-11(c)(2)* requires that at least one additional 20-ampere branch-circuit must be provided to supply the laundry receptacle outlet(s) as required by *Section 210-52(f)*.

A complete list of the branch-circuits for this residence is found in Unit 28. The circuit directories for Main Panel A and Subpanel B show clearly the number of lighting, appliance, and special-purpose circuits provided.

RECEPTACLE OUTLET BRANCH-CIRCUIT RATINGS

The receptacles in the kitchen, breakfast nook, workshop, powder room, bathrooms, and laundry room are connected to 20-ampere branch-circuits. All other receptacles are connected to 15-ampere branch-circuits.

SUMMARY OF WHERE RECEPTACLE AND LIGHTING OUTLETS MUST BE INSTALLED IN RESIDENCES

The following is a brief explanation of where receptacle outlets and lighting outlets must be installed in residential occupancies. Throughout this text, each room is discussed in detail regarding outlet location. The following recaps all of the *NEC®* rules that relate to receptacle and lighting outlet requirements.

Bear in mind that the Code provides *minimum* requirements for the placement of receptacles. A duplex receptacle located where there will be a concentration of such things as computers, printers, stereos, VCRs, CD players, television sets, radios, fax machines, answering machines, typewriters, adding machines, or lamps will probably not be adequate to serve these appliances without the use of a cord set having multiple receptacles. An electrician will do well to quiz the homeowner where this concentration might occur, then offer to install four receptacles at these locations. As an alternative, a quadplex receptacle (four receptacles) that can replace a standard duplex receptacle in a single-gang device box. This same quadplex receptacle can be

RECEPTACLE BEHIND CLOCK SOLELY TO SUPPLY AND SUPPORT ELECTRIC CLOCK

20-AMPERE SMALL APPLIANCE BRANCH CIRCUIT IN KITCHEN, DINING ROOM, PANTRY, AND BREAKFAST ROOM

RECEPTACLE INSTALLED TO PLUG IN GAS-FIRED APPLIANCE TO OPERATE IGNITION

Figure 3-3   A receptacle installed solely to supply and support an electric clock, or a receptacle that supplies power for clocks, clock timers, and electric ignition for gas ranges, ovens, and cooktops may be connected to a 20-ampere small appliance circuit or to a general-purpose lighting circuit, *Section 210-52(b)(2)*.

mounted on a 4-inch square or octagon outlet box, on a 1- or 2-gang device box, or on a 4¹¹⁄₁₆-inch square box using a special adapter plate.

See Unit 6 for a discussion of surge protection.

Receptacle Outlets (125-Volt, Single-Phase, 15- or 20-Ampere)

Ground-Fault Circuit Interrupters. As we discuss the locations where receptacle outlets are required, the letters GFCI stand for *Ground-Fault Circuit Interrupter*. This means that the receptacle must be of the GFCI type, or the receptacle must be protected with a GFCI circuit breaker. GFCIs are covered in detail in Unit 6.

- Wall receptacle outlets must be placed so that no point along the floor line is more than 6 feet (1.83 m) measured horizontally from an outlet. Fixed room dividers and railings are considered to be walls for the purpose of this requirement, *Section 210-52(a)(1)*.

- Nonsliding fixed glass panels on exterior walls are considered to be wall space and are to be figured in when applying the rule that no point along the floor line is more than 6 feet (1.83 m) from a receptacle. Sliding panels on exterior walls are not considered to be a wall, *Section 210-52(a)(2)*.

- Receptacle outlets located in the floor more than 18 inches (457 mm) from the wall are not to be counted as meeting the required number of wall receptacle outlets, *Section 210-52(a)(3)*. An example of this would be a floor receptacle installed under a dining room table for plugging in warming trays, coffee pots, or similar appliances.

- Hallways 10 feet (3.05 m) or longer in homes must have at least one receptacle outlet, *Section 210-52(h)*. See Figure 3-4.

- Any wall space 2 feet (610 mm) or more in width must have a receptacle outlet, *Section 210-52(a)(2)(a)*.

- Unfinished basement areas not generally "lived in," such as work areas and storage areas, are not included in the above 6-foot (1.83 m) rule.

Figure 3-4   Determination of hallway length. Hallways 10 feet or longer must have at least one receptacle outlet, *Section 210-52(h)*.

- Circuits supplying wall receptacle outlets for lighting loads are generally 15-ampere circuits, but are permitted by the Code to be 20 amperes.

- In crawl spaces, and in attics, a receptacle outlet must be installed in an accessible location within 25 feet (7.6 m) of heating, air-conditioning, or refrigeration equipment. Do *not* connect this receptacle outlet to the load side of the equipment disconnecting means, *Section 210-63*.

- A receptacle is *not* required on the rooftop of one- and two-family dwellings, *Section 210-63, Exception*.

- Circuits supplying electrical outlets in kitchens, dining rooms, and pantries are required to be 20 amperes where appliances will be plugged in, *Sections 210-11(c)(1) and 210-52(b)*.

Suggestion: In addition to installing receptacles per the *NEC®* spacing requirements, give serious thought to installing a receptacle below the wall switch in bedrooms, living rooms, family rooms, and similar rooms. In most cases, the required receptacles will be behind furniture, making it difficult to reach for plugging in a vacuum cleaner or other temporarily used appliance. A receptacle below the wall switch will be very easy to get to. A real plus!

See Figure 3-5 for explanation of *receptacle* and *receptacle outlet*.

Basements

- At least one receptacle outlet must be installed in addition to the laundry outlet, *Section 210-52(g)*.

- Connect to 15- or 20-ampere circuit.

- In unfinished basement areas that are not intended to be "lived in" such as storage areas or workshop areas, *all* 125-volt, single-phase, 15- or 20-ampere receptacles must have GFCI protection, *Section 210-8(a)(5)*. This includes receptacles that are an integral part of porcelain or plastic pull-chain or keyless lampholders.

- GFCI protection is *not* required for receptacles that are not readily accessible, or for a single receptacle or duplex receptacle for two appliances that are not easily moved, are cord- and plug-connected, and are located in a dedicated space, See *Section 210-8(a)(5), Exceptions No. 1* and *No. 2*. Examples of this include the receptacle for laundry equipment, a sump pump, a water softener, a freezer, or a refrigerator.

- In finished basements, **GFCI** protection for receptacle outlets is *not* required unless the receptacles are located within 6 feet (1.83 m) from the outer edge of a wet-bar sink and are intended to serve the countertop areas around the wet-bar sink, *Section 210-8(a)(7)*. The required number, spacing, and location of the receptacles would have to be in conformance to *Section 210-52(a)*.

- Unfinished basements, or unfinished portions of basements that are not generally "lived in" such as work areas and storage areas are not subject to the spacing requirements for receptacles as specified in *Section 210-52(a)*.

- ▶ Where a portion of the basement is finished, a GFCI receptacle must be installed in the unfinished portion of the basement, and this is in addition to the required receptacle for the laundry

A RECEPTACLE OUTLET WHERE ONE OR MORE RECEPTACLES WILL BE INSTALLED.

A RECEPTACLE OUTLET WITH A SINGLE RECEPTACLE.

A RECEPTACLE OUTLET WITH A MULTIPLE (DUPLEX) RECEPTACLE. THIS IS TWO RECEPTACLES.

A RECEPTACLE OUTLET WITH TWO MULTIPLE (DUPLEX) RECEPTACLES. THIS IS FOUR RECEPTACLES.

Figure 3-5   The branch-circuit wiring and the box where one or more receptacles will be installed is defined by the Code as a *receptacle outlet*. The receptacle itself, whether a single or multiple device (one, two, or three receptacles) on one strap (yoke) is defined as a *receptacle*. For house wiring, receptacle loads are considered to be part of the general lighting load calculations, *Section 220-3(b)(10)*. For commercial installations, single or multiple receptacles (one, two, or three) on one strap (yoke) must be figured at 180 volt-amperes, *Section 220-3(b)(11)*.

equipment. ◄ The finished portion of the basement generally does not require GFCI receptacles. Because the finished portion of a basement is considered to be habitable, the spacing and location of receptacle, switches, and lighting outlets must be installed according to *Sections 210-52(a)* and *210-70*.

- Refer to *Sections 210-8(a)(5)*, *210-52(f)*, and *210-52(g)*.

Small Appliance Receptacle Outlets

- At least two 20-ampere small appliance circuits must supply the receptacle outlets in the kitchen, dining areas, or pantry.

- See *Sections 210-11(c)(1)*, *210-52*, and *220-16(a)*.

Countertops *(See Figure 3-6)*

- In kitchens and dining rooms, receptacles must be installed above all countertops 12 inches (305 mm) or wider.

- Receptacles that serve countertops in kitchens and dining rooms must be installed above, but not more than 18 inches (458 mm) above the countertop.

This basic rule discourages installing receptacles below a countertop because of the hazards created when someone pulls on a cord that is plugged into a below-the-countertop receptacle. However, the authority having jurisdiction, usually the electrical inspector, can give special permission to mount receptacles not more than 12 inches (305 mm) below a countertop if:

1. the receptacles must be located so that the physically handicapped can reach them, or

▶ 2. in the case of islands and peninsulas, where there is no way to mount a receptacle within 18 inches (458 mm) above the countertop. ◄

Figure 3-6   Floor plan showing placement of receptacles that serve countertop surfaces in kitchens and wet-bar areas in conformance to *Section 210-52(c)*. All 120-volt receptacles that serve countertops must be GFCI protected, *Section 210-8(a)(6)* and *(7)*. Receptacles that are located behind the refrigerator, under the sink for plugging in a food waste disposer, or inside the cabinets for plugging in a microwave oven do not have to be GFCI protected because they do not serve the countertop surfaces. Drawing is not to scale. See text for detailed explanation.

- Receptacles mounted below the countertop must not be installed where the countertop extends more than 6 inches (153 mm) beyond the base cabinet.

- Receptacles that serve countertops must be supplied by at least two 20-ampere small appliance branch-circuits. These circuits are permitted to supply other receptacles in the kitchen, dining room, pantry, breakfast room, or similar locations.

- Receptacles that serve countertop surfaces in kitchens must be GFCI protected. This includes receptacles installed along the wall, and on peninsulas and freestanding islands.

- Receptacles located behind a refrigerator or under the sink serving a food waste disposer and/or dishwasher do not have to be GFCI protected because these receptacles do not serve countertop surfaces.

- Peninsulas and freestanding islands that have a long dimension of 24 inches (619 mm) or greater and a short dimension of 12 inches (305 mm) or greater must have at least one receptacle outlet. Measure the peninsula length from the edge where the peninsula connects to the wall base cabinet unit. The outlet(s) must be GFCI protected.

- If an island or peninsula has a range or sink, it is quite possible that the island may require two receptacles if the placement of the range or sink results in a countertop space of 12 inches or more on both sides of the range or sink.

- Receptacle outlets installed within 6 feet (1.8 m) from the outside edge of a wet-bar sink must be GFCI protected.

- Receptacle outlets in kitchen and dining areas located above countertops shall be positioned so that no point along the wall line is more than 24 inches (610 mm), measured horizontally, from a receptacle outlet in that space. Example: A section of countertop measures 5'3" along the wall line. The minimum number of receptacles is: 5'3" ÷ 4 = 1 + (install two receptacles). Example: The countertop length from the edge of a sink to a range space measures 11'2" measured along the wall line. The minimum number of receptacles for this space is: 11'2" ÷ 4 = 2 + (install three receptacles).

Note: Dividing by 4 ensures that there will be a receptacle outlet for every 4 linear feet or fraction thereof of counter length.

- Receptacles are not permitted to be installed "face up" on countertops.

- Receptacle outlets behind refrigerators or appliances fastened in place do not meet the previous requirements for receptacle outlets above countertops, because these outlets would be impossible to use once the refrigerator or appliance was fastened in place.

- See Unit 6 for detailed data on GFCI protection.

- See *Sections 210-8(a)(6)* and *(7)*, *210-11(c)*, *210-52(c)*, and *220-16(a)*.

Laundry

- At least one receptacle outlet is required within 6 feet (1.83 m) of the intended location of the washing machine.

- The circuit must be a 20-ampere circuit.

- This receptacle *need not* be GFCI protected because, acceptable levels of leakage current of some clothes washing machines, as determined by U.L., could cause the GFCI to trip.

- This outlet and circuit are in addition to the basement receptacle outlet requirements already discussed, and must be in addition to the required two small appliance circuits that serve the receptacle outlets in the kitchen and dining areas. This circuit must *not* serve other outlets.

- See *Sections 210-8(a), 210-11(c)(2), 210-50(c), 210-52(f)*, and *220-16(b)*.

Bathrooms

- ▶ At least one receptacle must be installed within 36 inches (914 mm) from the outside edge of a basin. This receptacle shall be located on a wall that is adjacent to the basin. ◄

- A receptacle on a lighting fixture or medicine cabinet does not qualify for the required receptacle in a bathroom.

- Do not install a receptacle face up in a bathroom countertop near the basin location.

- Connect the receptacle(s) to at least one separate 20-ampere branch-circuit. Although this separate branch circuit is permitted to supply more than one bathroom, installing a separate 20-ampere branch-circuit to each bathroom is recommended because of the high wattage appliances commonly used in bathrooms, such as electric hair dryers.

- For service and feeder load calculations, no additional load needs to be included for this separate circuit.

▶• The separate 20-ampere branch-circuit must not supply other outlets. An exception is that if the separate 20-ampere branch-circuit is supplying only one bathroom, then other equipment such as an exhaust fan is permitted to be connected to the separate 20-ampere branch-circuit. ◀

- Receptacles in bathrooms must be GFCI protected.

- See *Sections 210-8(a)(1), 210-11(c)(3), 210-52(d),* and *220-16.*

Outdoors

- At least one receptacle outlet must be installed at grade level in the front and another in the back of all one-family dwellings and each unit of a two-family dwelling. These receptacles must be **accessible** and not more than 6½ feet (1.98 m) above grade level, Figure 3-7. A receptacle on a porch or deck is acceptable, provided there is easy access onto the porch from grade level and to the receptacle. A screened-in porch or deck, or a porch or deck that requires access by climbing up a number of steps would not meet the intent of the Code. This is a judgment call. One or two steps from grade level onto the porch is acceptable. More steps might make it difficult to access the receptacle.

- Connect to 15- or 20-ampere circuit.

- *All* outdoor receptacles must be GFCI protected. An exception is that personnel GFCI protection is *not* required for an outdoor receptacle(s) installed for snow and ice-melting equipment (i.e., heating cables) that is not **readily accessible**, and is supplied by a separate branch-circuit. GFCI devices for *personnel protection* trip in the range of 4 to 6 milliamperes.

- Ground fault protection must be provided for "fixed" outdoor de-icing and snow melting equipment. This protection is referred to as

AT LEAST ONE OUTDOOR RECEPTACLE OUTLET REQUIRED IN FRONT AND ANOTHER IN BACK OF HOUSE.

Figure 3-7 Outdoor receptacle requirements for residences, *Section 210-52(e),* require that for one-family dwellings, and each unit of a two-family dwelling that is at grade level, at least one receptacle outlet must be installed in the front and another in the back. These receptacles must not be mounted more than 6 feet, 6 inches (1.98 m) above grade. *Section 210-8(a)(3)* requires that *all* receptacle outlets installed outdoors be GFCI protected, except those receptacles for snow and ice melting equipment that are not readily accessible, and are supplied from a dedicated branch-circuit. Additional receptacles may be installed higher, such as receptacles used for plugging in ice and snow melting heating cables and/or decorative outdoor lighting.

GFPE, and can be a circuit breaker, a receptacle, or be self-contained in the equipment. GFPE devices for *equipment protection* trip at approximately 30 milliamperes. ▶ When used for de-icing and snow melting, mineral insulated metal-sheathed cable embedded in non combustible material such as concrete does not require GFP. ◀

- See *Sections 210-8(a)(3), 210-52(e),* and *426-28.*

Garages

- At least one receptacle outlet must be provided.

- Connect to 15- or 20-ampere circuit.

- Must be GFCI protected. An exception to this is receptacles that are not readily accessible, such as a receptacle mounted on the ceiling for plugging in a garage door opener. Also exempt from the GFCI requirement are single receptacles or duplex receptacles intended for cord- and plug-connected appliances that are not easily moved, and are located in dedicated spaces. Examples are a freezer, refrigerator, or central vacuum equipment.

- If the garage is detached, the Code does not require electric power, but if electric power is provided, then at least one GFCI receptacle outlet must be installed.
- See *Sections 210-8(a)(2)* and *210-52(g)*.

Receptacles in Other Locations. All of the required receptacles discussed above are *in addition* to any receptacles that are part of lighting fixtures or appliances, are located inside of cabinets, or are located more than 5½ feet (1.68 m) above the floor, *Section 210-52*.

GFCI Exemption Recap

- Under certain conditions, GFCI protection is *not* required for receptacle outlets in garages, in crawl spaces at or below ground level, or in unfinished basements if the receptacles are:
  - not readily accessible.
  - single-type receptacles or duplex-type receptacles located in dedicated spaces to be used for cord- and plug-connected appliances that are not easily moved.

Examples of these would be such things as overhead door openers, refrigerators, freezers, water softeners, central vacuum equipment or sump pumps. These would be the receptacles that are dedicated to specific plug-in appliances. The homeowner would normally not unplug the appliance to plug in an extension cord. These receptacles are exempt from the GFCI requirement because they are not intended to be used for plugging in extension cords for portable electric tools or portable appliances. These GFCI exempt receptacles are *in addition* to the receptacles required by *Section 210-52(g)* for garages and basements. See *Section 210-8(a)*.

Specific Loads

- Appliances such as electric ranges and ovens, fans, air conditioners, electric heat, heat pumps, electric water heaters, garage door openers, and similar loads are purchased as desired by the homeowner, *but* when included in the design of a residence the circuitry and installation must conform to the Code. See index for location of coverage for specific appliance or load.

Lighting Outlets *(Section 210-70)*

The *NEC®* does contain certain minimum requirements for providing lighting for dwellings.

Per *Section 210-70*, lighting outlets in dwellings must be installed as follows:

Switched Lighting Outlets *(See Figure 3-8.)*
▶ For dwellings, *Section 210-70(a)* shows three categories for lighting outlets.

Habitable Rooms. In every habitable room and in every bathroom, at least one wall switch-controlled lighting outlet must be installed. Switched receptacles meet the intent of this requirement except in kitchens and bathrooms. However, if plug-in swag lighting as illustrated in Figure 3-10 is used in a dining room, a switched receptacle connected to a general lighting branch-circuit must be installed. This receptacle must be in addition to the receptacles supplied by the required 20-ampere small appliance branch-circuits, *Section 210-52(b)(1)*, *Exception No. 1*, Figure 3-9.

Occupancy motion sensors may be used if: (1) they are in addition to the required wall switch, or (2) they have a manual override and are located where the wall switch would normally be located.

Additional Locations. At least one wall switch-controlled lighting outlet must be installed in hallways, stairways, attached garages, detached garages if they have electric power, on the exterior side of any outdoor entrances or exits that have grade level access. A sliding glass door is considered to be an outdoor entrance. A vehicle door is not considered to be an outdoor entrance.

Where a stairway has six or more steps (risers), switch control must be provided at each level. This includes permanent stairways to basements and attics, but does not include fold-down storable ladders commonly installed for access to residential attics.

Remote control, central control, motion sensors, photocells, and other automatic devices may be installed instead of wall switches for the control of lighting in hallways, stairways, and at outdoor entrances. For example, an outdoor post light controlled by a photocell does not require a separate switch.

Although the *NEC®* does not require lighting outlets in clothes closets, some building codes do. You should check this out in your locality.

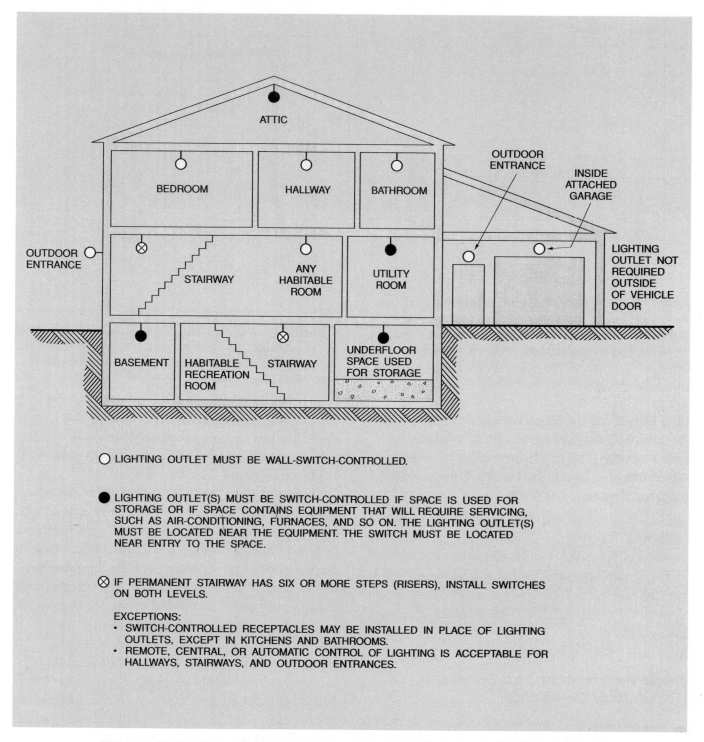

Figure 3-8 Lighting outlets required in a typical dwelling unit, *Section 210-70(a)*.

Storage or Equipment Spaces. At least one switch-controlled lighting outlet must be installed in spaces that are used for storage or contain equipment that requires servicing, such as attics, underfloor spaces, utility rooms, and basements. Examples of equipment that may need servicing are furnaces, air conditioners, heat pumps, sump pumps, etc. The control is permitted to be a wall switch or a lighting outlet that has an integral switch such as a pull chain switch. The lighting outlet must be at or near the equipment that may require servicing. The control must be at the usual point of entrance to the space.◀

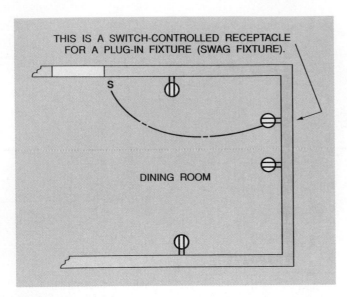

Figure 3-9 When a switch-controlled receptacle outlet is installed in rooms where the receptacle outlets normally would be 20-ampere small appliance circuits (breakfast nook, dining room, etc.), it must be in addition to the required small appliance receptacle outlets, *Section 210-52(b)(1), Exception No. 1.*

Figure 3-10 A receptacle controlled by a wall switch is acceptable as the required lighting outlet in rooms other than a kitchen or bathroom, *Section 210-70(a)(1), Exception.*

The Number of Branch-Circuits Required, and How to Determine the Number of Lighting and Receptacle Outlets to Be Connected to One Branch-Circuit

This Unit has explained the *NEC®* rules on where receptacle and lighting outlets are to be installed. The question of *how many* receptacle outlets and lighting outlets should be connected to each branch-circuit is covered in detail in Unit 8, where we begin the actual layout of the wiring for the residence discussed throughout this text.

Common Areas in Two-Family or Multifamily Dwellings

Section 210-25 contains specific requirements for circuits that serve common areas of two-family or multifamily dwellings. Be careful when wiring a two-family or multifamily dwelling. For areas such as a common entrance, common basement, or common laundry area, the branch-circuits for lighting, central fire/smoke/security alarms, communication systems, or any other electrical requirements that serve the common area *shall not* be served by circuits from any individual dwelling unit's power service. Common areas will require separate circuits, associated panels, and metering equipment.

The reasoning for this requirement is that power could be lost in one occupant's electrical system, resulting in no lights or alarms for the other tenant(s) using the common shared area.

For example, a two-family dwelling would probably require two separate stairways to the basement. If only one stairway is provided to serve the basement of a two-family dwelling, then lighting would have to be provided and served from a branch-circuit originating from each tenant's electrical system, so that each tenant would have control of the lighting. For a common entrance, two lighting fixtures would be installed and connected to each tenant's electrical system.

Likewise, fire/smoke/security systems in a two-family dwelling should be installed and connected to branch-circuits originating from each tenant's electrical panel. This becomes an architectural design issue. This is not generally a problem in two-family dwellings, but can become a problem in dwellings having more than two tenants. Rather than having to install a third watthour meter and associated electrical equipment on a two-family dwelling, the design of the structure can be such that there are no common areas, Figure 3-11. A four-plex will probably require a fifth watthour meter, four watthour meters for the tenants and one watthour meter to serve common area lighting and the fire/smoke/security systems.

Figure 3-11 This outline of a two-family dwelling shows a common rear entry, a common stairway to the basement, and a common front entry. Note that each entry has one lighting fixture. These common area lighting fixtures are not permitted to be connected to either of the tenant's electrical systems in the individual dwelling units, *Section 210-25*. The choices are (1) architect/designer must redo the layout of the building to eliminate the common entrances, (2) provide two lighting fixtures in each entry, with each fixture connected to the respective individual tenant's electrical circuit, or (3) install a third meter and panel to supply the electrical loads in the common areas.

REVIEW

Note: In the following questions the student may refer to the *NEC®* to obtain the answers. To help the student find these answers, references to Code sections are included after some of the questions. These references are not given in later reviews, where it is assumed that the student is familiar with the Code.

1. What is the meaning of computed load? _____

   _____

2. How are branch-circuits rated? *Section 210-3* _____

   _____

3. How is the rating of the branch-circuit protective device affected when the conductors used are of a larger size than called for by the Code? See *Section 210-3.* _____

   _____

4. What dimensions are used when measuring the area of a building? *Section 220-3(a)*

   _____

5. What spaces are not included in the floor area when computing the load in volt-amperes per square foot? *Section 220-3(a)* _____

   _____

6. What is the unit load per square foot for dwelling units? *Table 220-3(a)* _____

   _____

7. According to *Section 210-50(c)*, a laundry equipment outlet must be placed within _____ feet (_____ meters) of the intended location of the laundry equipment.

8. How is the total load in volt-amperes for lighting purposes determined? *Section 220-3(a)* _____

   _____

   _____

   _____

9. How is the total lighting load in amperes determined? _____

   _____

10. How is the required number of branch-circuits determined? _____

    _____

11. What is the minimum number of 15-ampere lighting circuits required if the dwelling has an occupied area of 4000 square feet (368 m^2)? Show all calculations.

12. How many lighting branch-circuits are provided in this dwelling? _____

13. a. What is the minimum load allowance for small appliance circuits for dwellings? *Section 220-16(a)* _____

    b. A separate 15-ampere branch-circuit is run to the receptacle outlet behind the refrigerator instead of connecting it to one of the two 20-ampere small appliance branch-circuits that are required in kitchens. For this separate circuit, an additional 1500 volt-ampere (must) (does not have to) be added to the load calculations for dwellings. Circle the correct answer.

14. What is the smallest size wire that can be used in a branch-circuit rated at 20 amperes?

    _____

15. How is the load determined for outlets supplying specific appliances? *Section 220-3(b)*

    _____

    _____

16. What type of circuits must be provided for receptacle outlets in the kitchen, pantry, dining room, and breakfast room? See *Section 210-11(c)(1)*. _____

    _____

17. How is the minimum number of receptacle outlets determined for most occupied rooms? See *Section 210-52(a)*. _____

_____

_____

_____

_____

_____

_____

_____

18. In a single-family dwelling, what types of overload protection for circuits are used? See *Section 240-6*. _____

_____

_____

_____

19. The *NEC*® in *Article 100* defines a continuous load as "a load where the maximum current is expected to continue for three hours or more." According to *Section 210-19(a)* and *210-20(a)*, the branch-circuit conductors and overcurrent protection for a continuous load shall be sized at not less than (100%) (125%) (150%) of the continuous load. Circle the correct answer.

20. The minimum number of outdoor receptacles for a residence is _____, *Section* _____ State the location. _____

21. The Code indicates the rooms in a dwelling that are required to have switched lighting outlets or switched receptacles. Write yes (switch required) or no (switch not required) for the following areas: See *Section 210-70(a)*.

    a. attic   _____

    b. stairway   _____

    c. crawl space (where
       used for storage)   _____

    d. hallway   _____

    e. bathroom   _____

22. Is a receptacle required in a bedroom on a 3-foot (914-mm) wall space behind the door? The door is normally left open._____

23. The Code requires that at least one 20-ampere circuit feeding a receptacle outlet must be provided for the laundry. May this circuit supply other outlets?_____

24. In a basement, at least one receptacle outlet must be installed in addition to the receptacle outlet installed for the laundry equipment. This additional receptacle outlet and any other receptacle outlets in unfinished areas must be _____ protected.

25. *Section 210-8(a)(2)* requires that *all* receptacles installed in garages must have GFCI protection, but there are certain exceptions. What are these exceptions?

_____

_____

_____

_____

_____

26. Define a branch-circuit. _____

_____

27. Although the Code contains many exceptions to the basic overcurrent protection requirements for conductors, in general, the rating of the branch-circuit overcurrent device must (not be less than) (not be more than) the ampacity of a conductor. Circle the correct answer.

28. The rating of a branch-circuit is based upon (circle the correct answer)
    a. the rating of the overcurrent device
    b. the length of the circuit
    c. the branch-circuit wire size

29. a. A 25-ampere branch-circuit conductor is derated to 70%. It is important to provide proper overcurrent protection for these conductors. The derated conductor ampacity is _____

_____

   b. If the connected load is a "fixed" nonmotor, noncontinuous load, the branch-circuit overcurrent device may be sized at (20) (25) amperes. Circle correct answer.

   c. If the above circuits supply receptacle outlets, the branch-circuit overcurrent device must be sized at not over (15) (20) (25) amperes. Circle correct answer.

30. Small appliance receptacle outlets are (included) (not included) in the 3 volt-amperes per square foot calculations. Circle correct answer.

31. If a homeowner wishes to have a switched receptacle outlet for a swag lamp in a dining room, may this switched receptacle outlet be considered to be one of the receptacle outlets required for the 20-ampere small appliance circuits? _____

32. How many receptacle outlets are required on a 13-foot (3.96 m) wall space between two doors? Refer to *Section 210-52(h)*. Draw the outlets in on the diagram.

33. Is a receptacle required in a hallway in a home when the hallway is 8 feet (2.44 m) long?

_____

34. Is a receptacle required in a hallway in a home when the hallway is 13 feet (3.96 m) long?

_____

35. A sliding glass door is installed in a family room that leads to an outdoor deck. The sliding door has one 4-foot (1.22 m) sliding section and one 4-foot (1.22 m) permanently mounted glass section. Inside the recreation room, what does the Code say about wall receptacle outlets relative to the receptacle's position near the sliding glass door?

_____

_____

_____

_____

_____

_____

_____

36. The Code permits certain other receptacles to be connected to a 20-ampere small appliance circuit. What are they? What *Section* of the *NEC*® supports your answer?

_____

_____

_____

_____

_____

_____

37. What is the minimum number of 20-ampere small appliance circuits required by the Code to supply the receptacle outlets that serve the countertop surfaces in the kitchen? In what section of the Code is this requirement found? _____

38. Does the Code allow a 120-volt receptacle located behind a gas range to be supplied by a small appliance circuit? (Yes) (No) Circle the correct answer. *Section* _____

39. When determining the location and number of receptacles required, fixed room dividers and railings (shall be) (need not be) considered to be wall space. Circle correct answer.

40. No point along the wall line shall be more than 6 feet (1.83 m) from a receptacle outlet. This requirement (does apply) (does not apply) to unfinished residential basements. Circle correct answer.

41. Receptacle outlets that serve countertop surfaces in kitchens are required to be _____ protected, and shall be installed so that no point along the wall line above the base cabinets is more than (18 inches) (24 inches) (36 inches) from a receptacle. Circle the correct answer.

42. In kitchens, the Code requires that at least (one) (two) receptacle(s) be installed to serve the countertop surfaces on an island or peninsula. Circle the correct answer. In what *Section* of the Code is this requirement found? *Section* _____

43. In the past, it was common practice to connect the lighting and receptacle(s) in a bathroom to the same circuit. Because of overloads caused by high wattage grooming appliances, the Code now requires that these receptacle(s) be supplied by a separate 20-ampere branch-circuit. In what *Section* of the Code is this requirement found? *Section* _____

44. If a residence has two bathrooms, the Code states that:

    a. the receptacles in each bathroom must be connected to a separate 20-ampere branch-circuit. For two bathrooms, this would mean two circuits. (True) (False) Circle the correct answer.

    b. the receptacles in both bathrooms are permitted to be served by the same 20-ampere branch-circuit. (True) (False) Circle the correct answer.

45. In your own words, explain the GFCI exemptions for receptacle outlets installed in unfinished basements and garages. Give some examples.

    _____

    _____

    _____

    _____

46. *Section* _____ of the Code prohibits connecting lighting and/or fire/smoke/ security systems to an individual tenant's electrical source where common (shared) areas are present, in which case the lighting provided for both tenants is connected on one tenant's electrical source.

47. When installing weatherproof outdoor receptacles high up under the eaves of a house intended to be used for plugging in de-icing and snow melting cable for the rain gutters, must these receptacles be GFCI protected for personnel protection? (Yes) (No) Circle the correct answer. *Section* _____

Conductor Sizes and Types, Wiring Methods, Wire Connections, Voltage Drop, Neutral Sizing for Services

OBJECTIVES

After studying this Unit, the student will be able to

- define the terms used to size and rate conductors.
- discuss the subject of aluminum conductors.
- describe the types of cables used in most dwelling unit installations.
- list the installation requirements for each type of cable.
- describe the uses and installation requirements for electrical conduit systems.
- describe the requirements for service grounding conductors.
- describe the use and installation requirements for flexible metal conduit.
- make voltage drop calculations.
- understand the fundamentals of the markings found on terminals of wiring devices and wire connectors.
- understand why the ampacity of high-temperature insulated conductors may not necessarily be suitable at that same rating on a particular terminal on a wiring device, breaker, or switch.
- understand why knob and tube wiring must not be buried in insulation.
- understand the use of reduced neutrals for residential feeders and services when certain criteria are met.

CONDUCTORS

Throughout this text, all references to conductors are for copper conductors, unless otherwise stated.

Wire Size

The copper wire used in electrical installations is graded for size according to the American Wire Gauge (AWG) Standard. The wire diameter in the AWG standard is expressed as a whole number. The higher the AWG number, the smaller the wire. AWG sizes vary from fine, hairlike wire used in coils and small transformers to very large diameter wire required in industrial wiring to handle heavy loads.

The wire may be a single strand (solid conductor), or it may consist of many strands. Each strand of wire acts as a separate conducting unit. The wire size used for a circuit depends on the maximum current to be carried. The *NEC®* in *Table 210-24* shows that the minimum conductor size for branch-circuit wiring is No. 14 AWG. Smaller size conductors are permitted for bell wiring, thermostat wiring, communications wiring, intercom wiring, fixture wires, and similar low-energy circuits.

The chart on the right shows typical applications for different size conductors.

Solid and Stranded Conductors

For residential wiring, sizes No. 14, 12, and 10 AWG are generally solid when the wiring method is nonmetallic-sheathed cable or armored cable.

In a raceway, the preference is to use stranded No. 10 AWG conductors because of the flexibility and ease of handling and pulling in the conductors.

Section 310-3 of the *NEC®* requires that conductors No. 8 AWG and larger must be stranded when installed in a raceway.

Ampacity

Ampacity means "the current in amperes a conductor can carry continuously under the conditions of use without exceeding its temperature rating." This value depends upon the conductor's cross-sectional area, whether the conductor is copper or aluminum, and the type of insulation around the conductor. Ampacity values, also referred to as current-carrying capacity, are found in *Article 310*. The most commonly used conductor ampacities are found in *Table 310-16*.

The ampacity values in the tables are "derated" when more than three current-carrying conductors are installed in the same raceway or cable. The ampacity values are further "corrected" when the conductors are installed in high-temperature applications.

More on conductor ampacities, correction factors, and derating factors is discussed in Unit 18.

Conductors must have an ampacity not less than the maximum load that they are supplying, Figure 4-1. All conductors of a specific branch circuit must

CONDUCTOR APPLICATIONS CHART		
Conductor Size	**Overcurrent Protection**	**Typical Applications** (Check wattage and/or ampere rating of load to select the correct size conductors based on *Table 310-16*.)
No. 18 AWG	7 amperes. See *Table 430-72(b)* and *Section 725-23*. Class 2 circuit transformers provide overcurrent protection.	Low-voltage control circuits such as thermostat wiring and chime wiring. Telephone wiring usually is No. 20 or 22 AWG. Also used for the LV wiring in "Advanced Home Systems" oftentimes referred to as "Home Automation."
No. 16 AWG	10 amperes. See *Table 430-72(b)* and *Section 725-23*. Class 2 circuit transformers provide overcurrent protection.	Same applications as above. Good for long runs to minimize voltage drop.
No. 14 AWG	15 amperes	Typical lighting branch-circuits.
No. 12 AWG	20 amperes	Small appliance branch-circuits for the receptacles in kitchens and dining rooms. Also laundry receptacles and workshop receptacles. Often used as the "home run" for lighting branch circuits. Some water heaters.
No. 10 AWG	30 amperes	Most clothes dryers, built-in ovens, cooktops, central air conditioners, some water heaters, heat pumps.
No. 8 AWG	40 amperes	Ranges, ovens, heat pumps, some large clothes dryers, large central air conditioners, heat pumps.
No. 6 AWG	50 amperes	Electric furnaces, heat pumps.
No. 4 AWG	70 amperes	Electric furnaces, feeders to subpanels.
No. 3 AWG and larger	100 amperes	Main service entrance conductors, feeders to subpanels, electric furnaces.

have an ampacity of the branch circuit's rating, Figure 4-2. There are exceptions to this rule, such as taps for electric ranges, Unit 20.

Figure 4-1   Branch-circuit conductors shall have an ampacity not less than the maximum load to be served, *Section 210-19(a).* See *Section 210-23* for permissible loads.

Figure 4-2   All conductors in this circuit supplying readily accessible receptacles shall have an ampacity of not less than the rating of the branch-circuit. In this example, 20-ampere conductors must be used, *Section 210-19(a)* and *Table 210-24.*

Conductor Sizing

The diameter of wire is usually given in a unit called a "mil." A *mil* is defined as one-thousandth of an inch (0.001 inch). Mils squared are known as "circular mils."

Table 8 in *Chapter 9* of the *NEC®* clearly shows that conductors are expressed in AWG numbers from No. 18 (1,620 circular mils) through No. 4/0 (211,600 circular mils). Wire sizes larger than No. 4/0 are expressed in circular mils.

Large conductors, such as 500,000 circular mils, are generally expressed as 500 kcmils. Since

the letter *k* designates 1,000, the term *kcmils* means "thousand circular mils." This is much easier to express in both written and verbal terms.

Older texts used the term MCM, which also means "thousand circular mils." The first letter *M* refers to the Roman numeral that represents 1,000. Thus, 500 MCM means the same as 500 kcmils. Roman numerals are no longer used in the electrical industry for expressing conductor sizes.

Overcurrent Protection for Conductors

Conductors must be protected against overcurrent by fuses or circuit breakers rated not more than the ampacity of the conductors, *Section 240-3.*

Section 240-3(b) permits the use of the next higher standard overcurrent device rating as listed in *Section 240-6.* For example, from *Table 310-16,* we find that a No. 6 Type TW 60°C conductor has an allowable ampacity of 55 amperes. In *Section 240-6,* we find that the next higher standard size overcurrent device is 60 amperes. ▶Nonstandard ampere rating of fuses and breakers are also permitted to be used. ◀

Section 240-3(d) shows the maximum overcurrent protection for the smaller sizes of branch circuit conductors. Here is what *Section 240-3(d)* says:

CONDUCTOR SIZE	MAXIMUM OVERCURRENT PROTECTION
No. 14 AWG copper	15 amperes
No. 12 AWG copper	20 amperes
No. 10 AWG copper	30 amperes

There are exceptions for motor branch-circuits and HVAC equipment. Motor circuits are discussed in Units 19 and 22. HVAC circuits are discussed in Unit 23.

Branch-Circuit Rating

Section 210-3 tells us that the rating of the overcurrent device (OCD) determines the rating of a branch-circuit, Figure 4-3.

Figure 4-3 All three of the above branch-circuits are rated as 15-ampere circuits, even though larger conductors were used for some other reason, such as solving a voltage drop problem. The rating of the overcurrent device (OCD) determines the rating of a branch circuit, *Section 210-3.*

PERMISSIBLE LOADS ON BRANCH-CIRCUITS *(SECTION 210-23)*

- the load shall not exceed the branch-circuit rating.

- the branch-circuit must be rated 15, 20, 30, 40, or 50 amperes when serving two or more outlets.

- an individual branch-circuit can supply any size load.

- 15 and 20 ampere branch-circuits:
  a. can supply lighting, other equipment, or both types of loads.
  b. cord- and plug-connected equipment shall not exceed 80% of the branch-circuit rating.
  c. equipment fastened in place shall not exceed 50% of the branch-circuit rating if the branch-circuit also supplies lighting, other cord- and plug-connected equipment, or both types of loads.

- the 20-ampere small appliance circuits in homes shall not supply other loads, *Section 210-11(c)(1).*

- 30-ampere branch-circuits can supply equipment such as dryers, cooktops, etc. Cord- and plug-connected equipment shall not exceed 80% of the branch-circuit rating.

- 40- and 50-ampere branch-circuits can supply cooking equipment that is fastened in place, such as an electric range.

- over 50-ampere-rated branch-circuits shall not supply lighting loads. These large circuits are for electric furnaces, large heat pumps, air-conditioning equipment, large double-ovens, and similar large loads.

Aluminum Conductors

The conductivity of aluminum wire is not as great as that of copper wire for a given size. For example, checking *Table 310-16*, it is found that a No. 8 THHN copper wire has an allowable ampacity of 55 amperes, whereas a No. 8 THHN aluminum or copper-clad aluminum wire has an ampacity of only 45 amperes. These ampacities are further reduced if the conductors are installed in high-ambient-temperature locations. (See ampacity correction factors below the *Table.*) Ampacities are also reduced when there are more than three current-carrying conductors in a raceway or cable. See *Section 310-15(b)(2)(a).*

Another example is found in *Section 240-3(d)* where the maximum overcurrent protection for No. 12 AWG copper wire is 20 amperes, but only 15 amperes for No. 12 AWG aluminum or copper-clad aluminum wire.

Resistance is an important consideration when installing aluminum conductors. An aluminum conductor has a higher resistance compared to a copper conductor for a given wire size, which therefore causes a greater voltage drop.

$$\text{Voltage Drop}(E_d) = \text{Amperes}(I) \times \text{Resistance}(R)$$

Common Connection Problems

Some common problems associated with aluminum conductors when not properly connected may be summarized as follows:

- a corrosive action is set up when dissimilar wires come in contact with one another when moisture is present.

- the surface of aluminum oxidizes as soon as it is exposed to air. If this oxidized surface is not penetrated, a poor connection results. When installing aluminum conductors, particularly in large sizes, an inhibitor is brushed onto the aluminum conductor, then the conductor is scraped with a stiff brush where the connection is to be made. The process of scraping the conductor breaks through the oxidation, and the inhibitor keeps the air from coming into contact with the conductor. Thus, further oxidation is prevented. Aluminum connectors of the compression type usually have an inhibitor paste already factory installed inside of the connector.

- aluminum wire expands and contracts to a greater degree than does copper wire for an equal load. This factor is another possible cause of a poor connection. Crimp connectors for aluminum conductors are usually longer than those for comparable copper conductors, thus resulting in greater contact surface of the conductor in the connector.

Proper Installation Procedures

Proper, trouble-free connections for aluminum conductors require terminals, lugs, and/or connectors that are suitable for the type of conductor being installed.

Terminals on receptacles and switches must be suitable for the conductors being attached. Table 4-1 shows how the electrician can identify these terminals. Listed connectors provide proper connection when properly installed.

See also *Sections 110-14, 380-14(c)*, and *410-56(b)*.

Wire Connections

When splicing wires or connecting a wire to a switch, fixture, circuit breaker, panelboard, meter socket, or other electrical equipment, the wires may be twisted together, soldered, then taped. Usually, however, some type of wire connector is required.

Wire connectors are known in the trade by such names as *screw terminal, pressure terminal connector, wire connector, Wing nut,® Wire-nut,® Scotchlok,® split-bolt connector, pressure cable connector, solderless lug, soldering lug, solder lug*, and others. Solder-type lugs are not often used today. In fact, connections that depend upon solder are *not* permitted for connecting service-entrance conductors to service equipment. The labor costs and time spent make the use of solder-type connections prohibitive.

Solderless connectors, designed to establish

TYPE OF DEVICE	MARKING ON TERMINAL OR CONDUCTOR	CONNECTOR PERMITTED
15- or 20-ampere receptacles and switches	CO/ALR	aluminum, copper, copper-clad aluminum
15- and 20-ampere receptacles and switches	NONE	copper, copper-clad aluminum
30-ampere and greater receptacles and switches	AL/CU	aluminum, copper, copper-clad aluminum
30-ampere and greater receptacles and switches	NONE	copper only
Screwless pressure terminal connectors of the push-in type	NONE	copper or copper-clad aluminum
Wire connectors	AL	aluminum
Wire connectors	AL/CU	aluminum, copper, copper-clad aluminum
Wire connectors	CC	copper-clad aluminum only
Wire connectors	CC/CU	copper or copper-clad aluminum
Wire connectors	CU	copper only
Any of the above devices	COPPER OR CU ONLY	copper only

Table 4-1 Terminal identification markings and the acceptable types of conductors permitted to be connected to a specific type of terminal.

connections by means of mechanical pressure, are quite common. Examples of some types of wire connectors, and their uses, are shown in Figure 4-4.

As with the terminals on wiring devices (switches and receptacles), wire connectors must be marked **AL** when they are to be used with aluminum conductors. This marking is found on the connector itself, or appears on or in the shipping carton.

Connectors marked **AL/CU** are suitable for use with aluminum, copper, or copper-clad aluminum conductors. This marking is found on the connector itself, or it appears on or in the shipping carton.

Connectors not marked *AL* or *AL/CU* are for use with copper conductors only.

Unless specially stated on or in the shipping carton, or on the connector itself, conductors made of copper, aluminum, or copper-clad aluminum *may not* be used in combination in the same connector.

CRIMP CONNECTORS USED TO SPLICE AND TERMINATE NO. 20 AWG TO 500 KCMILS ALUMINUM-TO-ALUMINUM, ALUMINUM-TO-COPPER, OR COPPER-TO-COPPER.

PROPERLY CRIMP THEN TAPE

CONNECTORS USED TO CONNECT WIRES TOGETHER ON COMBINATIONS OF NO. 18 AWG THROUGH NO. 6 AWG. THEY ARE TWIST-ON, SOLDERLESS, AND TAPELESS.

*WIRE-NUT® and WING-NUT® are registered trademarks of IDEAL INDUSTRIES, INC. Scotchlok® is a registered trademark of MINNESOTA MINING.

WIRE CONNECTORS VARIOUSLY KNOWN AS WIRE-NUT,® WING NUT,® AND SCOTCHLOK.®

CONNECTORS USED TO CONNECT WIRES TOGETHER IN COMBINATIONS OF NO. 16, NO. 14, AND NO. 12 AWG. THEY ARE CRIMPED ON WITH A SPECIAL TOOL, THEN COVERED WITH A SNAP-ON INSULATING CAP OR WRAPPED WITH INSULATING TAPE.

CRIMP-TYPE WIRE CONNECTOR AND INSULATING CAP

SOLDERLESS CONNECTORS ARE AVAILABLE IN SIZES NO. 14 AWG THROUGH 500 MCM. THEY ARE USED FOR ONE SOLID OR ONE STRANDED CONDUCTOR ONLY, UNLESS OTHERWISE NOTED ON THE CONNECTOR OR ON ITS SHIPPING CARTON. THE SCREW MAY BE OF THE STANDARD SCREWDRIVER SLOT TYPE, OR IT MAY BE FOR USE WITH AN ALLEN WRENCH OR SOCKET WRENCH.

SOLDERLESS CONNECTORS

COMPRESSION CONNECTORS ARE USED FOR NO. 8 AWG THROUGH 1000 MCM. THE WIRE IS INSERTED INTO THE END OF THE CONNECTOR, THEN CRIMPED ON WITH A SPECIAL COMPRESSION TOOL.

COMPRESSION CONNECTOR

SPLIT-BOLT CONNECTORS ARE USED FOR CONNECTING TWO CONDUCTORS TOGETHER, OR FOR TAPPING ONE CONDUCTOR TO ANOTHER. THEY ARE AVAILABLE IN SIZES NO. 10 AWG THROUGH 1000 MCM. THEY ARE USED FOR TWO SOLID AND/OR TWO STRANDED CONDUCTORS ONLY, UNLESS OTHERWISE NOTED ON THE CONNECTOR OR ON ITS SHIPPING CARTON.

SPLIT-BOLT CONNECTOR

Figure 4-4 Types of wire connectors.

When combinations are permitted, the connector will be identified for the purpose and for the conditions when and where they may be used. The conditions usually are limited to dry locations only. There are some "twist-on" wire connectors that have recently been listed for use with combinations of copper and aluminum conductors.

The preceding data is found in *Section 110-14* and in the Underwriters Laboratories (U.L.) Standards.

Insulation of Wires

The Code requires that all wires used in electrical installations be insulated, *Section 310-2(a)*. Exceptions to this rule are clearly indicated in the Code.

The most common type of insulation used on wires is a thermoplastic material, although rubber compound conductor insulations are still recognized by the Code. *Table 310-13* lists the various conductor insulations and applications. The insulation completely surrounds the metal conductor, has a uniform thickness, and runs the entire length of the wire or cable.

The allowable ampacities of copper conductors are given in *Table 310-16* for various types of insulation. *Table 310-15(b)(6)* is a table of conductor ampacities that are greater than the ampacities found in *Table 310-16*. This table is permitted to be used *only* for dwelling unit 120/240-volt, 3-wire, single-phase service-entrance conductors, service lateral conductors, and feeder conductors that serve as the main power feeder to a dwelling unit. This table is based upon the tremendous known load diversity found in homes. These special ampacities must not be used for feeder conductors that do not serve as the main power feeder to a dwelling. An example of this would be the feeder from the Main Panel A to Panel B in the residence discussed in this text.

Table 4-2 shows the special ampacities for types RH, RHH, RHW, THHW, THW, THWN, THHN, XHHW, and USE conductors. Table 4-2 is similar to the table found in *Table 310-15(b)(6)*.

The insulation covering wires and cables used in house wiring is usually rated at 600 volts or less. Exceptions to this statement are low-voltage wiring and fixture wiring.

Conductor Temperature Ratings

Conductors are also rated as to the temperature they can withstand. For example:

	Celsius	Fahrenheit
Type TW	60°	140°
Type THW	75°	167°
Type THHN	90°	194°

Without question, conductors with Type THHN insulation are the most popular and most commonly used, particularly in the smaller sizes, because of their small diameter (easy to handle; more conductors of a given size permitted in a given size raceway) and their ability to be installed where high temperatures are encountered, such as attics, buried in insulation, and supplying recessed fixtures.

The temperature rating of the conductors found in nonmetallic-sheathed cable, most commonly used for house wiring, is 90°C, *Section 336-30(b)*. However, the allowable ampacity of the conductors in nonmetallic-sheathed cable is that of the 60°C column of *Table 310-16*. See *Section 336-26*.

The Weakest Link of the Chain

A conductor has two ends!

Selecting a conductor ampacity based solely on the values found in *Table 310-16* is a very common mistake that can prove costly.

SPECIAL AMPACITY RATINGS FOR RESIDENTIAL 120/240-VOLT, 3-WIRE, SINGLE-PHASE SERVICE-ENTRANCE CONDUCTORS, SERVICE LATERAL CONDUCTORS, AND FEEDER CONDUCTORS THAT SERVE AS THE MAIN POWER FEEDER TO A DWELLING UNIT.

Copper Conductor AWG for Insulation of RH-RHH-RHW-THW-THWN-THHW-THHN-XHHW-USE	Aluminum or Copper-Clad Aluminum Conductors (AWG)	Allowable Special Ampacity
4	2	100
3	1	110
2	1/0	125
1	2/0	150
1/0	3/0	175
2/0	4/0	200
3/0	250 kcmil	225
4/0	300 kcmil	250
250 kcmil	350 kcmil	300
350 kcmil	500 kcmil	350
400 kcmil	600 kcmil	400

▶ **Note:** Where the conductors will be exposed to temperatures in excess of 30°C (86°F), apply the correction factors found at the bottom of *Table 310-16*.◀

Table 4-2

Conductors have maximum temperature ratings, *Tables 310-13, 310-16*, and U.L. 83. Receptacles covered under U.L. 498 and snap switches covered under U.L. 20 have maximum temperature ratings. Panelboards have maximum temperature ratings, U.L. 67. Circuit breakers have maximum temperature ratings, U.L. 489. Disconnect switches have maximum temperature ratings, U.L. 98. Wire splicing devices (connectors) have maximum temperature ratings, U.L. 486.

It is apparent that in any given circuit, we have an assortment of electrical components that have different maximum temperature ratings. After sorting this all out, the allowable ampacity of a conductor is based on the temperature limitations of the lowest temperature rating of any part of the circuit.

Most branch-circuits rated 15- and 20-amperes have receptacles and snap switches that have a maximum temperature rating of 60°C. Therefore, no matter what the temperature rating of the panelboard or conductors might be (i.e., 90°C for THHN), the conductor ampacity must be determined based on the 60°C column of *Table 310-16* because the "weakest link" of the chain are the receptacles and snap switches.

Section 110-14(c) and U.L. standards require that unless otherwise identified, terminals on wiring devices, switches, breakers, motor controllers, and so forth are based upon the ampacity of 60°C conductors where the circuit is rated 100 amperes or less, and the conductor sizes are No. 14 through No. 1 AWG. It is acceptable to install conductors having a higher temperature rating, such as 90°C THHN, but you must use the 60°C ampacity values.

For circuits rated over 100 amperes, or marked for use with conductors larger than No. 1 AWG, the conductors installed must have a minimum 75°C rating. The ampacity of the conductors would be based upon the 75°C values. It is acceptable to install conductors having a higher temperature rating, such as 90°C THHN, but you must use the 75°C ampacity values.

Conductors having a temperature rating higher than the rating of the termination may be used. When using high temperature insulated conductors where adjustment factors are to be applied, such as *derating* for more than three current-carrying conductors in a raceway or cable, or *correcting* where high temperatures are encountered, the final results of these adjustments (in amperes) must comply with the requirements of *Section 110-14(c)*. Examples of applying *derating factors* and *correcting factors* are found in Unit 18.

Individual conductors for most premise wiring are rated 60°C, 75°C, or 90°C. In the case of armored cable and nonmetallic-sheathed cable, we must use the 60°C ampacity column of *Table 310-16*. This requirement is found in *Sections 333-20* and *336-26*.

Snap switches and 15- and 20-ampere receptacles are rated for 60°C.

Some 30-ampere receptacles are rated for 60°C and some are rated for 75°C. Larger 50-ampere range receptacles are rated for 75°C.

Residential panelboards are generally marked 75°C. The temperature rating of the panelboard has been established as an assembly with circuit breakers in place. Do not use the circuit breaker's temperature marking by itself.

Circuit breakers and switches are generally marked "75°C only" or "60°/75°C."

It is interesting to note that in *Section 240-3(d)* there is somewhat of a built-in code compliance limitation for applying small branch-circuit conductors. Here we find that the maximum overcurrent protection for No. 14 is 15 amperes, for No. 12 is 20 amperes, and for No. 10 is 30 amperes.

From *Table 310-16* we find that:

No. 14 copper conductors' allowable ampacity is:

20 amperes in the 60°C column

20 amperes in the 75°C column

25 amperes in the 90°C column

No. 12 copper conductors' allowable ampacity is:

25 amperes in the 60°C column

25 amperes in the 75°C column

30 amperes in the 90°C column

No. 10 copper conductors' allowable ampacity is:

30 amperes in the 60°C column

35 amperes in the 75°C column

40 amperes in the 90°C column

But no matter what allowable ampacity the conductor has, the maximum size overcurrent protection for No. 14, No. 12, and No. 10 as stated in *Section 240-3(d)* would keep these conductors from becoming overloaded.

In summary, we cannot arbitrarily check *Table 310-16* to find the allowable ampacity of a given conductor. Conductor ampacity is not a simple "stand alone" issue. We must also consider the temperature limitations of the equipment, such as panelboards, receptacles, snap switches, etc. The lowest maximum temperature rated device in an electrical system is the "weakest link," and it is the "weakest link" that we must base our ultimate conductor ampacity decision on.

The following chart indicates the correct application for conductors and terminations.

TERMINATION RATING	CONDUCTOR INSULATION RATING		
	60°C	75°C	90°C
60°C	OK	OK (at 60°C ampacity)	OK (at 60°C ampacity)
75°C	NO	OK	OK (at 75°C ampacity)
60/75°C	OK	OK (at 60°C or 75°C ampacity)	OK (at 60°C or 75°C ampacity)
90°C	NO	NO	OK (Only if equipment has a 90°C rating.)

Derating Factors and Correction Factors

Allowable ampacities for various conductors are found in *Table 310-16* of the *NEC.*® *Table 310-16* is based on not more than three current-carrying conductors in a single raceway or cable, and at an ambient temperature of not over 86°F.

When more than three current-carrying conductors are installed in a single raceway or cable, and/or where high temperatures are encountered, penalties must be applied to these allowable ampacities. This is discussed in Unit 18.

It is recommended that the reader study *Article 310* in the *NEC,*® which is entitled *Conductors for General Wiring.*

Knob-and-Tube Wiring

Older homes were wired with open, individual, insulated conductors supported by porcelain *knobs* and *tubes.* In its day, knob-and-tube wiring served its purpose well. This text does not cover knob-and-tube wiring because this method is no longer used in new construction. On occasion, it is necessary to make some modifications to knob-and-tube wiring when doing remodel work. The *NEC*® covers this wiring method in *Article 324.*

Many fires have occurred because thermal insulation has been blown in attics of older homes, completely burying the knob-and-tube wiring. The conductors become overheated. *Section 324-4* in part states that "concealed knob-and-tube wiring shall not be used in hollow spaces of walls, ceilings, and attics where such spaces are insulated by loose, rolled, or foamed-in-place insulating material that envelops the conductors."

Keep an eye out for this potential fire hazard when working in older homes that have knob-and-tube wiring.

Neutral Size

Figure 4-5 shows how *line-to-line* loads and *line-to-neutral* loads are connected. The grounded neutral (white) conductor Ⓝ for residential services and feeders is permitted to be smaller than the "hot" ungrounded phase (black, red) conductors Ⓓ only when the neutral conductor is properly and adequately sized to carry the maximum unbalance loads computed according to *Sections 215-2, 220-22,* and *230-42.*

Section 215-2 relates to feeders and refers us to *Article 220* for computation requirements.

Section 230-42 relates to services, and refers us to *Article 220* for computation requirements.

Focusing on the neutral conductor, *Section 220-22* states that ▶"the feeder or service neutral load shall be the maximum unbalance of the load determined by this Article."◀

Section 220-22 further states that "the maximum unbalanced load shall be the maximum net computed load between the neutral and any one ungrounded conductor."

In this residence we find a number of loads that carry little or no neutral currents, such as the electric water heater, electric clothes dryer, electric oven and range, electric furnace, and air conditioner. These appliances for the most part are drawing their biggest current through the "hot" conductors because they are fed with 240-volt Ⓒ or 120/240-volt circuits Ⓔ. Loads Ⓑ are connected line to neutral only.

Thus we find logic in the *NEC*® which permits reducing the neutral size on services, and feeders where the computations prove that the neutral will be carrying less current than the "hot" phase conductors.

See Unit 20 for sizing neutral conductors for electric range branch-circuits, and Unit 29 for sizing neutral service-entrance conductors.

Figure 4-5 Diagram showing line-to-line and line-to-neutral loads.

VOLTAGE DROP

Low voltage can cause lights to dim, some television pictures to "shrink," motors to run hot, electric heaters to not produce their rated heat output, and appliances to not operate properly.

Low voltage in a home can be caused by:

- wire that is too small for the load being served,

- a circuit that is too long,

- poor connections at the terminals,

- conductors operating at high temperatures having higher resistance than when operating at lower temperatures.

A simple formula for calculating voltage drop on single-phase systems considers only the dc resistance of the conductors and the temperature of the conductor. See *Table 8, Chapter 9* in the *NEC®* for the DC resistance values. The more accurate formulas consider ac resistance, reactance, temperature, spacing, in metal conduit, and in nonmetallic conduit, *NEC® Table 9, Chapter 9*. Voltage drop is covered in great detail in *Electrical Wiring— Commercial* (Delmar Publishers). The simple voltage drop formula is more accurate with smaller conductors, and gets less and less accurate as conductor size increases. It is sufficiently accurate for voltage drop calculations necessary for residential wiring.

To Find Voltage Drop in a Single-Phase Circuit:

$$E_d = \frac{K \times I \times L \times 2}{CMA}$$

To Find Conductor Size for a Single-Phase Circuit:

$$CMA = \frac{K \times I \times L \times 2}{E_d}$$

In the above formulas:

E_d = allowable voltage drop in volts.

K = approximate resistance in ohms per circular-mil foot at 167°F (75°C).
- for uncoated copper wire use 12 ohms.*
- for aluminum wire use 20 ohms.*

I = current in amperes flowing through the conductors.

L = length in feet from beginning of circuit to the load.

CMA = cross-sectional area of the conductors in circular mils.**

---

*Derived values from *Table 8, Chapter 9, NEC®*

**From Table 4-3 in this text or from *Table 8, Chapter 9, NEC®*

We use the factor of 2 for single-phase circuits because there is voltage drop in both conductors . . . to and from the connected load.

To Find Voltage Drop and Conductor Size for a Three-Phase Circuit:

Although residential wiring does not use three-phase systems, commercial and industrial systems do. To make voltage drop and conductor sizing calculations for a three-phase system, substitute the factor of 1.73 in place of the factor 2 in the formulas on page 80.

CONDUCTOR SIZE	CROSS-SECTIONAL AREA IN CIRCULAR MILS
18	1,620
16	2,580
14	4,110
12	6,530
10	10,380
8	16,510
6	26,240
4	41,740
3	52,620
2	66,360
1	83,690
0	105,600
00	133,100
000	167,800
0000	211,600

Table 4-3

Code References to Voltage Drop

For branch-circuits, refer to *Section 210-19(a)*, *FPN No. 4*. The recommended maximum voltage drop is 3%, Figure 4-6.

For feeders, refer to *Section 215-2, FPN No. 2*. The recommended maximum voltage drop is 3%. See Figure 4-7.

When both branch-circuits and feeders are involved, the total voltage drop should not exceed 5%. This is shown in Figure 4-7.

▶ According to *Section 90-5(c)*, *FPNs* are informational only, and are not to be enforced as a Code requirement. ◀

Examples of Voltage Drop Calculations

EXAMPLE 1: What is the approximate voltage drop on a 120-volt, single-phase circuit consisting of No. 14 AWG copper conductors where the load is 11 amperes and the distance of the circuit from the panel to actual load is 85 feet (25.91 m)?

SOLUTION:

$$E_d = \frac{K \times I \times L \times 2}{CMA}$$

$$E_d = \frac{12 \times 11 \times 85 \times 2}{4110}$$

$$E_d = 5.46 \text{ volts drop}$$

Note: Refer to Table 4-3 for the CMA value.

THIS VOLTMETER READS 120 VOLTS.

BRANCH-CIRCUIT

LOAD

THIS VOLTMETER READS 116.4 VOLTS.

3% OF 120 VOLTS = 3.6 VOLTS
120 VOLTS − 3.6 VOLTS = 116.4 VOLTS

Figure 4-6 Maximum recommended voltage drop on a branch-circuit is 3%, *Section 210-19(a)*, *FPN, No. 4*.

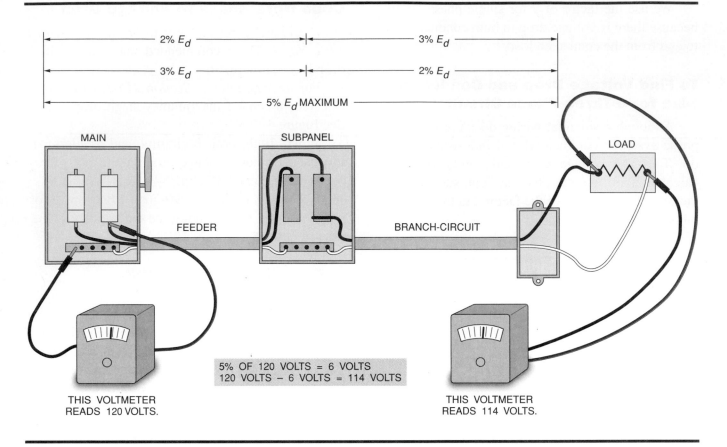

Figure 4-7 *Sections 210-19(a), FPN No. 4* and *215-2 FPN No. 2* of the Code state that the total voltage drop from the beginning of a feeder to the farthest outlet on a branch-circuit should not exceed 5%. In this figure, if the voltage drop in the feeder is 3%, then do not exceed 2% voltage drop in the branch-circuit. If the voltage drop in the feeder is 2%, then do not exceed 3% voltage drop in the branch-circuit.

This exceeds the voltage drop permitted by the Code, which is:

$$3\% \text{ of } 120 \text{ volts } = 3.6 \text{ volts}$$

Let's try it again using No. 12 AWG:

$$E_d = \frac{12 \times 11 \times 85 \times 2}{6530}$$

$$E_d = 3.44 \text{ volts drop}$$

This "Meets Code."

EXAMPLE 2: Find the wire size needed to keep the voltage drop to no more than 3% on a single-phase, 240-volt, air-conditioner circuit. The nameplate reads: "Minimum Circuit Ampacity 40 Amperes." The circuit originates at the main panel located approximately 65 feet (19.81 m) from the air-conditioner unit. No neutral is required.

SOLUTION: Checking *Table 310-16*, the conductors could be No. 6 Type TW copper or No. 8 Type THHN copper.

The permitted voltage drop is:

$$E_d = 240 \times 0.03 = 7.2 \text{ volts}$$

Let's see what voltage drop we might experience if we installed the No. 8 Type THHN conductors.

$$E_d = \frac{12 \times 40 \times 65 \times 2}{16510}$$

$$= 3.78 \text{ volts drop}$$

This is well below the permissible 7.2 volts drop and would make an acceptable installation.

A Word of Caution When Using High-Temperature Wire

Note in *Table 310-16* that conductors fall into three classes of temperature ratings—60°C, 75°C, and 90°C. For a given conductor size, we find that the allowable ampacity of a 90°C insulated conductor is greater than that of a 60°C insulated conductor. Therefore, be careful when selecting conductors

Table 310-16. Allowable Ampacities of Insulated Conductors Rated 0 through 2000 Volts, 60°C through 90°C (140°F through 194°F) Not More than Three Current-Carrying Conductors in Raceway, Cable, or Earth (Directly Buried), Based on Ambient Temperature of 30°C (86°F)

Size	Temperature Rating of Conductor (See Table 310-13)						Size
	60°C (140°F)	75°C (167°F)	90°C (194°F)	60°C (140°F)	75°C (167°F)	90°C (194°F)	
AWG or kcmil	Types TW, UF	Types FEPW, RH, RHW, THHW, THW, THWN, XHHW, USE, ZW	Types TBS, SA, SIS, FEP, FEPB, MI, RHH, RHW-2, THHN, THHW, THW-2, THWN-2, USE-2, XHH, XHHW, XHHW-2, ZW-2	Types TW, UF	Types RH, RHW, THHW, THW, THWN, XHHW, USE	Types TBS, SA, SIS, THHN, THHW, THW-2, THWN-2, RHH, RHW-2, USE-2, XHH, XHHW, XHHW-2, ZW-2	AWG or kcmil
	COPPER			ALUMINUM OR COPPER-CLAD ALUMINUM			
18	—	—	14	—	—	—	—
16	—	—	18	—	—	—	—
14*	20	20	25	—	—	—	—
12*	25	25	30	20	20	25	12*
10*	30	35	40	25	30	35	10*
8	40	50	55	30	40	45	8
6	55	65	75	40	50	60	6
4	70	85	95	55	65	75	4
3	85	100	110	65	75	85	3
2	95	115	130	75	90	100	2
1	110	130	150	85	100	115	1
1/0	125	150	170	100	120	135	1/0
2/0	145	175	195	115	135	150	2/0
3/0	165	200	225	130	155	175	3/0
4/0	195	230	260	150	180	205	4/0
250	215	255	290	170	205	230	250
300	240	285	320	190	230	255	300
350	260	310	350	210	250	280	350
400	280	335	380	225	270	305	400
500	320	380	430	260	310	350	500
600	355	420	475	285	340	385	600
700	385	460	520	310	375	420	700
750	400	475	535	320	385	435	750
800	410	490	555	330	395	450	800
900	435	520	585	355	425	480	900
1000	455	545	615	375	445	500	1000
1250	495	590	665	405	485	545	1250
1500	520	625	705	435	520	585	1500
1750	545	650	735	455	545	615	1750
2000	560	665	750	470	560	630	2000
CORRECTION FACTORS							

Ambient Temp. (°C)	For ambient temperatures other than 30°C (86°F), multiply the allowable ampacities shown above by the appropriate factor shown below.						Ambient Temp. (°F)
21–25	1.08	1.05	1.04	1.08	1.05	1.04	70–77
26–30	1.00	1.00	1.00	1.00	1.00	1.00	78–86
31–35	0.91	0.94	0.96	0.91	0.94	0.96	87–95
36–40	0.82	0.88	0.91	0.82	0.88	0.91	96–104
41–45	0.71	0.82	0.87	0.71	0.82	0.87	105–113
46–50	0.58	0.75	0.82	0.58	0.75	0.82	114–122
51–55	0.41	0.67	0.76	0.41	0.67	0.76	123–131
56–60	—	0.58	0.71	—	0.58	0.71	132–140
61–70	—	0.33	0.58	—	0.33	0.58	141–158
71–80	—	—	0.41	—	—	0.41	159–176

*See Section 240-3.

Reprinted with permission from NFPA 70-1999.

based on their ability to withstand high temperatures. For example:

- No. 8 THHN (90°C) copper has an ampacity of 55 amperes.

- No. 8 THW (75°C) copper has an ampacity of 50 amperes.

- No. 8 TW (60°C) copper has an ampacity of 40 amperes.

We learned earlier in this Unit that according to *Section 110-14(c)*, we use the 60°C column of *Table 310-16* for circuits rated 100 amperes or less, and the 75°C column for circuits rated over 100 amperes. However, when the equipment is marked as being suitable for 75°C, we can take advantage of the higher ampacity of 75°C conductors. This, oftentimes results in a smaller size conductor and a corresponding smaller size raceway.

But installing smaller conductors might result in excessive voltage drop.

Therefore, after selecting the proper size conductor for a given load, it is always a good idea to run a voltage-drop calculation to make sure the voltage drop is not excessive.

APPROXIMATE CONDUCTOR SIZE RELATIONSHIP

Rule One: For wire sizes up through 0000, every third size doubles or halves in circular mil area.

Thus, a No. 1 AWG conductor is 2X larger than a No. 4 AWG conductor (83,690 versus 41,470). Thus, a "0" wire is one-half the size of "0000" wire (105,600 versus 211,600).

Rule Two: For wire sizes up through 0000, every consecutive wire size is approximately 1.26 larger or smaller than the preceding wire size.

Thus, a No. 3 AWG conductor is approximately 1.26 larger than a No. 4 AWG conductor (41,740 × 1.26 = 52,592). Thus, a No. 2 AWG wire is approximately 1.26 smaller than a No. 1 AWG wire (83,690 ÷ 1.26 = 66,420).

Try to fix in your mind that a No. 10 AWG conductor has a cross-sectional area of 10,380 circular mils, and that it has a resistance of 1.2 ohms per 1000 feet (305 m). The resistance of aluminum wire is approximately 2 ohms per 1000 feet (305 m). By remembering these numbers, you will be able to perform voltage drop calculations without having the wire tables readily available.

EXAMPLE: What is the approximate cross-sectional area in circular mils, and resistance of a No. 6 AWG copper conductor?

SOLUTION:

WIRE SIZE	CMA (In Circular Mils)	OHMS PER 1000 FEET
10	10,380	1.2
9		
8		
7	20,760	0.6
6	26,158	0.0476

Note in this chart that when the CMA of a wire is doubled, then its resistance is cut in half. Inversely, when a given wire size is reduced to one-half, its resistance doubles.

NONMETALLIC-SHEATHED CABLE *(ARTICLE 336)*

Description

Most electricians still refer to nonmetallic-sheathed cable as *Romex*, a name chosen years ago by the Rome Wire and Cable Company.

Nonmetallic-sheathed cable is a factory assembly of two or more insulated conductors having an outer sheath of moisture-resistant, flame-retardant, nonmetallic material. This cable is available with two or three current-carrying conductors. The conductors range in size from No. 14 through No. 2 for copper conductors, and from No. 12 through No. 2 for aluminum or copper-clad aluminum conductors. Two-wire cables contain one black conductor, one white conductor, and one bare equipment grounding conductor. Three-wire cables contain one black, one white, one red, and one bare equipment grounding conductor. Equipment grounding conductors are permitted to have green insulation, but bare equipment grounding conductors are the most common.

Figure 4-8 clearly shows a bare equipment grounding conductor. Sometimes the bare equipment grounding conductor is wrapped with paper or fiberglass, which acts as a filler. The equipment grounding conductor is *not* permitted to be used as a current-carrying conductor.

U.L. lists three types of nonmetallic-sheathed cables:

- Type NM-B cable has a flame-retardant, moisture-resistant, nonmetallic outer jacket. The conductors are rated 90°C (194°F). The ampacity is based on 60°C (140°F).

- Type NMC-B cable has a flame-retardant, moisture-resistant, fungus-resistant, and corrosion-resistant nonmetallic outer jacket. The conductors are rated 90°C (194°F). The ampacity is based on 60°C (140°F).

Figure 4-8 A nonmetallic-sheathed cable showing (A) black "ungrounded" (hot) conductor, (B) bare equipment "grounding" conductor, and (C) white "grounded" conductor; *Courtesy of* Southwire Company.

- Type NMS cable contains the conventional insulated power conductors as well as telephone, coaxial, home entertainment, and signaling conductors all in one cable. This type of cable is used for home automation systems using the latest in digital technology. *Article 780* recognizes this wiring method for closed loop and programmed power distribution. Type NMS has a moisture-resistant, flame-retardant, nonmetallic outer jacket.

Although somewhat confusing, the *NEC®* in *Article 336* still refers to nonmetallic-sheathed cable as Type NM and Type NMC. In *Section 336-30(b)*, we find that the conductors in these cables must be rated 90°C (194°F).

Over the years, there have been many reported problems of conductor insulation becoming brittle, and breaking off of the conductors due to extreme high temperatures associated with recessed and surface-mounted lighting fixtures. Add to this the heat problem found in attics where the temperatures reach such high values, that when correctly applying the *Correction Factors* for the 60°C column of *Table 310-16*, we find that the allowable ampacity for the conductors in nonmetallic-sheathed cable in many instances is zero.

To solve this problem, U.L. required that all nonmetallic-sheathed cable manufactured after December 17, 1984, have 90°C insulated conductors, and required the manufacturers to mark the outer jacket with the suffix **B**. This was for ease in identifying the new 90°C cable from the older 60°C cable.

In the field today, you would find nonmetallic-sheathed cable marked Type NM-B, Type NMC-B, or Type NMS-B, Figure 4-8. Although the conductors found in nonmetallic-sheathed cable are insulated for 90°C, the actual ampacity of the conductors must be based on the ampacity of 60°C conductors. Refer to *Sections 336-26* and *336-30(b)*. According to *Section 336-26,* the 90°C rating ampacity is permitted to be used for derating purposes such as would be necessary in attics where the temperatures can reach extremely high values. The final derated ampacity must not exceed the ampacity for 60°C rated conductors. *Table 310-16* shows the allowable ampacities for conductors.

Where Nonmetallic-Sheathed Cable May Be Used

Table 4-4 shows the uses permitted for Type NM and Type NMC and NMS cables.

Equipment Grounding Conductors

Equipment grounding conductors are selected based upon the rating or setting of the overcurrent device protecting a given branch-circuit or feeder. See *Table 250-122*. The equipment grounding conductor in a nonmetallic-sheathed cable is sized according to this table.

Rating or Setting of Overcurrent Device	MINIMUM SIZE EQUIPMENT GROUNDING CONDUCTOR	
	Copper	Aluminum or Copper-Clad Aluminum
15	14	12
20	12	10
30	10	8
40	10	8
60	10	8
100	8	6
200	6	4

WHERE CAN YOU USE TYPES NM, NMC, AND NMS CABLE?

	TYPE NM	TYPE NMC	TYPE NMS
May be used on circuits of 600 volts or less.	Yes	Yes	Yes
May be run exposed or concealed in dry locations.	Yes	Yes	Yes
May be installed exposed or concealed in damp and moist locations.	No	Yes	No
Has flame-retardant and moisture-resistant outer covering.	Yes	Yes	Yes
Has fungus-resistant and corrosion-resistant outer covering.	No	Yes	No
May be used to wire one- and two-family dwellings, or multifamily dwellings that do not exceed three floors above grade.	Yes	Yes	Yes
May be embedded in masonry, concrete, plaster, adobe, fill.	No	No	No
May be installed or fished in the hollow voids of masonry blocks or tile walls where not exposed to excessive moisture or dampness.	Yes	Yes	Yes
May be installed or fished in the hollow voids of masonry blocks or tile walls where exposed to excessive moisture.	No	Yes	No
In outside walls of masonry block or tile.	No	Yes	No
In inside wall of masonry block or tile.	Yes	Yes	Yes
May be used as service-entrance cable.	No	No	No
Must be protected against physical damage.	Yes	Yes	Yes
May be run in shallow chase of masonry, concrete, or adobe if protected by a steel plate at least 1/16 inch (1.59 mm) thick, then covered with plaster, adobe, or similar finish.	No	Yes	No

Table 4-4 Uses permitted for nonmetallic-sheathed cable.

Figure 4-9 This stripper is used to remove the outer jacket from nonmetallic-sheathed cable. *Courtesy* of Seatek Co., Inc.

Installation

Nonmetallic-sheathed cable is probably the least expensive of the various wiring methods. It is relatively lightweight and easy to install. It is widely used for dwelling unit installations on circuits of 600 volts or less. Figure 4-9 shows an example of a stripper used for stripping nonmetallic-sheathed cable. The installation of nonmetallic-sheathed cable must conform to the requirements of *Article 336*. Refer to Figures 4-10 through 4-16.

- The cable must be strapped or stapled not more than 12 inches (305 mm) from a box or fitting.
- ▶• Do not staple flat nonmetallic-sheathed cables on edge, Figure 4-15.◀
- The intervals between straps or staples must not exceed 4½ feet (1.37 m).
- The cable must be protected against physical damage where necessary.
- The 4½ foot (1.37 m) securing requirement is not needed where nonmetallic-sheathed cable is run *horizontally* through holes in wood or metal framing members (studs, joists, rafters, etc.). The cable is considered to be adequately supported by the framing members, *Section 336-18*.
- The inner edge of the bend shall have a minimum radius not less than five times the cable diameter, *Section 336-16*.
- The cable must not be used in circuits of more than 600 volts.

Figure 4-10 Installation of nonmetallic-sheathed cable.

NONMETALLIC
SHEATHED CABLE

PROTECT CABLE
FROM ABRASION
ON BOTH ENDS,
*SECTION
300-15(b) Exc. 1.*

▶ PROTECT CABLE
WITH RIGID METAL
CONDUIT, INTERMEDIATE
METAL CONDUIT,
ELECTRICAL METALLIC
TUBING, SCHEDULE 80
PVC RIGID NONMETALLIC
CONDUIT, OR LISTED
SURFACE METAL OR
NONMETALLIC RACEWAY,
OR OTHER METAL PIPE.
PROTECT AT LEAST
6 INCHES (152 mm)
ABOVE FLOOR. ◀

Figure 4-11   Installation of exposed nonmetallic-sheathed cable where passing through a floor. See *Section 336-6(b)*.

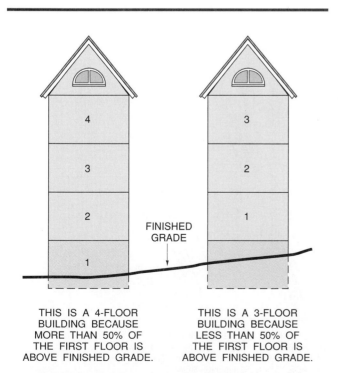

4

3

2

1

3

2

1

FINISHED
GRADE

THIS IS A 4-FLOOR
BUILDING BECAUSE
MORE THAN 50% OF
THE FIRST FLOOR IS
ABOVE FINISHED GRADE.

THIS IS A 3-FLOOR
BUILDING BECAUSE
LESS THAN 50% OF
THE FIRST FLOOR IS
ABOVE FINISHED GRADE.

Figure 4-12   Illustration of how *Section 336-5(a)(1)* defines the first floor of a building.

Figure 4-13   When cables are "bundled" for distances more than 24 inches (610 mm), their ampacities must be reduced (derated) according to the percentages found in *Table 310-15(b)(2)*.

A STRAP FOR ARMORED CABLE
OR LARGE NONMETALLIC
SHEATHED CABLE.

C INSULATED STAPLE
FOR NONMETALLIC
SHEATHED CABLE.

B STAPLE FOR
ARMORED CABLE.

Figure 4-14   (A) One hole strap used for conduit, EMT, large nonmetallic-sheathed cable and armored cable. (B) Metal staple for armored cable and nonmetallic-sheathed cable. (C) Insulated staple for nonmetallic-sheathed cable. Certain types of insulated and noninsulated staples can be applied with a staple gun.

- Nonmetallic-sheathed cable must be protected where passing through a floor by at least 6 inches (152 mm) ▶of rigid metal conduit, intermediate metal conduit, electrical metallic tubing, Schedule 80 PVC rigid nonmetallic conduit, listed surface metal or nonmetallic raceway, or other metal pipe, *Section 336-6(b)*.◀ A fitting (bushing or connector) must be used at both ends of the conduit to protect the cable from abrasion, Figure 4-11.

- When cables are "bundled" together for distances longer than 24 inches (610 mm), Figure 4-13, the heat generated by the conductors cannot easily dissipate. These conductors must be derated according to *Table 310-15(b)(2)*.

CODE VIOLATION
SECTION 336-18

• CABLE ON EDGE

"MEETS CODE"
SECTION 336-18

• CABLE LIES FLAT

▶Figure 4-15   It is a Code violation to staple flat nonmetallic-sheathed cables on edge, *Section 336-18.*◀

IT IS OK TO FASTEN CABLES NOT SMALLER
THAN TWO No. 6 OR THREE No. 8 CONDUCTORS
DIRECTLY TO BOTTOM OF JOISTS.

CABLES OF ANY SIZE MAY BE RUN THROUGH
BORED HOLES IN JOISTS, RAFTERS, AND STUDS.
THEY ARE CONSIDERED TO BE ADEQUATELY
SUPPORTED, *SECTION 336-18.*

CABLES OF ANY SIZE
MAY BE RUN ON THE
SIDES OF JOISTS.
THEY MUST BE
SECURED (STAPLED
OR STRAPPED).

THIS IS A
RUNNING BOARD.

CABLES OF ANY SIZE MAY BE RUN
PARALLEL TO SIDES OR FACE OF JOISTS.
KEEP CABLES AT LEAST 1 1/4" FROM EDGE
OF FRAMING MEMBER, *SECTION 300-4(d).*

CABLES SMALLER THAN TWO No. 6
OR THREE No. 8 MUST BE RUN THROUGH
BORED HOLES OR BE RUN ON THE
SURFACE OF A RUNNING BOARD.

Figure 4-16   *Section 336-6(c)* gives the rules for running nonmetallic-sheathed cable in unfinished basements.

▶• Nonmetallic-sheathed cable not over 4½ feet (1.37 m) may be used as a fixture whip above an accessible ceiling. It does not have to be secured within 12 inches (305 mm) of the lighting fixture. See *Section 336-18, Exception No. 3.*◀ Fixture whips are discussed in Units 7 and 17.

Figure 4-16 shows the installation of nonmetallic-sheathed cable in unfinished basements, *Section 336-6(c).*

See Unit 15 and Figure 15-13 for additional text and diagrams covering the installation of cables in attics.

Figure 4-10 shows the Code requirements for securing nonmetallic-sheathed cable when using nonmetallic boxes.

The Code in *Section 370-40(d)* requires that all metal boxes have provisions for the attachment of an equipment grounding conductor. Figure 4-17 shows a gang-type switch (device) box that is tapped for a screw by which the grounding conductor may be connected underneath. Figure 4-18 shows an outlet box, also with provisions for attaching grounding conductors. Figure 4-19 illustrates the use of a small grounding clip.

Section 410-20 of the Code requires that there be a provision whereby the equipment grounding conductor can be attached to the exposed metal parts of lighting fixtures.

Multistory Buildings

▶ Nonmetallic-sheathed cable may be installed in any one- and two-family dwelling regardless of how many floors, *Section 336-4(l).*◀

Nonmetallic-sheathed cable is not permitted in multifamily dwellings or other structures that exceed three floors above grade.

Section 336-5(a)(l) defines the first floor of a building as *"that floor that has 50% or more of the exterior wall surface area level with or above finished grade."* The *NEC®* allows one additional level for parking, storage, or similar use when such level is not used for human habitation. Walk around the entire building to determine the 50% criteria. Refer to Figure 4-12.

▶ Although nonmetallic-sheathed cables are usually secured to a panel, box, or cabinet, with a

Figure 4-18   Outlet box.

Figure 4-17   Gang-type switch (device) box.

Figure 4-19   Method of attaching ground clip to a metal switch (device) box. See *Section 250-148*.

A NONMETALLIC SHEATHED CABLE SECURED WITHIN 12 INCHES FROM WHERE IT EMERGES FROM RACEWAY.

B RACEWAY COMES OUT OF TOP OF PANEL AND MUST NOT PENETRATE STRUCTURAL CEILING.

C A FITTING NEEDED ON BOTH ENDS TO PREVENT ABRASION.

D RACEWAY MUST BE SEALED TO PREVENT DEBRIS FROM GETTING INTO PANEL.

E OUTER SHEATH OF CABLE MUST EXTEND INTO PANEL AT LEAST 1/4 INCH.

F SECURE RACEWAY.

G DO NOT EXCEED PERCENT FILL AS SHOWN IN *TABLE 1, CHAPTER 9, NEC.*® USE LARGEST DIMENSION OF CABLE AS ITS DIAMETER FOR CALCULATING THE CABLE'S CROSS-SECTIONAL AREA, *CHAPTER 9, TABLE 1, NOTE 9.*

H DEPENDING UPON THE NUMBER OF CONDUCTORS IN THE RACEWAY AND THE LENGTH OF THE RACEWAY, IT MAY BE NECESSARY TO DERATE THE AMPACITY OF THE CONDUCTORS, *SECTION 310-15(b)(2)(a).*

NONFLEXIBLE RACEWAY NOT LESS THAN 18 INCHES (457 mm) NOR MORE THAN 10 FEET (3.05 m).

PANEL

Figure 4-20   An alternate method for running nonmetallic-sheathed cables into a surface-mounted panel. This is permitted in *Section 373-5(c), Exception.*

cable clamp or connector, the *Exception* to *Section 373-5(c)* provides an alternate method for running nonmetallic-sheathed cables into surface-mounted enclosures. This exception is for surface-mounted enclosures only. This is illustrated in Figure 4-20. ◄

ARMORED CABLE *(ARTICLE 333)*

Most electricians refer to armored cable as BX, supposedly derived from an abbreviation of the Bronx in New York City, where armored cable was once manufactured. BX was a trademark owned by the General Electric Company, but over the years, has become a generic term.

The *NEC*® covers the installation requirements for armored cable in *Article 333.*

U.L. tests armored cable per their standard No. 4. Because U.L. standards have evolved in numerical order, it is apparent that the armored cable standard is one of the first and oldest standards to be developed.

Description

The *NEC®* describes Type AC cable as a "fabricated assembly of insulated conductors in a flexible metallic enclosure," Figure 4-21.

The armor may be galvanized steel or aluminum.

Armored cable with aluminum conductors is marked "Armored Aluminum Cable." Armored cable with aluminum armor is marked "Aluminum Armored Cable." The marking appears on the cable's marking tape and on the tag found on each coil.

▶ Type AC armored cables have conductors with *thermosetting* insulation. Thermosetting material becomes permanently hard and rigid when subjected to heat once. It will not melt when heated again. Type ACT armored cables have conductors with *thermoplastic* insulation. Thermosetting material will soften when heated. ◀

Type ACTH armored cable has 75°C (167°F) thermoplastic insulated conductors.

Type ACTHH armored cable has 90°C (194°F) thermoplastic insulated conductors.

Type ACHH armored cable has 90°C (194°F) thermosetting insulated conductors.

Armored cable buried in thermal insulation must have conductors rated for 90°C (194°F), and the ampacity of the conductors must be based upon the 60°C (140°F) column of *Table 310-16*. This is a requirement of *Section 333-20*.

Earlier in this unit, we discussed terminations and ampacities of conductors in conformance to *Section 110-14(c)*. You may want to read this again to be sure you fully understand it.

Effective Grounding Path

▶ *Section 250-2(d)* requires that *"the fault-current path shall be permanent and electrically continuous, shall be capable of safely carrying the maximum fault likely to be imposed on it, and shall have sufficiently low impedance to facilitate the operation of overcurrent devices under fault conditions."* ◀ This is accomplished by properly installed metal raceways, separate equipment grounding conductors, or cable armor.

▶ *Section 333-21* requires that armored cable provide an adequate path for equipment grounding. ◀

To ensure that the impedance of a given length of armored cable is low enough to meet these requirements, armored cable has an internal bonding strip (copper or aluminum) in direct contact with the metal armor for the entire length of the cable. In combination, the metal armor plus the internal bonding wire provides the required low impedance. Note in Figure 4-22 how the bonding strip is folded back over the armor. The metal armor also provides excellent mechanical protection for the conductors.

Never use the internal bonding strip as a grounded neutral conductor or as a separate equipment grounding conductor. It is an internal bonding strip—nothing more!

Figure 4-21   Flexible armored cable. *Courtesy* of AFC Cable Systems, Inc.

A

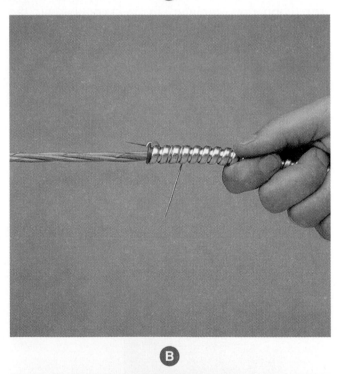

B

Figure 4-22 Antishort bushing prevents cutting of the conductor insulation by the sharp metal armor. *Courtesy of Arc Cable Systems, Inc.*

Connectors

Listed connectors for use with armored cable are permitted to be used on both steel and aluminum armored cable with one exception: Set-screw con-

nectors are *not* permitted to be used with aluminum armored cable. This is a U.L. requirement.

Listed connectors are suitable for grounding purposes.

Color Coding. For residential wiring, armored cable with two or three insulated conductors is most commonly used. Four-conductor, and even five-conductor, armored cable is available but is not commonly used for house wiring. In addition, armored cable could contain a separate green insulated equipment grounding conductor. This would be used if an isolated ground is required. Again, this is not very common in house wiring.

The current-carrying insulated conductors in armored cable can be No. 14 through No. 1 AWG copper, or No. 12 through No. 1 AWG aluminum. Two-conductor cable contains one black and one white conductor. Three-conductor cable contains one black, one white, and one red conductor. Four-conductor cable contains one black, one white, one red, and one blue conductor. Each insulated conductor is individually wrapped with a light brown kraft paper.

Use and Installation

Armored cable can be used in more applications than nonmetallic-sheathed cable.

Armored Cable:

- may be used on branch circuits and feeders for applications of 600 volts or less.

- may be used for open and concealed work in dry locations.

- may be run through walls and partitions.

- may be used where not subject to physical damage.

- may be embedded in the plaster finish on masonry walls or run through the hollow spaces of such walls if these locations are not considered damp or wet.

- may be fished in air voids of masonry block or tile walls where the walls are not exposed or subject to excessive moisture or dampness. Masonry in direct contact with the earth is considered to be a wet location.

- must be secured within 12 inches (305 mm) from every outlet box or fitting and at intervals not exceeding 4½ feet (1.37 m). This securing requirement is not necessary when the armored cable is fished through the walls and/or ceilings, *Section 333-7(b)(1)*.

- need not follow the 4½ foot (1.37 m) securing requirement where AC cable is run *horizontally* through holes in wood or metal framing members (studs, joists, rafters, etc.), or laid in notches in wood framing members. The cable is considered to be adequately supported by the framing members, *Section 333-7(a)*.

- must not be bent to a radius of less than five times the diameter of the cable, *Section 333-8*, Figure 4-10.

- may be used in unsupported lengths of not over 6 feet (1.83 m) to run between an outlet box and a lighting fixture (or other equipment) above an accessible ceiling. The 12 inch (305 mm) and 4½ feet (1.37 m) securing requirement mentioned above is waived when the armored cable is used as a fixture "whip" to connect lighting fixtures or equipment, *Section 333-7(b)(3)*.

- must have an insulating bushing (antishort) at the cable ends between the conductors and the armor, *Section 333-9*. See Figure 4-22.

- may be used where the dwelling or structure exceeds three floors above grade.

- when exposed, must closely follow the surface of the building or of running boards if used. You do not have to closely follow the surface when (1) hooking up equipment where flexibility is needed and the length of the cable is not more than 24 inches (610 mm), or (2) when the cable is supported on the underside of each joist and is not subject to physical damage, *Sections 333-7(b)(2)* and *333-11*.

- in accessible attics or roof spaces, must be protected by guard strips at least as high as the cable when run across the top of floor joists, or within 7 feet (2.13 m) of the floor or floor joists when the cable is run across the face of studs or rafters. If there is no permanent stairway or ladder, this protection is needed only within 6 feet (1.83 m) from the edge of the scuttle hole or entrance to the attic. The cable is considered to

be adequately protected from physical damage when run parallel to the sides of joists, rafters, or studs, in which case no guard strips or running boards are necessary. See *Section 333-12*.

▶• if buried in thermal insulation, must have 90°C (194°F) insulation. The ampacity must be based on the 60°C (140°F) values from *Table 310-16*.◀

- must not be buried underground.

- must not be buried in masonry, concrete, or fill during construction.

- must not be used anywhere directly exposed to the weather.

See Unit 15 for additional discussion and diagrams regarding physical protection of cable in attics.

To remove the outer metal cable armor, it is recommended that a tool similar to the one shown in Figure 4-23 be used. Otherwise, a hacksaw can be used, Figure 4-24, to cut through one of the raised convolutions of the cable armor. Be very careful not to cut too deep or you will cut into the conductor insulation. Then bend the cable armor at the cut. It will snap off easily, exposing the desired 8 to 10 inches (203 to 254 mm) of conductor.

To prevent the sharp metal armor from cutting the conductor insulation, an insulating (antishort)

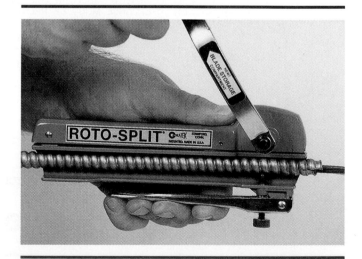

Figure 4-23 This is an example of a tool that precisely cuts the outer armor of armored cable, making it easy to remove the armor with a few turns of the handle. *Courtesy* of Seatek Co., Inc.

Figure 4-24 Using a hacksaw to cut through a raised convolution of the cable armor.

bushing is inserted at the cable ends as shown in Figure 4-22. Photo (A) shows an antishort being inserted. Photo (B) shows the recommended way of bending the bonding strip to hold the antishort in place.

INSTALLING CABLES THROUGH WOOD AND METAL FRAMING MEMBERS
(SECTION 300-4)

To complete the wiring of a residence, cables must be run through studs, joists, and rafters. One method of installing cables in these locations is to run the cables through holes drilled at the approximate centers of wood building members, or at least 1¼ inches (31.8 mm) from the nearest edge. Holes bored through the center of standard 2 × 4s meet the requirements of *Section 300-4(a)(1)*. Refer to Figure 4-25.

If the 1¼-inch (31.8 mm) distance cannot be maintained, or if the cables are to be laid in a notch, a metal plate at least ¹⁄₁₆ inch (1.59 mm) thick must cover the notch to protect the cable from nails, *Section 300-4(a)(1)* and *(2)*. Refer to Figure 4-25.

Where nonmetallic-sheathed cables pass through holes in metal framing members, the cable must be protected by a bushing or grommet securely fastened in place prior to installing the cable. ▶The bushing or grommet must cover all metal edges, *Section 300-4(b)(1)*.◀

Bushings that snap into a hole sized for the specific size bushing (½-inch, ¾-inch, etc.), or a two-piece snap in bushing that fits just about any size

1 CABLES RUN THROUGH FRAMING MEMBERS (JOISTS, RAFTERS, STUDS) MUST NOT BE LESS THAN 1 1/4″ (31.8 mm) FROM THE NEAREST EDGE OF THE FRAMING MEMBER. IF THIS CLEARANCE CANNOT BE MAINTAINED, THE CABLE MUST BE PROTECTED BY A STEEL PLATE AT LEAST 1/16″ (1.59 mm) THICK.

LESS THAN 1 1/4″ (31.8 mm)

1 1/4″ (31.8 mm)

3 AN ALTERNATIVE TO USING STEEL PLATES IS TO USE CABLE "STAND OFF" DEVICES. SEE FIGURE 4-27.

CABLES

NOTCHED STUDS

2 x 2

2 x 4

2 METAL PLATE, 1/16″ (1.59 mm) THICK NEEDED TO PROTECT CABLES LAID IN NOTCH IN STUD.

Figure 4-25 Methods of protecting nonmetallic and armored cables so nails and screws that "miss" the studs will not damage the cables, *Section 300-4(a)(1)* and *(2)*. Because of their strength, intermediate and rigid metal conduit, rigid nonmetallic conduit, and electrical metallic tubing are exempt from this rule per the *Exceptions* found in *Section 300-4*. See Figure 4-27 for alternative methods of protecting cables.

precut hole in the metal framing members is acceptable to use, Figure 4-26. These types of bushings (insulators) are not required in metal framing members when the wiring method is armored cable, electrical metallic tubing (EMT), flexible metal conduit, or electrical nonmetallic tubing.

To reduce the possibility of nails being driven into electrical cables and certain types of raceways, Figure 4-27 illustrates some devices that can be used to *stand off* the cables from a framing member. Use

of this type of device meets the requirements of *Section 300-4(d)*.

When running nonmetallic-sheathed cable or electrical nonmetallic tubing through metal framing members where there is a likelihood that nails or screws could be driven into the cables or tubing, a steel plate, steel sleeve, or steel clip at least $\frac{1}{16}$-inch thick must be installed to protect the cables and/or tubing. This is spelled out in *Section 300-4(b)(2)*. See Figures 4-26 and 4-28.

METAL
STUDS

SHARP
EDGES

Figure 4-26 *Section 300-4(b)(1)* requires protection of nonmetallic-sheathed cable where run through holes in metal framing members. Shown are two types of bushings (grommets) that can be installed to protect nonmetallic-sheathed cable from abrasion. Electricians generally avoid using nonmetallic-sheathed cable through metal studs and joists. They use electrical metallic tubing, flexible metal conduit, armored cable (Type AC or MC), or electrical nonmetallic tubing. Where there is a likelihood that nails or screws might penetrate the nonmetallic-sheathed cable, $\frac{1}{16}$-inch (1.59 mm) thick steel plates, sleeves, or clips must be installed to protect the cable or electrical nonmetallic tubing.

Where cables and certain types of raceways are run concealed or exposed, parallel to a stud, joist, or rafter (any building framing member), the cables and raceways must be supported and installed in such a manner that there is not less than 1¼ inches (31.8 mm) from the edge of the stud or joist (framing member) to the cable or raceway. If the 1¼-inch (31.8 mm) clearance cannot be maintained, then a steel plate, sleeve, or equivalent that is at least ¹⁄₁₆-inch (1.59 mm) thick must be provided to protect the cable or raceway from damage that can occur when nails or screws are driven into the wall or ceiling, *Section 300-4(d)*, Figures 4-25 and 4-28.

Section 300-4(e) addresses the wiring problems associated with cutting grooves into building material because there is no hollow space for the wiring.

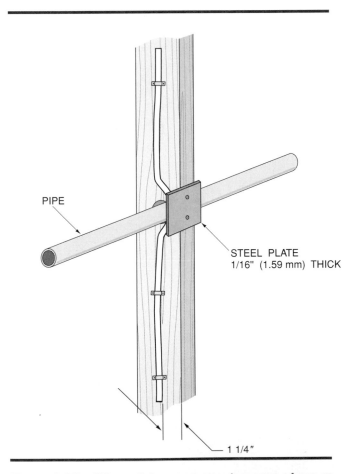

PIPE

STEEL PLATE
1/16" (1.59 mm) THICK

1 1/4"

Figure 4-28   Where it is necessary to cross pipes or other similar obstructions, making it impossible to maintain at least 1¼-inch (31.8 mm) clearance from the framing member to the cable, install a steel plate at least ¹⁄₁₆-inch (1.59 mm) thick to protect the cable from possible damage from nails or screws. See *Section 300-4(d)*.

Figure 4-27   Devices that can be used to meet the Code requirement to maintain a 1¼-inch (31.8 mm) clearance from framing members.

Nonmetallic-sheathed cable and some raceways must be protected when laid in a groove. Protection must be provided with a ¹⁄₁₆-inch (1.59 mm) steel plate, sleeve, or equivalent, or the groove must be deep enough to allow not less than 1¹⁄₄-inch (31.8 mm) free space for the entire length of the groove. This situation can be encountered where styrofoam insulation building blocks are grooved to receive the electrical cables, then covered with wall-board, wood paneling, or other finished wall material. Another typical example is where solid wood planking is installed on top of wood beams. The top side of the planking is covered with roofing material, and the bottom side is exposed and serves as the finished ceiling. The only way to install wiring for ceiling fixtures and fans is to groove the planking in some manner, Figure 4-29.

This additional protection is not required when the raceway is intermediate metal conduit *(Article 345)*, rigid metal conduit *(Article 346)*, rigid non-metallic conduit *(Article 347)*, or electrical metallic tubing *(Article 348)*, Figure 4-26.

Additional protection obviously cannot be installed inside the walls or ceilings of an existing building where the walls and ceilings are already closed in.

Sections 336-18, Exception No. 1, and *300-4(d), Exception 2* allow cables to be "fished" between outlet boxes or other access points without additional protection.

In the recreation room of the residence plans, precaution must be taken so that there is adequate protection for the wiring. The walls in the recreation room are to be paneled. Therefore, if the carpenter uses 1 × 2-inch or 2 × 2-inch furring strips, Figure 4-30, nonmetallic-sheathed cable would require the additional ¹⁄₁₆-inch (1.59 mm) steel plate protection or the use of *stand off* devices similar to that illustrated in Figure 4-27.

Some building contractors attach 1-inch insulation to the walls, then construct a 2 × 4-inch wall in front of the basement structural foundation wall, leaving 1-inch spacing between the insulation and the back side of the 2 × 4-inch studs. This makes it easy to run cables or conduits behind the 2 × 4-inch studs, and eliminates the need for the additional mechanical protection, Figure 4-31.

Watch out for any wall partitions where 2 × 4-inch studs are installed "flat," Figure 4-32. In this case, the choice is to provide the required mechanical protection as required by *Section 300-4*, or to install the wiring in electrical metallic tubing.

When running nonmetallic-sheathed cable and armored cable through or parallel to studs, joist, and

Figure 4-29   Example of how a solid wood framing member or styrofoam insulating block might be grooved to receive the nonmetallic-sheathed cable. After the cable is laid in the groove, a steel plate not less than ¹⁄₁₆-inch (1.59 mm) thick must be installed over the groove to protect the cable from damage, such as from driven nails or screws. See *Section 300-4(e)*.

FOUNDATION WALL

1 x 2-INCH OR
2 x 2-INCH
FURRING STRIPS

DRY WALL
OR PANELING

Figure 4-30   When 1 × 2-inch or 2 × 2-inch furring strips are installed on the surface of a basement wall, additional protection as shown in Figures 4-25, 4-26, 4-27, and 4-29 must be provided.

Figure 4-31 This sketch shows how to insulate a foundation wall and how to construct a wall that provides space to run cables and/or raceways that will not require the additional protection as shown in Figures 4-25, 4-26, 4-27, and 4-29.

Figure 4-32 When 2 × 4-inch studs are installed "flat," such as might be found in nonbearing partitions, cables running parallel to or through the studs must be protected against the possibility of having nails or screws driven through the cables. This means protecting the cable with ¹⁄₁₆-inch (1.59 mm) steel plates or equivalent for the entire length of the cable, or installing the wiring using intermediate metal conduit, rigid metal conduit, rigid nonmetallic conduit, or electrical metallic tubing. See Figures 4-25, 4-26, 4-27, and 4-28.

rafters, refer to *Sections 333-10* and *336-9*. These sections direct us to *Section 300-4* for more detailed information.

See Unit 15 for the installation of cables in attics.

Building Codes

When complying with the *National Electrical Code,* be sure to also comply with the Uniform Building Code. Building codes are concerned with the weakening of framing members when drilled or notched. Meeting the requirements of the *NEC* can be in conflict with building codes. Here is a recap of some of the key building code requirements that electricians should be aware of.

Studs:

- Cuts or notches in exterior or bearing partitions shall not exceed 25 percent of the stud's width.

- Cuts or notches in nonbearing partitions shall not exceed 40 percent of the stud's width.

- Bored or drilled holes in any stud:
  - shall not be greater than 40 percent of the stud's width, and
  - shall not be closer than ⅝ inch to the edge of the stud, and
  - shall not be located in the same section as a cut or notch.

Joists:

- Notches on the top or bottom of joists shall not exceed one-sixth the depth of the joist, and shall not be located in the middle third of the span.

- Holes bored in joists shall not be within 2 inches of the top or bottom of the joist, and the diameter of any such hole shall not exceed one-third the depth of the joist.

Engineered Wood Products

Today, we find many homes constructed with engineered wood I-joists and beams. The strength of these factory-manufactured framing members conform to various building codes. It is critical that certain precautions be taken relative to drilling and cutting these products so as not to weaken them. In general,

- Do not drill or notch beams.

- Do not cut, drill, or notch flanges. Flanges are the top and bottom pieces.

- Do not cut holes too close to supports or to other holes.

- For multiple holes, the amount of web to be left between holes must be at least twice the diameter of the largest adjacent hole. The web is the piece between the flanges.

- Holes not over 1½" usually can be drilled anywhere in the web.

- Do not hammer on the web except to remove knockout prescored holes.

- Check and follow the manufacturer's instructions!

INSTALLATION OF CABLES THROUGH DUCTS

Section 300-22 is extremely strict as to what types of wiring methods are permitted for installation of cables through ducts or plenum chambers. These stringent rules are for fire safety.

▶ The Code requirements are somewhat relaxed by *Section 300-22(c), Exception*. In this exception, permission is given to run nonmetallic-sheathed and armored cables in joist and stud spaces (i.e., cold air returns), but only if they are run perpendicular to the long dimensions of such spaces as illustrated in Figure 4-33. ◀

CONNECTORS FOR INSTALLING NONMETALLIC-SHEATHED AND ARMORED CABLE

The connectors shown in Figure 4-34 are used to fasten nonmetallic-sheathed cable and armored cable to the boxes and panels in which they terminate. These connectors clamp the cable securely to each outlet box. Many boxes have built-in clamps and do not require separate connectors.

The question continues to be asked: May more than one cable be inserted in one connector? Unless the U.L. listing of a specific cable connector indicates that the connector has been tested for use with more than one cable, the rule is *one cable, one connector*. This will be marked on the carton.

ELECTRICAL METALLIC TUBING *(ARTICLE 348)*, INTERMEDIATE METAL CONDUIT *(ARTICLE 345)*, RIGID METAL CONDUIT *(ARTICLE 346)*, AND RIGID NONMETALLIC CONDUIT *(ARTICLE 347)*

Some communities do not permit the installation of cable of any type in residential buildings. These communities require the installation of a raceway system of wiring, such as electrical metallic tubing (thinwall), intermediate metal conduit, rigid metal conduit, or rigid nonmetallic conduit.

Where the building construction is cement block, cinder block, or poured concrete, it will be necessary to make the electrical installation in conduit.

Sufficient data is given in this text for both cable and conduit wiring methods. Thus, the student will be able to complete the type of installation permitted or required by local codes.

According to installation requirements for a raceway system of wiring, electrical metallic tubing (EMT), intermediate metal conduit, and rigid metal conduit:

- may be buried in concrete or masonry and may be used for open or concealed work.

- may *not* be installed in cinder concrete or cinder fill unless (1) protected on all sides by a layer of noncinder concrete at least 2 inches (50.8 mm) thick, or (2) the conduit is at least 18 inches (457 mm) below the fill.

In general, conduit must be supported within

FACTORY
MANUFACTURED
I-JOISTS

FACTORY
MANUFACTURED
I-JOISTS

COLD AIR
SPACE
(RETURN)

NONMETALLIC-SHEATHED
CABLE *MAY PASS*
THROUGH THIS
SPACE – ONLY IF
RUN PERPENDICULAR
(RIGHT ANGLES) TO
THE LONG DIMENSION.

STUD

STUD

FOR OTHER THAN THE SPECIAL
PERMISSION TO RUN TYPE NM CABLE
THROUGH THIS SPACE PERPENDICULAR
TO THE LONG DIMENSION, WIRING
METHOD IN THIS SPACE MUST BE
A METAL RACEWAY, TYPE MI, MC, OR
AC CABLE OR OTHER FACTORY
ASSEMBLED MULTICONDUCTOR CONTROL
OR POWER CABLE SPECIFICALLY LISTED
FOR THE USE.

NONMETALLIC-
SHEATHED CABLE
NOT PERMITTED
TO BE RUN
PARALLEL TO THE
LONG DIMENSION
IN THIS SPACE.

ELECTRICAL EQUIPMENT NOT PERMITTED
IN THIS SPACE UNLESS LISTED FOR THIS
TYPE OF USE. METAL BOXES PERMITTED.
NONMETALLIC BOXES PERMITTED ONLY IF
LISTED FOR FIRE-RESISTANCE AND LOW
SMOKE PRODUCING CHARACTERISTICS.

Figure 4-33 Nonmetallic-sheathed cable may pass through cold air return, joist, or stud spaces in dwellings, but only if it is run at right angles to the long dimension of the space, *Section 300-22(c), Exception.*

USED FOR NONMETALLIC-
SHEATHED CABLE

USED FOR ARMORED CABLE

Figure 4-34 Cable connectors.

NOT OVER 3 FEET (914 mm) BETWEEN SUPPORT
AND OUTLET BOX, JUNCTION BOX, CABINET,
CONDUIT BODY, OR CONDUIT TERMINATION

NOT OVER 10 FEET (3.05 m)
BETWEEN SUPPORTS

Figure 4-35 Code requirements for the support of raceways are found in *Sections 345-12* (intermediate metal conduit), *346-12* (rigid metal conduit), *347-8* (rigid nonmetallic conduit), and *348-13* (electrical metallic tubing). Horizontal runs through bored holes in framing members (studs, joists, rafters) are considered to be adequately supported. *Section 348-13, Exception No. 1* permits unbroken lengths of EMT to be secured 5 feet (1.52 m) from boxes, cabinets, or conduit bodies if structural members do not readily permit fastening within 3 feet (914 mm). *Section 348-13, Exception No. 2* permits unbroken lengths of EMT to be "fished" through finished or concealed partitions where it is impractical to secure the EMT such as conditions found in remodel work.

3 feet (914 mm) of each box or fitting and at 10-foot (3.05 m) intervals along runs, Figure 4-35.

The number of conductors permitted in conduit and tubing is found in *Chapter 9* of the Code. Heavy-wall rigid conduit provides greater protection against mechanical injury to conductors than does either EMT or intermediate metal conduit, which have much thinner walls.

The residence specifications indicate that a meter pedestal is to be furnished by the utility and installed by the electrical contractor. Main Panel A is located in the workshop. The meter pedestal is located on the outside of the house near the air-conditioning unit. The electrical contractor must furnish and install a 1½-inch (38.1 mm) electrical metallic tubing between the meter pedestal and Main Panel A.

Sections 300-4(f) and *373-6(c)* require that smoothly rounded insulating fittings or equivalent be used where No. 4 or larger conductors enter a raceway, box, cabinet, auxiliary gutter, or other enclosure. Various types of fittings are shown in Figure 4-37. Examples of insulated fittings that protect conductors from abrasion are illustrated in Figures 28-28 and 28-29. Where smoothly rounded or flared threaded hubs and bosses such as found on meter sockets are encountered, insulating fittings are not required, Figure 28-16.

Figure 4-36 (A) Flexible metal conduit. (B) Flexible liquid-tight metal conduit. (C) Flexible liquidtight non-metallic conduit. *Courtesy* of Electri-Flexco.

Figure 4-37 Various fittings for (A) flexible metal conduit, (B) liquidtight flexible metal conduit, and (C) liquidtight flexible nonmetallic conduit.

In addition to these insulating fitting requirements, a bonding type bushing (Figure 28-23) must be used where the service-entrance conduit enters Main Panel A. Unit 28 covers grounding and bonding of service equipment.

The plans also indicate that electrical metallic tubing is to be used in the workshop. Because the recreation room and basement stairwell have finished ceilings and walls, they will be wired with cable.

Many of the *NEC*® rules are applicable to both cable and conduit installations. Some Code requirements are applicable only to cable installations. For example, there are special requirements for the attachment of the grounding conductors and when protection from physical damage must be provided. Many of the Code requirements governing the installation of cables do not apply to a raceway system. Properly installed electrical metallic tubing provides good continuity for equipment grounding, a means of withdrawing or pulling in additional conductors, and excellent protection against physical damage to the conductors.

RIGID NONMETALLIC CONDUIT

Rigid nonmetallic conduit is made of polyvinyl chloride (PVC), and oftentimes referred to as rigid PVC, is covered in *Article 347* of the *National Electrical Code.*®

Rigid PVC comes in 10-foot (3.05 m) lengths. It comes in two types:

- Schedule 40 (heavy wall thickness)
- Schedule 80 (extra-heavy wall thickness)

The outside diameter is the same for both Schedule 40 and Schedule 80. The inside diameter of Schedule 40 is greater than that of Schedule 80.

Schedule 40 has a thinner wall than Schedule 80 and is used where it is not subject to physical damage. Schedule 80 is stronger and is used where exposed to physical damage.

The fittings used with rigid PVC are attached with a solvent-type cement, much the same as what plumbers use. Plumbing PVC pipe may look the same as electrical PVC—but they are not the same. One is designed to withstand water pressure from the inside—the other is designed to withstand forces from the outside.

The applications where rigid PVC are permitted are found in *Article 347.* For residential use, here are some of the key requirements for rigid PVC.

- may be used concealed in walls, ceilings, and floors.
- may be used in wet locations, such as outdoors.
- may be used underground.
- may not be used to support lighting fixtures.
- may be used to support nonmetallic conduit bodies.
- if used to support nonmetallic conduit bodies, the conduit body must not contain wiring devices or support lighting fixtures.
- is nonmetallic, therefore, an equipment grounding conductor may be needed.
- shall not have more than the equivalent of four 90° bends between pull points. Simply stated, the sum of all bends and offsets must not exceed 360°.

- must be supported within 3 feet (914 mm) of each outlet box, junction box, device box, conduit body, or other termination.
- must be supported at intervals as follows:

RIGID NONMETALLIC CONDUIT SIZE	MAXIMUM SPACING BETWEEN SUPPORTS
½", ¾", 1"	3 Feet
1¼", 1½", 2"	5 Feet

- must have expansion fittings on long runs, *Section 347-9.* Instructions furnished with the rigid PVC will provide the necessary details. *Table 10* in *Chapter 9* of the *NEC®* has dimensional requirements for anticipated expansion. Long runs of PVC in direct sunlight can and will pull out of the hubs of boxes, or will pull the boxes off the wall if proper attention is not given to thermal expansion.
- when all conductors are the same size and type, the number of conductors permitted in rigid PVC Schedule 80 is found in *Table C9* in *Appendix C* of the *NEC®.*
- when conductors are different sizes, refer to *Chapter 9* in the *NEC®.* The number of conductors permitted is determined using the percentage fill shown in *Table 1*, the dimensions of rigid PVC found in *Table 4*, and the dimensions of the conductors as shown in *Tables 5*, *5A*, and *8*.
- conductors emerging from rigid PVC must be protected from abrasion, either by the rounded edges of the box, fitting, or other enclosure.
- *Table 346-10* shows the minimum radius for bends in rigid nonmetallic conduit.
- bends can be made with a proper bender that electrically heats the PVC to just the right temperature for bending. Manufactured 90, 45, and 30 elbows are available.

FLEXIBLE CONNECTIONS

The installation of certain equipment requires flexible connections, both to simplify the installation and to stop the transfer of vibrations. In residential wiring, flexible connections are used to wire attic fans, food waste disposers, dishwashers, air conditioners, heat pumps, recessed fixtures, and similar equipment.

The three types of flexible conduit used for these connections are: flexible metal conduit, flexible liquidtight metal conduit, and flexible liquidtight nonmetallic conduit, Figure 4-36. Figure 4-37 shows many of the types of connectors used with flexible metal conduit, flexible liquidtight metal conduit, and flexible liquidtight nonmetallic conduit.

FLEXIBLE METAL CONDUIT
(ARTICLE 350)

Article 350 of the Code covers the use and installation of flexible metal conduit. This wiring method is similar to armored cable, except that the conductors are installed by the electrician. For armored cable, the cable armor is wrapped around the conductors at the factory to form a complete cable assembly. Flexible metal conduit is often referred to as *Greenfield*, named after Harry Greenfield, the individual who submitted the product for listing back in 1902.

Common installations using flexible metal conduit are shown in Figure 4-38. Note that the flexibility required to make the installation is provided by the flexible metal conduit. Figure 4-38 calls

FOOD WASTE DISPOSER

EQUIPMENT GROUNDING CONDUCTOR

FLEXIBLE METAL CONDUIT

Figure 4-38   Some of the more common places where flexible metal conduit may be used.

attention to the *NEC®* and U.L. restrictions on the use of flexible metal conduit with regard to relying on the metal armor as a grounding means.

The following are some of the common uses, limitations, and applications for flexible metal conduit, combining the material found in *Article 350* of the Code, and in the U.L. White Book:

- do not install in wet locations unless the conductors are suitable for use in wet locations (TW, THW, THWN), and the installation is made so water will not enter the enclosure or other raceways to which the flex is connected.

- do not bury in concrete.

- do not bury underground.

- do not use in locations subject to corrosive conditions.

- is acceptable as an equipment grounding conductor if:
  - the flex is listed for grounding.
  - the fittings are listed for grounding.
  - the overcurrent protection does not exceed 20 amperes.
  - the total length of the return ground path through the flex does not exceed 6 feet (1.83 m), Figure 4-39.

- if longer than 6 feet (1.83 m), regardless of trade size, is not acceptable as grounding means. You must install a separate equipment grounding conductor sized per *Table 250-122*.

MAXIMUM 6 FEET (1.83 m)

4 FEET (1.22 m) PLUS 2 FEET (.61 m)

Figure 4-39 If a flexible metal conduit is serving as an equipment ground return path, the total length shall not exceed 6 feet (1.83 m). Refer to *Sections 250-118(6)(c), (7)(d), (8)(b)*.

- fittings larger than ¾ inch (19 mm) that have been listed as acceptable for grounding means will be marked "GRND" or some equivalent marking.

- ⅜-inch trade size not permitted longer than 6 feet (1.83 m). Fixture "whips" are a good example. See Recreation Room, Unit 17, for a discussion of fixture whips.

- must be supported every 4½ feet (1.37 m).

- must be supported within 12 inches (305 mm) of box, cabinet, or fitting.

Note: The 4½ feet (1.37 m) and the 12-inch (305 mm) requirements are waived when the flexible metal conduit is fished through walls and ceilings, where flexibility is needed and the length of the flex is not over 3 feet (914 mm), or for fixture whips where the length of the flex is not over 6 feet (1.83 m).

- horizontal runs through holes and is supported by framing members not more than 4½ feet (1.37 m) apart and is secured within 12 inches (305 mm) of a box, cabinet, fitting, or enclosure are considered to be adequately supported.

- if flexibility is needed, support within 36 inches (914 mm) of termination.

- if the flexible metal conduit is not acceptable as a grounding means, then install a bonding jumper inside or outside the flexible metal conduit. If run outside of the flexible metal conduit, the bonding jumper must not be longer than 6 feet (1.83 m) and it must be routed with the flexible metal conduit, *Section 250-102(e)*. Size the bonding jumper per *Table 250-122*. For mechanical protection and to keep the impedance of the circuit to a minimum, it is best to run the bonding jumper inside the flexible metal conduit.

- do not have more than four quarter bends (360°) between boxes.

- do not conceal angle-type fittings because of the difficulty that would present itself when pulling wires in.

- if flexibility is needed, always install a separate equipment grounding conductor regardless of trade size. Size equipment grounding conductor per *Table 250-122*.

- may be used for exposed or concealed installations.

- conductor fill same as regular conduit. See *Chapter 9, NEC, Tables 1* and *4.*

- conductor fill for ⅜-inch (9.5 mm) size. See *Table 350-12.*

- fittings used with flexible metal conduit must be listed.

- be sure to remove rough edges at "end-cuts" unless fittings are used that thread into the convolutions of the flex.

See Figure 4-42 when the installation has combined lengths of flex.

Sections 250-118(5), (6), and *(7)* also discuss flexible metal conduit installations.

LIQUIDTIGHT FLEXIBLE METAL CONDUIT *(ARTICLE 351, PART A)*

The use and installation of liquidtight flexible metal conduit are described in *Article 351* of the Code. Liquidtight flexible metal conduit has a "tighter" fit of its spiral turns as compared to standard flexible metal conduit. Liquidtight conduit also has a thermoplastic outer jacket that is liquidtight. Liquidtight flexible metal conduit is commonly used as the flexible connection to a central air-conditioning unit located outdoors, Figure 4-40.

Figure 4-41 shows the Code rules for the use of liquidtight flexible metal conduit as a grounding means. These limitations are given in the Underwriters Laboratories Standards.

Listed here are some of the common uses, limitations, and applications for liquidtight flexible metal conduit, combining *NEC* rules and U.L. White Book requirements:

- may be used for exposed and concealed installations.

- may be buried directly in the ground if so listed and marked.

- may not be used where subject to physical abuse.

- may not be used if ambient temperature and heat from conductors will exceed the temperature limitation of the nonmetallic outer jacket. U.L. listing will indicate maximum temperature.

LIQUIDTIGHT FLEXIBLE METALLIC CONDUIT OR LIQUIDTIGHT FLEXIBLE NONMETALLIC CONDUIT

AIR CONDITIONER

DISCONNECT SWITCH

Figure 4-40   Use of liquidtight flexible metal conduit or flexible liquidtight nonmetallic conduit.

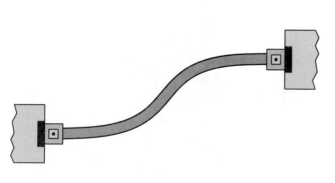

Ⓐ LIQUIDTIGHT FLEXIBLE METAL CONDUIT MAY BE USED AS A GROUNDING MEANS IF IT IS NOT OVER 1 1/4 INCH TRADE SIZE, IS NOT OVER 6 FEET (1.83 m) LONG, AND IS CONNECTED BY FITTINGS LISTED FOR GROUNDING PURPOSES.

Ⓑ MINIMUM SIZE 1/2 INCH, EXCEPT 3/8 INCH WHEN NOT LONGER THAN 6 FEET (1.83 m) PERMITTED ONLY AS PART OF AN APPROVED ASSEMBLY OR TO CONNECT LIGHTING FIXTURES. WHEN USED AS THE GROUNDING MEANS, THE MAXIMUM RATED OVERCURRENT DEVICE IS 20 AMPERES FOR 3/8- AND 1/2-INCH SIZES, AND 60 AMPERES FOR 3/4-INCH, 1-INCH, AND 1 1/4-INCH SIZES.

Ⓒ LIQUIDTIGHT FLEXIBLE METAL CONDUIT MAY NOT BE USED AS A GROUNDING MEANS IN ANY SIZE WHEN LONGER THAN 6 FEET (1.83 m), OR IN SIZES 1 1/2 INCH AND LARGER OF ANY LENGTH. IN SUCH CASES, A BONDING JUMPER IS REQUIRED. SEE FIGURE 4-39.

Figure 4-41   Liquidtight flexible metal conduit Code rules, *Article 351.*

- must not be used in sizes smaller than ½ inch except that ⅜-inch trade size is acceptable for enclosing the leads to a motor, or as a fixture "whip."
- for conductor fill, see *Chapter 9, NEC,® Tables 1 and 4.*
- conductor fill for ⅜-inch size. See *Table 350-12.*
- the flex must be "listed."
- all fittings must be "listed."
- must be supported every 4½ feet (1.37 m).
- must be supported within 12 inches (305 mm) of box, cabinet, fitting, or enclosure.
- horizontal runs through holes and supported by framing members not more than 4½ feet (1.37 m) apart and secured within 12 inches (305 mm) of a box, cabinet, fitting, or enclosure are considered to be adequately supported.
- the 4½ foot (1.37 m) and 12-inch (305 mm) securing requirements are waived when the flex is used as a fixture "whip" whose length is not over 6 feet (1.83 m).
- the 12-inch (305 mm) securing requirement is waived when the flex is used for a flexible connection when length is not over 3 feet (914 mm).
- the 4½ foot (1.37 m) and 12-inch (305 mm) securing requirements are waived when the flex is "fished."
- if flexibility is needed, support within 3 feet (914 mm) of termination.

- ⅜-inch and ½-inch trade sizes are acceptable for grounding purposes if the flex and the fittings are "listed" for grounding, the total length of the return ground path through the flex does not exceed 6 feet (1.83 m), and the overcurrent device is not over 20 amperes.
- ¾-inch, 1-inch, and 1¼-inch trade sizes are acceptable for grounding purposes if the flex and the fittings are "listed" for grounding, the total length of the return ground path through the flex does not exceed 6 feet (1.83 m), and the overcurrent device is not over 60 amperes.
- is not suitable for grounding if 1½-inch trade size and larger.
- when the flex is not acceptable as a grounding means, then a separate equipment grounding conductor must be installed, sized per *Table 250-122.*
- if flexibility is needed, always install a separate equipment grounding conductor regardless of the trade size of the flex. Size equipment grounding conductor per *Table 250-122.*
- do not conceal angle-type fittings.
- do not have more than four quarter bends (360°) between boxes.

See Figure 4-42 when the installation has combined lengths of flex.

Section 250-118(7) also discusses liquidtight flexible metal conduit installations.

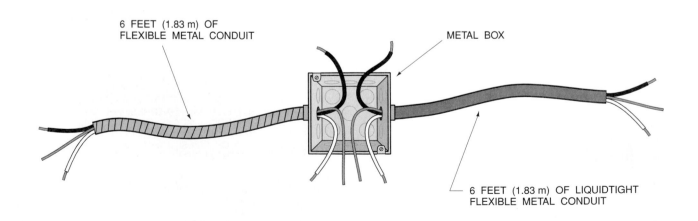

Figure 4-42 The "combined" length of flexible metal conduit, liquidtight flexible metal conduit, and flexible metal tubing shall not exceed 6 feet (1.83 m) if it is to serve the same ground return path, *Sections 250-118(6)(c), (7)(d), and (8)(b).* In the above illustration, a separate equipment grounding conductor (EGC) would have to be provided.

LIQUIDTIGHT FLEXIBLE NONMETALLIC CONDUIT (ARTICLE 351, PART B)

The following are some of the important Code and U.L. rules pertaining to liquidtight flexible nonmetallic conduit:

- do not use in direct sunlight unless specifically marked for use in direct sunlight.

- may be used in exposed or concealed installations.

- can become brittle in extreme cold applications.

- may be used outdoors when listed and marked as suitable for this application.

- may be buried directly in the earth when listed and marked as suitable for this application.

- may not be used where subject to physical abuse.

- may not be used if ambient temperature and heat from the conductors will exceed the temperature limitation of the nonmetallic material. U.L. listing will indicate this.

- do not have more than four quarter bends (360°) between boxes.

- must not be used in sizes smaller than ½ inch except that ⅜-inch trade size is acceptable for enclosing the leads to a motor, or as a fixture "whip."

- must be secured every 3 feet (914 mm) and within 12 inches (305 mm) of a box, cabinet, fitting, or enclosure.

- horizontal runs through holes and supported by framing members not more than 3 feet (914 mm) apart and secured within 12 inches (305 mm) of a box, cabinet, fitting, or enclosure are considered to be adequately supported.

- the 3-foot (914 mm) and 12-inch (305 mm) securing requirements are waived when the flex is used as a fixture "whip" whose length is not over 6 feet (1.83 m).

- the 12-inch (305 mm) securing requirement is waived when the flex is used for a flexible connection whose length is not over 3 feet (914 mm).

- the 3-foot (914 mm) and 12-inch (305 mm) securing requirements are waived when the flex is "fished."

- conductor fill same as for regular conduit. See *Table 1, Chapter 9, NEC*®.

- conductor fill for ⅜-inch size flex. See *Table 350-12*.

- do not conceal angle-type fittings.

- the flex must be "listed."

- must be used with listed fittings.

- if an equipment grounding conductor is needed, the equipment grounding conductor may be run inside or outside of the flex. The equipment grounding conductor shall not be longer than 6 feet (1.83 m) if run on the outside of the flex. Size equipment grounding conductor per *Table 250-122*.

- must not be used in lengths longer than 6 feet (1.83 m) unless absolutely necessary for flexibility, such as when connecting a motor.

- must not be used in lengths longer than 6 feet (1.83 m) unless it is of the integral reinforcement type and is secured at intervals not over 3 feet (914 mm) and within 12 inches (305 mm) from a box or other similar enclosure.

ELECTRICAL NONMETALLIC TUBING

Electrical nonmetallic tubing, commonly referred to as ENT is covered in *Article 331* of the *NEC*®. It is a pliable corrugated raceway made of PVC, and requires special fittings made specifically for the ENT.

ENT is not often used in house wiring, but is mentioned here to emphasize that it is a recognized wiring method that could be used.

SERVICE-ENTRANCE CABLE

Service-entrance cable is covered in *Article 338* of the *National Electrical Code*®.

Service-entrance cable is defined as *"a single conductor or multi-conductor assembly provided with or without an overall covering, primarily used for services."*

Type SE service-entrance cable is for above-ground use. Where permitted by local electrical codes, SE cable is most often run from the utility's service-point to the meter base to the main panel. The outer jacket is suitable where the cable is

Figure 4-43 A Type SE service-entrance cable. This cable contains two insulated conductors, and one "wrap-around" bare neutral conductor. There is a reinforcement tape plus the final sunlight-resistant outer jacket.

Figure 4-44 A Type USE service-entrance conductor. The insulation is abrasion, moisture, heat, and sunlight resistant suitable for direct burial. Utility companies generally use Type USE conductors that are prebundled cable assemblies having the proper number of conductors.

exposed to the sun. This information is found in the U.L. White Book. Figure 4-43 shows a Type SE service-entrance cable.

Type USE service-entrance conductors are for underground use. Type USE service-entrance conductors are not permitted for premise wiring, or aboveground wiring except to terminate in service-entrance equipment or metering equipment. Where emerging from the ground to terminate in a meter base, Type USE conductors are protected from physical damage by running the conductors in a raceway, such as from the bottom of a meter through to a point below the ground line (Figure 28-10), or from the bottom of a meter pedestal as illustrated in Figure 28-18. See *Section 338-2.* Multiple Type USE conductors are available prebundled in a cable assembly. Figure 4-44 shows a USE conductor.

Some service-entrance cables have a reduced size neutral in which case the cable is so marked. This information is found in *Section 338-5.* The neutral conductor for services and feeders in residential installations is permitted to be smaller than the ungrounded conductors, *Section 310-15(b)(4).* The U.L. standards state that where an uninsulated neutral conductor is smaller than the insulated conductors in a service-entrance cable, the current-carrying capacity of the smaller uninsulated neutral conductor is considered to be the same as for the insulated conductor.

Section 338-3 permits Type SE cable for interior wiring to hook up major electrical appliances, such as a range, a wall-mounted oven, a counter-mounted cooktop, or clothes dryer. Type SE cable may also be used as a feeder to a subpanel. For these applications, all of the circuit conductors must be insulated. The equipment grounding conductor in these cables is permitted to be bare.

For many years, there was a cost advantage of using SE cable with two insulated conductors and a bare neutral conductor for the branch-circuits supplying ranges and dryers. The *NEC®* permitted the frames of ranges and dryers to be grounded to the neutral conductor. Since the 1996 edition of the *NEC,®* this is no longer permitted. Now, SE cable and nonmetallic-sheathed cable with insulated grounded and ungrounded conductors is required. The bare equipment grounding conductor in these cables is used to ground the frames of these appliances.

The choice of whether to install SE cable or NMC cable becomes one of cost. Type SE cable with aluminum conductors costs less than Type NMC cable with copper conductors. In some parts of the country, nonmetallic-sheathed cable is used for hooking up these appliances. In other parts of the country, Type SE cable is used.

Section 338-4 requires that when SE cable is used for interior wiring, the installation must comply with the same requirements as nonmetallic-sheathed cable regarding securing and physical protection.

When installing SE cable for interior wiring, the ampacity of SE conductors is found using the 75°C column of *Table 310-16*, assuming the equipment is marked as suitable for use with 75°C conductors. See *Section 110-14(c)*.

For 120/240-volt, 3-wire, single-phase dwelling services and feeders, the ampacity of the SE conductors is permitted to be the special values found in *Table 310-15(b)(6)*.

If the service-entrance cable has an uninsulated grounded conductor (neutral), a final outer non-metallic covering and no conductor in the cable operates over 150 volts to ground as is the case in house wiring—its use is restricted to services or to supply another building on the same property.

To avoid damage to the cable where the cable is bent, the inner edge of the cable shall have a radius of not less than five times the diameter of the cable, *Section 338-6*.

Where service-entrance cable is run from the outside to the inside, passing through a wall, the penetration must be sealed and the cable must be protected using a *sill plate* such as illustrated in Figure 28-3.

REVIEW

Note: Refer to the Code or the plans where necessary.

1. The largest size solid conductor permitted to be installed in a raceway is (No. 10) (No. 8) (No. 6). Circle the correct answer.

2. What is the minimum branch-circuit wire size that may be installed in a dwelling?

   _____

3. What exceptions, if any, are there to the answer for question 2? _____

   _____

   _____

   _____

   _____

4. What determines the ampacity of a wire? _____

   _____

   _____

   _____

   _____

5. What unit of measurement is used for the diameter of wires? _____

6. What unit of measurement is used for the cross-sectional area of wires? _____

   _____

7. What is the voltage rating of the conductors in Type NMC cable? _____

8. Indicate the allowable ampacity of the following Type THHN (copper) conductors. Refer to *Table 310-16*.

   a. 14 AWG _____ amperes   d. 8 AWG _____ amperes

   b. 12 AWG _____ amperes   e. 6 AWG _____ amperes

   c. 10 AWG _____ amperes   f. 4 AWG _____ amperes

9. What is the maximum operating temperature of the following conductors? Give the answer in degrees Fahrenheit and degrees Celsius.

   a. Type XHHW _____   c. Type THHN _____

   b. Type RH   _____   d. Type TW   _____

10. What are the colors of the conductors in nonmetallic-sheathed cable for

    a. two-wire cable? _____ , _____

    b. three-wire cable? _____ , _____ , _____

11. For nonmetallic-sheathed cable, may the uninsulated conductor be used for purposes other than grounding? _____

12. What size equipment grounding conductor is used with the following sizes of non-metallic-sheathed cable?

    a. 14 AWG _____   c. 10 AWG _____

    b. 12 AWG _____   d.   8 AWG _____

13. Under what condition may nonmetallic-sheathed cable (Type NM) be fished in the hollow voids of masonry block walls? _____

    _____

    _____

14. a. What is the maximum distance permitted between straps on a cable installation?

       _____

    b. What is the maximum distance permitted between a box and the first strap in a cable installation? _____

    c. Does the *NEC*® permit 2-wire Romex stapled on edge? _____

    d. When Type NMC cable is run through holes in studs and joists, must additional support be provided? _____

15. What is the difference between Type AC and Type ACT armored cables? _____

    _____

    _____

    _____

    _____

    _____

    _____

    _____

16. Type ACT armored cable may be bent to a radius of not less than _____ times the diameter of the cable.

17. When armored cable is used, what protection is provided at the cable ends?

_____

_____

_____

_____

18. What protection must be provided when installing a cable in a notched stud or joist, or when a cable is run through bored holes in a stud or joist where the distance is less than 1¼ inches (32 mm) from the edge of the framing member to the cable, or where the cable is run parallel to a stud or joist and the distance is less than 1¼ inches (32 mm) from the edge of the framing member to the cable? _____

_____

_____

_____

_____

19. For installing directly in a concrete slab, (armored cable, nonmetallic-sheathed cable, conduit) may be used. Circle the correct method of installation.

20. Circle the correct answer to the following statements:

    a. Type SE service-entrance cable is for aboveground use.   True   False

    b. Type USE service-entrance cable is suitable for direct burial in the ground.   True   False

    c. Type SE cable with an uninsulated neutral is permitted for hooking up an electric range.   True   False

    d. Type SE cable with an insulated neutral is permitted for hooking up an electric range.   True   False

    e. Are Types SE and USE service-entrance cables used in your area?   Yes   No

21. The edge of a bored hole in a stud shall not be less than _____ inches from the edge of the stud.

22. Where is the main service entrance panel located in this residence? _____

_____

23. a. Is nonmetallic-sheathed cable permitted in your area for residential wiring?

      _____

    b. From what source is this information obtained? _____

24. Is it permitted to use flexible metal conduit over 6 feet (1.83 m) in length as a grounding means? (Yes) (No)　Circle the correct answer.

25. Liquidtight flexible metal conduit may serve as a grounding means in sizes up to and including _____ inches where used with approved fittings.

26. The allowable current-carrying capacity (ampacity) of aluminum wire, or the maximum overcurrent protection in the case of No. 14, No. 12, and No. 10 AWG conductors, is less than that of copper wire for a given size, insulation, and temperature of 86°F. Refer to *Table 310-16* and *Section 240-3(d)* and complete the following table. Important: where the ampacity of the conductor does not match the rating of a standard fuse or circuit breaker as listed in *Section 240-6(a)*. *Section 240-3(b)* permits the selection of the next "higher" standard rating of fuse or circuit breaker if the next higher standard rating does not exceed 800 amperes. For ratings above 800 amperes, select the next "lower" rated over-current device. ▶Nonstandard ampere ratings are also permitted.◀

WIRE	COPPER		ALUMINUM	
	Ampacity	Overcurrent Protection	Ampacity	Overcurrent Protection
No. 12 THHN				
No. 10 THHN				
No. 3 THW				
0000 THWN				
500 KCMIL THWN				

27. It is permissible for an electrician to connect aluminum, copper, or copper-clad aluminum conductors together in the same connector. True or False. Circle the correct answer.

28. Terminals of switches and receptacles marked CO/ALR are suitable for use with _____ , _____ , and _____ conductors.

29. Wire connectors marked AL/CU are suitable for use with _____ , _____ , and _____ conductors.

30. A wire connector bearing no marking or reference to AL, CU, or ALR is suitable for use with (copper) (aluminum) conductors only. Underline the correct answer.

31. When Type NM or NMC cable is run through a floor, it must be protected by at least _____ inches (_____ mm) of _____

_____

_____

32. When nonmetallic-sheathed cables are "bunched" or "bundled" together for distances longer than 24 inches (610 mm), what happens to their current-carrying ability?

_____

33. In diagrams A and B, nonmetallic-sheathed cable is run through the cold air return. Which diagram "Meets Code"? A _____ B _____   Check one.

34. The marking on the outer jacket of a nonmetallic-sheathed cable indicates the letters NMC-B. What does the letter **B** signify?_____

_____

_____

35. A 120-volt branch-circuit supplies a resistive heating load of 10 amperes. The distance from the panel to the heater is approximately 140 feet. Calculate the voltage drop using (a) No. 14, (b) No. 12, (c) No. 10, (d) No. 8 AWG copper conductors. See *Sections 210-19(a), FPN No. 4*, and *215-2, FPN No. 2*.

36. In question 35, it is desired to keep the voltage drop to 3% maximum. What minimum size wire would be installed to accomplish this 3% maximum voltage drop? See *Sections 210-19(a)*, FPN No. 4, and *215-2, FPN No. 2*.

_____

_____

_____

_____

_____

37. *Section 310-15(b)(6)* states that the neutral conductor of a residential service or feeder can be smaller than the phase conductors, but only

_____

_____

_____

38. The allowable ampacity of a No. 4 THHN from *Table 310-16* is 95 amperes. What is this conductor's ampacity if connected to a terminal listed for use with 60°C wire?

_____

39. If, because of some obstruction in a wall space, it is impossible to keep an NMC cable at least 1¼ inches (32 mm) from the edge of the stud, then it shall be protected by a metal plate at least _____ inch thick.

40. The recessed fluorescent fixtures installed in the ceiling of the recreation room of this residence are connected with ⅜-inch flexible metal conduit. These flexible connections are commonly referred to as "fixture whips." Does the flexible metal conduit provide adequate grounding for the fixtures, or must a separate equipment grounding conductor be installed? _____

_____

_____

_____

41. Flexible liquidtight metal conduit will be used to connect the air-conditioner unit. It will be ¾-inch trade size. Must a separate equipment grounding conductor be installed in this flex to ground the air conditioner properly?_____

_____

_____

_____

_____

42. What size overcurrent device protects the air-conditioning unit?_____

_____

43. May the 20-ampere small appliance branch-circuits in the kitchen of a residence also supply the lighting above the kitchen sink and under-cabinet lighting? Check the correct answer. Yes _____ No _____

44. A 30-ampere branch-circuit is installed for an electric clothes dryer. What is the maximum ampere rating of a dryer permitted by the *NEC®* to be cord and plug connected to this circuit? Check the correct answer.

a._____ 30 amperes

b._____ 24 amperes

c._____ 15 amperes

45. In many areas, metal framing members are being used in residential construction. When using nonmetallic-sheathed cable, what must be used where the cable is run through the metal framing members?_____

46. Are set-screw-type connectors permitted to be used with armored cable that has an aluminum armor? _____

47. Most armored cable today has 90°C conductors. What is the correct designation for this type of armored cable?_____

Switch Control of Lighting Circuits, Receptacle Bonding, and Induction Heating Resulting from Unusual Switch Connections

OBJECTIVES

After studying this Unit, the student will be able to

- identify the grounded and ungrounded conductors in cable or conduit (color coding).

- identify the various types of toggle switches for lighting circuit control.

- select a switch with the proper rating for the specific installation conditions.

- describe the operation that each type of toggle switch performs in typical lighting circuit installations.

- demonstrate the correct wiring connections for each type of switch per Code requirements.

- analyze some unusual three-way switch connections.

- understand the various ways to bond wiring devices to the outlet box.

- understand how to design circuits to avoid heating by induction.

The electrician installs and connects various types of lighting switches. To do this, both the operation and method of connection of each type of switch must be known. In addition, the electrician must understand the meanings of the current and voltage ratings marked on lighting switches, as well as the *NEC®* requirements for the installation of these switches.

CONDUCTOR IDENTIFICATION
(ARTICLES 200 AND 210)

Before making any wiring connections to devices, the electrician must be familiar with the ways in which conductors are identified. ▶For alternating-current circuits, the *NEC®* requires that the grounded (identified) conductor have an outer finish that is either continuous white or natural gray, or have an outer finish (not green) that has three continuous white stripes along the conductor's entire length.◀ In multiwire circuits, the grounded conductor is also called a *neutral* conductor.

▶ An ungrounded conductor must have an outer finish that is a color other than green, white, natural gray, or three continuous white stripes.◀ This conductor generally is called the "hot" conductor. A shock is felt if this "hot" conductor and a grounded conductor or grounded surface such as a metal water pipe are touched at the same time.

The hazards associated with electric shock are discussed in detail in Unit 6.

Grounded Neutral Conductor

For residential wiring, the white grounded conductor is commonly referred to as the *neutral*.

For branch circuits and feeders, the neutral conductor must be insulated, *Section 310-2(a)*.

For services, the neutral conductor is permitted to be a bare conductor, *Section 230-41*.

For residential wiring, the 120/240-volt electrical system is grounded by the electric utility at their transformer, and again by the electrician at the main service. This is in accordance with *Section 250-20(b)(1)* requiring that the system be grounded so the maximum voltage to ground on the ungrounded conductors does not exceed 150 volts. *Section 250-26(2)* requires that the neutral conductor be grounded for single-phase, three-wire systems.

Electricians often use the term "neutral" when-

ever they refer to a white grounded circuit conductor. Proposals have been made to include a definition of a neutral in the *NEC*®—with no success. Even electrical engineering textbooks are evasive as to whether the definition of a neutral refers only to the actual source (transformer), or down into the system such as a two-wire branch-circuit.

A neutral conductor is:

- the conductor that carries only the unbalanced current from the other conductors, as in the case of multiwire circuits of three or more conductors, and

- the conductor where the voltage from every other conductor to it is equal under normal operating conditions.

By these definitions, the white grounded conductor in a two-wire circuit is not truly a neutral conductor. In a multiwire three-wire branch-circuit, the white grounded conductor is a neutral conductor. See Figure 5-1.

For residential wiring, if you are comfortable

Figure 5-1   Definition of a true neutral conductor.

calling the white grounded conductor a neutral, continue to do so. Do not get hung up on technicalities.

For those of you who will broaden your training into commercial and industrial wiring, you will find that there are electrical systems such as a 3-phase Delta corner grounded system where the grounded conductor is a phase conductor and is not a neutral conductor. This is covered in detail in the *Electrical Wiring—Commercial* text.

Multiwire Branch-Circuit

The Code defines a multiwire branch-circuit as *"A branch circuit consisting of two or more ungrounded conductors having a potential difference between them, and a grounded conductor having equal potential difference between it and each ungrounded conductor of the circuit and that is connected to the neutral or grounded conductor of the system."*

For additional information relative to the hazards involved with multiwire circuits, refer to Figures 17-5, 17-6, 17-7, 17-8, and 17-9 in Unit 17.

Color Coding (Cable Wiring)

The conductors in nonmetallic-sheathed cable (Romex) are color coded as follows.

Two-wire: one black ("hot" phase conductor)
               one white (grounded "identified" conductor)
               one bare (equipment grounding conductor)

Three-wire: one black ("hot" phase conductor)
               one white (grounded "identified" conductor)
               one red ("hot" phase conductor)
               one bare (equipment grounding conductor)

Four-wire: one black ("hot" phase conductor)
               one white (grounded "identified" conductor)
               one red ("hot" phase conductor)
               one blue ("hot" phase conductor)
               one bare (equipment grounding conductor)

Four-wire nonmetallic sheathed cable and armored cable are not commonly used for house wiring. In fact, it may be impossible to find this type of cable in stock at electrical supply houses, hardware stores, or home centers. It may need to be ordered special. The cost is very high when compared to two-wire and three-wire cables. It is generally easier and more economical to lay out the wiring using two-conductor and three-conductor cables.

Color Coding (Raceway Wiring)

When the wiring method is a raceway such as EMT, the choice of colors for the ungrounded conductors is virtually unlimited, with the following restrictions:

Green: reserved for use as an equipment grounding conductor only. See *Section 250-119.*

White, Natural Gray, or ►*an insulated conductor (not green) with three continuous white stripes* ◄: reserved for use as a grounded circuit conductor (neutral) only. The striping is done by the manufacturer—not by the electrician on the job site. See *Sections 200-6(a) and (b) and 200-7(a)(b) and (c).*

Changing Colors When Conductors Are in a Cable

The basic rule in *Section 200-7* is that grounded conductors must be white, natural gray, ►or have three continuous white stripes on other than green insulation. ◄

For cable wiring such as nonmetallic-sheathed cable or armored cable, *Section 200-7(c)(1) and (2)* permits the white or natural gray conductor to be used for single-pole, three-way, or four-way switch loops. ►These sections require that when used as a switch "loop," the conductor that is white, natural gray, or marked with three continuous white stripes is to be used for the supply to the switch, and not as the return conductor from the switch to the switched outlet. Also, the conductor must be permanently reidentified to indicate its use by painting or other effective means, at its terminations, and at each location where the conductor is visible and accessible. ◄ See Figure 5-2.

In the past, merely attaching a white conductor to the terminal of a switch was acceptable as proper identification of an ungrounded conductor at the switch location. This practice is no longer permitted.

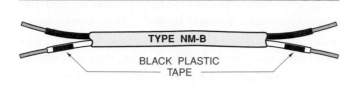

Figure 5-2 A white wire may be wrapped with a piece of black plastic tape so that everyone will know that this wire is *not* a grounded conductor. Use red tape if it is desired to mark the conductor red.

Much confusion resulted and inadvertent electric shocks happened because we always think of a white conductor as being a grounded conductor. Now when wiring with cable, we must reidentify a white conductor that is used as an ungrounded conductor wherever it is visible and accessible, and whenever it is used as a switch leg. The intent of the Code is that we do not want to see white conductor insulation when, in fact, the conductor is really a "hot" ungrounded conductor.

The wiring diagrams that follow are good examples of where permanent reidentification of the white conductor is necessary.

Changing Colors When Conductors Are in a Raceway

Occasionally, it may be necessary to change the actual color of the conductor's insulation. The following table provides guidelines for doing this.

TO CHANGE FROM THIS	TO THIS	DO THIS
red, black, blue, etc.	an *equipment grounding* conductor	**The Basic Rule:** No reidentification permitted for conductors No. 6 AWG and smaller. An equipment grounding conductor must be bare, covered, or insulated. If covered or insulated, the outer finish must be green or green with yellow stripes. The color must run the entire length of the conductor. ►**Conductors larger than No. 6 AWG:** At time of installation, at both ends and at every point where the conductor is accessible, do any of the following, For the entire exposed length: • strip off the covering or insulation • paint the covering or insulation green • use green tape, adhesive labels, or green heat shrink tubing. ◄ See *Sections 210-5(b), 250-119, 310-12(b).*
red, black, blue, etc.	a *grounded* conductor	**Conductors No. 6 AWG or smaller:** For the conductor's entire length, the insulation must be white, natural gray, ►or have three continuous white stripes on other than green insulation.◄ Reidentification in the field is not permitted when the wiring method is a raceway. **Conductors larger than No. 6 AWG:** For the conductors entire length, the insulation must be white, natural gray, ►or have three continuous white stripes on other than green insulation.◄ At terminations, reidentification may be white paint, white tape, or white heat shrink tubing. ►This marking must encircle the conductor or insulation.◄ Do this marking at time of installation. See *Sections 200-2, 200-6(a) and (b), 200-7(a)(b) and (c), 210-5(a), 310-12(a).*
White, natural gray	an *ungrounded* (hot) conductor	Insulated conductors that are white, natural gray, ►or have three continuous white stripes are to be used as grounded conductors only.◄ There is no provision in the *NEC®* for reidentifying white or natural gray conductors for use as ungrounded conductors in raceways. The only exception to use white or natural gray insulated conductors as ungrounded conductors is for cable wiring. See *Sections 200-7(a)(b) and (c), and 310-12(c).*

CONNECTING WIRING DEVICES

Table 4-1 shows in detail the terminal identification markings found on wiring devices (switches and receptacles) indicating the types of conductors permitted to be connected to them.

Most commonly used wiring devices have screw terminals.

Some wiring devices have back-wiring holes only. For these, the conductor insulation is stripped off for the desired length, inserted into the hole, and held in place by the device's internal spring pressure mechanism. These wiring devices are referred to as having *push-in* terminals.

For convenience, some wiring devices have both screw terminals and back-wiring holes as illustrated in Figures 5-3 and 5-4. These can be connected or hooked up in different ways. One way is the conductor insulation is stripped off for the desired length and inserted into the back hole, then the screw terminal is properly tightened. In another way, the conductor insulation is stripped off for the desired length and terminated under the screw terminal, in which case the back-wiring holes are not used.

Using both the back-wiring terminals and the screw terminals, it would be possible to terminate as many as four conductors to the white terminal, and four conductors to the brass terminal. But this would be considered poor workmanship. So many conductors on the wiring device would make it extremely difficult to work with. A much better way is shown in Figure 5-5 where all of the necessary splicing is done independent of the receptacle. Short lengths of wire called *pigtails* are included in the splice, and these pigtails in turn connect to the receptacle. Removal of the receptacle would not have any effect on the splice.

Figure 5-4 Exploded view of a grounding-type receptacle, showing all internal parts. *Section 200-10(b)* **requires that terminals on receptacles, plugs, and connectors that are intended for the ground circuit conductor be substantially white (usually silver colored), or marked with the word "white" or the letter "W" adjacent to the terminal. The terminal for ungrounded "hot" conductor is usually brass colored.**

Figure 5-3 Grounding-type receptacle detailing various parts of the receptacle.

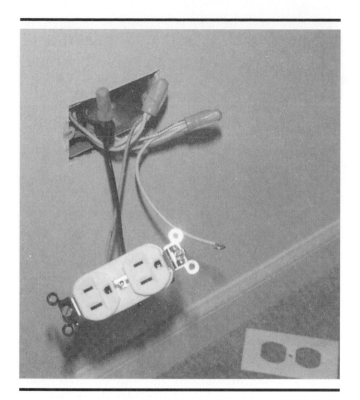

Figure 5-5 A receptacle connected with pigtails. This particular installation has one outlet switched, and one outlet on continuously. Six conductors enter the box because the wiring runs from receptacle to receptacle around the room. A 4-inch square outlet box trimmed with a single-gang plaster (dry-wall) cover is used. Removal of the receptacle does not affect the circuit to the other receptacles.

PUSH-IN TERMINATIONS

Be careful when using screwless push-in terminals.

Screwless push-in terminals on receptacles are "listed" by U.L. for use *only* with solid No. 14 AWG copper conductors. They *are not* to be used with:

- aluminum or copper-clad aluminum conductors.

- stranded conductors.

- No. 12 AWG conductors. By design, the holes are only large enough to take a No. 14 AWG solid conductor.

Prior to January 1, 1995, screwless push-in terminals on receptacles accepted both No. 14 and No. 12 AWG conductors. Manufacturers are no longer permitted to produce these.

Push-in terminals for No. 12 AWG solid copper conductors are still permitted on snap switches.

TOGGLE SWITCHES *(ARTICLE 380)*

U.L. refers to them as snap switches. Manufacturers and electricians refer to them as toggle switches. They are one and the same.

The most frequently used switch in lighting circuits is the toggle flush switch, Figure 5-6. When mounted in a flush switch box, the switch is concealed in the wall with only the insulated handle or toggle protruding.

Figure 5-7 shows how switches and receptacles are weatherproofed.

Toggle Switch Ratings

Underwriters Laboratories (U.L.) lists toggle switches used for lighting circuits as *general-use snap switches*. The U.L. requirements are the same as *Section 380-14(a)* and *(b)* of the *NEC.*

Figure 5-6 Toggle flush switches. *Courtesy* Hubbell, Incorporated.

Figure 5-7 Receptacle outlet and toggle switch are protected by a weatherproof cover.

AC/DC General-Use Snap Switches

The requirements for general-use snap switches include the following:

- alternating-current (ac) or direct-current (dc) circuits.

- resistive loads not to exceed the ampere rating of the switch at rated voltage.

- inductive loads not to exceed one-half the ampere rating of the switch at rated voltage unless otherwise marked.

- tungsten filament lamp loads not to exceed the ampere rating of the switch at 125 volts when marked with the letter **T**.

- for switches marked with a horsepower rating, a motor load shall not exceed the rating of the switch at rated voltage.

A tungsten filament lamp draws a very high momentary inrush current at the instant the circuit is energized. This is because the *cold resistance* of tungsten is very low. For instance, the cold resistance of a typical 100-watt lamp is approximately 9.5 ohms. This same lamp has a *hot resistance* of 144 ohms when operating at 100% of its rated voltage.

Normal operating current would be

$$I = \frac{E}{R} = \frac{120}{144} = 0.83 \text{ ampere}$$

But, *maximum* instantaneous inrush current could be as high as

$$I = \frac{E}{R} = \frac{170 \text{ (peak voltage)}}{9.5} = 17.9 \text{ amperes}$$

This instantaneous inrush current drops off to normal operating current in about 6 cycles (0.10 second). The contacts of T-rated switches are designed to handle these momentary high inrush currents. See Unit 13 for more information pertaining to inrush currents.

The ac/dc general-use snap switch normally is not marked ac/dc. However, it is always marked with the current and voltage rating, such as 10A-125V or 5A-250V-T.

AC General-Use Snap Switches

Ac general-use snap switches are marked "ac only," in addition to identifying their current and voltage ratings. A typical switch marking is 15A,

120-277V ac. The 277-volt rating is required on 277/480-volt systems. Other requirements include:

- alternating-current (ac) only.

- resistive and inductive loads not to exceed the ampere rating of the switch at the voltage involved.

- tungsten-filament lamp loads not to exceed the ampere rating of the switch at 120 volts.

- motor loads not to exceed 80% of the ampere rating of the switch at rated voltage. A U.L. requirement is that the load shall not exceed 2 horsepower.

Conductor Color Coding for Switch Connections

The insulated conductors in nonmetallic-sheathed cable or armored cable are *black-white*, or *black-white-red*, and, in some cases, *black-white-red-blue*. We usually think of the black insulated conductor as the feed to a switch, and a red insulated conductor as the switch leg. However, when wiring with nonmetallic-sheathed cable or armored cable, we are "stuck" with the colors found in the cables. ▶We do not want to use a white conductor as the return conductor from a switch, *Section 200-7(c)(2).*◀ A color-coding scheme needs to be established for the feeds, switch legs, and travelers. The wiring diagrams that follow show the recommended color schemes when wiring with cable.

When the wiring method is a raceway or tubing (EMT), the choices of insulation colors are virtually unlimited.

Always connect a white wire to the white (silverish) terminal or to the white wire of a lampholder or receptacle, *Sections 200-9* and *200-10*. Always connect the black switch leg conductor (or red in some cases) to the black wire (or dark brassy terminal) of a lampholder or receptacle.

▶ In cables, always reidentify white conductors when they are used as ungrounded (hot) conductors. This reidentification must be done wherever the conductors are visible and accessible. These requirements are found in *Section 200-7(c)(2).*◀

Never use a green-colored insulation for a grounded or ungrounded conductor. Green is reserved for equipment grounding conductors. See *Sections 210-5* and *250-119*.

Connecting Switches, Receptacles, and Lighting Outlets

To prevent induction heating and to keep the impedance of a circuit as low as possible where conductors are run in metal raceways, metal jacketed cables, and in metal enclosures, the installation must follow certain *NEC®* requirements.

▶ *Section 300-3(b)* requires that "*All conductors of the same circuit and, where used, the grounded conductor and all equipment grounding conductors shall be contained within the same raceway, auxiliary gutter, cable tray, trench, cable, or cord.*" ◀

This section has a few exemptions: for paralleled installations, for grounding conductors, for nonferrous wiring methods, and certain enclosures.

Section 300-20(a) requires that "*Where conductors carrying alternating current are installed in metal enclosures or metal raceways, they shall be so arranged as to avoid heating the surrounding metal by induction. To accomplish this, all phase conductors and, where used, the grounded conductor and all equipment grounding conductors shall be grouped together.*"

Section 380-2(a) requires that "*Three-way and*

Figure 5-8 A few examples of arranging circuitry to avoid induction heating according to *Section 300-20* when metal raceway or metal cables (BX) are used. Always ask yourself, "Is the same amount of current flowing in both directions in the metal raceway?" If the answer is "no," there will be induction heating of the metal that can damage the insulation on the conductors.

four-way switches shall be so wired that all switching is done only in the ungrounded circuit conductor. Where in metal raceways or metal-jacketed cables, wiring between switches and outlets shall be in accordance with Section 300-20(a)." Never switch a grounded circuit conductor except in the case of a 2-pole switch that breaks both the ungrounded and grounded conductors simultaneously.

The above Code requirements mean that the same amount of current going to the load through one conductor will return through another conductor in the same raceway or cable. This Code requirement ensures that there will be no induction heating because the flux surrounding each conductor is equal and opposite, thereby canceling one another out. The impedance of the circuit is kept to a minimum.

Figures 5-8, 5-9, 5-10, and 5-11 illustrate the intent of the *NEC®* to prevent induction heating.

Unusual switch connections may or may not conform to the *NEC,®* depending upon how the circuitry is run. Some electricians have found that they can save substantial lengths of three-wire cable when hooking up three-way and four-way switches by using two-wire cable. Connections as shown in Figure 5-12 are clearly in violation of *Sections 300-3(b), 300-20(a),* and *380-2(a).*

Switch Types

Switches are available in single-pole, three-way, four-way, double-pole, dimmer, and speed control types. They are available in snap (toggle), push, slider, rotary, electronic, with and without pilot lights. Faceplates are commonly available in white, brown, and ivory plastic, stainless steel, brass, chrome, and wood.

In all of the following wiring diagrams, the equipment grounding conductors are not shown as this is a separate issue covered elsewhere in this text.

Single-Pole Switches. A single-pole switch is marked with an "ON" "OFF" position. Used when the load is to be controlled from one switching

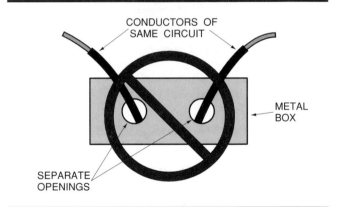

Figure 5-10   Another typical example of how induction heating can occur is to run conductors of the same circuit through different openings of a metal box.

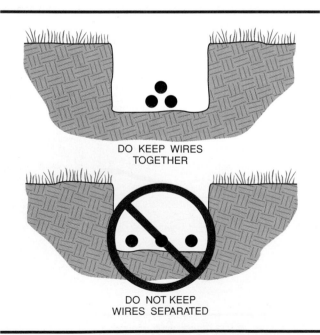

Figure 5-9   Arrangement of conductors in trenches. Refer to Unit 16 for complete information pertaining to underground wiring.

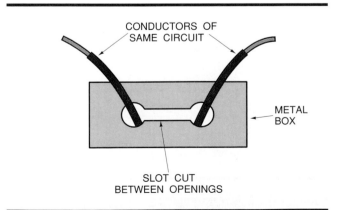

Figure 5-11   Inductive effects can be minimized by cutting slots between the openings.

Figure 5-12   Example of violations of *Sections 300-3(b)*, *300-20(a)*, and *380-2(a)*.

point, a single-pole switch is connected in series with the ungrounded (hot) conductor.

Single-Pole Switch, Feed at Switch.   In Figure 5-13, the 120-volt source enters the switch box where the white grounded conductor is spliced directly through to the lamp. The black conductor from the supply circuit connects to one terminal on the switch. The black conductor in the cable coming from the outlet box at the lamp is connected to the other terminal on the switch. Because a single-pole switch connection is simply a series circuit, it makes no difference which terminal is the supply and which terminal is the load. Although not a Code requirement, try to use the top terminal as the supply and the bottom terminal as the load. This line/load arrangement is similar to that of a standard disconnect switch. This layout needs no reidentification of the white conductor.

Single-Pole Switch, Feed at Light.   In Figure 5-14, the 120-volt source enters the outlet box at the lamp. The switch loop is a two-wire cable. *Section*

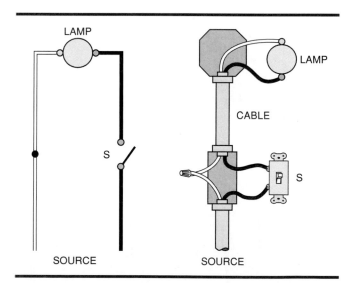

Figure 5-13   Single-pole switch in circuit with feed at switch.

200-7(c)(2) permits the use of the white wire of a cable to be used as a switch loop. ▶When so used, the white conductor must be reidentified in the outlet box and in the switch box, *Section 200-7(c)(2)*.◀

Figure 5-14 Single-pole switch in circuit with feed at light.

Color plastic tape is most often used to reidentify a conductor. For this diagram, black tape would be suitable.

Single-Pole Switch, Feed at Switch, Receptacle "On" Continuously. Figure 5-15 shows the 120-volt source entering the switch box where the switch is located. The switch controls the lamp. The receptacle remains "hot" at all times. The white conductor in the switch box is spliced straight through to the outlet box. The black conductor in the switch box is spliced through to the outlet box and also feeds the switch. The red conductor in the three-wire cable is the switch leg, connecting to the switch and to the lamp. A two-wire cable from the outlet box carries the circuit to the receptacle. The circuitry is such that no reidentification of the white conductor is necessary.

Three-Way Switches. Three-way switches are used when a load is to be controlled from two locations. The term "three-way" is very misleading. Three-way switches have three terminals. Maybe they should have been named "three-terminal" switches. One terminal is called the "common" and is darker in color than the other two terminals. The other two terminals are called "traveler terminals."

Figure 5-16 shows the two positions of a three-way switch. Actually, a three-way switch is a single-pole, double-throw switch.

There is no "ON" "OFF" marking on a three-way switch as can be seen in Figure 5-17.

Three-Way Switch Control, Feed at Switch. Figure 5-18 shows the 120-volt source entering at

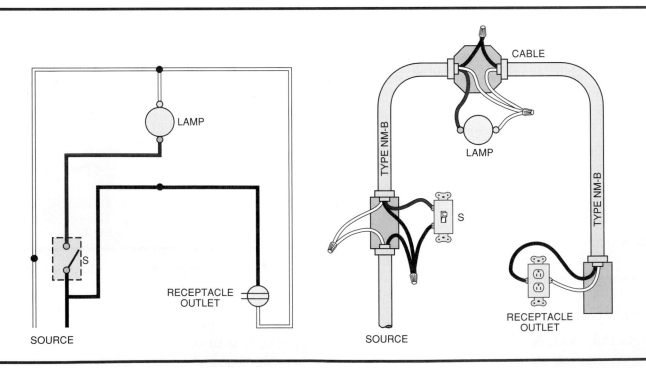

Figure 5-15 Ceiling outlet controlled by single-pole switch with live receptacle outlet and feed at switch.

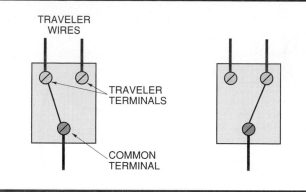

Figure 5-16   Two positions of a three-way switch.

Figure 5-17   Toggle switch: three-way flush switch. *Courtesy of* **Pass & Seymour.**

Figure 5-18   Circuit with three-way switch control. The feed is at first switch. The load is connected to the second switch. The black and red conductors are used for the travelers. The black conductor has been reidentified with red tape.

the first switch location. A three-wire cable is run between the two switches. A two-wire cable is run from the second switch to the lamp. The white conductors in both switch boxes are spliced directly through to the lamp. The red and black conductors in the three-wire cable are the "travelers." The black conductor in the three-wire cable is reidentified with red tape, consistent with keeping both "travelers" red in color. The black conductor from the source is connected to the "common" terminal of the first switch. The black conductor in the cable coming from the outlet box at the lamp is connected to the "common" terminal of the second switch. The cir-

cuitry is such that no reidentification of the white conductor is necessary.

Three-Way Switch Control, Feed at Light. Figure 5-19 shows the 120-volt source feeding into the outlet box at the lamp. A two-wire cable is run from the outlet box to the first switch location. A three-wire cable is run from the first switch to the second switch. ▶In all locations where the white wire is to be used as an ungrounded "hot" conductor, it must be reidentified, *Section 200-7(c)(2).*◀ In this diagram, reidentification of the white conductor in the outlet box is with black tape.

The black conductor of the two-wire cable is connected to the "common" terminal of the first switch. ▶At the first switch, the white conductor in the two-wire cable is reidentified with black tape.◀ The white conductor of the three-wire cable

Figure 5-19   Circuit with three-way switch control and feed at light.

is reidentified with red tape. The red conductor and the re-identified white conductor in the three-wire cable are the "travelers." The white conductor is re-identified with red tape, consistent with keeping both "travelers" red in color.

At the second switch location, the black conductor of the three-wire cable is connected to the "common" terminal of the switch. ▶The white conductor of the three-wire cable is reidentified with red tape.◀ The red conductor and the reidentified white conductor in the three-wire cable are the "travelers." Here again, the white conductor is re-identified with red tape, so that the "travelers" are red in color.

Three-Way Switch Control, Feed at Light (Alternate Connection).   Figure 5-20 shows the 120-volt source entering at the outlet box. The white conductor from the 120-volt source at the outlet box goes directly to the lampholder. ▶In all locations where the white wire is to be used as an ungrounded "hot" conductor, it must be reidentified, *Section 200-7(c)(2)*.◀ From the outlet box to the switch on the left, the black conductor in the three-wire cable serves as the switch return to the lamp, and the red and reidentified white conductor are the "travelers." From the outlet box to the switch on the right, the black conductor serves as the feed to the three-way switch, and the red and reidentified white conductor are the "travelers." For uniformity, re-identification of the white conductors used for "travelers" through the outlet box and in the switch boxes is with red tape.

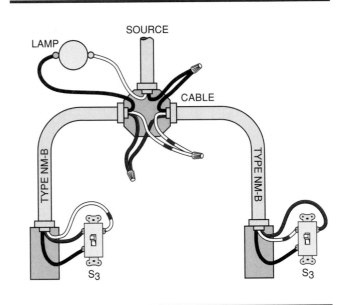

Figure 5-20   Alternative circuit with three-way switch control and feed at light.

Four-Way Switches.   Four-way switches are used where switch control is needed from more than two locations. The term "four-way" is misleading. Four-way switches have four terminals. Maybe four-way switches should have been called "four-terminal" switches. A four-way switch does not have "ON" "OFF" markings. Four-way switches are connected to the "travelers" *between* two three-way switches. One three-way switch is fed by the source—the other three-way switch is connected to the load. Make sure that the "travelers" are connected to the

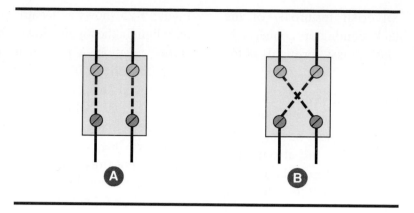

Figure 5-21   Two positions of a four-way switch.

proper "pairs" of terminals on a four-way switch; otherwise the switching will not work.

Figure 5-21 shows the internal switching of a four-way switch.

Figure 5-22 shows a lamp controlled from three switching points. Because the white conductor from the source is connected straight through all of the switch boxes and on to the lamp, no reidentification of the white wire is necessary. The black conductor from the source is connected to the common termi-

nal on the first three-way switch. The black conductor at the third switch is connected to the common terminal on the three-way switch. The black and red conductors in the three-wire cables are the "travelers" and are connected to the proper "traveler" terminals of the three-way and four-way switches. The black conductors in the three-wire cables have been reidentified with red tape, consistent with keeping both "travelers" red in color. At the lamp, the white conductor of the two-wire cable connects to the

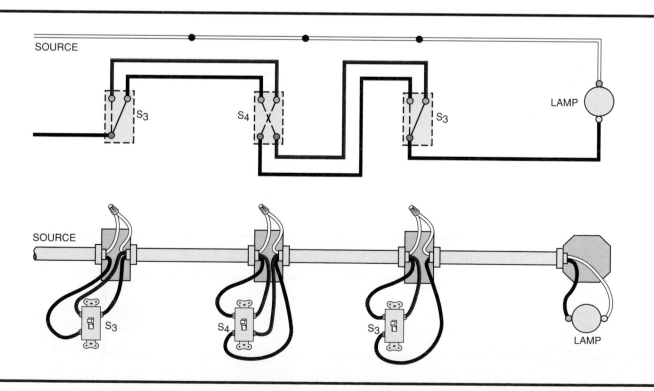

Figure 5-22   Circuit with switch control at three different locations.

white conductor (or silverish terminal) of the lampholder, and the black conductor connects to the black conductor (or dark brassy terminal) of the lampholder.

Double-Pole Switches. Double-pole switches are not common in residential work. They can be used when two separate circuits must be controlled with one switch, Figure 5-23. They are also used to control 240-volt loads, such as electric heat, motors, electric clothes dryers, and similar 240-volt loads, Figure 5-24. When line-voltage thermostats for electric heat are double-pole, they switch both ungrounded "hot" conductors, and are marked with "ON" "OFF" positions.

Switches with Pilot Lights. There are instances when a pilot light is desired at the switch location.

Figure 5-25 shows a "locator" type switch where the pilot light in the toggle glows when the switch is in the "OFF" position. A resistor inside the switch is connected in series with the pilot lamp. Be careful when working on this type of switch because when the switch is in the "OFF" position, the load and the pilot lamp are actually connected in "series." Even though these switches are marked with an "OFF" position, they are not truly off. The circuit's path through the switch and connected load is still complete. There is a voltage drop (almost full open circuit voltage) across the pilot lamp/resistor, and an infinitesimal voltage drop across the connected load. Although there is not enough voltage at the lamp to light the lamp, there may be enough voltage and current present to cause a person to "flinch," that might result in the person jerking back and falling off a ladder.

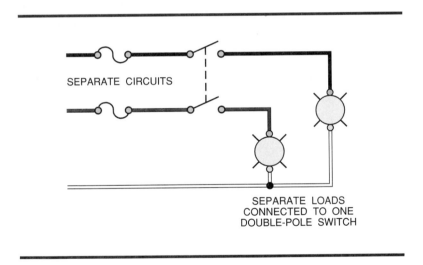

Figure 5-23 Application of a double-pole switch.

Figure 5-24 Double-pole (2-pole) disconnect switch.

Figure 5-25   In (A) the switch is in the OFF position. The neon lamp in the handle of the switch glows and the load is off. This is a series circuit. The neon lamp has extremely high resistance. In a series circuit voltage divides proportionately to the resistance. Therefore, the neon lamp "sees" line voltage (120 volts) for all practical purposes, and the load "sees" zero voltage. The neon lamp glows.

In (B) the switch is in the ON position. The neon lamp is shunted out and therefore has zero voltage across it. Thus, the neon lamp does not glow. Full voltage is supplied to the load. This type of switch might be referred to as a *locator* because it glows when the load is off and does not glow when the load is on.

Certain electronic controls such as motion (occupancy) sensors need the voltage drop across the device to operate. Again, an ever so slight amount of voltage is present at the load because the load and switching device are in "series." For a typical human, a current flow of ½ milliampere cannot be felt. A current of 1 milliampere will cause a slight tingling sensation. U.L. is addressing this issue. Refer to Unit 6 for additional information relating to the effects of an electrical shock.

Figure 5-26 shows a true pilot light that is "on" when the light is on.

Figure 5-26   A true pilot light. The pilot lamp is an integral part of the switch. When the load is turned on, the pilot light is also on. When the load is turned off, the pilot light is also off. This switch has three terminals because it requires a grounded neutral circuit conductor.

COMBINATION AND INTERCHANGEABLE WIRING DEVICES

In the space of one standard wiring device, it is possible to install one, two, or three wiring devices using combination or interchangeable wiring devices. These are available in single-pole, three-way, four-way, pilot lights, grounding-type receptacles, and GFCI receptacles. Using these types of wiring devices can offer a neater appearance than having two-gang or three-gang wallplates at a given location. This arrangement is also used where the wall space is limited, such as between a door casing and a window casing where there is not enough space for a multi-gang wallplate.

When combination or interchangeable wiring devices are installed, standard sectional switch boxes generally are not used because of the lack of wiring space in the box. Instead, a 4-inch square box with a raised single-gang cover is used. Another pos-

sibility is to install 3-inch or 3½-inch deep device boxes. The maximum number of conductors and wiring devices permitted in a box is critical. Crowding of conductors in a box can only lead to problems such as short-circuits and ground-faults. Box fill calculations were discussed in Unit 2 and elsewhere in this text.

Figure 5-27(A) shows two switches and a pilot light of the interchangeable type. Figure 5-27(B) shows an interchangeable grounding-type receptacle, a three-hole mounting bar, and a single-pole switch. Figure 5-27(C) shows three types of combination wiring devices.

Miscellaneous Connections. Figures 5-28 and 5-29 are diagrams that somewhat combine parts of the previous wiring diagrams. These connections become rather complicated using cable because of the limitation on the available insulation colors.

Figure 5-27 (A) Two switches and one pilot light of the interchangeable type. (B) Interchangeable type single grounding receptacle and single-pole switch and a three-hole mounting bar. (C) Three types of combination wiring devices. *Photos courtesy* of Leviton Manufacturing Co., Inc.

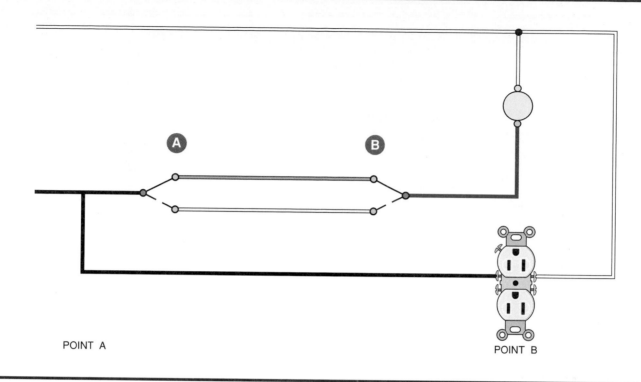

POINT A   POINT B

Figure 5-28 A lighting outlet controlled by two three-way switches. The receptacle is "live" at all times. Four conductors are run between Point A and Point B. A typical example of this might be between a house and a detached garage.

These connections are more suited to raceway wiring.

Do You Need the Grounded Circuit Conductor at the Switch?
In most instances, the answer is no, *Section 380-2(a), Exception.* However, a grounded conductor is needed when a true pilot light is connected at the switch location.

Some electronic motion sensors need the grounded circuit conductor to operate. When the grounded circuit conductor is not present at the switching location, uninformed individuals have been known to connect the motion sensor to the equipment grounding conductor. This is wrong and constitutes a serious Code violation. Equipment grounding conductors are not intended to be used as a current-carrying conductor.

If there is any indication when wiring a home that certain switching points might be a candidate for an occupancy motion sensor or other electronic switching device, give consideration to running the grounded circuit conductor to that location.

Unusual Switch Connections. Over the years, "creative" electricians have come up with some very unusual ways to connect three-way switches. Figures 5-30, 5-31, and 5-32 are shown here only as a challenge for the student. These unusual connections are difficult to troubleshoot, are not recommended, and in some cases are not in conformance to the *NEC.*® But if you ever run across one of these hookups, you now have some wiring diagrams that might help you out of a jam.

Bonding and Grounding at Receptacles and Switches

A metal box is considered to be adequately grounded when the wiring method is armored cable, nonmetallic-sheathed cable with ground, or a metal raceway such as electrical metallic tubing (EMT), *Section 250-118.* A separate equipment grounding conductor can also provide the required grounding, *Section 250-134(b).* Equipment grounding conductor may be bare or green.

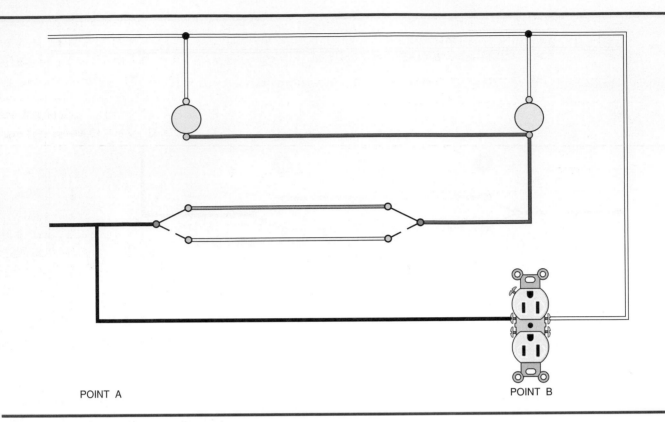

POINT A   POINT B

Figure 5-29 Two three-way switches controlling two lighting outlets. The receptacle outlet is "live" at all times. Five conductors are needed between Point A and Point B. An example of this hookup might be between a house and a detached garage, where one lighting outlet is on the house, and another lighting outlet is on the garage.

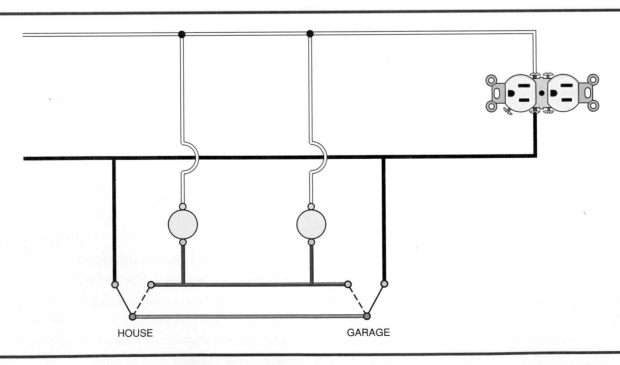

HOUSE   GARAGE

Figure 5-30 A set of three-way switches controlling one light on the side of the house, and another light on the side of the garage. The receptacle is "hot" at all times. The lights can be connected anywhere between the white and red wire. Normally, five wires are needed, whereas this hookup gets the job done with only four wires. Insofar as the Code is concerned, there are no Code violations. This is a very confusing hookup, and is not a recommended way to make the connections.

Figure 5-31   This hookup is *not* in conformance to the Code. Here we have a set of three-way switches controlling one light. The receptacle is "hot" at all times. The switches as shown have the light turned off. If you were to snap any one of the three-way switches to its other position, the light would go on. But note that the reason the light went on is that you are now connecting both terminals of the light with the same circuit conductor. Snapping the other switch to its other position turns the light off again. Each time a three-way switch is snapped, the wires on the light socket reverse. Sometimes the "shell" of the lampholder is connected to the grounded conductor as it should be, and other times the shell is connected to the "hot" circuit conductor. This is a violation of *Section 200-10(c)* and *200-11* of the Code. This is a very unsafe hookup. Here are the four possible three-way switch positions and the respective polarity of the lampholder:

	SHELL	CENTER CONTACT	LAMP
POSITION 1	grounded	"live"	on
POSITION 2	"live"	grounded	on
POSITION 3	grounded	grounded	off
POSITION 4	"live"	"live"	off

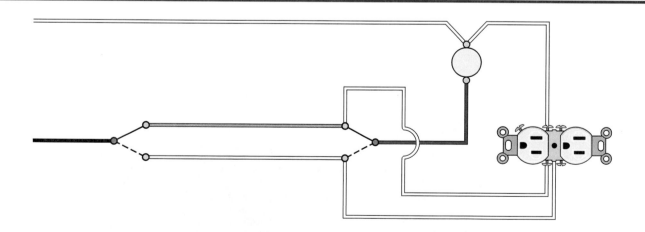

Figure 5-32   Here we have a set of three-way switches controlling a light. This connection "meets Code." At a later date, a receptacle was installed. Note that the receptacle is of the "split circuit" type, so the white grounded conductor is common to both outlets on the receptacle, but a different "hot" conductor connects to each of the "hot" terminals on the receptacle outlet. Depending upon the position of the three-way switches, one or the other outlet is "hot." But both outlets cannot be "live" at the same time. In the diagram, the left outlet on the receptacle is "live." Snap any one of the three-way switches to another position, and the right outlet on the receptacle becomes "live." This is a very unusual "creative" hookup.

Figure 5-33 Single-pole toggle switch. Note that this switch has a terminal for the connection of an equipment grounding conductor. Thus, when a metal faceplate is installed, it will be grounded per the requirement of *Section 380-9. Courtesy of* Pass & Seymour.

Figure 5-33 illustrates a switch that has an equipment grounding conductor terminal. Using a switch of this type provides an easy means for providing the required grounding of a metal faceplate when using nonmetallic boxes.

Grounding and bonding of the equipment grounding conductor to a metal box, switch, or receptacle is important. Figures 5-34, 5-35, and 5-36 show how an equipment grounding conductor of a nonmetallic-sheathed cable can be attached to a metal box and to the equipment grounding screw of a receptacle. Most metal boxes have a No. 10/32 tapped hole for securing a green hexagon-shaped equipment grounding screw.

Other than for special "isolated ground" receptacles, the metal yoke of a receptacle is integrally bonded to the equipment grounding terminal of the receptacle. Isolated ground receptacles are discussed in Unit 6.

Figures 5-3 and 5-4 show a receptacle that has a special "self-grounding" feature for the No. 6/32 mounting screws that results in effective grounding

SEE *SECTIONS 210-7, 250-146,* AND *250-148*

GROUNDED CONDUCTOR

EQUIPMENT GROUNDING CONDUCTOR

HOT CONDUCTOR

GROUNDING CONDUCTOR ATTACHED TO BOX AND GROUNDING TERMINAL ON RECEPTACLE *SECTION 250-146*

NONMETALLIC-SHEATHED CABLE WITH GROUNDING CONDUCTOR

GROUNDING-TYPE RECEPTACLE *SECTION 210-7*

SWITCH BOX WITH GROUNDING WASHERHEAD SCREW

Figure 5-34 Connections for grounding-type receptacles. See *Sections 210-7, 250-146,* and *250-148.*

Figure 5-35 An approved method of connecting the grounding conductor using a special grounding-type wire connector. See *Sections 250-146 and 250-148.*

Figure 5-36 Method of attaching grounding clip to switch (device) box. See *Sections 250-146 and 250-148.*

of the metal yoke and equipment grounding terminal without the use of a separate equipment grounding conductor.

To ensure the continuity of the equipment grounding conductor path, *Section 250-148* requires that where more than one equipment grounding conductor enters a box, they shall be spliced with devices "suitable for the use." Splices shall not depend on solder. The splicing must be done so that if a receptacle or other wiring device is removed, the continuity of the equipment grounding path shall not be interrupted. This is clearly shown in Figure 5-35. Splicing must be done in accordance with *Section 110-14(b)*.

Grounding Metal Faceplates and Switches

Section 380-9(b) requires that switches and dimmers be grounded. Metal switch boxes where switches will be installed must be grounded, *Section 380-12*. The No. 6/32 screws that attach the switch or receptacle to a metal box provide effective grounding where the metal yoke of the switch has direct metal-to-metal contact with a metal box. Many switches and receptacles have small pieces of cardboard holding the No. 6/32 mounting screws from falling out. These pieces of cardboard must be removed when metal-to-metal contact is needed.

A separate equipment grounding conductor may have to be brought out to the receptacle if there in no metal-to-metal contact, Figures 5-34 and 5-35.

Metal faceplates must be grounded, *Section 410-56(d)*. *Section 380-9* requires that snap switches and dimmer switches provide a means to ground metal faceplates. *Section 370-25* states that metal faceplates shall be grounded according to the requirements described in *Section 250-110*. A metal faceplate is considered grounded when fastened to the switch or receptacle with the faceplate's No. 6/32 mounting screws.

Section 380-12 requires that nonmetallic boxes where switches are used must be installed with a wiring method that provides or includes an equipment grounding conductor.

According to *Section 380-9(b)(1) and (2)*, when using nonmetallic boxes, a metal faceplate is considered grounded if the box has integral means for grounding wiring devices. If the circuit's equipment grounding conductor is connected to the wiring device, then the No. 6/32 screws that attach the faceplate to the device provide the necessary grounding of the metal faceplate.

In existing installations where an equipment ground does not exist in the switch box, and where located within reach of a conductive floor (i.e., concrete, tile) or other conductive surfaces (i.e., metal pipes, metal ducts), use nonmetallic faceplates, *Section 380-9(b), Exception.*

COMMON CODE VIOLATION– MAKING "TAPS"

Section 240-3(d) shows the maximum overcurrent protection for the smaller sizes of branch-circuit conductors. Here is what *Section 240-3(d)* says:

CONDUCTOR SIZE	MAXIMUM OVERCURRENT PROTECTION
No. 14 AWG copper	15 amperes
No. 12 AWG copper	20 amperes
No. 10 AWG copper	30 amperes

There are some exceptions to this basic rule, such as for motor circuits, fixture conductors, and for "taps" to electric surface-mounted cooking units and wall-mounted ovens.

▶ A "tap" is a conductor that has overcurrent protection ahead of its point of supply that exceeds the ampacity of the conductor, *Section 240-3(e).*◀

Figure 5-37 illustrates a code violation. The switch leg (loop) conductors are part of the branch-

Figure 5-37 *Violation:* **The electrician has run a 20-ampere branch-circuit consisting of No. 12 AWG conductors to the outlet box where a lighting fixture will be installed. No. 14 AWG conductors are run as the switch legs between the lighting outlet and the switch. The switch legs are part of the branch-circuit and must have an ampacity equal to that of the branch-circuit ampere rating, 20 amperes. Switch legs are not "taps." For the same reasons, it is a Code violation to use short "pigtails" of No. 14 AWG wire to connect receptacles to the No. 12 AWG conductors fed from a 20-ampere branch-circuit.**

circuit wiring and are not to be considered a "tap." Although the connected load may be well within the ampacity of the switch leg, the switch leg must also be capable of handling short-circuits and ground-faults. The switch leg shown in Figure 5-30 must be the same size as the branch-circuit conductors, which are No. 12 AWG 20-ampere conductors protected by a 20-ampere circuit breaker.

To minimize voltage drop, sometimes No. 12 AWG conductors are used for long "home runs" (50 feet or longer) even though the branch-circuit overcurrent device is rated 15 amperes. A "home run" is the branch-circuit wiring from the panel to the first outlet or junction box in that circuit. This is permitted by *Section 210-3*, which states that *"where conductors of higher ampacity are used for any reason, the ampere rating or setting of the specified overcurrent device shall determine the circuit rating."* In this case, No. 14 AWG conductors are permitted for the switch legs. But this leads to confusion later when someone removes the cover

from the panel, and wonders why No. 12 AWG branch-circuit conductors are connected to 15-ampere circuit breakers. For house wiring, it is best to use the same size conductor throughout a given branch-circuit.

Specifications for commercial and industrial installations oftentimes will call for a "home run" of 50 feet or more to be No. 10 AWG conductors, even though the branch-circuit rating is 20 amperes.

Another very common Code violation is where short No. 14 AWG pigtails connect receptacles to the 20-ampere small appliance branch-circuit conductors in a kitchen or dining room. These short pigtails must be rated 20 amperes because they are an *extension* of the 20-ampere branch-circuit wiring. Figure 5-4A illustrates this type of connection.

Table 210-24 is a summary of branch-circuit requirements. This table lists branch-circuit ratings, branch-circuit conductor sizes, tap conductor sizes, overcurrent protection, outlet (lampholder and receptacle) ratings, maximum loads, and permissible loads.

REVIEW

Note: Refer to the Code or the plans where necessary.

1. The identified grounded circuit conductor must be _____ or _____ in color.

2. Explain how lighting switches are rated. _____
_____
_____
_____
_____

3. A T-rated switch may be used to its _____ current capacity when controlling an incandescent lighting load.

4. What switch type and rating is required to control five 300-watt tungsten filament lamps on a 120-volt circuit? Show calculations. _____
_____
_____
_____

5. List four types of lighting switches.

   a. _____   c. _____

   b. _____   d. _____

6. To control a lighting load from one control point, what type of switch would be used?

   _____

7. Single-pole switches are always connected to the _____ wire.

8. Complete the connections in the following arrangement so that both ceiling light outlets are controlled from the single-pole switch. Assume the installation is in cable.

WHITE   WHITE   WHITE

        LAMP   LAMP   SWITCH

120-VOLT   RED
SOURCE

BLACK   BLACK   BLACK

9. Complete the connections for the diagram. Installation is cable.

WHITE   WHITE

              LAMP

120-VOLT   SWITCH
SOURCE

BLACK   BLACK

10. A three-way switch may be compared to a _____ switch.

11. What type of switch is installed to control a lighting fixture from two different control points? How many switches are needed and what type are they? _____

   _____

12. Complete the connections in the following arrangement so that the lamp may be controlled from either three-way switch.

WHITE   WHITE   WHITE

                              LAMP

120-VOLT S_3 RED   RED S_3
SOURCE

BLACK   BLACK   BLACK

13. When connecting four-way switches, care must be taken to connect the "travelers" to the _____ terminals.

14. Show the connections for a ceiling outlet that is to be controlled from any one of three switch locations. The 120-volt feed is at the light. Label the conductor colors. Assume the installation is in cable.

15. Match the following switch types with the correct number of terminals for each.

    | Three-way switch | Two terminals |
    | Single-pole switch | Four terminals |
    | Four-way switch | Three terminals |

16. When connecting single-pole, three-way, and four-way switches, they must be wired so that all switching is done in the _____ circuit conductor.

17. What section of the Code emphasizes the fact that all circuiting must be done so as to avoid the damaging effects of induction heating? _____

18. If you had to install an underground three-wire feeder to a remote building using three individual conductors, which of the following installations "meets Code"? Circle correct installation.

19. Is it always necessary to attach the bare equipment grounding conductor of a nonmetallic-sheathed cable to the green hexagon-shaped grounding screw on a receptacle? Explain.

_____

_____

_____

_____

20. List the methods by which an equipment grounding conductor is connected to a device box.

_____

_____

_____

_____

_____

21. When two nonmetallic-sheathed cables (Romex) enter a box, is it permitted to bring both of the bare equipment grounding conductors directly to the grounding terminal of a receptacle, using the terminal as a splice point? _____

22. This installation is in electrical metallic tubing. Because receptacle outlet "B" is only a few feet from switch "C," and switch "D" is only a few feet from receptacle outlet "A," the electrician saved a considerable amount of wire by picking up the feed for switch "C" from receptacle "B." The switch leg return is at switch "D." Please comment on this installation.

_____

_____

_____

_____

23. Define an equipment grounding conductor. _____

_____

_____

_____

24. When metal toggle switchplates are used with nonmetallic boxes, the faceplate must be grounded according to *Section 380-9(b)*. How is this accomplished?

_____

25. Does the Code permit the ampacity of switch legs to be less than the ampere rating of the branch-circuit? _____

26. Does the Code permit connecting receptacles to a 20-ampere branch-circuit, using short lengths of No. 14 AWG as the "pigtail" between the receptacle terminals and the No. 12 branch-circuit conductors? _____

27. Do receptacles that have back-wired "push-in" terminals accept No. 12 AWG conductors? Give a brief explanation for your answer. _____

_____

_____

_____

Ground-Fault Circuit Interrupters, Transient Voltage Surge Suppressors, Isolated Ground Receptacles, Immersion Detection Circuit Interrupters

OBJECTIVES

After studying this Unit, the student will be able to

- understand the theory of operation of ground-fault circuit-interrupting (GFCI) devices.
- explain the operation and connection of GFCIs.
- discuss the do's and don'ts relative to the use of GFCI devices.
- explain why GFCIs are required.
- discuss locations where GFCIs must be installed in homes.
- understand the proper way to install and utilize feedthrough GFCI receptacles.
- understand testing, recording, and identification of GFCI receptacles.
- discuss the Code rules relating to the replacement of existing receptacles with GFCI receptacles.
- understand the Code requirements for GFCI protection on construction sites.
- understand the logic of the exemptions to GFCI mandatory requirements for receptacles in certain locations in kitchens, garages, and basements.
- discuss and understand the basics of transient voltage surge suppressors, and isolated ground receptacles.
- discuss and understand the principles of *immersion detection circuit interrupters*.

ELECTRICAL HAZARDS

There can be no compromise where human life is involved!

Many injuries have occurred and many lives have been lost because of electrical shock. Coming in contact with live wires or with an appliance or other equipment that is "hot" spells danger. Problems arise in equipment when there is a breakdown of insulation because of wear and tear, defective construction, or misuse of the equipment. Insulation failure can result when the "hot" ungrounded conductor in the appliance comes in contact with the metal frame of the appliance. If the equipment is not properly grounded, the potential for an electrical shock is present.

The shock hazard exists whenever the user can touch both the defective equipment and grounded surfaces, such as water pipes, stainless steel sinks, metal faucets, grounded metal lighting fixtures, earth, concrete in contact with the earth, water, or any other grounded surface.

A severe shock can cause considerably more damage to the human body than is visible. A person may suffer internal hemorrhages, destruction of tissues, nerves, and muscles. Further injury can result from a fall, cuts, burns, or broken bones.

The effect of an electric current passing through a human body is determined by *how much current is flowing and the length of time the current will flow*. Figure 6-1 shows a time/current curve that indicates the amount of current that a normal healthy adult can stand for a certain time. The EFFECT OF ELECTRIC SHOCK chart shows expected sensations at various milliamperes current levels.

Experts say that the resistance of the human body varies from a few hundred ohms to many thousand ohms, but generally agree on values of 800 to 1000 ohms as typical. Body internal resistance, body contact resistance, moisture, duration, voltage, as well as the path of current: finger to finger, hand to hand, hand holding "hot" wire, hand holding a pair of pliers, hand holding an electric drill, hand grasping a grounded metal pipe, two hands grasping a grounded metal pipe, hand in water, standing in water, hand to foot, foot to foot, etc. all have an effect on the severity of the shock.

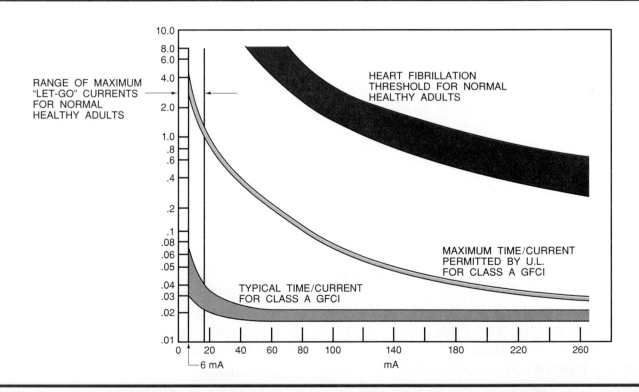

Figure 6-1   The time/current curve shows the tripping characteristics of typical Class A GFCI. Note that if you follow the 6-mA line vertically to the crosshatched typical time/current curve, you will find that the GFCI will open in from approximately 0.035 second to just less than 0.1 second. One electrical cycle is 1/60 of a second (0.0167 second). An air bag in an automobile inflates in approximately 1/20 of a second.

EFFECT OF ELECTRIC SHOCK		
	Current in milliamperes @ 60 hertz	
	Men	Women
• cannot be felt	0.4	0.3
• a little tingling—mild sensation	1.1	0.7
• shock—not painful—can still let go	1.8	1.2
• shock—painful—can still let go	9.0	6.0
• shock—painful—just about to point where you can't let go—called "threshold"—you may be thrown clear	16.0	10.5
• shock—painful—severe—can't let go—muscles immobilize—breathing stops	23.0	15.0
• ventricular fibrillation (usually fatal) length of time . . . 0.03 sec. length of time . . . 3.0 sec.	1000 100	1000 100
• heart stops for the time current is flowing—heart may start again if time of current flow is short	4 amperes	4 amperes
• burning of skin—generally not fatal unless heart or other vital organs burned	5 or more amperes	5 or more amperes
• cardiac arrest, severe burns, and probable death	10,000	10,000

Generally, voltages of less than 40 volts to ground are considered safe. Voltages 50 volts and greater are considered lethal. *Section 250-20(b)* requires that (with a few exceptions) all ac systems of 50 volts to 1,000 volts that supply premise wiring be grounded.

Some circuits of less than 50 volts must be grounded according to *Section 250-20(a)*:

- if the circuit is supplied by a transformer where the primary exceeds 150 volts to ground, or

- if the circuit is supplied by a transformer where the primary supply is ungrounded, or

- if the circuit is overhead wiring outside of a building.

CODE REQUIREMENTS FOR GROUND-FAULT CIRCUIT INTERRUPTERS *(SECTION 210-8)*

To protect against electric shocks, *Section 210-8(a)* requires that personnel ground-fault protection be provided for *all* 125-volt, 15- and 20-ampere receptacle outlets in dwellings as follows:

- in bathrooms. See Figure 10-7 for the definition of a bathroom.

- outdoors.

- in kitchens, for all receptacles that serve the countertop surfaces. This includes receptacles installed on islands and peninsulas, Figure 6-2. GFCI protection is *not* required for:
  - a receptacle installed solely for a clock.
  - a receptacle installed inside of an upper kitchen cabinet for plugging in a microwave oven that is fastened to the underside of the cabinet.
  - a receptacle installed below the kitchen sink for plugging in a food-waste disposer or dishwasher.

- within 6 feet (1.83 m) from the edge of a wet-bar sink, Figure 6-2.

- in attached or detached garages. GFCI protection is *not* required for:
  - receptacles that are not readily accessible, such as a receptacle installed on the ceiling for an overhead garage door opener.
  - a single receptacle, or a duplex receptacle is installed for cord- and plug-connected appliances in a dedicated space where the appliance is not easily moved from one location to another, such as a freezer or a refrigerator.

- in other buildings (sheds, workshops, or storage buildings). GFCI protection is *not* required for:
  - receptacles that are not readily accessible.
  - a single receptacle, or a duplex receptacle is installed for cord- and plug-connected appliances in a dedicated space where the appliance is not easily moved from one location to another. Examples are a freezer or a refrigerator.

- in crawl spaces that are at or below grade. Porcelain or plastic lampholders that have a plug on them also must be GFCI protected. See the exceptions listed below.

- in unfinished basements. Porcelain or plastic lampholders that have a plug on them also must be GFCI protected. See the exceptions listed below.

Exceptions:
  - GFCI protection is *not* required for receptacles that are not readily accessible. The Code defines "readily accessible" as being able to be reached without the use of ladders or chairs.

Figure 6-2 In kitchens, *all* 125-volt, single-phase, 15- and 20-ampere receptacles that serve countertop surfaces (walls, islands, or peninsulas) must be GFCI protected. The key words are "that serve countertop surfaces." GFCI personnel protection is *not* required for the refrigerator receptacle, a receptacle under the sink for plugging-in a food waste disposer, or a receptacle installed inside cabinets above the countertops for plugging in a microwave oven attached to the underside of the cabinets. Where wet-bar sinks are installed outside of the kitchen area, *all* 125-volt, single-phase, 15- and 20-ampere receptacles that serve the countertops must be GFCI protected if they are within 6 feet (1.83 m) measured from the outer edge of the wet-bar sink. See *Section 210-8(a)(6)* and *(7)*. Drawing is not to scale.

– GFCI protection is *not* required for a single receptacle, or a duplex receptacle installed for cord- and plug-connected appliances in a dedicated space where the appliance is not easily moved from one location to another. Examples are a freezer, refrigerator, water softener, and a gas-fired electric-controlled water heater. Do not plug sump pumps into a GFCI receptacle because of the possibility of the GFCI tripping off, unknown to the homeowner. Result—a flooded basement. Since a single receptacle is exempt from the GFCI requirement, it is better to install a single receptacle for a sump pump. Also, plan on installing a single receptacle for the laundry receptacle that is required by *Section 210-52(f)*. The laundry receptacle must be supplied by a separate 20-ampere branch-

circuit, and must not supply any other loads, *Section 210-11(5)(a)*.

Some individuals try to get around the requirement for GFCI personnel protection in basements by calling the basement "finished," even though it is clearly evident that it is not "finished." If the basement is intended to be "finished," then the Code requirements for the spacing of receptacles, required lighting outlets, and switch controls would have to be followed. This is covered in Unit 3.

• there is a special rule for residential underground branch-circuit wiring that can result in a labor cost savings when the underground branch-circuit wiring is GFCI protected. *Table 300-5* of the *NEC®* indicates that for residential underground 120-volt or less branch-circuits that are rated 20 amperes or less, the underground wiring need only be buried 12 inches below grade when the circuit is GFCI protected. An example of this would be the Type UF cable supplying the post light in front of the residence. Type UF cable must be buried at least 24 inches below grade if the circuit is not GFCI protected. If the underground circuit is installed in rigid or intermediate metal conduit, a depth of not less than 6 inches is required. Refer to Unit 16 for more detailed information regarding residential underground wiring.

• for swimming pools, see Unit 30.

The Code requirements for ground-fault circuit protection can be met in many ways. Figure 6-3 illustrates a GFCI circuit breaker installed on a branch circuit. A ground fault in the range of 4 to 6 milliamperes or more shuts off the entire circuit. If for example, a GFCI circuit breaker were to be installed on Circuit B14, a ground-fault at any point in the circuit would shut off the garage lighting, garage receptacles, rear garage door outside bracket fixture, post light, and overhead door opener.

When a GFCI receptacle is installed, then only that receptacle is shut off when a ground fault greater than 6 milliamperes occurs, Figure 6-4.

GFCI receptacles break both the ungrounded (hot) and grounded conductors.

GFCI circuit breakers break only the ungrounded (hot) conductor.

Figure 6-5 shows the effect of a GFCI installed as part of a feeder supplying 15- and 20-ampere receptacle branch-circuits. This is rarely used in residential wiring.

Figure 6-6 is a pictorial view of how a ground-fault circuit interrupter operates. Figure 6-7 shows the actual internal wiring of a GFCI.

Occasionally a GFCI will trip for seemingly no reason. This could be "leakage" in extremely long runs of cable from a GFCI circuit breaker in a panel to the protected branch-circuit wiring. The manufacturer of the GFCI circuit breaker may have recommendations relative to the total length of the branch-circuit wiring.

The allowable "leakage" for U.L. listed appliances is so small that the appliances should not contribute to nuisance tripping of GFCIs. However, if nuisance tripping can be traced to an appliance, the appliance should be checked by a qualified appliance repairman.

Nuisance tripping occasionally can be traced to moisture somewhere in the circuit wiring, receptacles, or lighting outlets.

Some electricians are of the opinion that using nonmetallic staples (see Figure 4-14(C)) will reduce the possibility of nuisance tripping. No actual testing has been done to prove or disprove this opinion.

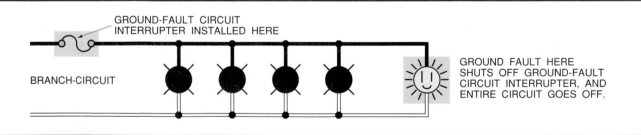

Figure 6-3   Ground-fault circuit interrupter as a part of the branch-circuit overcurrent device.

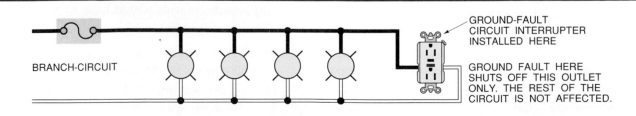

Figure 6-4   Ground-fault circuit interrupter as an integral part of a receptacle outlet.

Figure 6-5   Ground-fault circuit interrupter as part of the feeder supplying a panel that serves a number of 15- and 20-ampere branch-circuits. This is permitted by *Section 215-9*, but is rarely—if ever—used in residential wiring.

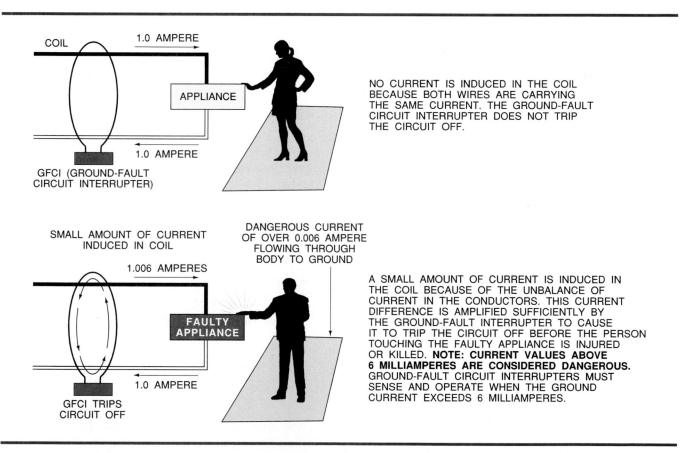

Figure 6-6   Basic principle of how a ground-fault circuit interrupter operates.

What a GFCI Does

A GFCI monitors the current balance between the ungrounded "hot" conductor and the grounded "neutral" conductor. As soon as the current flowing through the "hot" conductor is in the range of 4 to

6 milliamperes or more than the current flowing in the "return" grounded conductor, the GFCI senses this unbalance and trips (opens) the circuit off. The unbalance indicates that part of the current flowing in the circuit is being diverted to some path other

Figure 6-7 **Ground-fault circuit interrupter internal components and connections. Receptacle-type GFCIs switch both the phase (hot) and grounded conductors. Note that when the test button is pushed, the test current passes through the test button, the sensor, then back around (bypasses, outside of) the sensor, then back to the opposite circuit conductor. This is how the "unbalance" is created, then monitored by the solid-state circuitry to signal the GFCI's contacts to open. Note that because both normal "load" currents pass through the sensor, no unbalance is present.**

than the normal return path along the grounded return conductor. If the "other" path is through a human body, as illustrated in Figure 6-6, the outcome could be fatal.

U.L. Standard No. 943 covers ground-fault circuit interrupters.

- Class "A" GFCI devices are designed to trip when a ground fault in the range of 4 to 6 milliamperes ($^4/_{1000}$ to $^6/_{1000}$ of an ampere) or more occurs.

- Class "B" GFCI devices are pretty much obsolete. They were designed to trip on ground faults of 20 milliamperes ($^{20}/_{1000}$ of an ampere) or more. They were used only for underwater swimming pool lighting installed before the adoption of the 1965 *NEC.* For this application, Class "A" devices were too sensitive and would nuisance trip.

What a GFCI Does *Not* Do

- It *does not* protect against electrical shock when a person touches both circuit conductors at the same time (two "hot" wires, or one "hot" wire and one grounded neutral wire) because the current flowing in both conductors is the same. Thus, there is no unbalance of current for the GFCI to sense and trip.

- It *does not* limit the magnitude of ground-fault current. It *does* limit the length of time that a ground fault will flow. In other words, you will still receive a severe shock during the time it takes the GFCI device to trip "off." See Figure 6-1.

- It *does not* sense solid short-circuits between the "hot" conductor and the grounded "neutral" conductor. The branch-circuit fuse or circuit breaker provides this protection.

- It *does not* sense solid short-circuits between two "hot" conductors. The branch-circuit fuse or circuit breaker provides this protection.

- It *does not* sense "arcing" faults between the "hot" conductor and the grounded "neutral" conductor. The branch-circuit fuse or circuit breaker provides this protection. An example of an arcing fault is common with frayed extension cords. The heat generated by the arc can easily result in a fire, yet the amount of current flowing can be much less than the rating of the branch-circuit fuse or circuit breaker.

- A GFCI receptacle *does not* provide overload protection for the branch-circuit wiring. It provides *ground-fault protection only*.

GROUND-FAULT CIRCUIT INTERRUPTER IN RESIDENCE CIRCUITS

In this residence, GFCI personnel protection must be provided for receptacle outlets installed outdoors, in the bathrooms, in the garage, in the workshop, specific receptacles in the basement area, in the kitchen, and adjacent to the wet-bar in the recreation room. The GFCI protection can be provided by GFCI circuit-breakers or by GFCI receptacles of the type shown in Figure 6-8. This is a design issue that confronts the electrician. The electrician must decide how to provide the GFCI personnel protection required by *Section 210-8*. These receptacle outlets are connected to the circuits listed in Table 6-1.

Swimming pools also have special requirements for GFCI protection. These requirements are covered in Unit 30.

GFCIs operate properly only on grounded electrical systems, as is the case in all residential,

CIRCUITS	GFCI RECEPTACLE LOCATION
A15	Front porch receptacle
A16	Outdoor receptacle on rear of residence outside of Front Bedroom
A18	Workbench receptacles
A19	Outdoor receptacle on rear of residence outside of Master Bedroom
A20	Workshop receptacles on window wall
A22	Master Bath receptacle
A23	Bath (off hall) receptacle
B9	Wet-bar (Recreation Room) receptacles
B13	Kitchen receptacles
B14	Garage receptacles
B15	Kitchen receptacles
B16	Kitchen receptacles
B17	Outdoor receptacle outside of Living Room next to sliding door
B20	Outdoor receptacle outside of laundry
B21	Powder Room receptacle

Table 6-1 Circuit number and general location of GFCI receptacles in this residence.

A

GROUND-FAULT CIRCUIT INTERRUPTER AS AN INTEGRAL PART OF A DUPLEX GROUNDING-TYPE CONVENIENCE RECEPTACLE.

B

A GROUND-FAULT CIRCUIT-INTERRUPTER CIRCUIT BREAKER.

Figure 6-8 A ground-fault circuit-interrupter receptacle (A) *courtesy of* Pass & Seymour and a ground-fault circuit-interrupter circuit breaker (B). The switching mechanism of a GFCI receptacle opens both the ungrounded "hot" conductor and the grounded conductor. The switching mechanism of a GFCI circuit breaker opens the ungrounded "hot" conductor only. *Courtesy of* Square D Company.

condo, apartment, commercial, and industrial wiring. The GFCI will operate on a two-wire circuit even though an equipment grounding conductor is not included with the circuit conductor. In this residence, equipment grounding conductors are in the cables. In the case of metal conduit, the equipment grounding conductor is the metal conduit.

Never ground a system neutral conductor except at the service equipment. A GFCI would be inoperative.

Never connect the neutral of one circuit to the neutral of another circuit.

When a GFCI feeds an isolation transformer (separate primary winding and separate secondary winding), as might be used for swimming pool underwater fixtures, the GFCI *will not* detect any ground faults on the secondary of the transformer.

Ground-fault circuit interrupters may be installed on other circuits, in other locations, and even when rewiring existing installations where the Code does not specifically call for GFCI protection,

Do not connect fire and smoke detector systems to a circuit that is protected by a GFCI breaker. A nuisance tripping of the GFCI would render the fire alarm and smoke detectors inoperative. This requirement is found in *NFPA 72, The National Fire Alarm Code.*

FEEDTHROUGH GROUND-FAULT CIRCUIT INTERRUPTER

The decision to use more GFCIs rather than trying to protect many receptacles through one GFCI becomes one of economy and practicality. GFCI receptacles are more expensive than regular receptacles. Here is where knowledge of material and labor costs comes into play. The decision must be made separately for each installation, keeping in mind that GFCI protection is a safety issue recognized and clearly stated in the Code. However, the actual circuit layout is left up to the electrician.

Figure 6-9 illustrates how a feedthrough GFCI receptacle supplies many other receptacles. Should a ground fault occur anywhere on this circuit, all 11 receptacles lose power—not a good circuit layout. Attempting to locate the ground-fault problem, unless obvious, can be very time consuming. The use of more GFCI receptacles is generally the more practical approach.

Figure 6-10 shows how a feedthrough receptacle is connected into a circuit. Figure 6-11 shows a feedthrough GFCI receptacle connected midway into a circuit. In both diagrams, the feedthrough GFCI receptacle and all downstream outlets are ground-fault protected.

FEEDTHROUGH GFCI

Figure 6-9   Illustration showing one GFCI feedthrough receptacle protecting ten other receptacles. Although this might be an economical way to protect many receptacle outlets with only one GFCI feedthrough receptacle, there is great likelihood that the GFCI will nuisance trip because of leakage currents. Use common sense and manufacturer's recommendations as to how many devices should be protected from one GFCI.

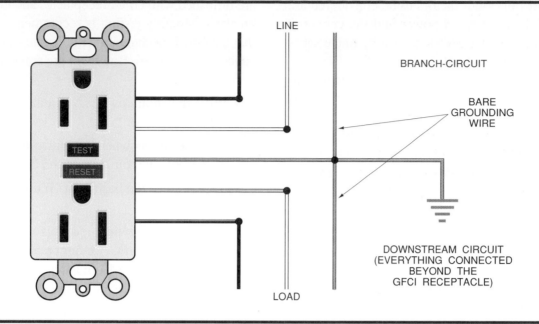

Figure 6-10   Connecting feedthrough ground-fault circuit interrupter into circuit.

Figure 6-11   Feedthrough ground-fault circuit interrupter as an integral part of the convenience receptacle outlet.

In Underwriters Standard 943, we find that a GFCI receptacle may have black and white line wire leads, red and gray load wire leads, and a green lead for the equipment grounding conductor as illustrated in Figure 6-10. GFCI receptacles may have ordinary screw terminals: brass color for the "hot" conductors, silver color for the grounded conductors, and a green hexagon-shaped screw for the equipment grounding conductor. The line and load connections are clearly marked.

Where the wiring method is a grounded method such as electrical metallic tubing, flexible metal conduit, or armored cable, the grounding terminal of the GFCI receptacle is considered adequately grounded by the metallic grounded wiring method, and a separate connection to the green lead or to the green hexagon-shaped screw is not necessary. This was discussed and illustrated in Unit 5.

A "safety yellow" adhesive label must cover the load terminals, or be wrapped around the load wire leads. The wording on the label must be marked with something like this: *"Attention! The load terminals under this label are for feeding additional receptacles. Miswiring can leave this outlet without ground-fault protection. Read instructions prior to wiring."*

The most important factors to be considered are the continuity of electrical power and the economy of the installation. The decision must be made separately for each installation.

Warning: Do not reverse the line and load connections on a feedthrough GFCI receptacle. This would result in the feedthrough GFCI receptacle itself still being "live" even though the GFCI mechanism has tripped "off" and has shut off the circuit downstream from the feedthrough GFCI receptacle. It is easy to tell if a GFCI receptacle has the line and load leads reversed. Push the test button. The GFCI will trip. If the GFCI receptacle is still "hot," it has been wired incorrectly.

TESTING AND RECORDING OF TEST DATA FOR GFCI RECEPTACLES

Refer to Figure 6-8 and note that the GFCI receptacle has "T" (Test) and "R" (Reset) buttons. Pushing the test button places a small ground fault on the circuit. If operating properly, the GFCI receptacle should trip to the OFF position. Pushing the reset button will restore power. The operation is similar for the GFCI circuit breakers.

Detailed information and testing instructions are furnished with GFCI receptacles and GFCI circuit breakers. Monthly testing is recommended to ensure that the GFCI mechanism will operate properly when called upon to do so. But, there are some questions to keep in mind:

- Will the homeowner remember to do the monthly testing?
- Will the homeowner recognize and understand the purpose and function of the GFCI receptacle?
- Should all of the basement GFCI receptacles be identified?
- Will the homeowner recognize a receptacle if protected by a GFCI circuit breaker located in an electrical panel far from the actual receptacle?

These are important questions and a possible reason why Underwriters Laboratories (U.L.) requires detailed installation and testing instructions to be included in the packaging for the GFCI receptacle or circuit breaker. Instructions are not only for the electrician but also for homeowners, and must be left in a conspicuous place so homeowners can familiarize themselves with the receptacle, its operation, and its need for testing. Figure 6-12 shows a chart homeowners can use to record monthly GFCI testing.

OCCUPANT'S TEST RECORD

TO TEST, depress the "TEST" button, the "RESET" button should extend. Should the "RESET" button not extend, the GFCI will not protect against electrical shock. Call qualified electrician.

TO RESET, depress the "RESET" button firmly into the GFCI unit until an audible click is heard. If reset properly, the "RESET" button will be flush with the surface of the "TEST" button.

This label should be retained and placed in a conspicuous location to remind the occupants that for maximum protection against electrical shock, each GFCI should be tested monthly.

Year	Jan	Feb	Mar	Apr	May	Jun	Jul	Aug	Sep	Oct	Nov	Dec

Figure 6-12 Homeowner's testing chart for recording GFCI testing dates.

REPLACING EXISTING RECEPTACLES

The *NEC*® is very specific on the type of receptacle permitted to be used as a replacement for an existing receptacle. These Code rules are found in *Sections 210-7(d)* and *250-130(c)*.

Section 210-7(d)(2) is a retroactive ruling requiring that when an existing receptacle is replaced in *any* location where the present code requires GFCI protection (bathrooms, kitchens, outdoors, unfinished basements, garages, etc.), the replacement receptacle must be GFCI protected. This could be a GFCI-type receptacle, or the branch-circuit could be protected by a GFCI-type circuit breaker.

It is important to note that on existing wiring systems, such as "knob and tube," or older nonmetallic-sheathed cable wiring that did not have an equipment grounding conductor, that a GFCI receptacle will still properly function. It does not need an equipment grounding conductor. Thus, protection against lethal shock is still provided. See Figure 6-7.

Replacing Existing Two-Wire Receptacles Where Grounding Means Does Exist

When replacing an existing receptacle (Figure 6-13(A) where the wall box is properly grounded (Figure 6-13(E)) or where the branch-circuit wiring contains an equipment grounding conductor (Figure 6-13(D)), at least two easy choices are possible for the replacement receptacle:

1. The replacement receptacle must be of the grounding type (Figure 6-13(B)) unless . . .

2. The replacement receptacle is of the GFCI type (Figure 6-13(C)).

If the enclosure is not grounded, the *Exception* to *Section 250-130* permits an equipment grounding conductor to be run from the enclosure to be grounded, or from the green hexagon grounding screw of a grounding-type receptacle, to any accessible point on the grounding electrode system or to any accessible point on the grounding electrode conductor. Note however, that this permission is only for existing installations. See Figure 6-16.

Be sure that the equipment grounding conductor of the circuit is connected to the receptacle's green hexagon-shaped grounding terminal.

Do not connect the white grounded circuit conductor to the green hexagon-shaped grounding terminal of the receptacle.

Do not connect the white grounded circuit conductor to a metal box.

Figure 6-13 Where box (E) is properly grounded by any of the methods specified in *Section 250-118*, an existing receptacle of the type shown in (A) *must* be replaced with a grounding-type receptacle (B) or with a GFCI receptacle (C). See *Section 210-7(d)*.

Replacing Existing Two-Wire Receptacles Where Grounding Means Does Not Exist

When replacing an existing two-wire non-grounding-type receptacle (Figure 6-14(A)), where the box is not grounded (Figure 6-14(E)), or where an equipment grounding conductor has not been run with the circuit conductors (Figure 6-14(D)), four choices are possible for selecting the replacement receptacle:

1. The replacement receptacle may be a non-grounding type (Figure 6-14(A)).

2. The replacement receptacle may be a GFCI type (Figure 6-14(C)).

   • The GFCI replacement receptacle must be marked "*No Equipment Ground*," Figure 6-15.

   • The green hexagon grounding terminal of the GFCI replacement receptacle does not have to be connected to any grounding means. It can be left "unconnected." The GFCI's trip mechanism will operate properly when ground faults occur anywhere on the load side of the GFCI replacement receptacle. Ground-fault protection is still there. Refer to Figure 6-7.

• Do not connect an equipment grounding conductor from the green hexagon grounding terminal of a replacement GFCI receptacle (Figure 6-14(C)) to any other downstream receptacles that are fed through the replacement GFCI receptacle.

The reason this is not permitted is that if at a later date someone saw the conductor connected to the green hexagon grounding terminal of the downstream receptacle, there would be an immediate assumption that the other end of that conductor had been properly connected to an acceptable grounding point of the electrical system. The fact is that

REPLACING A RECEPTACLE WHERE GROUNDING MEANS DOES NOT EXIST.

A — OLD TWO-WIRE RECEPTACLE

B — GROUNDING-TYPE RECEPTACLE — GRND

C — GFCI RECEPTACLE — GRND — TEST / RESET

NO EQUIPMENT GROUNDING CONDUCTOR RUN WITH CIRCUIT CONDUCTORS

GROUNDED CONDUCTOR

HOT CONDUCTOR

D

E — NONGROUNDED BOX

Figure 6-14 Where box (E) is *not* grounded, or where an equipment grounding conductor has *not* been run with the circuit conductors (D), an existing nongrounding-type receptacle (A) may be replaced with a nongrounding-type receptacle (A), a GFCI-type receptacle (C), a grounding-type receptacle (B) if supplied through a GFCI receptacle (C), or a grounding-type receptacle (B) if a separate equipment grounding conductor is run from the receptacle to any accessible point on the grounding electrode system, *Section 250-130(c)*. Most electricians would replace an existing nongrounding-type receptacle with a GFCI type rather than install an equipment grounding conductor, which could be very labor-intensive. See *Section 210-7(d)*.

the so-called equipment grounding conductor had been connected to the replacement GFCI receptacle's green hexagon grounding terminal that was not grounded in the first place. We now have a false sense of security, a real shock hazard.

3. The replacement receptacle may be a grounding type (Figure 6-14(B)) if it is supplied through a GFCI-type receptacle (Figure 6-14(C)).

    • The replacement receptacle must be marked "*GFCI Protected*" and "*No Equipment Ground,*" Figure 6-15.

    • The green hexagon equipment grounding terminal of the replacement grounding-type receptacle (Figure 6-14(B)) need not be connected to any grounding means. It may be left "unconnected." The upstream feedthrough GFCI receptacle (Figure 6-14(C)) trip mechanism will work properly when ground faults occur anywhere on its load-side. Ground-fault protection is still there. Refer to Figure 6-7.

    • Do not connect an equipment grounding conductor from the green hexagon grounding terminal of a replacement GFCI receptacle (Figure 6-14(C)) to any other downstream receptacles that are fed through the replacement GFCI receptacle.

4. The replacement receptacle may be a grounding type (Figure 6-14(B)) if:

    • an equipment grounding conductor, sized per *Table 250-122*, is run from the replacement receptacles green hexagon grounding screw and properly connected to a listed ground clamp at some point on the first 5 feet (1.5 m) of the metal water piping as it enters the building. You are *not* permitted to make the equipment ground connection to the interior metal water piping anywhere beyond the first 5 feet (1.5 m). See *Section 250-130(c)* and the *FPN*. Also read *Section 250-50* for the explanation of a "grounding electrode system." This permission is only for replacing a receptacle in existing installations. See Figure 6-16.

IMPORTANT:
WHERE AN EXISTING NON-GROUNDING TYPE RECEPTACLE IS REPLACED WITH A GROUNDING TYPE RECEPTACLE, AND IS SUPPLIED THROUGH ANOTHER GFCI RECEPTACLE, IT MUST BE MARKED "GFCI PROTECTED" AND "NO EQUIPMENT GROUND." SEE *SECTION 210-7(d)*. WITHOUT THIS REQUIRED MARKING, IT WOULD BE DIFFICULT TO KNOW IF THAT RECEPTACLE IS GFCI PROTECTED, OR IF THE GROUNDING TERMINAL ON THAT RECEPTACLE IS TRULY GROUNDED. MANUFACTURERS OF RECEPTACLES PROVIDE PRESSURE-SENSITIVE LABELS FOR THIS PURPOSE.

REPLACEMENT GFCI RECEPTACLE WHERE NO GROUNDING MEANS EXISTS. MUST BE MARKED "NO EQUIPMENT GROUND."

DOWNSTREAM REPLACEMENT GROUNDING TYPE RECEPTACLE FED THROUGH A REPLACEMENT GFCI RECEPTACLE. MUST BE MARKED "GFCI PROTECTED" AND "NO EQUIPMENT GROUND."

Figure 6-15   Where a grounding means *does not* exist, there are certain marking requirements as presented in this diagram. *Do not* connect an equipment grounding conductor between these receptacles.

REPLACEMENT RECEPTACLE

GROUNDING SCREW

NO. 14 OR NO. 12 AWG GREEN OR BARE EQUIPMENT GROUNDING CONDUCTOR.

GROUNDING ELECTRODE SYSTEM. SEE *SECTION 250-50*

BECAUSE OF THE POSSIBLE HIGH COST OF INSTALLING AN EQUIPMENT GROUNDING CONDUCTOR PER THE *NEC®* REQUIREMENTS, MOST ELECTRICIANS WOULD OPT TO MAKE THE REPLACEMENT WITH A GFCI-TYPE RECEPTACLE.

LISTED GROUNDING CLAMP

MAKE CONNECTION ON FIRST FIVE FEET OF METAL WATER PIPE, OR TO ANY ACCESSIBLE POINT ON THE GROUNDING ELECTRODE CONDUCTOR.

Figure 6-16 **In existing installations that do not have an equipment grounding conductor as part of the branch-circuit wiring, *Section 210-7(d)* permits replacing an old-style nongrounding-type receptacle with a grounding-type receptacle. The grounding terminal on the receptacle must be properly grounded. One of the acceptable ways to do this is to run an equipment grounding conductor to any accessible point on the grounding electrode system. Another way is to run an equipment grounding conductor to any accessible point on the grounding electrode conductor. See *Section 250-130(c)*.**

And remember, whenever you replace a two-wire nongrounding-type receptacle in a location where the present *NEC®* requires GFCI protection, such as in kitchens, bathrooms, garages, outdoors, and basements, the replacement receptacle must be a GFCI receptacle, or the branch-circuit must be protected with a GFCI circuit breaker. This is in conformance to *Section 210-7(d)(2)*.

PERSONNEL GROUND-FAULT PROTECTION FOR ALL TEMPORARY WIRING

Because of the nature of construction sites, there is a continual presence of shock hazard that can lead to serious personal injury or death through electrocution. Workers are standing in water, standing on damp or wet ground, or in contact with steel framing members. Electric cords and cables are lying on the ground, subject to severe mechanical abuse. All of these conditions spell danger.

Therefore, *Section 305-6(a)* requires that all 125-volt, single-phase, 15- and 20-ampere receptacle outlets that are not part of the permanent wiring of the building and that will be used by the workers on the construction site must be GFCI protected. Receptacle outlets that are part of the actual permanent wiring of a building and are used by personnel for temporary power are also required to be GFCI protected.

For receptacles *other* than the 125-volt, single-phase, 15- and 20-ampere receptacles, *Section 305-6(b)* permits a written *Assured Equipment Grounding Conductor Program* that must be continuously enforced on the construction site. A designated person must keep a written log, ensuring that all electrical equipment is properly installed and maintained according to the applicable requirements of *Sections 210-7(c)*, *250-114*, *250-138*, and *305-4(d)*. This option has proved to be difficult to enforce. Thus, most "authorities having jurisdiction" require GFCI protection as stated in *Section 305-6(a)*.

IMMERSION DETECTION CIRCUIT INTERRUPTERS

The danger of electrical shock is always present when using electrical appliances. GFCIs are required in specific locations in new homes. But something had to be done to protect users from this shock hazard in older homes that did not require GFCIs at the time they were built. The major culprit was found to be personal grooming appliances.

To address the problem, a new type of circuit interrupter was developed. This device is called an **Immersion Detection Circuit Interrupter** (IDCI). It usually is an integral part of the attachment plug cap found on personal grooming appliances that are commonly used near water and basins, such as hair dryers, heated stylers, heated air combs, heated air curlers, curling iron–hair dryer combinations, and so on.

These electrical appliances are most commonly used at a vanity having a wash basin with water in it, or even worse, next to a water-filled bathtub with a person (child or adult) in it. Accidentally dropping or pulling the appliance into the tub when reaching for a towel can be lethal.

Electricity and water do not mix! The shock hazard protection of these appliances is permitted to be in the form of a GFCI device or in the form of an immersion detection device. If any of these appliances fall into a basin or tub that is filled with water, the IDCI will sense the leakage current and will trip to the OFF position. Both GFCI and IDCI devices disconnect both conductors of the branch-circuit. These devices are also required to open the circuit regardless of whether the appliance switch is in the ON or OFF position.

The operation of an IDCI depends upon a "third" wire in the cord, referred to as a *probe*, which is connected to a sensor within the appliance. This probe will detect leakage current flow when conductive liquid enters the appliance and makes contact with any live part inside the appliance and the internal "sensor." When this happens, the IDCI is instructed to trip to the OFF position.

It is not permitted to replace the special GFCI or IDCI attachment plug cap on these appliances because that would make the appliance absolutely unsafe. The U.L. requirement is that the plug cap shall be marked: "Warning: To Reduce The Risk Of Electric Shock, Do Not Remove, Modify, Or Immerse This Plug."

An IDCI must open the circuit when the sensor detects a leakage current to ground in the range of 4 to 6 milliamperes. An IDCI on a hair dryer or similar appliance must be manually reset. *Section 422-41* requires protection against electrocution for freestanding hydromassage appliances and hand-held hair dryers. This is intended to include heated stylers, heated air combs, heated air curlers, and curling iron–hair dryer combinations.

ARC-FAULT CIRCUIT INTERRUPTERS

Arc-fault circuit interrupters are the "new kid on the street."

Electrical arcing is one of the leading causes of electrical fires in homes. The heat of an arc is extremely high—enough to ignite most combustible material. Arcing faults can be line-to-ground, line-to-neutral, or line-to-line. When a conductor that is carrying current breaks, or when a loose connection occurs, it is a "series" arcing fault because the arcing is traveling in the intended path. A broken or frayed cord is a good example of this.

An *arc-fault circuit interrupter (AFCI)* is designed to sense the rapid fluctuations of current flow typical of an arcing condition. It will open the circuit when arcing occurs. The sensing range of an AFCI is 5 amperes up through the expected instantaneous tripping current value of a typical circuit breaker. However, the design of the electronic circuitry is such that it will not nuisance trip because of the normal momentary arc that occurs when a switching device is turned on or off.

▶ Effective January 1, 2002, AFCI protection must be provided for all 125-volt, single-phase, 15- and 20-ampere receptacles in dwelling bedrooms. This protection can be provided with AFCI circuit breakers or AFCI receptacles. See *Section 210-12.* ◀

Other Types of "People Protectors"

Most older homes do not have GFCI protection. There are other devices that manufacturers of appliances are required to design into certain electrical products in conformance to various U.L. standards. In addition to GFCIs and IDCIs, other devices currently being used are *appliance leakage circuit interrupters (ALCI)* that trip at 6 milliamperes, *equipment leakage circuit interrupters (ELCI)* that trip at 50 milliamperes, and *leakage monitoring receptacles (LMR)* that trip at 50 milliamperes.

TRANSIENT VOLTAGE SURGE SUPPRESSOR (TVSS)

In today's homes, we find many electronic appliances (television, stereo, personal computers, fax equipment, word processors, VCRs, CD players, digital stereo equipment, microwave ovens), all of which contain many sensitive, delicate electronic components (printed circuit boards, chips, microprocessors, transistors, etc.).

Voltage transients, called *surges* or *spikes*, can stress, degrade, and/or destroy these components. They can cause loss of memory in the equipment or "lock up" the microprocessor.

The increased complexity of electronic integrated circuits makes this equipment an easy target for "dirty" power that can and will affect the performance of the equipment.

Voltage transients cause abnormal current to flow through the sensitive electronic components. This energy is measured in joules. A *joule* is the unit of energy when one ampere passes through a one-ohm resistance for one second (like wattage, only on a much smaller scale).

Line surges can be line-to-neutral, line-to-ground, and line-to-line.

Transients

Transients are generally grouped into two categories:

- Ring wave—These transients originate within the building and can be caused by copiers, computers, printers, HVAC cycling, spark igniters on furnaces, water heaters, dryers, gas ranges, ovens, motors, and other inductive loads.

- Impulse—These transients originate outside of the building and are caused by utility company switching, lightning, and so on.

To minimize the damaging results of these transient line surges, service equipment, panels, load centers, feeders, branch-circuits, and individual receptacle outlets can be protected with a device called a *transient voltage surge suppressor* (TVSS), Figure 6-17.

A TVSS contains one or more metal-oxide varistors (MOV) that clamp the transients by absorbing the major portion of the energy (joules) created by the surge, allowing only a small, safe amount of energy to enter the actual connected load.

The MOV clamps the transient in times of less than 1 nanosecond, which is one billionth of a second, and keeps the voltage spike passed through to the connected load to a maximum range of from 400 to 500 peak volts.

Typical TVSS devices for homes are available as an integral part of receptacle outlets that mount in the same wall boxes as do regular receptacles. They look the same as a normal receptacle, and may have an audible alarm that sounds when a MOV has failed. The TVSS device may also have a visual indication, such as an LED (light-emitting diode) that glows continuously until an MOV fails. Then the LED starts to flash on and off, Figure 6-18.

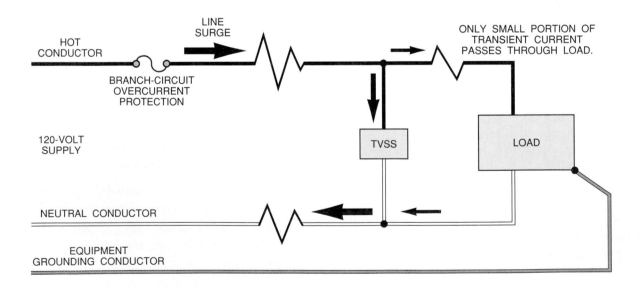

Figure 6-17 Transient voltage surge suppressor absorbs and bypasses (shunts) transient currents around the load. The MOV dissipates the surge in the form of heat.

Figure 6-18   Surge suppressor protector. *Courtesy* of Hubbell, Incorporated.

TVSS devices are available in plug-in strips (Figure 6-19), and as part of desktop computer hardware power direction stands commonly placed under a monitor for plugging in a monitor, printer, scanner, and other peripheral equipment. These minimize the accumulation of the multitude of cords under and behind the equipment.

A TVSS on a branch-circuit will provide surge suppression for all of the receptacles on the same circuit. These surges are classified as Category A low-level surges.

Whole house surge protectors are available that offer surge protection for the entire house. These are generally available at electrical supply houses and are installed at the main electrical panel of a home. They take up the same room as a 2-pole circuit breaker and offer protection for Category B, medium-level surges.

When a whole house surge protector is installed, it is still a good idea to install surge protectors at the plug-in locations to more closely protect against low level surges at the computer or other delicate electronic equipment location.

Noise

"Noise" is recognized as snow on a TV screen or as static on a radio or telephone. Low-level nondamaging transients can also be present. These can be caused by fluorescent lamps and ballasts, other electronic equipment in the area, x-ray equipment, motors running or being switched on and off, improper grounding, and so forth. Although not physically damaging to the electronic equipment, computers can lose memory or perform wrong calculations. Their intended programming can malfunction.

"Noise" comes from *electromagnetic interference* (EMI) and *radio frequency interference* (RFI). EMI is usually caused by ground currents of very low values from motors, utility switching loads, lightning, and so forth, and are transmitted through metal conduits. RFI is noise, like the buzzing heard on a car radio when driving under a high-voltage transmission line. This interference "radiates" through the air from the source and is picked up by the grounding system of the building.

This undesirable noise can be reduced by reducing the number of ground reference points on a system. This can be accomplished by installing an *isolated ground receptacle*.

Figure 6-19   Surge suppressors. *Courtesy* of Hubbell, Incorporated.

ISOLATED GROUND RECEPTACLE

In a standard conventional receptacle outlet, Figure 6-20, the green grounding hexagon screw, the grounding contacts, the yoke (strap), and the metal wall box are all "tied" together to the building's equipment grounding system. Picture then the many receptacles in a building, all having and creating the multiple-ground situation.

In an **isolated ground receptacle**, the green grounding hexagon screw and the grounding contacts of the receptacle are isolated from the metal yoke (strap) of the receptacle and also from the building's equipment grounding system. A separate green insulated grounding conductor is then installed from the green hexagon screw on the receptacle all the way back to the main service disconnect. This separate green grounding conductor does *not* connect to any panels, load centers, or other ground reference points in between. The system grounding is now "clean," and the result is less transient noise (disturbances) transmitted to the connected load.

Equipment grounding conductors and isolated equipment grounding conductors are two different items. The *FPN* to *Section 250-146(d)* calls attention to the fact that metal raceways and outlet boxes may still need to be grounded with an equipment grounding conductor. An additional isolated equipment grounding conductor may then be installed to minimize "noise" as discussed earlier.

When nonmetallic boxes are used in conjunction with an isolated ground receptacle, use nonmetallic faceplates. This is because the metal yoke on the receptacle is not connected to the receptacle's equipment grounding conductor terminal, in which case a metallic faceplate would not be properly grounded.

In this residence, a TVSS and/or isolated ground-type receptacle could be installed in the study/bedroom at the request of the owner. This might be desirable should the owner have a personal computer (PC), a word processor, or other electronic equipment subject to the problems that arise when transients appear on the wiring.

Figure 6-20   (A) Standard conventional receptacle and (B) isolated ground receptacle. *Courtesy* of Hubbell Incorporated.

As more and more electronic equipment is brought onto the marketplace, the subject of transient voltage surge suppression and isolated ground receptacles will be given even more attention than it is today.

To reduce "noise," isolated ground receptacles are permitted by *Section 250-146(d)*. They are identified by an orange triangle imprinted on the face of the receptacle, *Section 410-56(c)*.

Permission is granted in *Section 384-20, Exception* to pass the separate insulated equipment grounding conductor through one or more panels without connecting it to that panel's equipment grounding terminal bar. The *FPN* of this Code reference reminds us, however, that the metal raceway and metal outlet both must still be grounded.

This separate insulated grounding conductor is carried all the way back to the main service, where it is connected to the point where the neutral of the system is bonded to the grounding electrode conductor and the equipment grounding bus. In a residence, this would be the terminal bus provided for the connection of equipment grounding conductors as shown in Figure 28-12.

When a transformer is installed between the equipment and the main service, the isolated grounding conductor is then run to the panelboard connected to the secondary of the transformer where the conductor is connected to the terminal. At the terminal, the grounding electrode conductor, equipment grounding conductor, and neutral conductors are bonded together.

Here are some related U.L. Standards:

- No. 1449: covers transient voltage surge suppressors.
- No. 498 covers attachment plugs and receptacles, including isolated ground receptacles.
- No. 1283 covers noise suppression.
- No. 1363 covers temporary power taps.

REVIEW

1. Explain the operation of a ground-fault circuit interrupter. Why are GFCI devices used? Where are GFCI receptacles required? _____

_____

_____

_____

_____

_____

2. Residential GFCI devices are set to trip a ground-fault current above _____ milliamperes.

3. Where must GFCI receptacles be installed in residential garages? _____

_____

_____

_____

_____

_____

4. The Code requires GFCI protection for certain receptacles in the kitchen. Explain where these are required. _____

_____

_____

5. Is it a Code requirement to install GFCI receptacles in a fully carpeted, finished recreation room in the basement? _____

6. A homeowner calls in an electrical contractor to install a separate circuit in the basement (unfinished) for a freezer. Is a GFCI receptacle required? _____

7. GFCI protection is available as (a) branch-circuit breaker GFCI, (b) feeder circuit breaker GFCI, (c) individual GFCI receptacles, (d) feedthrough GFCI receptacles. In your opinion, for residential use, what type would you install? _____

8. Extremely long circuit runs connected to a GFCI branch-circuit breaker might result in
   a. nuisance tripping of the GFCI
   b. loss of protection
   c. the need to reduce the load on the circuit
   Circle the correct answer.

9. If a person comes in contact with the "hot" and grounded conductors of a two-wire branch-circuit that is protected by a GFCI, will the GFCI trip "off"? Why?

   _____

   _____

   _____

   _____

   _____

10. What might happen if the line and load connections of a feedthrough GFCI receptacle were reversed? _____

   _____

   _____

   _____

   _____

   _____

11. May a GFCI receptacle be installed as a replacement in an old installation where the 2-wire circuit has no equipment grounding conductor? _____

12. What two types of receptacles may be used to replace a defective receptacle in an older home that is wired with knob-and-tube wiring where no equipment grounding means exists in the box? _____

   _____

13. You are asked to replace a receptacle. Upon checking the wiring, you find that the wiring method is conduit and that the wall box is properly grounded. The receptacle is of the older style two-wire type that does not have a grounding terminal. You remove the old receptacle and replace it with (Circle the letter of the correct answers.)
   a. the same type of receptacle as the type being removed.
   b. a receptacle that is of the grounding type.
   c. a GFCI receptacle.

14. What color are the terminals of a standard grounding-type receptacle? _____

   _____

15. What special shape are the grounding terminals of receptacle outlets and other devices? _____

_____

16. Construction sites can be dangerous because of the manner in which extension cords, portable electrical tools, and other electrical equipment are used and abused. To reduce this hazard, *Section 305-6(a)* or the *NEC®* requires that all 125-volt, single-phase, 12-, 20-, and 30-ampere receptacle outlets used for temporary wiring during construction must be _____ protected.

17. In your own words, explain why the Code does not require certain receptacle outlets in kitchens, garages, and basements to be GFCI protected. _____

_____

_____

_____

_____

_____

_____

_____

_____

_____

18. Circuit A1 supplies GFCI receptacles ① and ②. Circuit A2 supplies GFCI receptacles ③ and ④. Receptacles ① and ③ are feedthrough type. Using colored pencils or marking pens, complete all connections.

19. The term TVSS is becoming quite common. What do the letters stand for? _____

_____

_____

20. Transients (surges) on a line can cause spikes or surges of energy that can damage delicate electronic components. A TVSS device contains one or more _____ _____ _____ that bypass and absorb the energy of the transient.

21. Undesirable "noise" on a circuit can cause computers to lock up, lose their memory, and/or cause erratic performance of the computer. This "noise" does not damage the equipment. The two types of this "noise" are EMI and RFI. What do these letters mean?

_____

_____

_____

_____

22. Can TVSS receptacles be installed in standard device boxes? _____

23. Some line transients are not damaging to electronic equipment, but can cause the equipment to operate improperly. The effects of these transients can be minimized by installing _____ _____ _____ .

24. When an isolated ground receptacle is installed, the Code permits carrying the separate equipment grounding conductor back through the raceway, through one or more panels, back to the point in the panel where the grounding electrode conductor, the equipment grounding conductors, and the neutral circuit conductors are connected together.

What sections of the Code cover this topic? _____

_____

_____

25. Briefly explain the operation of an immersion detection circuit interrupter (IDCI).

_____

_____

_____

_____

_____

_____

26. What range of leakage current must trip an IDCI? _____

_____

27. What amount of current flowing through a male human being will cause muscle contractions that will keep him from letting go of the live wire? _____ milliamperes

28. Other than for a few exceptions, the *NEC*® requires that all systems having a voltage of over _____ volts shall be grounded.

29. In an old house, an existing nongrounding non-GFCI-type receptacle in a bathroom needs to be replaced. According to *Section 210-7(d)*, this replacement receptacle shall be (circle the correct letter):
    a. a nongrounding-type receptacle.
    b. a GFCI-type receptacle or be GFCI protected in some other way, such as changing the breaker that supplies the receptacle to a GFCI type.

30. Do all receptacle outlets in kitchens have to be GFCI protected? Explain. _____

    _____

    _____

    _____

UNIT 7

Lighting Fixtures, Ballasts, Lamps

OBJECTIVES

After studying this Unit, the student will be able to

- understand fixture terminology, such as Type IC, Type NON IC, suspended ceiling fixtures, recessed fixtures, and surface-mounted fixtures.
- connect recessed fixtures, both prewired and nonprewired types, according to Code requirements.
- specify insulation clearance requirements.
- discuss the importance of temperature effects while planning recessed fixture installations.
- describe thermal protection for recessed fixtures.
- understand the cautions to be observed when installing outdoor lighting.
- understand Class P and other types of fluorescent ballasts.

TYPES OF LIGHTING FIXTURES

There are literally thousands of different types of lighting fixtures from which to choose to satisfy certain needs, wants, desires, space requirements, and last but not least, price considerations.

As more and more lighting fixtures are imported into the United States, you will see cartons marked with the term "Luminaire," the international term for a lighting fixture.

FLUORESCENT	INCANDESCENT
• Surface	• Surface
• Recessed	• Recessed
• Suspended Ceiling	• Suspended Ceiling

The Code Requirements

Article 410 sets forth the requirements for installing lighting fixtures. The electrician must "Meet Code" when mounting, supporting, grounding, live-parts exposure, insulation clearances, supply conduc-

tor types, maximum lamp wattages, and so forth.

Probably the two biggest contributing factors to fires caused by lighting fixtures are installing lamp wattages that exceed that for which the lighting fixture has been designed, and burying recessed lighting fixtures under thermal insulation when the fixture has not been designed for such an installation.

U.L. tests, lists, and labels lighting fixtures that are in conformance to their standards. Always install lighting fixtures that bear the U.L. label.

In addition to the *NEC*® the U.L. *Electrical Construction Materials Directory* (Green Book) and the U.L. *General Information for Electrical Equipment* (White Book), and manufacturer's catalogs and literature are excellent sources of information about lighting fixtures.

Read the Label

Always carefully read the label on a lighting fixture. Many Code requirements can be met by simply doing what the label says. It will provide such information as:

- For wall mount only.
- Ceiling mount only.
- Maximum lamp wattage.
- Type of lamp.
- Access from above ceiling required.
- Access from behind wall required.
- Suitable for air handling only.
- For chain or hook suspension only.
- Suitable for operation in ambient temperature not exceeding _____°F (°C).
- Suitable for installation in poured concrete.
- For line volt-amperes, multiply lamp wattage by 1.25.
- Suitable for use in suspended ceilings.
- Suitable for use in noninsulated ceilings.
- Suitable for use in insulated ceilings.
- Suitable for damp locations.
- Suitable for wet locations.
- Suitable for use as a raceway.
- Suitable for mounting on low-density cellulose fiberboard.
- For supply connections, use wire rated for at least _____°F (°C).
- Not for use in dwellings.
- Thermally protected.
- Type-IC.
- Type Non-IC.
- Inherently protected.

Surface-Mounted Lighting Fixtures

These are easy to install. It is simply a matter of following the manufacturers' instructions furnished with the fixtures. The lighting fixture is attached to a ceiling outlet box or wall outlet box using fixture studs, hickeys, bar straps, or fixture extensions. Do not exceed the maximum lamp wattage marked on the fixture.

Special attention must be given when installing surface-mounted lighting fixtures on low-density cellulose fiberboard, Figure 7-1. Because of the potential fire hazard, *Section 410-76(b)* states that

Figure 7-1 When mounted on *low-density cellulose fiberboard*, surface-mounted fluorescent fixtures must be marked "Suitable for Surface Mounting on Low-Density Cellulose Fiberboard."

fluorescent lighting fixtures that are surface mounted on this material must be spaced at least 1½ inches (38 mm) from the fiberboard surface unless the fixture is marked "Suitable for Surface Mounting on Low-Density Cellulose Fiberboard."

The *Fine Print Note* to *Section 410-76(b)* explains combustible low-density cellulose fiberboard as sheets, panels, and tiles that have a density of 20 pounds per cubic foot or less that are formed of bonded plant fiber material. It does not include fiberboard that has a density of over 20 pounds per cubic foot or material that has been integrally treated with fire-retarding chemicals to meet specific standards. Solid or laminated wood does not come under the definition of "combustible low-density cellulose fiberboard."

To obtain fire-resistance ratings, most acoustical ceiling panels today are made from mineral wool fibers or fiberglass.

Recessed Lighting Fixtures

The Code requirements for the installation of recessed lighting fixtures are found in *Sections 410-64* through *410-72*. Installing recessed lighting fixtures requires more attention than simple surface-mounted lighting fixtures.

Recessed lighting fixtures as shown in Figure 7-2 have an inherent heat problem. They must be suitable for the application and must be installed properly.

Figure 7-2   Typical recessed fixture.

It is absolutely essential for the electrician to know early in the roughing-in stages of wiring a house what types of lighting fixtures are to be installed. This is particularly true for recessed-type fixtures. To ensure that such factors as location, proper and adequate framing, and possible obstructions have been taken into consideration, the electrician must work closely with the general building contractor, carpenter, plumber, heating contractor, the thermal insulation installer, and the other building trades.

Figure 7-3(A) shows a Type Non-IC (insulated ceiling) recessed lighting fixture where a minimum 3-inch (76 mm) clearance between the sides of the fixture and thermal insulation is needed. Above the fixture, thermal insulation must be installed so as not to entrap heat in the cavity. How far should the thermal insulation be above the fixture? It is a judgment call. Three inches (76 mm) should suffice. Check the manufacturer's instructions.

Figure 7-3(B) shows a Type IC recessed lighting fixture that may be completely buried in thermal insulation.

Figure 7-3(C) is a sketch that summarizes the minimum clearances for a Type Non-IC recessed lighting fixture that is not listed for direct burial in thermal insulation, *Section 410-66(b)*.

Recessed lighting fixtures are available for both new work and remodel work. Remodel work recessed lighting fixtures can be installed from below an existing ceiling by cutting a hole in the ceiling,

Figure 7-3   (A) is a Type Non-IC recessed lighting fixture that requires clearance between the fixture and the thermal insulation. (B) is a Type IC recessed lighting fixture that may be completely covered with thermal insulation. (C) shows the required clearances from thermal insulation for a Type Non-IC fixture. See *Section 410-66*.

bringing power to the fixture, making the electrical connections in the integral junction box on the fixture, then installing the housing through the hole.

Energy Efficient Housings

Some recessed fixtures have energy efficient housings that restrict the airflow (leakage) through the fixture to less than 2 cubic feet of air per minute. These fixtures meet energy codes that have been adopted by some states. You should check with your local electrical inspector and/or building official to see if such laws are applicable in your area for one- and two-family dwellings.

Thermal Protection

To protect against the hazards of overheating, U.L. and the *NEC®* in *Section 410-65(c)* require that recessed lighting fixtures be equipped with an integral thermal protector, Figure 7-4. These devices

OPEN   CLOSED

Figure 7-4   Recessed fixture thermal cutout.

will cycle on and off repeatedly until the heat problem has been resolved. There are two exceptions to this requirement. Thermal protection is not required for recessed incandescent fixtures designed for installation directly in a concrete pour, or for recessed incandescent fixtures that are constructed so as to not exceed "temperature performance characteristics" equivalent to thermally protected fixtures. These fixtures are so identified.

What the Junction Box on the Recessed Lighting Fixture Housing Is for

Most recessed lighting fixtures come equipped with a junction box that is an integral part of the fixture housing. This is clearly shown in Figures 7-2 and 7-5. For most residential-type recessed lighting fixture installations, nonmetallic-sheathed cable or armored cable can be run directly into this junction box. These cables have conductors rated 90°C. Check the marking on the lighting fixture for supply conductor temperature limitations.

Figure 7-6 shows the clearances normally needed between recessed lighting fixtures and combustible framing members. This figure also summarizes the *NEC®* requirements for connecting the supply conductors to a recessed lighting fixture. Note in one case that the branch-circuit wiring is brought directly into the integral junction box on the fixture. In the other case, a "fixture whip" is run

Figure 7-5   Roughing-in box of a recessed fixture with mounting brackets and junction box.

Figure 7-6   Requirements for installing recessed lighting fixtures.

from an adjacent junction box to the lighting fixture. In some instances, the fixture comes with a flexible metal "fixture whip" and in other situations, the "fixture whip" is supplied by the electrician.

▶ *Section 410-11* states that "branch-circuit wiring, other than 2-wire or multiwire branch-circuits supplying power to fixtures connected together, shall not be passed through an outlet box that is an integral part of a fixture unless the fixture is identified for through-wiring."◀

▶ *Section 410-31* states that "fixtures shall not be used as a raceway for circuit conductors unless listed and identified for use as a raceway."◀

Recessed lighting fixtures that are marked "Identified for through-wiring" will have a label that looks something like this:

Maximum of No. ___ AWG through branch-circuit conductors suitable for at least ____°C (___°F) permitted in a box (____ in ____ out).

These fixtures have been tested for the added heat generated by the additional branch-circuit conductors.

Figure 7-7 shows three recessed lighting fixtures, each with an integral outlet box. If these fixtures are marked "Identified for through-wiring," then it would be permitted to run conductors through the outlet boxes *in addition* to the conductors that supply the fixtures. If the "Identified for through-wiring" marking is not found on the fixture, then it is a Code violation to run any conductors through the box, other than the conductors that supply the fixture itself.

Some electricians prefer to run the branch-circuit wiring to the junction box on the recessed fixture, then drop the switch loop to the switch box location. Others prefer to run the branch-circuit to the switch location, then to the junction box on the fixture. Either way is acceptable according to the *NEC*® and U.L. Standards.

Figure 7-7 ►The only conductors permitted to run into or through the junction boxes on the recessed lighting fixtures are those that supply the lighting fixtures. Conductors other than those that supply the lighting fixtures are permitted to run through the outlet boxes if the fixture is marked "Identified For Through-Wiring." See *Sections 410-11* and *410-31.*◄

Recessed Lighting Fixture Trims

Recessed incandescent lighting fixtures are listed by Underwriters Laboratories with specific trims. Fixture/trim combinations are marked on the label. Do not use trims that are not listed for use with the particular recessed lighting fixture. Installing a trim not "Listed" for use with a specific recessed lighting fixture can result in overheating and possible on-off cycling of the fixture's thermal protective device. Mismatching fixtures and trims is a violation of *Section 110-3(b)* of the *National Electrical Code,*® which states that "Listed or labeled equipment shall be used or installed in accordance with any instructions included in the listing or labeling."

Suspended Ceiling Lay-In Lighting Fixtures

Figure 7-8 illustrates lighting fixtures installed in a typical suspended ceiling.

Figure 7-8   Suspended ceiling fixtures.

The recreation room of this residence has a "dropped" suspended acoustical paneled ceiling. Fixtures installed in a "dropped" ceiling must bear a label stating "Suspended Ceiling Fixture." These types of fixtures are not classified as recessed fixtures because in most cases, there is a great deal of open space above and around them.

Support of Suspended Ceiling Lighting Fixtures

Section 410-16(c) requires that when framing members of a suspended ceiling grid are used to support lighting fixtures, all framing members must be securely fastened together and to the building structure. It is the responsibility of the installer of the ceiling grid to do this.

All lay-in suspended ceiling lighting fixtures must be securely fastened to the ceiling grid members with bolts, screws, or rivets. ►"Listed" clips supplied by the manufacturer of the fixture are also permitted to be used.◄ Clips that meet this criteria are shown in Figure 17-1.

The logic behind all of the "securely fastening" requirements is that in the event of a major problem such as an earthquake or fire, the lighting fixtures will not fall down and injure someone.

Connecting Suspended Ceiling Lighting Fixtures

The most common way to connect suspended ceiling lay-in lighting fixtures is to complete all of the wiring above the suspended ceiling using standard wiring methods such as electrical metallic tubing, nonmetallic-sheathed cable, or armored cable.

Then, from outlet boxes strategically placed above the ceiling near the intended location of the lay-in lighting fixtures, a flexible connection is made between the outlet boxes and the lighting fixtures. In the electrical trade, these flexible connections are referred to as "fixture whips." ▶*Section 410-67(c)* permits the conductors in "fixture whips" to be in a suitable raceway, Type AC, or Type MC cables at least 18 inches (450 mm) but not more than 6 feet (1.83 m) in length.◀

A "fixture whip" might be:

- a 6-foot (1.83 m) length of ⅜-inch ▶listed◀ flexible metal conduit (FMC), using conductors suitable for the temperature requirement as stated on the fixture's label. Usually 90°C conductors satisfy the temperature requirement. This metal flex is acceptable as the grounding means when it is not longer than 6 feet (1.83 m), the overcurrent protection does not exceed 20 amperes, and the fittings are "listed" as being acceptable for grounding purposes, *Sections 250-118(5)* and *350-14*. See Figure 7-9. The requirements for securing the flex every 4½ feet (1.37 m) and within 12 inches (305 mm) of the fixture are waived, *Section 350-18, Exception No. 3.*

- a 6-foot (1.83 m) length of armored cable (AC). The metal armor and the internal bonding wire provide the necessary equipment grounding for the lighting fixture, *Sections 250-118(9)* and *333-21*. The requirements for securing the cable every 4½ feet (1.37 m) and within 12 inches (305 mm) of the fixture are waived, *Section 333-7(3).*

- a 6-foot (1.83 m) length of ⅜-inch liquidtight flexible nonmetallic conduit (LFNC), *Section 351-24(b)(2)*. A separate equipment grounding conductor must be installed in this flex to provide the necessary equipment grounding for the lighting fixture. The 3-foot (914 mm) and the 12-inch (305 mm) supporting requirements are waived, *Section 351-27(6).*

- a 6-foot (1.83 m) length of ⅜-inch liquidtight flexible metallic conduit (LFMC), *Section 350-10(a)(2)(c)*. This metal flex is acceptable as

FLEXIBLE METAL CONDUIT AND LIQUIDTIGHT FLEXIBLE METAL CONDUIT SUITABLE AS EQUIPMENT GROUNDING MEANS IF BRANCH-CIRCUIT NOT OVER 6 FEET (1.83 m) LONG BRANCH-CIRCUIT OVERCURRENT DEVICE IS NOT OVER 20 AMPERES, AND FITTINGS ARE LISTED FOR GROUNDING, *SECTIONS 250-118(6)* AND *250-118(7)*.

TIE WIRES

THIS SPACE ACCESSIBLE FROM BELOW CEILING BY LIFTING OUT PANELS.

SUSPENDED CEILING FIXTURE

OUTLET BOX

SUSPENDED CEILING

CONDUCTORS IN FLEXIBLE CONNECTION (FIXTURE WHIP) MUST BE SUITABLE FOR TEMPERATURE REQUIREMENTS AS SPECIFIED ON FIXTURE LABELING, *SECTION 410-67.*

▶3/8-INCH FLEXIBLE METAL CONDUIT OR ARMORED CABLE NOT LESS THAN 18 INCHES (450 mm) LONG NOR LONGER THAN 6 FEET (1.83 m), *SECTION 410-67(C).*◀

Figure 7-9   A suspended ceiling fixture connected with a flexible "fixture whip." Fixture whips may be ⅜-inch flexible metal conduit, *Section 350-18, Exception No. 3*, armored cable, *Section 333-7(b)(3)*, liquidtight flexible metal conduit, *Section 351-8, Exception No. 3*, liquidtight flexible nonmetallic conduit if part of a listed assembly, *Section 351-24(b)(2)*, ▶and nonmetallic-sheathed cable, *Section 336-18, Exception No. 3*. The maximum length of a fixture whip is 6 feet (1.83 m) except for nonmetallic-sheathed cable, which has a maximum permitted length of 4½ feet.◀

the grounding means when it is not longer than 6 feet (1.83 m), the overcurrent protection does not exceed 20 amperes, and the fittings are "listed" as being acceptable for grounding purposes, *Sections 351-9, 351-8(6), 250-118(6), and 250-128(7)*. The 4½-foot (1.37 m) and the 12-inch (305 mm) supporting requirements are waived, *Section 351-8, Exception No. 3*.

- a nonmetallic-sheathed cable connection to the fixture may also be used. Grounding of the lighting fixture is accomplished with the equipment grounding conductor in the cable. ▶The requirements for securing the cable every 4½ feet (1.37 m) and within 12 inches (305 mm) of the fixture are waived, *Section 336-18.*◀

- a 6-foot (1.83 m) electrical nonmetallic tubing (ENT). A separate equipment grounding conductor must be installed in this flex to provide the necessary equipment grounding for the lighting fixture. The 3-foot (914 mm) support requirement is waived, *Section 331-11, Exception*.

In Unit 17, Figures 17-1 and 17-2 have additional information regarding recessed lighting fixture installations as are installed in the recreation room of this residence.

Lighting Fixtures in Closets

There are special requirements for installing lighting fixtures in closets. This is covered in Unit 8.

Lighting Fixtures in Bathrooms

Because of the electrical shock hazards associated with electricity and water, there are special restrictions for lighting fixtures installed in bathrooms. This is covered in Unit 10.

FLUORESCENT BALLASTS AND LAMPS, INCANDESCENT LAMPS

The following text introduces the basics of fluorescent ballasts and lamps. This topic is extremely complex. The student is urged to send for the technical brochures, bulletins, and booklets mentioned in the Instructor's Guide. These brochures contain a wealth of information on ballasts, lamps, wiring diagrams, troubleshooting hints, application suggestions, and so on.

Fluorescent Ballasts

An incandescent lamp contains a filament that has a specific "hot" resistance value when operating at rated voltage. The current flowing through the lamp filament is governed by Ohm's law, where Amperes = Volts/Ohms. When energized, the lamp will provide the light output for which the lamp was designed.

Fluorescent lamps do not have a filament running from end-to-end. Instead, they have a filament at each end. The filaments are connected to a ballast, Figure 7-10. A ballast is needed to control the energy, control voltage to heat the filaments, control voltage across the lamp to start the arc within the

Figure 7-10 Simple series circuit.

tube (end-to-end) needed to ionize the gas and vaporize the droplets of mercury inside the lamp, and to "limit" the current flowing through the lamp. Without a ballast, the fluorescent lamp would probably not start. But, if it did, then the internal resistance (impedance) would get lower and lower, and the current flowing through the lamp would get higher and higher, and in a very short time, the lamp would be destroyed.

Ballast Types (Circuitry)

Preheat. Preheat ballasts are connected in a simple series circuit configuration. They are easily identified because the fixture will have a "starter." One type of starter is automatic, and looks like a small roll of life-savers with two "buttons" on one end. When the fixture is turned on, the starter is in series with the lamp's filaments and the ballast. The lamp filaments glow for a few seconds. The starter's bimetal element heats up and "warps," causing the starter contacts to automatically open the series circuit. At the instant the starter contacts open, an arc is established inside the lamp, end-to-end, and the lamp starts.

Another type of starter is a manual ON/OFF switch that has a momentary "make" position just beyond the ON position. When you push the switch on, and hold it there for a few seconds, the lamp filaments glow. When the switch is released, the start contacts open, an arc is initiated within the lamp— and the lamp lights up.

The fluorescent lamps for use with preheat ballasts have two pins (bi-pin) on each end of the lamp.

Preheat lamps and ballasts are not used for dimming applications.

Rapid Start. Rapid start lamps do not require a starter. This is probably the most common type used today. The lamps start in less than 1 second. The filaments remain energized with a low-voltage circuit from the ballast. The lamps start up much faster than the preheat type. To ensure proper starting, ballast manufacturers recommend that there be a grounded reflector within ½ inch (12.7 mm) to the lamp and running the full length of the lamp, that the ballast be grounded, and that the supply circuit originate from a grounded system. Residential electrical systems in the United States are grounded systems and operate at 240 volts between phases and 120 volts to

ground from either phase. Commercial installations might be 208/120 volts or 480/277 volts.

Fluorescent lamps for use with rapid start ballast have two pins (bi-pin) on each end of the lamp.

A *modified* rapid start ballast operates just like a standard rapid start ballast, but has a delay feature that shuts off or reduces the filament heat after the lamps have started.

Another version of a rapid start ballast is the *trigger start* ballast. Trigger start ballasts are used with preheat fluorescent lamps up to 32 watts. The trigger start ballast was the forerunner of the rapid start ballast.

Rapid start lamps can be dimmed using special dimming ballast. See Unit 13.

Instant Start. Instant start lamps do not require a starter. These ballasts provide a high voltage "kick" to start the lamp instantly. They require special fluorescent lamps that do not require preheating of the lamp filaments. Because instant start fluorescent lamps are started by brute force, they have a shorter life (as much as 40% less) than rapid start lamps. Most instant start fluorescent lamps have a single pin on each end of the lamp, although some have two pins (bi-pin), in which case a short-circuiting arrangement is designed into the lampholder.

Instant start ballasts/lamps cannot be dimmed.

Mismatching of Fluorescent Lamps and Ballasts

Be sure to use the proper lamp in a given ballast. Mismatching a lamp and ballast may result in poor starting, poor performance, as well as shortened lamp and/or ballast life. The manufacturer's ballast and/or lamp warranty may be null and void.

T8 lamps are bi-pin as they have two-pins on each end. They are designed to be used interchangeably on magnetic or electronic rapid start ballasts, or electronic instant start ballasts. Lamp life is reduced slightly when used with an instant start ballast.

Ballast Types (Applications)

The three types of ballast discussed previously are available in:

High Efficiency. High efficiency ballasts are highly recommended because they save energy. Energy-saving ballasts can be of the *electronic type* or *magnetic type* (core & coil). If they are of the magnetic type, then they will probably have steel laminations and copper wire instead of iron laminations and aluminum wire. High efficiency, energy-saving ballasts have a high power factor rating. Power factor is the ratio of watts to volt-amperes.

For example, a 40-watt fluorescent lamp is installed in a fixture that has a ballast marked 42 VA. The power factor of the complete fixture is:

$$\text{Power Factor} = \frac{\text{Watts}}{\text{Volt-amperes}} = \frac{40}{42} = 0.95\ (95\%)$$

Energy-saving ballasts cost more to buy initially than low power factor ballasts, but over time, the savings in energy consumption far exceeds the higher initial costs. A standard fluorescent ballast might consume 14 to 16 watts. An energy-efficient ballast consumes 8 to 10 watts. Combined with an energy-efficient fluorescent lamp that uses 32 or 34 watts instead of 40 watts, there is a considerable energy savings. You are buying lighting—not heat. Always install fixtures that have a high power factor rating. This data is found on the nameplate of the ballast. Today, electronic ballasts are the most energy efficient when compared to magnetic ballasts containing coils, capacitors, and transformers.

The federal government's National Appliance Energy Conservation Amendment of 1988, Public Law 100-357, prohibited manufacturers from producing certain types of commonly used low power factor (less than 90%) ballasts as of January 1990. Among these were ballasts for single and two-lamp ballasts for 4-foot T12 rapid start lamps. Fixture manufacturers had to start using high power factor ballasts (90% or greater) as of April 1991. Ballasts that "meet or exceed" the requirements of the federal law are marked with an "E" in a circle.

Dimming ballasts and ballasts designed specifically for residential use were exempted.

Following the ballast legislation, the National Energy Policy Act of 1992 enacted restrictions on lamps. Lamp manufacturers could no longer produce certain 8-foot fluorescent lamps as of April 1994. In October 1995, the common 4-foot T12 linear and U-tube fluorescent lamps, and the R30, R40, and PAR38 incandescent lamps were eliminated. These were replaced by energy-efficient lamps that are direct replacements for the discontinued lamps. For example, the typical PAR38 reflector flood lamp could be replaced with a 60PAR/HIR incandescent lamp. An F40T12WW fluorescent lamp could be replaced with an F40SP30RS/WM fluorescent lamp that consumes 34 watts instead of 40 watts for just about the same amount of light (lumen output).

It has been estimated that the total electric bill savings across the country will exceed $250 billion over the next 15 years. Figure 7-11 clearly illustrates how much energy saving is possible when using high-efficiency (high power factor) ballasts. It compares a standard F40T12 fluorescent lamp when used with different types of ballasts.

Watts versus Amperes

It is possible to overload a branch-circuit if too many low power factor ballasts are connected to the circuit. In this residence, the use of low cost, low power factor ballasts for the recessed fluorescent lighting fixtures in the recreation room could result in the need for two branch-circuits instead of one—as well as higher energy (bigger light bills) consumption. High-efficiency, high power factor ballasts should always be used.

BALLAST	LINE CURRENT	LINE VOLTAGE	LINE VOLT/AMPERES	LAMP WATTAGE	LINE POWER FACTOR
No. 1	0.35	120	42	40	0.95 (95%)
No. 2	0.45	120	54	40	0.74 (74%)
No. 3	0.55	120	66	40	0.61 (61%)
No. 4	0.85	120	102	40	0.39 (39%)

Figure 7-11 Table showing various line currents, volt-amperes, wattages, and overall power factor for various single-lamp fluorescent ballasts. These values were taken from actual ballast manufacturers' data. For two-lamp ballasts, double the values in this table.

Fluorescent loads are a good example of why volt-amperes, rather than watts, must be considered in order to find the current draw on the circuit.

Let's make a comparison of possibilities for the recreation room lighting where there are six recessed fluorescent fixtures, each containing two 2-lamp ballasts. The nameplate on each ballast indicates a line current rating of 0.70 ampere at 120 volts, which is 84 volt-amperes. The lamps are marked 40-watts.

The total current in amperes is:

$$12 \times 0.70 = 8.4 \text{ amperes}$$

The total power in volt-amperes is:

$$8.4 \times 120 = 1008 \text{ volt-amperes}$$

The total lamp wattage is:

$$40 \times 24 = 960 \text{ watts}$$

If we had calculated the load for these fixtures based upon wattage, the result would have been:

$$\frac{W}{E} = \frac{960}{120} = 8 \text{ amperes}$$

This result is not much different than the 8.4 amperes using data from the ballast nameplate.

Had we used low-cost, low-efficiency ballasts like No. 4 shown in Figure 7-11, the result would be entirely different. The ballasts in Figure 7-11 are single-lamp ballasts. We double the current values for a good approximation of a similar 2-lamp ballast. Using the low-power factor ballast, we find that:

The total current in amperes is:

$$12 \times 0.85 \times 2 = 20.4 \text{ amperes}$$

The total power in volt-amperes is:

$$20.4 \times 120 = 2,448 \text{ volt-amperes}$$

or

$$12 \times 102 \times 2 = 2,448 \text{ volt-amperes}$$

The total lamp wattage is:

$$24 \times 40 = 960 \text{ watts}$$

As previously shown, if we use lamp wattage to calculate the current draw:

$$I = \frac{W}{E} = \frac{960}{120} = 8 \text{ amperes}$$

But we really have a current draw of 20.4 amperes, which would overload the 15-ampere branch-circuit B12. In fact, a load of 20.4 amperes is too much for a 20-ampere branch-circuit. Two 15-ampere branch-circuits would have been needed to hook-up the recreation room recessed fluorescent fixtures. Remember, *do not* load any circuit to more than 80% of the branch-circuit's rating. That is 16 amperes for a 20-ampere branch-circuit and 12 amperes for a 15-ampere branch-circuit. And that is the maximum.

It is apparent that the *installed* cost using cheap fixtures is greater and more complicated than if high-quality fixtures using energy-efficient ballasts and lamps had been used. However, after the initial installation, the energy savings adds up significantly.

Dimming Ballast. Dimming ballasts are discussed in Unit 13. They are used where the dimming of rapid start fluorescent lamps is desired. Instant start lamps and ballasts cannot be used for dimming

Figure 7-12 In-line fuseholder for ballast added protection. *Photo courtesy* of Cooper Bussmann, Inc.

applications. Dimming ballasts are more costly and are less energy efficient than standard ballasts.

Noise Levels. All ballasts will hum, but some do so more than others. Ballasts containing transformers and choke coils hum when the metal laminations vibrate because of the alternating current reversals. This hum can be magnified by the fixture and/or the surface the fixture is mounted on. Electronic ballasts have little if any hum.

Do not insert spacers, washers, or shims between a ballast and the fixture in attempts to make the ballast more quiet. This will cause the ballast to run much hotter, and could result in shortened ballast life, and possible fire hazard. Instead, replace the noisy ballast with a quiet, sound-rated ballast. Sometimes checking and tightening the many nuts, bolts, and screws of the fixture will solve the problem.

Ballasts are sound rated. Again, cost is a consideration. Ballasts are marked with letters *A* through *F*. *A* is the quietest, and *F* is the noisiest.

> *A*—homes, churches, libraries, hospital rooms.
>
> *B*—classrooms, private offices.
>
> *C*—open office areas.
>
> *D*—retail stores.
>
> *E & F*—outdoors and in factories.

Lamp Diameters. There is a simple way to determine the diameter of a lamp (light bulb or tube) at its widest measurement. The industry uses an "eighths of an inch rule." For example, a T5 fluorescent lamp is $5/8$ inch in diameter. A T8 fluorescent lamp is $8/8$ inches, or 1 inch in diameter. A T12 lamp is $12/8$ inches, or $1\frac{1}{2}$ inches in diameter. Incandescent lamps follow the same system. A PAR38 lamp is $38/8$ inches, or $4\frac{3}{4}$ inches in diameter. An R30 lamp is $30/8$ inches, or $3\frac{3}{4}$ inches in diameter.

Other Considerations

- Because heat trapped by insulation around and on top of a fixture can shorten the life of a ballast, always follow the manufacturer's installation requirements and the requirements found in *Article 410*, discussed earlier in this Unit. A 10°C temperature rise above the ballast's rated temperature (90°C) can reduce the ballast's life by one-half. This "half-life rule" is true for conductors, motors, transformers, and other electrical equipment.

- Fluorescent lamps that are intensely blackened on both ends should be replaced. Operating a two lamp ballast with only one lamp working will cause the ballast to run hot, and will shorten the life of the ballast. Severe blackening of one end of the lamp can also ruin the ballast. A flickering lamp should be replaced.

- Poor starting of a fluorescent lamp can be caused by poor contact in the lamp holder, poor grounding, excessive moisture on the outside of the tube, cold temperature (approximately 50°F), as well as dirt, dust, and grime on the lamp.

- Most ballasts will operate satisfactorily within a range of +5% to −7% of their rated voltage. The higher quality CBM certified ballasts will operate satisfactorily within a range of ±10%.

- Most T12 fluorescent lamps sold today are of the energy-efficient type. Generally, ballasts should not be used with energy saving lamps unless so marked. Check the markings on the ballast, on the lamp carton, or in the instructional literature available at the point-of-sale of the ballast or lamp.

- Lamp life generally is rated in "X" number of hours of operation based on 3 hours per start. Frequent switching results in shortened expected lamp life. Inversely, leaving the lamps on for long periods of time extends the expected lamp life. Vibration, rough handling, cleaning, etc., shortens lamp life.

- Be careful what type of conductors you use to connect fluorescent fixtures. Branch-circuit conductors within 3 inches (76 mm) of a ballast must have an insulation temperature rating of not less than 90°C (194°F), *Section 410-33*. Type THHN conductors, the conductors in non-metallic-sheathed cable, and in Type ACTHH armored cable are rated 90°C (194°F).

LIGHTING FIXTURE VOLTAGE
Voltage Limitations

The maximum voltage allowed for residential lighting fixtures is 120 volts between conductors per *Section 210-6(a)*.

120-VOLT SUPPLY   TRANSFORMER   NEON TUBING

OPEN CIRCUIT SECONDARY VOLTAGE: 15,000 VOLTS

Figure 7-13 It is *not* permitted to install lighting in or on residences where the open-circuit voltage is over 1,000 volts. This is a violation of *Section 410-80(b).*

Section 410-80(b) makes a further restriction stating that for dwelling occupancies, no lighting equipment shall be used if it operates with an open-circuit voltage over 1000 volts, Figure 7-13. This section more or less focuses on the desire to use neon lighting for more decorative lighting purposes.

Incandescent Lamp Life at Different Voltages

Operating an incandescent lamp at other than rated voltage will result in longer—or shorter—lamp life. The following formula can predict the approximate expected lamp life at different voltages. For example, assume that a gas-filled 120-volt incandescent lamp has a published lamp life of 1,000 hours. The calculations show the expected lamp life of this lamp when operated at 130 volts, and the expected lamp life when operated at 110 volts.

- At 130 volts:

$$\text{Expected Life} = \left(\frac{\text{Rated Volts}}{\text{Actual Volts}}\right)^{13.1} \times \text{Rated Life}$$

$$= \left(\frac{120}{130}\right)^{13.1} \times 1,000$$

$$= 0.350 \times 1,000$$

$$= 350 \text{ hours}$$

- At 110 volts:

$$\text{Expected Life} = \left(\frac{\text{Rated Volts}}{\text{Actual Volts}}\right)^{13.1} \times \text{Rated Life}$$

$$= \left(\frac{120}{110}\right)^{13.1} \times 1,000$$

$$= 3.126 \times 1,000$$

$$= 3,126 \text{ hours}$$

Incandescent Lamp Lumen Output at Different Voltages

Notice that operating an incandescent lamp at a voltage lower than the lamp's voltage rating will result in longer lamp life. A strong case might be made to install 130-volt lamps, particularly where they are hard to reach, such as flood lights located high up. However, the lumen output is reduced. For example, calculate the approximate lumen output of a 100-watt, 130-volt incandescent lamp that has an initial lumen output of 1,750 lumens at rated voltage. The lamp is to be operated at 120 volts.

$$\frac{\text{Expected}}{\text{Lumens}} = \left(\frac{\text{Actual Volts}}{\text{Rated Volts}}\right)^{3.38} \times \text{Rated Lumens}$$

$$= \left(\frac{120}{130}\right)^{3.38} \times 1,750$$

$$= 0.763 \times 1,750$$

$$= 1,335 \text{ Lumens}$$

Formulas such as these are found in lamp manufacturer's catalogs. These formulas are useful in determining the affect of applied voltage to lamp wattage, line current, lumen output, lumens per watt, and lamp life. A calculator that has a y^x power function is needed to solve these equations.

LAMP EFFICIENCY

Lighting efficiency is measured in *Lumens Per Watt*. A lumen is a unit of light measurement. One lumen falling on one square foot of surface produces one *foot-candle*. The following is a very brief comparison of different types of lamps used in residences.

The more lumens per watt, the greater the efficiency of the lamp, Figure 7-14.

Class P Ballasts

Section 410-73(e) states that all fluorescent ballasts installed indoors (except simple reactance-type ballasts), both for new and replacement installations, must have thermal protection built into the ballast by

the manufacturer of the ballast, Figure 7-15. Ballasts provided with built-in thermal protection are listed by the U.L. as Class P ballasts. Under normal conditions, the Class P ballast has a case temperature not exceeding 90°C. The thermal protector must open within two hours when the case temperature reaches 110°C.

Some Class P ballasts also have a nonresetting fuse integral with the capacitor to protect against capacitor leakage and violent rupture. The Class P ballast's internal thermal protector will disconnect the ballast from the circuit in the event of excessive temperature. Excessive temperatures can be caused by abnormal voltage and improper installation, such as being covered with insulation.

The reason for thermal protection is to reduce the hazards of possible fire due to an overheated ballast when the ballast becomes shorted, grounded, covered with insulation, lacking in air circulation,

	LUMENS/WATT	DIMMING	COLOR AND APPLICATION	LIFE (Approx. Hours)	TYPICAL SHAPES
Incandescent	18	Yes	Warm and Natural. Great for general lighting.	500, 1000, 1500 Depends on type of lamp.	Standard, spots, floods, decorative, tubes, globes, PAR (similar to standard spots and floods but much stronger). Use rough service where there is vibration, like garage door openers and ceiling fans.
Halogen	22	Yes	Brilliant white. Excellent for accent and task lighting. Are filled with halogen gas and have an inner lamp allowing the filament to run hotter (whiter).	2000 to 4000	PAR spots and floods, mini-reflector spots and floods, double and single end quartz.
Fluorescent	105	Yes, but only 40-watt rapid start lamps using special dimming ballast. See Unit 13.	Warm and deluxe warm white, cool and deluxe cool white plus many other shades of white. Great for general lighting, like the Recreation Room in this residence.	6000 to 24000. Compact fluorescents can last 9 to 13X longer than incandescent lamps. Compact fluorescents are complete units having both ballast and lamp. When lamp burns out, ballast is also replaced. Compacts with screw-in bases can replace most incandescents.	Straight tubular, U-shaped tubular, circular tubular, compacts. Compact screw-in can have twin, double twin, and triple twin tubes. This refers to the number of "loops" in the lamp. Compacts can save $20 to $30 over its lifetime when compared to an equivalent incandescent lamp. Compacts last longer than incandescent lamps.

Figure 7-14 Comparisons of various lamps' characteristics.

Figure 7-15 An electronic Class P ballast, thermally protected as required in *Section 410-73(e)*. Electronic ballasts provide improved performance in fluorescent lighting installations. Electronic ballast lighting systems are 25–40% more energy efficient than conventional magnetic (core & coil) ballast fluorescent systems. *Courtesy* of Motorola Lighting.

and so on. Ballast failure has been a common cause of electrical fires.

Additional backup protection can be provided by an in-line fuseholder connected in series with the black line lead and the black ballast lead, Figure 7-12. The in-line fuseholder acts as a disconnect for the ballast. Should a ballast fail (short out), the individual fuse opens, thus isolating the faulty ballast but not affecting the rest of the circuit. When the ballast is to be replaced, the fuse is removed to disconnect the ballast, but the entire circuit need not be turned off. In general, the ballast manufacturer will recommend the correct type and ampere rating of the fuse to be used for the particular ballast.

See Figure 7-16 for some important requirements for fluorescent and incandescent lighting fixtures.

Incandescent Tungsten Filament Lamp Load Inrush Currents

On rare occasions, a branch-circuit breaker may nuisance trip where a single dimmer controls a substantial incandescent tungsten filament lamp load. Why?

The cold resistance of a tungsten filament is very low. When the lamp is connected to the proper voltage, the filament resistance increases very quickly — within $1/240$ second or one-quarter of an electrical cycle after the circuit is energized. This momentary inrush current might be 10 to 20 times greater than the normal operating current of the lamp.

T-rated switches have this momentary high inrush current taken into consideration. T-rated switches were discussed in Unit 5. Branch-circuit conductors are not affected by this momentary high surge of current. However, a circuit breaker might nuisance trip as a result of this momentary high inrush of current.

In cases where a dimmer is set at a low (dim) setting and the circuit is turned on with a separate or integral switch, the lower voltage will heat up the lamp filament slower, and the inrush current will last slightly longer.

Nuisance tripping problems will generally not occur in residential installations because the lamp loads controlled by a dimmer are usually small. But nuisance tripping can happen in commercial dimming applications where the connected lamp load controlled by a single dimmer is quite large. Manufacturers can furnish high magnetic trip circuit breakers if the problem happens repeatedly.

Branch-circuit fuses are not affected by this momentary inrush current phenomena because they have sufficient time-delay to override the inrush.

Outdoor Lighting

Before installing outdoor lighting, check with the local electrical inspector and/or building official to find out if there are any restrictions regarding outdoor lighting.

A virtually unlimited array of outdoor lighting

FIXTURES

- SEE *ARTICLE 410, NEC*[®]
- WILL BE MARKED WITH INSULATION TEMPERATURE RATING REQUIRED FOR SUPPLY CONDUCTORS IF OVER 60°C.
- READ THE LABEL AND INSTRUCTIONS FURNISHED WITH ALL FIXTURES.

FLUORESCENT

- IF BRANCH-CIRCUIT CONDUCTORS ARE WITHIN 3 INCHES (76.2 mm) OF BALLAST, USE 90°C. CONDUCTORS, *SECTION 410-33, NEC.*[®]
- DO NOT USE AS RACEWAY UNLESS PERMITTED, *SECTION 410-31, NEC.*[®]
- ALL FLUORESCENT BALLASTS INSTALLED INDOORS MUST BE CLASS P TYPE, *SECTION 410-73(e), NEC.*[®]

SURFACE

- NOT SUITABLE FOR MOUNTING WITHIN 1½ INCHES (38.1 mm) OF THE SURFACE OF LOW-DENSITY CEILING FIBERBOARD UNLESS MARKED "SUITABLE FOR SURFACE MOUNTING ON LOW-DENSITY CELLULOSE FIBERBOARD." SEE *SECTION 410-76(b), NEC.*[®]
- READ THE LABEL ON THE FIXTURE FOR SPECIAL REQUIREMENTS OR LIMITATIONS.
- SEE FIGURE 7-1.

RECESSED

- SUITABLE FOR RECESSED INSTALLATION.
- MAY BE MOUNTED IN SUSPENDED CEILINGS IF PROVIDED WITH APPROPRIATE HARDWARE FOR MOUNTING TO OR IN SUSPENDED CEILINGS.
- READ FIXTURE LABEL FOR SPECIAL REQUIREMENTS OR LIMITATIONS.
- SEE FIGURES 7-3, 7-6, 7-7.

SUSPENDED

- FOR INSTALLATION IN A SUSPENDED GRID ONLY WHERE THE LAY-IN TILES ARE NOT FASTENED IN PLACE, AND WHERE THE TIE WIRES, T BARS, CEILING TILES AND OTHER COMPONENTS DIRECTLY ASSOCIATED WITH THE GRID ARE NOT PART OF THE BUILDING STRUCTURE.
- THE FIXTURES ARE INTENDED TO BE MOUNTED IN THE CEILING OPENINGS.
- READ FIXTURE LABEL OR INSTRUCTIONS FURNISHED WITH THE FIXTURE FOR SPECIAL REQUIREMENTS OR LIMITATIONS.
- SEE FIGURE 7-9.

INCANDESCENT

SURFACE

- FOR SURFACE MOUNTING ONLY.
- READ FIXTURE LABEL OR INSTRUCTIONS FURNISHED WITH THE FIXTURE FOR SPECIAL REQUIREMENTS OR LIMITATIONS.

RECESSED

TYPE IC
- ARE MARKED TYPE IC
- MAY BE INSTALLED IN INSULATED CEILINGS WHERE INSULATION AND OTHER COMBUSTIBLE MATERIALS MAY BE IN DIRECT CONTACT WITH AND OVER THE TOP OF THE FIXTURE.
- HAS INTEGRAL THERMAL PROTECTION THAT DEACTIVATES THE LAMP IF THE FIXTURE IS MIS-LAMPED, FIGURE 7-4.
- IS MARKED "NOTICE—THERMALLY PROTECTED FIXTURE. BLINKING LIGHT MAY INDICATE IMPROPER LAMP WATTAGE OR IMPROPER LAMP SIZE." MAY ALSO BE MARKED WITH OTHER CONDITIONS THAT WILL CAUSE OVERHEATING AND WILL RESULT IN THE LAMPS LINKING.
- MAY BE USED IN NONINSULATED CEILINGS.
- USUALLY ARE LOW-WATTAGE FIXTURES.
- READ FIXTURE LABEL OR INSTRUCTIONS FURNISHED WITH THE FIXTURE FOR SPECIAL REQUIREMENTS OR LIMITATIONS.
- SEE FIGURES 7-3, 7-6, 7-7.

TYPE NON-IC
- FOR INSTALLATION IN UNINSULATED CEILINGS.
- IF INSTALLED IN AN INSULATED CEILING, KEEP INSULATION AT LEAST 3 INCHES (76.2 mm) FROM SIDES AND NOT PLACED OVER THE FIXTURE SUCH THAT IT WOULD ENTRAP THE HEAT PRODUCED BY THE FIXTURE.
- HAS AN INTEGRAL THERMAL PROTECTION THAT WILL DEACTIVATE THE LAMP IF INSULATION COVERS THE FIXTURE, RESULTING IN AN OVERHEATING SITUATION, FIGURE 7-4.
- UNLESS OTHERWISE MARKED, KEEP FIXTURE AT LEAST ½ INCH (12.7 mm) FROM COMBUSTIBLE MATERIAL (LIKE WOOD JOISTS) EXCEPT AT SUPPORT POINTS.
- READ FIXTURE LABEL OR INSTRUCTIONS FURNISHED WITH THE FIXTURE FOR SPECIAL REQUIREMENTS OR LIMITATIONS.
- SEE FIGURES 7-3, 7-6, 7-7.

INHERENTLY PROTECTED
- IF MARKED INHERENTLY PROTECTED, THE FIXTURE IS SO DESIGNED THAT THE SURFACE TEMPERATURE WILL NOT EXCEED 90°C EVEN IF THE FIXTURE IS COVERED WITH INSULATION, IS MIS-LAMPED OR OVER-LAMPED, AN EXAMPLE MIGHT BE "DOUBLE-WALLED" CONSTRUCTION.
- THESE FIXTURES ARE NOT THERMALLY PROTECTED.
- READ FIXTURE LABEL OR INSTRUCTIONS FURNISHED WITH THE FIXTURE FOR SPECIAL REQUIREMENTS OR LIMITATIONS.

SUSPENDED

- FOR INSTALLATION IN A SUSPENDED GRID ONLY WHERE THE LAY-IN TILES ARE NOT FASTENED IN PLACE, AND WHERE THE TIE WIRES, T BARS, CEILING TILES, AND OTHER COMPONENTS DIRECTLY ASSOCIATED WITH THE GRID ARE NOT PART OF THE BUILDING STRUCTURE.
- THE FIXTURES ARE INTENDED TO BE MOUNTED IN CEILING OPENINGS.
- READ FIXTURE LABEL OR INSTRUCTIONS FURNISHED WITH THE FIXTURE FOR SPECIAL REQUIREMENTS OR LIMITATIONS. SEE FIGURES 7-8, 7-9.

Figure 7-16 Some of the most important U.L. and *NEC*[®] requirements for fluorescent and incandescent fixtures. Always refer to the U.L. Standards, the *NEC*,[®] and the label and/or instructions furnished with the fixture.

fixtures is available that provide uplighting, downlighting, diffused lighting, moonlighting, shadow and texture lighting, accent lighting, silhouette lighting, and bounce lighting. After the lighting fixtures are selected, the type and color of the lamp is then selected. Some lighting fixtures have specific light "cut-off" data that is useful in determining whether the emitting light will spill over onto the neighbor's property. The method of control must also be considered. Switch control, timer control, dusk-to-dawn control, and motion sensors are ways of turning outdoor lighting on and off.

In recent years, more and more complaints are coming from neighbors claiming that they are being bothered by glare, brightness, and light spillover from their neighbor's outdoor lighting fixtures. Security lighting, yard flood lights, driveway lighting, "moonlighting" in trees, and shrub lighting are examples of sources of light that might cross over the property line and be a "nuisance" to the next door neighbor.

Outdoor "lightscape" lighting considered by a homeowner to be aesthetically wonderful might be offensive to the neighbor. Nuisance lighting is also referred to as *light pollution, trespassing, intrusion, glare, spill-over,* and *brightness*. Some quiet residential neighborhoods are beginning to look like commercial areas, used car lots, and airport runways because no restrictions govern outdoor lighting methods. This is not a safety issue, and is not addressed in the *NEC.*® However, the issue might be found in local building codes.

Many communities are being forced to legislate strict outdoor lighting laws, specifying various restrictions on the location, type, size, wattage, and/or footcandles for outdoor lighting fixtures. Checking building codes in your area might reveal requirements such as:

Light Source: The source of light (the lamp) must not be seen directly.

Glare: Glare, whether direct or reflected, such as from floodlights, and as differentiated from general illuminations, shall not be visible at any property line.

Exterior Lighting: Any lights used for exterior illuminations shall direct light away from adjoining properties.

Before installing exterior lighting fixtures, be sure to check with the building authorities in your community to see if there are any requirements or restrictions for outdoor lighting.

REVIEW

1. Is it permissible to install a recessed fixture directly against wood ceiling joists when the label on the fixture does not indicate that the fixture is suitable for insulation to be in direct contact with the fixture? This is a Type Non-IC fixture.

_____

2. If a recessed fixture without an integral junction box is installed, what extra wiring must be provided? _____

_____

_____

_____

_____

3. Thermal insulation shall not be installed within _____ inches (millimeters) of the top or _____ inches (millimeters) of the side of a recessed fixture unless the fixture is identified for use in direct contact with thermal insulation. This is a Type Non-IC fixture.

4. Recessed fixtures are available for installation in direct contact with thermal insulation. These fixtures bear the U.L. mark "Type _____ ."

5. Unless specifically designed, all recessed incandescent fixtures must be provided with factory-installed _____ .

6. Plans require the installation of surface-mounted fluorescent fixtures on the ceiling of a recreation room that is finished with low-density ceiling fiberboard. What sort of mark would you look for on the label of the fixture? _____

   _____

   _____

   _____

7. A recessed lighting fixture bears no marking indicating that it is "Identified for Through Wiring." Is it permitted to run branch-circuit conductors *other* than the conductors that supply the lighting fixture through the integral junction box on the fixture?

   _____

8. Fluorescent ballasts for all indoor applications must be _____ type. These ballasts contain internal _____ protection to protect against overheating.

9. Additional backup protection for ballasts can be provided by connecting a(an) _____ with the proper size fuse as recommended by the ballast manufacturer.

10. You are called upon to install a number of lighting fixtures in a suspended ceiling. The ceiling will be dropped approximately 8 inches (203 mm) from the ceiling joists. Briefly explain how you might go about wiring these fixtures. _____

    _____

    _____

    _____

    _____

    _____

    _____

    _____

    _____

    _____

11. The Code places a maximum open-circuit voltage on lighting equipment in or on homes. This maximum voltage is (600) (750) (1000). (Circle the correct answer.) Where in the *NEC*® is this voltage maximum referenced?

    _____

    _____

12. The letter **E** in a circle on a ballast nameplate indicates that the ballast _____

_____

_____

13. A 2-lamp fluorescent ballast for two 40-watt lamps is marked 85 volt-amperes. What is the power factor of the ballast? _____

_____

_____

14. Can an incandescent lamp dimmer be used to control a fluorescent lamp load? _____

15. A good "rule-of-thumb" to estimate the expected life of a motor, a ballast, or other electrical equipment is that for every _____°C above rated temperature, the expected life will be cut in _____ .

16. A post light has a 120-volt, 60-watt lamp installed. The lamp has an expected lamp life of 1,000 hours. The homeowner installed a dimmer ahead of the post light, and leaves the dimmer set so the output voltage is 100 volts. The lamp burns slightly dimmer when operated at 100 volts, but this is not a problem. What is the expected lamp life when operated at 100 volts? _____

_____

_____

_____

17. Does your community have exterior outdoor lighting restrictions? _____

    If yes, what are they? _____

_____

_____

18. a. What is meant by incandescent lamp load inrush? _____

    _____

    _____

    b. What type of snap switch is required to control an incandescent load? _____

    _____

    c. What problems could occur when a single dimmer controls a large incandescent tungsten filament lamp load? _____

    _____

    _____

19. Circle the correct answer (True) or (False) for the following statements.

    a. Always match a fluorescent lamp's wattage and type designation to the type of ballast in the fixture. (True) (False)

    b. It really makes no difference what type of fluorescent lamp is used, just so the wattage is the same as marked on the ballast. (True) (False)

20. Circle the correct answer (True) or (False) for the following statements.

    a. When selecting a trim for a recessed incandescent fixture, select any trim that physically fits and can be attached to the fixture. (True) (False)

    b. When selecting a trim for a recessed incandescent fixture, select a trim that the manufacturer indicates may be used with that fixture. (True) (False)

Lighting Branch-Circuit for Front Bedroom

OBJECTIVES

After studying this Unit, the student will be able to

- explain the factors that influence the grouping of outlets into circuits.
- understand the meaning of general, accent, task, and security lighting.
- estimate loads for the outlets of a circuit.
- draw a cable layout and a wiring diagram based on information given in the residence plans, the specifications, and Code requirements.
- select the proper wall box for a particular installation.
- explain how wall boxes can be grounded.
- list the requirements for the installation of fixtures in clothes closets.

This Unit is longer than the other Units because it is the first exposure to real-life installations. Most issues confronting the electrician for a typical residential installation are covered in this Unit. Repetition in later Units is kept to a minimum.

RESIDENTIAL LIGHTING

Residential lighting is a personal thing. The homeowner, builder, and electrical contractor must meet to decide on what types of lighting fixtures are to be installed in the residence. Many variables (cost, personal preference, construction obstacles, etc.) must be taken into consideration.

Residential lighting can be segmented into four groups.

General Lighting

General lighting provides overall illumination for a given area, such as the front hall, the bedroom hall, the workshop, or the garage lighting in this res-

idence. General lighting can be very basic, which means just getting the job done by providing adequate lighting for the area involved. Or, it can take the form of decorative lighting, such as a chandelier over a dining room table or other decorative fixtures that can be attached to ceiling paddle fans.

Accent Lighting

Accent lighting provides focus and attention to an object or area in the home. Examples are the recessed "spots" over the fireplace. Another example is the track lighting in the living room of this residence, which can accent a picture, photo, painting, or sculpture that might be hung on the wall. To be effective, accent lighting should be at least five times that of the surrounding general lighting.

Task Lighting

This is sometimes referred to as "activity" lighting. Task lighting provides proper lighting where "tasks" or "activities" are performed. The fluorescent fixtures above the workbench, the kitchen range hood, and the recessed fixture over the kitchen sink are all examples of "task" lighting. To avoid too much contrast, task lighting should not exceed three times that of the surrounding general lighting.

Security Lighting

Security lighting generally includes outdoor lighting, such as post lamps, wall fixtures, walkway lighting, and all other lighting that serves the purpose of providing lighting for security and safety reasons. Security lighting in many instances is provided by the normal types of lighting fixtures found in the typical residence. In the residence discussed in this text, the outdoor bracket fixtures in front of the garage and next to the entry doors, as well as the post light, might be considered to be security lighting even though they add to the beauty of the residence. Security lighting can be controlled manually with regular wall switches, timers, light sensors, or motion detectors.

An organization called the American Lighting Association offers excellent reading materials with residential lighting suggestions and other materials relating to residential lighting. This organization comprises the many manufacturers of lighting fixtures, as well as electrical distributors that maintain extensive lighting showrooms across the country. A visit to one of these showrooms offers the prospective buyer of lighting fixtures an opportunity to select from hundreds (in some cases thousands) of lighting fixtures. Most of the finer lighting showrooms are staffed with individuals highly qualified to make recommendations to homeowners so that they will get the most value for their money.

The lighting provided throughout this residence conforms to the residential lighting recommendations of the American Lighting Association. Certainly many other variations are possible. For the purposes of studying an entire wiring installation for this residence, all of the load calculations, wiring diagrams, and so on, have been accomplished using the fixture selection as indicated on the plans and in the specifications.

GROUPING LIGHTING AND RECEPTACLE OUTLETS

The grouping of outlets into circuits must conform to *NEC®* standards and good wiring practices. There

are many possible combinations or groupings of outlets. In most residential installations, circuit planning is usually done by the electrician, who must ensure that the circuits conform to Code requirements. In larger, more costly residences, the circuit layout may be completed by the architect and included on the plans.

Because many circuit arrangements are possible, there are few guidelines for selecting outlets for a particular circuit. An electrician plans circuits that are economical without sacrificing the quality of the installation.

For example, some electricians prefer to have more than one circuit feed a room. Should one circuit have a problem, the second circuit would continue to supply power to the other outlets in that room, Figure 8-1.

Some electricians keep receptacles on different branch circuits than the lighting outlets. Again, it is a matter of choice.

Another excellent way to provide good wiring while economizing on wiring materials is to connect receptacle outlets back-to-back, as illustrated in Figure 8-2.

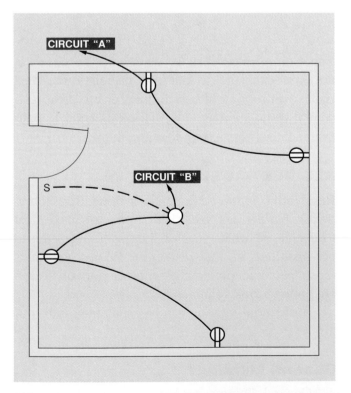

Figure 8-1 Wiring layout showing one room that is fed by two different circuits. If one of the circuits goes out, the other circuit will still provide electricity to the room.

Figure 8-2 Receptacle outlets connected back-to-back. This can reduce the cost of the installation because of the short distance between the outlets.

Although not a Code issue, it is poor practice to include outlets on different floors on the same circuit. Some local building codes limit this type of installation to lights at the head and foot of a stairway.

Cable Runs

A student studying the great number of wiring diagrams throughout this text will find many different ways to run the cables and make up the circuit connections. Sizing of boxes that will conform to the Code for the number of wires and devices is covered in great detail in this text. When a circuit is run "through" a recessed fixture like the type shown in Figure 7-7, that recessed fixture must be identified for "through wiring." This subject is discussed in detail in Unit 7. When in doubt as to whether the fixture is suitable for running wires to and beyond it to another fixture or part of the circuit, it is recommended that the circuitry be designed so as to end up at the recessed fixture with only the two wires that will connect to that fixture.

ESTIMATING LOADS FOR OUTLETS

Although this text is not intended to be a basic electrical theory text, it should be mentioned that, when calculating loads,

$$\text{Volts} \times \text{Amperes} = \text{Volt-amperes}$$

Yet, many times we say that

$$\text{Volts} \times \text{Amperes} = \text{Watts}$$

What we really mean to say is that

$$\text{Volts} \times \text{Amperes} \times \text{Power factor} = \text{Watts}$$

In a pure resistive load such as a simple light bulb, a toaster, an iron, or a resistance electric heating element, the power factor is 100%. Then

$$\text{Volts} \times \text{Amperes} \times 1 = \text{Watts}$$

With transformers, motors, ballasts, and other "inductive" loads, wattage is not necessarily the same as volt-amperes.

EXAMPLE: Calculate the wattage and volt-amperes of a 120-volt, 10-ampere resistive load.

SOLUTIONS:
    a. $120 \times 10 \times 1 = 1{,}200$ watts
    b. $120 \times 10 \quad = 1{,}200$ volt-amperes

EXAMPLE: Calculate the wattage and volt-amperes of a 120-volt, 10-ampere motor load at 50% power factor.

SOLUTIONS:
    a. $120 \times 10 \times 0.5 = 600$ watts
    b. $120 \times 10 \quad = 1{,}200$ volt-amperes

Therefore, to be sure that adequate ampacity is provided for in branch-circuit wiring, feeder sizing, and service-entrance calculations, the Code requires that we use the term *volt-amperes*. This allows us to ignore power factor, and address the *true* current draw that will enable us to determine the correct ratings of electrical equipment.

However, in some instances the terms *watts* and *volt-amperes* can be used interchangeably without creating any problems. For instance, *Section 220-19* recognizes that for electric ranges and other cooking equipment, the kVA and kW ratings shall be considered to be equivalent for the purpose of branch-circuit and feeder calculations.

The load calculation examples in *Appendix D* of the *NEC*® show kW values in the list of connected loads, but the actual calculations are done using volt-ampere values. Throughout this text, *wattage* and *volt-amperes* are used when calculating and/or estimating loads.

Building plans typically do not specify the ratings in watts for the outlets shown. When planning circuits, the electrician must consider the types of fixtures that may be used at the various outlets. To do this, the electrician must know the general uses of the receptacle outlets in the typical dwelling.

Unit 3 shows that the general lighting load of a residence is determined by allowing a load of 3 volt-amperes for each square foot (0.093 m²) of floor area, *Section 220-3(a)*. For the residence in the plans, it is shown that six lighting circuits meet the minimum standards set by the Code. However, to provide sufficient capacity, 13 lighting circuits are to be installed in this residence.

Number of Outlets Permitted on One Branch-Circuit

The *NEC*® does not specify the maximum number of receptacle outlets or lighting outlets that may be connected to one 120-volt lighting or small appliance branch-circuit in a residence. It may seem to be ridiculous and illogical that 10, 20, or more receptacle outlets and lighting outlets can be connected to one branch-circuit and not be in violation of the Code. Consider the fact, however, that having many "convenience" receptacles is "safer" because many receptacle outlets will virtually eliminate the use of extension cords, one of the most frequent causes of electrical fires. Rarely, if ever, would all receptacle outlets and lighting outlets be fully loaded at the same time. There is much diversity of load in residential occupancies.

Circuit Loading Guidelines

Recessed and surface-mounted incandescent lighting fixtures are marked with their maximum lamp wattage. Fluorescent fixtures are marked with lamp wattage and the ballast is marked in volt-amperes. But for new construction, usually the exact type of fixture that will be installed at each lighting outlet is unknown. Also unknown is the exact load that will be connected to the receptacle outlets.

Checking a lighting fixture manufacturer's catalog provides excellent guidelines for lamp wattages recommended for various types of lighting fixtures.

So it comes down to approximating and estimating what the connected loads might be.

In Unit 3, we discussed how typical 15-ampere lighting branch-circuits in homes can be figured at one 15-ampere branch-circuit for every 600 square feet. Although not often used in residential wiring, if you choose to install 20-ampere lighting branch-circuits, figure one 20-ampere branch-circuit for every 800 square feet. This results in a good approximation for the minimum number of lighting branch-circuits needed.

Another guideline to follow in residential wiring is *never load the circuit to more than 80% of its rating*.

A 15-ampere, 120-volt lighting branch-circuit would be calculated as:

$$15 \times 0.80 = 12 \text{ amperes}$$

or

$$12 \text{ amperes} \times 120 \text{ volts} = 1{,}440 \text{ watts}$$
$$\text{(volt-amperes)}$$

A 20-ampere, 120-volt lighting branch-circuit would be calculated as:

$$20 \times 0.80 = 16 \text{ amperes}$$

or

$$16 \text{ amperes} \times 120 \text{ volts} = 1{,}920 \text{ watts}$$
$$\text{(volt-amperes)}$$

Estimating Number of Outlets by Assigning an Amperage Value to Each

One method of determining the number of lighting and receptacle outlets to be included on one circuit is to assign a value of 1 to 1½ amperes to each outlet to a total of 15 amperes. Thus, a total of 10 to 15 outlets can be included in a 15-ampere circuit.

As stated previously, the *NEC*® does not limit the number of outlets on one circuit. However, many local building codes do specify the maximum number of outlets per circuit. Before planning any circuits, the electrician must check the local building code requirements.

All outlets will not be required to deliver 1½ amperes. For example, closet lights, night lights, and clocks will use only a small portion of the allowable current. A 60-watt closet light will draw less than 1 ampere.

$$I = \frac{W}{E} = \frac{60}{120} = 0.5 \text{ ampere}$$

If low-wattage fixtures are connected to a circuit, it is quite possible that 15 or more lighting outlets (many times referred to as "openings") could be connected to the circuit without a problem. On the other hand, if the load consists of high-wattage lamps, the number of outlets would be less. In this residence, estimated loads were figured at 120 volt-amperes (1 ampere) for those receptacle outlets obviously intended for general use. Receptacle outlets in the garage and workshop were figured at 180 volt-amperes (1½ amperes) because of the use these outlets will be supplying (i.e., tools, drills, table saws, etc.).

Receptacle outlets in the kitchen and laundry have been included in the small appliance circuit calculations.

If an electrician were to *estimate* the number of 15-ampere lighting branch-circuits desired for a new residence where the total count of lighting and receptacle outlets is, for example, 80, then

$$\frac{80}{10} = 8 \text{ (eight 15-ampere lighting circuits)}$$

If the circuit consists of low-wattage loads, then

$$\frac{80}{15} = 5.3 \text{ (six 15-ampere lighting circuits)}$$

In residential occupancies there is great diversity in the loading of lighting branch-circuits.

CAUTION: The foregoing procedure for estimating the number of lighting and receptacle outlets is for branch-circuits only. Obviously, where the small appliance circuits in the kitchen, laundry area, and other areas supply heavy concentrations of plug-in appliances, the 1½ amperes per receptacle outlet would not be applicable. These small appliance circuits are discussed later on in this text.

Review of How to Determine the Minimum Number of Lighting Branch-Circuits in a Residence

To determine the minimum number of lighting circuits:

1. Use the 3 volt-amperes per square foot method as required by the *NEC.*®

2. Figure a minimum of one 15-ampere lighting branch-circuit for every 600 square feet of floor area.

3. Figure a minimum of one 20-ampere lighting branch-circuit for every 800 square feet of floor area.

4. Estimate the probable loading for each lighting outlet and receptacle outlet (do not include the small appliance receptacle outlets). Try to have a total VA not over 1440 for a 15-ampere lighting circuit, and not over 1920 for a 20-ampere lighting circuit.

5. Connect approximately 10 to 15 outlets per circuit—more than 10 if the probable connected load is small, less than 10 if the probable connected load is large. For instance, the wattage of a 120-volt smoke detector is approximately ½ watt.

Divide Loads Evenly

Section 210-11(b) makes it mandatory to divide loads evenly between the various circuits. The obvious reason is to avoid overload conditions on some circuits when other circuits might be lightly loaded. This is common sense. At the Main Service, don't connect the branch-circuits and feeders to result in, for instance, 120 amperes on phase A and 40 amperes on phase B. Look at the probable and/or calculated loads to attain as close a balance as possible, such as 80 amperes on each A and B phase.

SYMBOLS

The symbols used on the cable layouts in this text are the same as those found on the actual electrical plans for the residence. Refer to Figures 2-4, 2-5, 2-6, 2-7, and 2-8.

Pictorial illustrations are used on all wiring diagrams in this text to make it easy for the reader to complete the wiring diagrams, Figure 8-3.

DRAWING THE WIRING DIAGRAM OF A LIGHTING CIRCUIT

The electrician must take information from the building plans and convert it to the forms that will be most useful in planning an installation that is economical, conforms to all Code regulations, and follows good wiring practice. To do this, the electrician

DUPLEX RECEPTACLE
OUTLET GROUNDING TYPE

DUPLEX SPLIT-CIRCUIT
RECEPTACLE OUTLET
GROUNDING TYPE

CEILING OR
WALL FIXTURE
OUTLET

PILOT
LIGHT

SINGLE-POLE
SWITCH

THREE-WAY
SWITCH

FOUR-WAY
SWITCH

Figure 8-3   Pictorial illustrations used on wiring diagrams in this text.

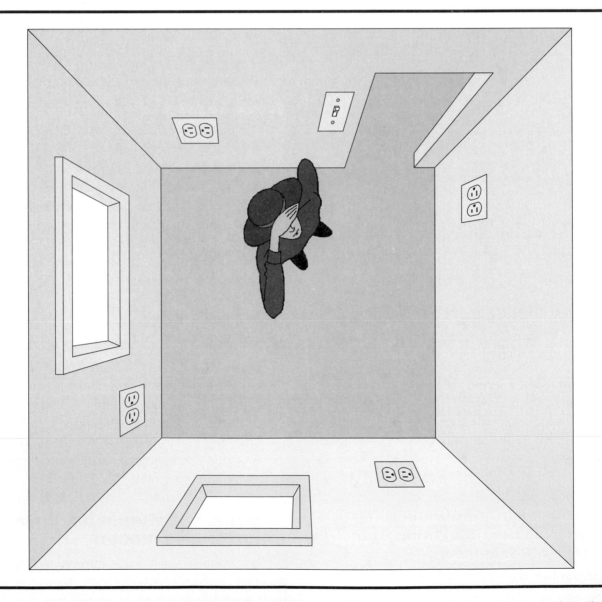

Figure 8-4   This top view of a room shows the relative positioning of receptacle outlets so that the wide grounded terminal slot is at the top. This provides for added safety if a metal faceplate loosens or if a metal clip or other object falls on top of the blades of an attachment plug cap that is not plugged in all the way. All of the wiring diagrams throughout this text will illustrate the receptacles in this manner.

first makes a cable layout. Then a wiring diagram is prepared to show clearly each wire and all connections in the circuit.

A skilled electrician, after many years of experience, will not prepare wiring diagrams for most residential circuitry because he has the ability to "think" the connections through mentally. An unskilled individual should make detailed wiring diagrams.

The following steps will guide the student in the preparation of wiring diagrams. The student is required to draw wiring diagrams in later Units of this text. As the student works on the wiring dia-

grams of the many circuits provided in this text, note that the receptacles are positioned exactly as though you were standing in the center of the room. Then, as you turn to face each wall, you will find the receptacles positioned in that wall with the grounded conductor slot on the top, as shown in Figure 8-4.

1. Refer to plans and make a cable layout of all lighting and receptacle outlets, as in Figure 8-5.

2. Draw a wiring diagram showing the traveler conductors for all three-way switches, if any, as in Figure 8-6.

3. Draw a line between each switch and the outlet or outlets it controls. For three-way switches, do this for one switch only. This is the "switch leg."

4. Draw a line from the grounded terminal on the lighting panel to each current-consuming outlet. This line may pass through switch boxes, but must not be connected to any switches. This is the white grounded circuit conductor, often called the "neutral" conductor.

NOTE: An exception to step 4 may be made for double-pole switches. For these switches, all conductors of the circuit are opened simultaneously. They are rarely used in residential wiring.

Figure 8-5   Typical cable layout.

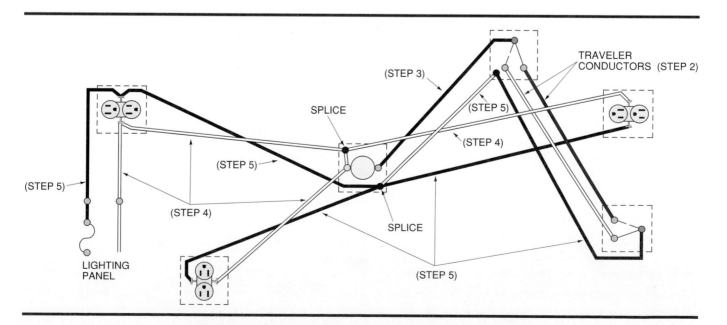

Figure 8-6   Wiring diagram of circuit shown in Figure 8-5.

5. Draw a line from the ungrounded "hot" terminal on the panel to connect to each switch and unswitched outlet. Connect to one three-way switch only.

6. Show splices as small dots where the various wires are to be connected together. In the wiring diagram, the terminal of a switch or outlet may be used for the junction point of wires. In actual wiring practice, however, the Code does not permit more than one wire to be connected to a terminal unless the terminal is a type identified for use with more than one conductor. The standard screw-type terminal is not acceptable for more than one wire, *Section 110-14*.

7. The final step in preparing the wiring diagram is to mark the color of the conductors, as in Figure 8-6. Note that the colors selected— black (B), white (W), and red (R)—are the colors of two- and three-conductor cables (refer to Unit 5). It is suggested that the reader use colored pencils or markers for different conductors when drawing the wiring diagram.

Use a fine line for the white conductor.

LIGHTING BRANCH-CIRCUIT A16 FOR FRONT BEDROOM

Circuit A16 is a 15-ampere branch-circuit. As we look at this circuit, we find five split-circuit receptacle outlets. These will provide the general lighting through the use of table or other lamps that will be plugged into the switched receptacles, Figure 8-7.

Radio, television, clocks, and other electrical and electronic items not intended to be controlled by the wall switch will be plugged into the "hot continuously" receptacle.

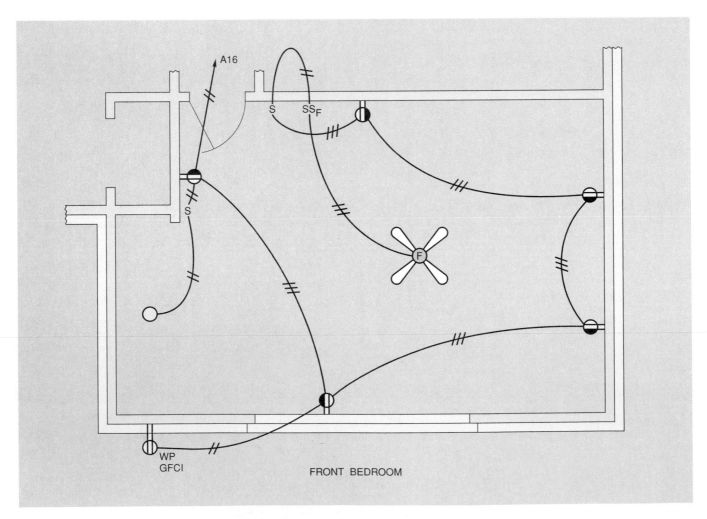

Figure 8-7   Cable layout for front bedroom circuit A16.

Next to the wall switch we find the control for the ceiling fan, which also has a lighting fixture as an integral part of the fan, Figure 8-8. A single-pole switch controls the ON/OFF and speed of the fan motor.

The closet light is controlled by a single-pole switch outside and to the right of the closet door.

Note that one outside weatherproof receptacle is also connected to circuit A16. Outdoor receptacles must have GFCI protection. See Unit 6 for a complete discussion of GFCI protection.

Checking the actual electrical plans, we find one television outlet and one telephone outlet are to be installed in the front bedroom. Television and telephones are discussed in Unit 25. Table 8-1 summarizes the outlets in the front bedroom and the estimated load.

Figure 8-8   Conceptual view of how the front bedroom switch arrangement is to be accomplished.

DESCRIPTION	QUANTITY	WATTS	VOLT-AMPERES
Receptacles @ 120 watts each	5	600	600
Weatherproof receptacle	1	120	120
Closet fixture	1	75	75
Ceiling fan/light	1		
3 - 50-W lamps		150	150
Fan motor (0.75 A @ 120 V)		80	90
TOTALS	8	1,025	1,035

Table 8-1   Front bedroom: outlet count and estimated load. Circuit A16.

DETERMINING THE SIZE OF OUTLET, DEVICE, JUNCTION BOXES, AND CONDUIT BODIES

One of the most common problems found in residential wiring are wires jammed into electrical boxes. Factors that need to be considered are the size and number of conductors, the number of wiring devices, cable clamps, fixture studs, hickeys, and splices. *Section 370-16* provides the rules for sizing boxes.

Experience has shown that *minimum* does not always make for easy working the splices, conductors, wiring devices, and carefully putting them all into the box without crowding. Do not skimp on box sizing. The small additional cost of larger boxes is worth much more than the grief and added time it will take to trim out the box using *minimum* size boxes. You will be much happier with boxes that have more than the *minimum* size. When it comes to box sizing, think big—not small.

The following example shows how to calculate the *minimum* size box for a particular installation, Figure 8-9.

1. Add the No. 14 AWG circuit conductors:
   2 + 3 + 3 =   8
2. Add equipment grounding wires (count one only)   1
3. Add two conductors for the receptacle   2

   Total   11   Conductors

Note that the pigtails connected to the receptacle, Figure 8-9, need not be counted when determining the correct box size. *Section 370-16(b)(1)* states, "A conductor, no part of which leaves the box, *shall not* be counted."

Once the total number of conductors, plus the volume count required for wiring devices, fixture studs, hickeys, and clamps is known, refer to *Table*

THE "PIGTAILS" CONNECTED TO THE RECEPTACLE NEED NOT BE COUNTED WHEN DETERMINING THE CORRECT BOX SIZE. THE CODE IN *SECTION 370-16(b)(1)* STATES, "A CONDUCTOR, NO PART OF WHICH LEAVES THE BOX, SHALL NOT BE COUNTED."

Figure 8-9 Determining size of box according to number of conductors.

Figure 8-10 Split-circuit wiring for receptacles. The top receptacle is "hot" at all times. The bottom receptacle is controlled by the switch.

370-16(a) and *(b)* and Figure 2-19 to select a box that has sufficient space. Also refer to Figure 2-34 for a summary of box fill requirements. For example, a 4" × 2⅛" square box with a suitable plaster ring can be used.

The volume of the box plus the space provided by plaster rings, extension rings, and raised covers may be used to determine the total available volume. In addition, it is desirable to install boxes with external cable clamps. Remember that if the box contains one or more devices, such as cable clamps, fixture studs, or hickeys, the number of conductors permitted in the box shall be one less than shown in *Table 370-16(a)* for *each type* of device contained in the box.

GROUNDING OF WALL BOXES

The specifications for the residence state that *all* metal boxes are to be grounded. The means of grounding is armored cable or nonmetallic-sheathed cable containing an extra equipment grounding conductor. This conductor is used only to ground the metal box. This bare equipment grounding conductor must *not* be used as a current-carrying conductor because severe shocks can result.

According to *Section 210-7(a)*, grounding-type receptacles must be installed on 15-ampere and 20-ampere branch-circuits. The methods of attaching the equipment grounding conductor to the proper terminal on the receptacle are covered in Unit 5.

POSITIONING OF SPLIT-CIRCUIT RECEPTACLES

The receptacle outlets shown in the front bedroom are called two-circuit or split-circuit receptacles. The top portion of such a receptacle is "hot" at all times, and the bottom portion is controlled by the wall switch, Figure 8-10. It is recommended that the electrician wire the bottom section of the receptacle as the switched section. As a result, when the attachment plug cap of a lamp is inserted into the bottom switched portion of the receptacle, the cord does not hang in front of the unswitched section. This unswitched section can be used as a receptacle for clock, vacuum cleaner, radio, stereo, compact disc player, personal computer, television, or other appliances where a switch control is not necessary or desirable.

When split-circuit receptacles are horizontally mounted, which is common when 4" square boxes are used because the plaster ring is easily fastened to the box in either a vertical or horizontal position, locate the switched portion to the right.

POSITIONING OF RECEPTACLES NEAR ELECTRIC BASEBOARD HEATING

Electric heating for a home can be accomplished with an electric furnace, electric baseboard heating units, heat pump, or resistance heating cables embedded in plastered ceilings or sandwiched between two layers of drywall sheets on the ceiling.

The important thing to remember at this point is that some baseboard electric heaters may not be permitted to be installed below a receptacle outlet, *Section 210-52 FPN*.

See Unit 23 for a detailed discussion of installing electric baseboard heating units and their relative positions below receptacle outlets.

The front bedroom does have a ceiling fan/light on the ceiling. Ceiling fans are discussed in Unit 9.

FIXTURES IN CLOTHES CLOSETS

The *NEC®* does not require lighting fixtures in clothes closets, *Section 210-70*. However, some local codes and some building codes do. When light-

Figure 8-11 Typical closet with one shelf and one rod. The shaded area defines "storage space." Dimension A is width of shelf or 12 inches (305 mm) from wall, whichever is greater. Dimension B is below rod, 24 inches (610 mm) from wall. See *Section 410-8(a)*.

ing fixtures are installed in clothes closets, they must be of the correct type and must be installed properly.

Clothing, boxes, and other material normally stored in clothes closets are a potential fire hazard. These items may ignite on contact with the hot surface of exposed lamps. Incandescent lamps have a hotter surface temperature than fluorescent lamps. *Section 410-8* gives very specific rules for the location and types of lighting fixtures permitted to be installed in clothes closets. Figures 8-11 through 8-15 illustrate the requirements of *Section 410-8*.

Figure 8-12 Large walk-in closet where there is access to the center rod from both sides. The shaded area defines "storage space." Dimension A is width of shelf or 12 inches (305 mm), whichever is greater. Dimension B is 24 inches (610 mm) from wall. See *Section 410-8(a)*.

In this residence, the closets in the bedrooms and front hall are 24 inches (619 mm) deep. The storage closet in the recreation room is 30 inches (762 mm) deep. The shelving is approximately 12 inches (305 mm) wide. Because of the required clearances between lighting fixtures and the storage space, locating lighting fixtures in closets can be difficult. In fact, some clothes closets are so small that lighting fixtures of any kind are not permitted.

Figure 8-14 shows the possibilities: recessed incandescents, recessed fluorescents, surface-mounted incandescents, and surface-mounted fluorescents. These can be on the ceiling, or on the wall space above the closet door. Whichever type is chosen, be sure that the minimum clearances are provided as required by *Section 410-8*.

Figure 8-16 shows typical bedroom-type lighting fixtures.

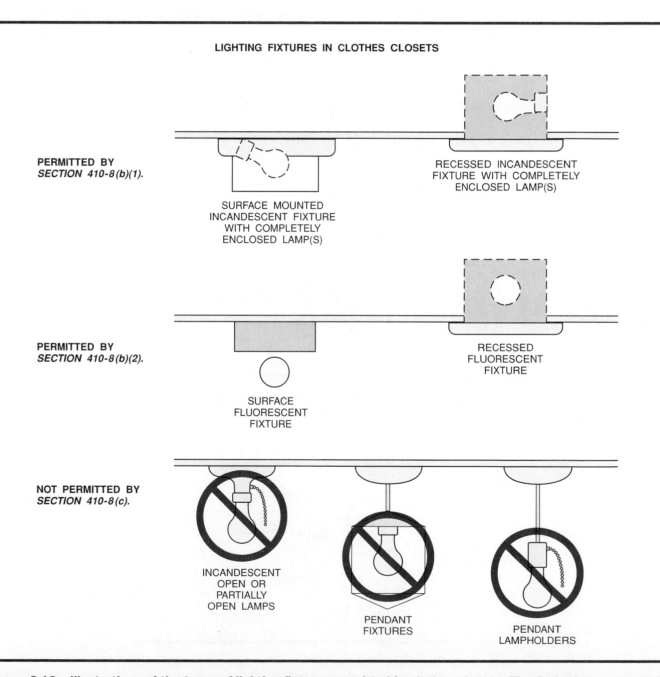

Figure 8-13 **Illustrations of the types of lighting fixtures permitted in clothes closets. The Code does not permit bare incandescent lamps, pendant fixtures, or pendant lampholders to be installed in clothes closets.**

LOCATION OF LIGHTING FIXTURES IN CLOTHES CLOSETS

MINIMUM CLEARANCES FOR SURFACE-MOUNTED INCANDESCENT FIXTURES, *SECTION 410-8 (d)(1).*

STORAGE SPACE

MINIMUM CLEARANCES FOR SURFACE-MOUNTED FLUORESCENT FIXTURES, *SECTION 410-8 (d)(2).*

STORAGE SPACE

MINIMUM CLEARANCES FOR RECESSED COMPLETELY ENCLOSED INCANDESCENT FIXTURES, *SECTION 410-8 (d)(3)* **AND RECESSED FLUORESCENT FIXTURES,** *SECTION 410-8 (d)(4).*

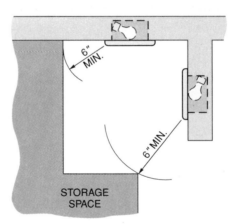

STORAGE SPACE

Figure 8-14   The above illustrations show the minimum clearances required between lighting fixtures and the storage space in clothes closets. See Figures 8-11 and 8-12 for a definition of storage space.

Figure 8-15   Recessed incandescent closet lighting fixture with pull-chain switch that may be used where a separate wall switch is not installed. *Courtesy* Progress Lighting.

Figure 8-16   Typical bedroom-type lighting fixtures. *Courtesy* Progress Lighting.

REVIEW

Note: Refer to the Code or the plans where necessary.

1. Can the outlets in a circuit be arranged in different groupings to obtain the same result? Why? _____

   _____

   _____

   _____

   _____

2. Is it good practice to have outlets on different floors on the same circuit? Why?

   _____

   _____

   _____

   _____

3. A good rule to follow is to never load a circuit to more than _____ percent of the branch-circuit rating.

4. The *NEC*® does not limit the number of lighting and receptacle outlets permitted on one branch-circuit for residential installations. A guideline that is often used by electricians for house wiring is to allow _____ to _____ amperes per outlet. For a 15-ampere circuit, this results in _____ to _____ outlets on the branch-circuit.

5. For residential wiring, not less than one 15-ampere lighting branch-circuit should be provided for every _____ square feet of floor area. If 20-ampere lighting branch-circuits are used, provide one 20-ampere lighting branch-circuit for every _____ square feet of floor area.

6. For this residence, what are the estimated wattages used in determining the loading of branch-circuit A16?

   Receptacles _____ watts (volt-amperes)

   Closet fixture _____ watts (volt-amperes)

7. What is the ampere rating of circuit A16? _____

8. What size wire is used for the lighting circuit in the front bedroom? _____

9. How many and what type of receptacles are connected to this circuit? _____

   _____

   _____

   _____

10. What main factors influence the choice of wall boxes? _____

    _____

    _____

11. How is a wall box grounded? _____

    _____

    _____

12. What is a split-circuit receptacle? _____

    _____

    _____

    _____

    _____

    _____

13. Is the switched portion of an outlet mounted toward the top or the bottom? Why?

    _____

    _____

    _____

    _____

14. The following questions pertain to lighting fixtures in clothes closets.
    a. Does the Code permit bare incandescent lamp fixtures such as porcelain keyless or porcelain pull-chain lampholders to be installed? _____
    b. Does the Code permit bare fluorescent lamp fixtures to be installed? _____
    c. Does the Code permit pendant fixtures or pendant lampholders to be installed?

       _____

    d. What is the minimum clearance from the storage area to surface-mounted incandescent fixtures? _____
    e. What is the minimum clearance from the storage area to surface-mounted fluorescent fixtures? _____
    f. What is the minimum distance between recessed incandescent or recessed fluorescent fixtures and the storage area? _____

g. Define the "storage area."

_____

_____

_____

_____

_____

_____

h. If a clothes hanging rod is installed where there is access from both sides, such as might be found in a large walk-in closet, define the storage area under that rod.

_____

_____

_____

_____

15. How many switches are in the bedroom circuit, what type are they, and what do they control? _____

_____

16. The following is a layout of the lighting circuit for the front bedroom. Using the cable layout shown in Figure 8-8, make a complete wiring diagram of this circuit. Use colored pencils or pens.

17. When planning circuits, what common practice is followed regarding the division of loads? _____

_____

18. The Code uses the terms *watts, volt-amperes, kW,* and *kVA*. Explain their significance in calculating loads. _____

_____

_____

_____

19. How many No. 14 AWG conductors are permitted in a device box that measures 3" × 2" × 2¾"? _____

20. A 4" × 1½" octagon box has one cable clamp and one fixture stud. How many No. 14 AWG conductors are permitted? _____

_____

_____

_____

Lighting Branch-Circuit for Master Bedroom

OBJECTIVES

After studying this unit, the student will be able to

- draw the wiring diagram of the cable layout for the master bedroom.
- study Code requirements for the installation of ceiling fans.
- estimate the probable connected load for a room based on the number of fixtures and outlets included in the circuit supplying the room.
- gain more practice in determining box sizing based upon the number of conductors, devices, and clamps in the box.
- make the connections for three-way switches.

The discussion in Units 3 and 8 of grouping outlets, estimating loads, selecting wall box sizes, and drawing wiring diagrams can also be applied to the circuit for the master bedroom.

The residence panel schedules show that the master bedroom is supplied by Circuit A19. Because Panel A is located in the basement below the wall switches next to the sliding doors, the home run for Circuit A19 is brought into the outdoor weatherproof receptacle. This results in six conductors in the outdoor receptacle box. The home run could have been brought into the corner receptacle outlet in the bedroom. Again, it is a matter of studying the circuit to determine the best choice for conservation of cable or conduit in these runs and to economically select the correct size of wall boxes.

LIGHTING BRANCH-CIRCUIT A19 FOR MASTER BEDROOM

Figure 9-1, Table 9-1, and the electrical plans show that Circuit A19 has four split-circuit receptacle outlets in this bedroom, one outdoor weatherproof GFCI receptacle outlet, two closet fixtures, each on a separate switch, plus a ceiling fan/light fixture, one telephone outlet, and one television outlet.

In addition, an outdoor bracket fixture is located adjacent to the sliding door and is controlled by a single-pole switch just inside the sliding door.

The split-circuit receptacle outlets are controlled by two three-way switches. One is located next to the sliding door. As in the front bedroom, living room, and study/bedroom, the use of split-circuit receptacles offers the advantage of having switch

DESCRIPTION	QUANTITY	WATTS	VOLT-AMPERES
Receptacles @ 120 watts each	4	480	480
Weatherproof receptacle	1	120	120
Outdoor bracket fixture	1	150	150
Closet fixtures One 75-W lamp each	2	150	150
Ceiling fan/light	1		
Three 50-W lamps		150	150
Fan motor (0.75 @ 120 V)		80	90
TOTALS	9	1130	1140

Table 9-1   Master bedroom outlet count and estimated load. Circuit A19.

control of one of the receptacles at a given outlet, while the other receptacle remains "live" at all times. See Figures 3-5 and 12-14 for definition of a receptacle.

Next to the bedroom door are one three-way switch, plus the ceiling fan/light controls, which are installed in a separate two-gang box, Figure 9-2.

SLIDING GLASS DOORS AND FIXED GLASS PANELS

Receptacle outlets must be installed so that no point along the floor line in any wall space is more than 6 feet (1.83 m), measured horizontally, from an outlet in that space, including any wall space 2 feet (610 mm) or more in width. Wall space occupied by

Figure 9-1   Cable layout for master bedroom.

fixed glass panels in exterior walls are considered to be wall space. Sliding glass panels are not considered to be wall space, the same as any other interior or exterior door. This is covered in *Section 210-52(a)(2)*.

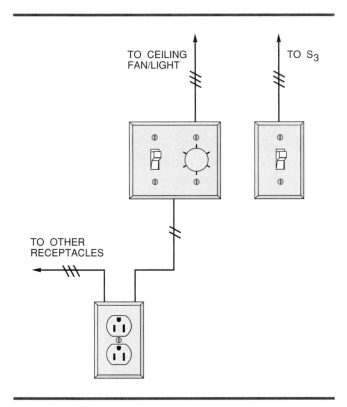

Figure 9-2 Conceptual view of how the switching arrangement is to be accomplished in the master bedroom Circuit A19.

Because fixed glass panels are considered to be wall space, in the master bedroom, a receptacle must be located so as to be not more than 6 feet (1.83 m) from the left-hand edge of the fixed glass panel.

SELECTION OF BOXES

As discussed in Unit 2, the selection of outlet boxes and switch boxes is made by the electrician. These decisions are based on Code requirements, space allowances, good wiring practices, and common sense.

For example, in the master bedroom, the electrician may decide to install a 4-inch square box with a two-gang raised plaster cover at the box location next to the sliding door. Or two sectional switch boxes ganged together may be installed at that location, Figures 9-3 and 9-4.

The type of box to be installed depends upon the number of conductors entering the box. The suggested cable layout, Figure 9-1, shows that four cables enter this box for a total of 14 conductors (10 No. 14 AWG circuit conductors and 4 equipment grounding conductors).

The requirements for calculating box fill are found in *Section 370-16*. These requirements are summarized in Figure 2-34. Also refer to *Tables 370-16(a)* and *(b)*, and Figure 2-19.

In this example, we have ten circuit conductors, four equipment grounding conductors (counted as

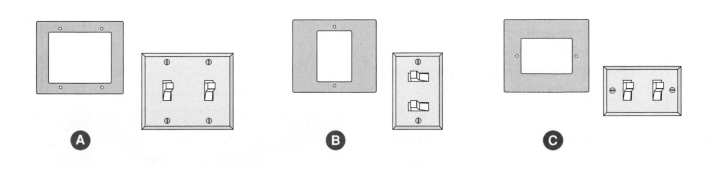

Figure 9-3 (A) shows two switches and wall plate that attach to a two-gang, 4"-square raised plaster cover or to two device boxes that have been ganged together. (B) and (C) show two interchangeable-type switches and wall plate that attach to a one-gang, 4"-square raised plaster cover or to a single device (switch) box. These types of wiring devices may be mounted vertically or horizontally.

NO. OF GANGS	HEIGHT	WIDTH
1	4½" (114.3 mm)	2¾" (69.85 mm)
2	4½" (114.3 mm)	4⁹⁄₁₆" (115.9 mm)
3	4½" (114.3 mm)	6⅜" (161.9 mm)
4	4½" (114.3 mm)	8³⁄₁₆" (207.9 mm)
5	4½" (114.3 mm)	10" (254 mm)
6	4½" (114.3 mm)	11¹³⁄₁₆" (300 mm)

Figure 9-4 Height and width of standard wall plates.

one conductor), four cable clamps (counted as one conductor), one single-pole switch (counted as two conductors), and one three-way switch (counted as two conductors), for a total of 16 conductors.

1. Count the circuit conductors
   2 + 2 + 3 + 3 =   10
2. Add one for one or more equipment grounding conductors   1
3. Add two for each switch (2 + 2)   4
4. Add one for one or more cable clamps   1

                        Total   16

Now look at the Quik-Chek Box Selection Guide, Figure 2-19, and select a box or a combination of gangable device boxes that are permitted to hold 16 conductors.

Possibilities are:

1. Gang two 3" × 2" × 3½" device boxes together (9 × 2 = 18 No. 14 AWG conductors allowed).
2. Install a 4" × 4" × 2⅛" square box (15 No. 14 AWG conductors allowed). Note that the raised plaster cover, if marked with its cubic-inch volume, can increase the maximum number of conductors permitted for the combined box and raised cover. See Figure 2-35.

CEILING-SUSPENDED (PADDLE) FANS

For appearance and added comfort, ceiling-suspended (paddle) fans have become extremely popular. Figure 9-5 is a typical ceiling-suspended (paddle) fan. Light kits are available for attaching to the bottom of the fan. These fans rotate slowly, in the range of approximately 60 r/min to 250 r/min for home-type fans.

Residential ceiling-suspended (paddle) fans are available that hug the ceiling as well as those with short stems so the fan will be approximately 12 inches (305 mm) below the ceiling. Light kits in many styles are available for most ceiling-suspended (paddle) fans.

Control possibilities are to use the pull-chain on off/speed control switch that is an integral part of the fan, or to use wall switches. Upscale models feature remote on-off and speed control.

Regardless of the speed of the air movement, the temperature of the air has not changed. However, the faster speed will carry heat away from an object, person, or area more rapidly than slow-moving air. Since air movement results in increased evaporation from our skin, we feel cooler.

Direction a Ceiling-Suspended (Paddle) Fan Should Run

Most ceiling-suspended (paddle) fans have a small switch for changing the fan's rotation.

There really is no right or wrong direction to run a ceiling-suspended (paddle) fan. Some people like air blowing directly on them—others do not. Air blowing on you makes you feel cooler. No matter which direction the ceiling-suspended (paddle) fan runs, the circulating air will make you feel more comfortable.

Without air movement, warm air rises to the ceiling and cool air drops to the floor.

Blowing downward, the fan will pull the air across the ceiling toward the fan, and blow it downward. The air will then move toward the outside of the room, upward along the walls of the room, and again be pulled toward the ceiling fan, repeating the process.

Figure 9-5 Typical home-type ceiling fan. *Courtesy NuTone.*

Blowing upward, the fan will pull the air upward, and will move the air across the ceiling away from the fan. The air will then move downward along the outside walls, across the floor, then repeat the process.

Some manufacturers of ceiling-suspended (paddle) fans suggest directing the air downward in the summer and upward in the winter.

Upward or downward, what really matters is that the air is moving instead of being stratified and still. If the air conditioner is running, or the furnace fan is running, a ceiling fan provides additional movement of the air.

The best recommendation is to try different speeds and direction, and do whatever feels best for you.

Ceiling-suspended (paddle) fans that are mounted on short stems are slightly more efficient in moving air than fans that are mounted directly on the ceiling.

Supporting Ceiling-Suspended (Paddle) Fans

This is a very important issue! There have been many reports of injuries (contusions, concussions, lacerations, fractures, head trauma, etc.) caused by the falling of a ceiling-suspended (paddle) fan. There is also the possibility of starting an electrical fire when the wires in the outlet box and fan canopy "short-circuit" as the fan falls from the ceiling. These problems are generally the result of an improper installation.

The safe supporting of a ceiling-suspended (paddle) fan involves three issues:

- the actual weight of the fan

- the twisting and turning motion when started

- vibration caused by unbalanced fan blades

Pay careful attention to the instructions furnished with the fan, and the marking *on the carton* of metallic and nonmetallic boxes to see if they are "listed" as suitable for the support of lighting fixtures and/or ceiling-suspended (paddle) fans.

Most residential-type ceiling-suspended (paddle) fans weigh 35 pounds (15.8 kg) or less, and may be supported by an outlet box that is clearly marked as being "Acceptable For Fan Support," *Sections 370-27(c)* and *422-18*. This is shown in Figure 9-6.

For ceiling-suspended (paddle) fans that weigh more than 35 pounds (15.8 kg), the fan must be supported independently of the box.

The combined weight of the fan and any accessories such as a light kit must be considered when determining compliance with the 35-pound requirement.

▶ The exception to *Section 422-18(b)* allows listed outlet boxes or outlet box systems that are identified for the purpose to be used to support a ceiling suspended fan, with or without accessories, that weigh 70 lb. (31.76 kg) or less. ◀

Where ceiling-suspended (paddle) fans are likely to be installed, make sure that the outlet box is marked "Acceptable For Fan Support." This would include all habitable rooms such as kitchens, family rooms, dining rooms, parlors, libraries, dens, sun rooms, bedrooms, recreation rooms, some stairways, or similar rooms or areas. Boxes installed close to a wall, or boxes installed specifically for

Figure 9-6 Typical mounting of a ceiling-suspended (paddle) fan.

Figure 9-7   A, B, C, and D show how a listed ceiling fan hanger and box assembly is installed through a properly sized and carefully cut hole in the ceiling. This is done for existing installations. Similar hanger/box assemblies are used for new work. The hanger adjusts for 16-inch (406 mm) and 24-inch (610 mm) joist spacing but can be cut shorter if necessary. (E) shows a type of box listed and identified for the purpose where the fan is supported from the joist, independent of the box, as required by *Section 422-18* for fans that weigh more than 35 pounds. *Courtesy* Reiker Enterprises Inc.

such things as security systems, fire alarms, and smoke detectors are examples of where the outlet box would not be required to be suitable for fan support. Where will the homeowner decide to install a ceiling fan after all of the wiring has been completed? This becomes an important judgment call on the part of the electrician.

Listed assemblies that include a hanger and an outlet box can be installed where a ceiling fan or a particularly heavy lighting fixture is to be installed. Figure 9-7 shows one type of box/hanger assembly that can be installed for new work or for existing installations.

A recent entry into the market place is an outlet box that is "listed" for both fixture and fan support for fans of 35 lb. or less, including all accessories. The outlet box has No. 8-32 holes for fixture mounting, and No. 10-32 holes for fan support. The box itself must be securely fastened to framing members of the building structure. Figure 9-8 shows this dual-purpose outlet box.

Section 410-16(a) requires that lighting fixtures weighing more than 50 pounds (22.7 kg) must be supported independently of an outlet box.

Figure 9-8   A special outlet box for the support of a lighting fixture or a ceiling-suspended (paddle) fan. There are no ears on the box. Instead, special strong mounting brackets are provided. Note that the mounting holes on the brackets are of two sizes—one set of holes is tapped for No. 8-32 screws (for a lighting fixture), and the other set of holes is tapped for No. 10-32 screws (for a ceiling-suspended [paddle] fan). *Photo courtesy* Reiker Enterprises.

Requirements relating to the support of ceiling-suspended (paddle) fans are also found in the U.L. White Book, and in the U.L. Standards 507 (electric fans), 514A (metallic boxes), and 514C (nonmetallic boxes).

Pay careful attention to the marking on metallic and nonmetallic boxes to see if they are "listed" as suitable for the support of ceiling-suspended (paddle) fans. This marking is required to be on the product.

Figure 9-9 shows the wiring for a fan/light combination with the supply at the fan/light unit. Figure 9-10 shows the fan/light combination with the sup-

ply at the switch. Figure 9-7 shows two types of ceiling fan hanger/box assemblies. Many other types are available.

Figure 9-11 illustrates a typical combination light switch and three-speed fan control. The electrician is required to install a deep 4-inch box with a two-gang raised plaster ring for this light/fan control.

Figure 9-12 illustrates a "slider" switch for speed control of a fan, and a "dual control" that provides 4-speed fan control and 2-level "hi/lo" dimming for the incandescent lamps. Figure 9-13 shows two other styles of fan/light controls. Dual controls

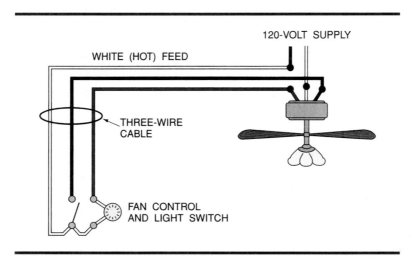

Figure 9-9 Fan/light combination with the supply at the fan/light unit.

Figure 9-10 Fan/light combination with the supply at the switch.

Figure 9-11 Combination light switch and three-speed fan switch.

Figure 9-12 The "slider" switch is for speed control of a fan. The "dual-control" provides 4-speed fan control and a 2-level "hi/lo" dimming for an incandescent lamp. *Courtesy* Pass & Seymour.

Figure 9-13 Two types of fan/light controls. One has a toggle switch for turning a ceiling fan off and on, plus a "slider" that provides 3-speed control of the fan. The other has a toggle switch for turning a light off and on, plus two "sliders," one for ON/OFF and 3-speed control of a ceiling fan, the other for full range dimming of lighting. *Courtesy* of Lutron.

are needed where the ceiling fan also has a lighting fixture. These controls fit into a deep single-gang device box, or preferably into a 4-inch square box with single-gang plaster cover. Be careful of "box fill." Make sure the wall box is large enough. Dimmer and speed control devices are quite large when compared to regular wall switches. Do not jam the control into the wall box.

A typical home-type ceiling fan motor draws 50 to 100 watts (50 to 100 volt-amperes). This is approximately 0.4 to slightly over 0.8 ampere.

Ceiling fan/light combinations increase the load requirements somewhat. Some ceiling fan/light units have one lamp socket, whereas others have four or five lamp sockets.

For the ceiling fan/light unit in the master bedroom, 240 volt-amperes were included in the load calculations. The current draw is

$$I = \frac{VA}{V} = \frac{240}{120} = 2 \text{ amperes}$$

For most fans, switching on and off may be accomplished by the pull-switch, an integral part of the fan (OFF-HIGH-MED-LOW), or with a solid-state switch having an infinite number of speeds. Some of the more expensive models offer a remote control similar to that used with television, VCRs, stereos, and cable channel selectors.

Most ceiling fan motors have an integral reversing switch to blow air downward or upward.

Do not use a standard incandescent lamp dimmer to control fan motors, fluorescent ballasts, transformers, low-voltage lighting systems (they use transformers to obtain the desired low voltage), motor-operated appliances, or other "inductive" loads. Serious overheating and damage to the motor can result. Fan speed controllers have internal circuitry that is engineered specifically for use on inductive load circuits. Fan speed controllers virtually eliminate any fan motor "hum." Read the label and the instructions furnished with a dimmer to be sure it is suitable for the application.

REVIEW

Note: Refer to the Code or the plans where necessary.

1. What circuit supplies the master bedroom? _____

2. What lighting besides the bedroom is supplied by this circuit? _____

    _____

3. What type of receptacles are provided in this bedroom? How many receptacles are there? _____

4. How many ceiling outlets are included in this circuit? _____

    _____

5. What wattage for the closet fixtures was used for calculating their contribution to the circuit? _____

6. What is the current draw for the ceiling fan/light fixture? _____

7. What is the estimated load in volt-amperes for the circuit supplying the master bedroom? _____

8. Which is installed first, the switch and outlet boxes or the cable runs? _____

    _____

9. How many conductors enter the ceiling fan/light wall box? _____

10. What type and size of box may be used for the ceiling fan/light wall box? _____

_____

_____

_____

11. What type of covers are used with 4-inch square outlet boxes? _____

_____

_____

12. a. Does the circuit for the master bedroom have a grounded circuit conductor?

    b. Does it have an equipment grounding conductor? _____

    c. Explain the difference between a "grounded" conductor and a "grounding" conductor. _____

_____

_____

_____

_____

_____

13. Approximately how many feet (meters) of two-wire cable and three-wire cable are needed to complete the circuit supplying the master bedroom? Two-wire cable _____ feet (_____ meters). Three-wire cable _____ feet (_____ meters).

14. If the cable is laid in notches in the corner studs, what protection for the cable must be provided? _____

_____

_____

15. How high are the receptacles mounted above the finished floor in this bedroom?

_____

16. Approximately how far from the bedroom door is the first receptacle mounted? (See the plans.) _____

17. What is the distance from the finished floor to the center of the wall switches in this bedroom? _____

18. The master bedroom features a sliding glass door. For the purpose of providing the proper receptacle outlets, answer the following statements true or false.

    • Sliding glass panels are considered to be wall space. _____

    • Fixed panels of glass doors are considered to be wall space. _____

19. What type of receptacle will be installed outdoors, just outside of the master bedroom?

_____

_____

20. Show your calculations of how to select a proper wall box for the closet fixture switch. Keep in mind that the available space between the wood casings is small. See Figure 9-1. _____

_____

_____

_____

_____

_____

_____

_____

21. When an outlet box is to be installed to support a ceiling fan, how must it be marked?

_____

_____

22. The following is a layout of the lighting circuit for the master bedroom. Using the cable layout shown in Figure 9-1, make a complete wiring diagram of this circuit. Use colored pencils or pens.

A19

GFCI
WP

MASTER BEDROOM

FAN/LIGHT

L

F

FAN   LIGHT

23. Connect the 3-wire cable, fan, light, fan speed control and light switch. Refer to *Section 200-7* to review the permitted use of the white conductor in the cable.

24. What is the width of a two-gang wall plate? _____

25. May a standard electronic dimmer be used to control the speed of a fan motor?

   _____

26. Outlet boxes that pass the tests at Underwriters Laboratories for the support of ceiling-suspended (paddle) fan are listed and marked "Acceptable For Fan Support" or "Suitable For Fan Support." When roughing-in the electrical wiring for a new home, this type of outlet box must be installed wherever there is a likelihood of installing a ceiling-suspended (paddle) fan. Name those locations in homes that you consider likely to support a ceiling-suspended (paddle) fan.

   _____

   _____

   _____

Lighting Branch-Circuit–Bathrooms, Hallway

OBJECTIVES

After studying this Unit, the student will be able to

- list equipment grounding requirements for bathroom installations.
- draw a wiring diagram for the bathroom and hallway.
- understand Code requirements for receptacles installed in bathrooms.
- understand Code requirements for receptacle outlets in hallways.
- discuss fundamentals of proper lighting for bathrooms.

Circuit A14 supplies the lighting outlets and receptacle outlet in the bedroom hall and both bathrooms. The receptacle outlets in the bathrooms are supplied by separate circuits A22 and A23. This is a Code requirement, discussed later in this Unit and in Unit 15.

Table 10-1 summarizes outlets and estimated load for the bathrooms and bedroom hall.

DESCRIPTION	QUANTITY	WATTS	VOLT-AMPERES
Receptacles @ 120 W each	1	120	120
Vanity fixtures @ 200 W each	2	400	400
Hall fixture	1	100	100
TOTALS	4	620	620

Table 10-1 Bathroom and bedroom hall: outlet count and estimated load. Circuit A14. The receptacles in the master bathroom and hall bathroom are not included in this table as they are connected to separate 20-ampere branch-circuits A22 and A23.

Note that each bathroom shows a ceiling heater/light/fan that is connected to separate circuit Ⓐ J and Ⓐ K. These are discussed in detail in Unit 22.

A hydromassage tub is located in the bathroom serving the master bedroom. It is connected to a separate circuit Ⓐ A and is also discussed in Unit 22.

The attic exhaust fan in the hall is supplied by a separate circuit Ⓐ L, also covered in Unit 22.

LIGHTING BRANCH-CIRCUIT A14 FOR HALLWAY AND BATHROOMS

Figure 10-1 and the electrical plans for this area of the home show that each bathroom has a fixture above the vanity mirror. Some typical fixtures are shown in Figure 10-2. Of course, the homeowner might decide to purchase a medicine cabinet complete with a self-contained lighting fixture. See Figure 10-3, which illustrates how to rough-in the wiring for each type. These fixtures are controlled by single-pole switches at the doors.

Figure 10-1 Cable layout for the master bathroom, hall bathroom, and hall. Layout includes lighting Circuit A14, two circuits (A11 and A12) for the heat/vent/lights, and two circuits (A22 and A23) for the receptacles in the bathrooms. The special purpose outlets for the hydromassage tub, attic exhaust fan, and smoke detector are covered elsewhere in this text.

Figure 10-2 Typical vanity (bathroom) lighting fixtures of the side bracket and strip types. *Courtesy* of Progress Lighting.

Figure 10-3 Two methods most commonly used for roughing-in the wiring for lighting above a vanity.

You must remember that a bathroom (powder room) should have proper lighting for shaving, combing hair, grooming, and so on. Mirror lighting can accomplish this, because a mirror will reflect what it "sees." If the face is poorly lit, with shadows on the face, that is precisely what will be reflected in the mirror.

A lighting fixture directly overhead will light the top of one's head, but will cause shadows on the face. Mirror lighting and/or adequate lighting above and forward of the standing position at the vanity can provide excellent lighting in the bathroom. Figures 10-4 through 10-6 show pictorial as well as section views of typical soffit lighting above a bathroom vanity.

Bathroom Receptacles

The *NEC®* in *Article 100* defines a bathroom as "an area including a basin, with one or more of the following: a toilet, a tub, or a shower." See Figure 10-7.

Bathroom receptacles are required to be GFCI protected per *Section 210-8(a)(1)* as discussed in Unit 6.

▶ At least one receptacle must be installed within 36 inches (914 mm) of the outside edge of the basin, and must be located on the wall adjacent to the basin, *Section 210-52(d)*.◀

Receptacles must not be installed face-up on counters or work surfaces, *Section 210-52(d)*.

Section 210-11(c)(3) requires that at least one 20-ampere branch-circuit must be provided for

Figure 10-4 **Positioning of bathroom lighting fixtures. Note the wrong way and the right way to achieve proper lighting.**

bathroom receptacles. ▶This separate 20-ampere branch-circuit shall not supply any other outlets, except that if the circuit supplies a single bathroom, then that 20-ampere branch-circuit is permitted to supply other electrical equipment in the same bathroom.◀

The intent of *Section 210-11(c)(3)* is to take bathroom receptacles off of the lighting branch-circuits in homes. In many instances, overloads caused by plugging in high-wattage hair dryers and similar high-wattage appliances have resulted in total loss of power to that particular branch-circuit.

The present wording of *Section 210-11(c)(3)* permits more than one bathroom to be connected to the one required separate 20-ampere branch-circuit.

In this text, we have chosen to run a separate 20-ampere branch-circuit A22 to the receptacle in the master bedroom bathroom, a separate 20-ampere branch-circuit A23 to the receptacle in the hall bathroom, and another separate 20-ampere branch-circuit B21 to the receptacle in the powder room located near the laundry. These separate circuits are included in the general lighting load calculations, so no additional load need be added.

Receptacles in Bathtub and Shower Spaces

Because of the obvious hazards associated with water and electricity, receptacles are *not* permitted to be installed in bathtub and shower spaces, *Section 410-57(c)*.

GENERAL COMMENTS ON LAMPS AND COLOR

Incandescent lamps (light bulbs) provide pleasant color tones, bringing out the warm red flesh tones similar to those of natural light. This is particularly true for the "soft" white lamps. Fluorescent lamps available today provide a wide range of "coolness" to "warmth." Rated in degrees Kelvin (i.e., 2500°K, 3000°K, 3500°K, 4000°K, 5000°K, etc.), the lower the Kelvin degrees, the "warmer" the color tone. Conversely, the higher the Kelvin degrees, the

TYPICAL INCANDESCENT RECESSED SOFFIT LIGHTING
OVER BATHROOM VANITY

END CUTAWAY OF SOFFIT SHOWING RECESSED
INCANDESCENT FIXTURES IN TYPICAL SOFFIT ABOVE
BATHROOM VANITY. TWO OR THREE FIXTURES
GENERALLY INSTALLED TO PROVIDE PROPER LIGHTING.

Figure 10-5 Incandescent soffit lighting.

TYPICAL FLUORESCENT RECESSED SOFFIT LIGHTING
OVER BATHROOM VANITY. NOTE ADDITIONAL
INCANDESCENT "SIDE-OF-MIRROR" LIGHTING.

END CUTAWAY VIEW OF SOFFIT ABOVE BATHROOM
VANITY SHOWING RECESSED FLUORESCENT FIXTURES
CONCEALED ABOVE TRANSLUCENT ACRYLIC LENS

Figure 10-6 Combination fluorescent and incandescent bathroom lighting.

"cooler" the color tone. Warm fluorescent lamps bring out the red tones, whereas cool fluorescent lamps tend to give a person's skin a pale appearance. Thus, you will find many types of fluorescent lamps, each providing different color tones and different efficiencies. These might be marked daylight **D** *(very cool)*, cool white **CW** *(cool)*, white **W** *(moder-*

ate), and warm white **WW** *(warm)*. These categories break down further into a *deluxe* **X** series, *specification* **SP** series, and *specification deluxe* **SPX** series. Unit 7 of this text discusses some basics of lighting. The Instructor's Guide lists some lamp manufacturers' publications that cover the subject of lighting in greater detail.

Figure 10-7 The Code in *Article 100* defines a *bathroom* as an area including a basin with one or more of the following: a toilet, a tub, or a shower.

HANGING FIXTURES IN BATHROOMS

Section 410-4(d) states that no parts of cord-connected fixtures, hanging fixtures, lighting track, pendants, or ceiling-suspended (paddle) fans shall be located within a zone measuring 3 feet (914 mm) horizontally and 8 feet (2.44 m) vertically from the top of the bathtub rim. Figure 10-8 projects the top view of the 3-foot (914 mm) restriction. Figure 10-9 shows both the allowable installation and Code violations of the given dimensions. Recessed or surface

Figure 10-8 No parts of cord-connected lighting fixtures, hanging fixtures, lighting track, pendants, or ceiling-suspended (paddle) fans shall be located above the tub and the area within 3 feet (914 mm) measured horizontally from the top of the bathtub rim (top view), *Section 410-4(d)*.

mounted lighting fixtures and recessed exhaust fans may be located within the restricted zone.

HALLWAY LIGHTING

The hallway lighting is provided by one ceiling fixture that is controlled with two three-way switches located at either end of the hall. The home run to Main Panel A has been brought into this ceiling outlet box.

RECEPTACLE OUTLETS IN HALLWAYS

One receptacle outlet has been provided in the hallway as required in *Section 210-52(h)*, which states that "for hallways of 10 feet (3.05 m) or more in length, at least one receptacle outlet shall be required."

For the purpose of determining the length of a hallway, the measurement is taken down the centerline of the hall, turning corners if necessary, but not passing through a doorway, Figure 10-10.

EQUIPMENT GROUNDING REQUIREMENTS FOR A BATHROOM CIRCUIT

All exposed metal equipment, including fixtures, electric heaters, faceplates, and similar items must be grounded. Equipment grounding requirements are found in *Section 250-110*.

Figure 10-9 No parts of cord-connected fixtures, hanging fixtures, lighting track, pendants, or ceiling-suspended (paddle) fans shall be allowed in shaded areas, *Section 410-4(d)*.

Figure 10-10 The centerline measurement of the bedroom hallway in this residence is 12'6" (3.8 m), which requires at least one wall receptacle outlet, *Section 210-52(h)*. This receptacle may be installed anywhere in the hall. The Code does not specify a location.

In general, all exposed noncurrent-carrying metal parts of electrical equipment must be grounded:

- if they are within 8 feet (2.44 m) vertically or 5 feet (1.52 m) horizontally of the ground or other grounded metal objects.

- if the ground and the electrical equipment can be touched at the same time.

- if they are located in wet or damp locations, such as in bathrooms, showers, and outdoors.

- if they are in electrical contact with metal. This requirement includes such things as metal lath, aluminum foil insulation, and metal sidings.

Equipment is considered grounded when it is properly and permanently connected to metal raceway, the armor of armored cable, the equipment grounding conductor in nonmetallic-sheathed cable, or a separate equipment grounding conductor. Of course, the means of grounding (the metal raceway, the armored cable, or the equipment grounding conductor in the nonmetallic-sheathed cable) must itself be properly grounded.

Section 250-114(3) lists the types of cord- and plug-connected appliances in residences that must be grounded. The second paragraph of *Section 250-114, Exception* accepts "double insulation" or its equivalent in lieu of grounding. Many hand-held appliances, such as electric drills, electric razors, and electric toothbrushes make use of the "double insulation" technique. Such appliances have a 2-wire cord and 2-wire plug cap instead of 3-wire cord that has an equipment grounding conductor and a 3-wire plug cap that has a ground prong.

Double-insulated appliances are clearly marked to indicate that they are double insulated.

Immersion Detection Circuit Interrupters

Another way to protect people from electrical shock is to use grooming appliances that are equipped with a special attachment plug-cap that is a liquid immersion detection circuit interrupter (IDCI). IDCIs are discussed in Unit 6.

Switches in Wet Locations

Do *not* install switches in wet locations, such as in bathtubs or showers, unless they are part of a listed tub or shower assembly, in which case the manufacturer has taken all of the proper precautions, and submitted the assembly to a recognized testing laboratory to undergo exhaustive testing to establish the safety of the equipment. See *Section 380-4.*

REVIEW

1. List the number and types of switches and receptacles used in Circuit A14.

2. There is a three-way switch in the bedroom hallway leading into the living room. Show your calculation of how to determine the box size for this switch. The box contains cable clamps.

3. What wattage was used for each vanity fixture to calculate the estimated load on Circuit A14? _____

4. What is the current draw for the answer given in question 3?

5. Exposed noncurrent-carrying metallic parts of electrical equipment must be grounded if installed within _____ feet (_____ meters) vertically or _____ feet (_____ meters) horizontally of bathtubs, plumbing fixtures, pipes, or other grounded metal work or grounded surfaces.

6. What color are the faceplates in the bathrooms? Refer to the specifications.

   _____

7. Most appliances of the type commonly used in bathrooms, such as hair dryers, electric shavers, and curling irons, have two-wire cords. These appliances are _____ insulated or _____ _____ _____ _____ protected.

8. a. The *NEC*® in *Section* _____ requires that all receptacles in bathrooms be _____ protected.

   b. The *NEC*® in *Section* _____ requires that all receptacles in bathrooms be connected to one or more separate 20-ampere branch-circuits that serve no other outlets.

   c. The *NEC*® in *Section* _____ permits the additional required 20-ampere branch-circuit for bathroom receptacles to supply more than one bathroom.

   d. The *NEC*® in *Section* _____ permits other electrical equipment to be connected to the additional required 20-ampere branch-circuit for bathroom receptacles, but only if the branch-circuit supplies a single bathroom, and the other equipment is located in that same bathroom.

   e. The *NEC*® in *Section* _____ prohibits mounting receptacles in bathrooms face-up in the countertops and work surfaces near basins.

9. Hanging lighting fixtures must be kept at least _____ feet (_____ meters) from the edge of the tub as measured horizontally. In bathrooms with high ceilings, where the hanging fixture is installed directly over the tub, it must be kept at least _____ feet (_____ meters) above the edge of the tub.

10. The following is a layout of a lighting circuit for the bathroom and hallway. Using the cable layout shown in Figure 10-1, make a complete wiring diagram of this circuit. Use colored pencils or pens.

A14

VANITY

VANITY

NOTE:
CEILING HEAT/VENT/LIGHT
AND RECEPTACLE OUTLET
IN BATHROOM NOT ON THIS
CIRCUIT. THESE ARE ON
CIRCUITS A22 AND A23.

BATHROOMS – HALLWAY

11. Circle the correct answer as to whether a receptacle outlet is required in the following hallways.

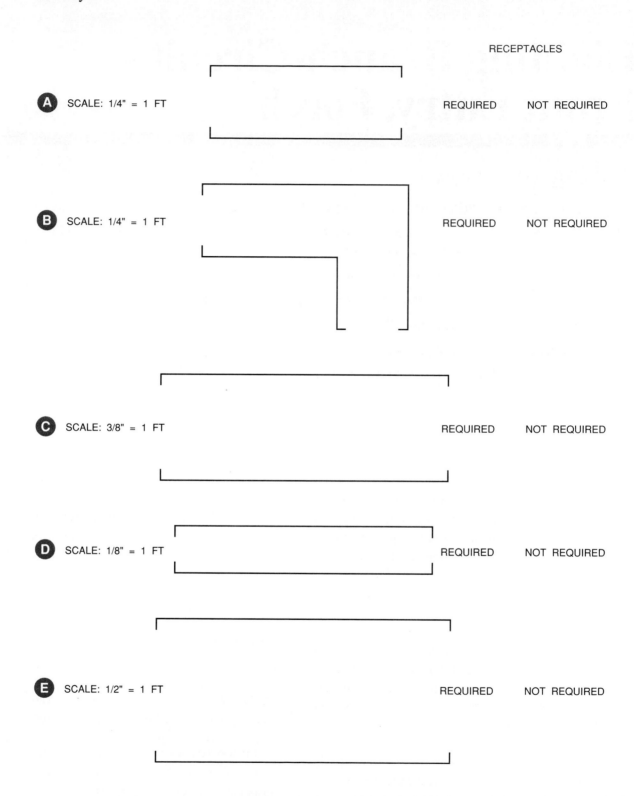

RECEPTACLES

A SCALE: 1/4" = 1 FT   REQUIRED   NOT REQUIRED

B SCALE: 1/4" = 1 FT   REQUIRED   NOT REQUIRED

C SCALE: 3/8" = 1 FT   REQUIRED   NOT REQUIRED

D SCALE: 1/8" = 1 FT   REQUIRED   NOT REQUIRED

E SCALE: 1/2" = 1 FT   REQUIRED   NOT REQUIRED

Lighting Branch-Circuit–Front Entry, Porch

OBJECTIVES

After studying this Unit, the student will be able to

- understand how to install a switch in a doorjamb for automatic ON/OFF when the door is opened or closed.
- discuss types of fixtures recommended for porches and entries.
- complete the wiring diagram for the entry-porch circuit.
- discuss the advantages of switching outdoor receptacles from indoors.
- define wet and damp locations.
- understand box fill for sectional ganged device boxes.

The front entry-porch is connected to Circuit A15. The home run enters the ceiling box in the front entry. From this box, the circuit spreads out, feeding the closet light, the porch bracket fixture, the two bracket fixtures on the front of the garage, one receptacle outlet in the entry, and one outdoor weatherproof GFCI receptacle on the porch, Figure 11-1. Table 11-1 summarizes the outlets and estimated load for the entry and porch.

Note that the receptacle on the porch is controlled by a single-pole switch just inside the front door. This allows the homeowner to plug in outdoor lighting, such as strings of ornamental Christmas lights, or decorative lighting, and have convenient switch control of that receptacle from inside the house, a nice feature.

Typical ceiling fixtures commonly installed in front entryways, where it is desirable to make a good first impression on guests, are shown in Figure 11-2.

Typical outdoor wall-bracket-style porch and entrance lighting fixtures are shown in Figure 11-3.

According to the Code, any location exposed to the weather is considered to be a wet location. *Section 410-4(a)* states that outdoor fixtures must be constructed so that water cannot enter or accumulate in lampholders, wiring compartments, or other electrical parts. These fixtures must be marked "Suitable for Wet Locations."

DESCRIPTION	QUANTITY	WATTS	VOLT-AMPERES
Receptacles @ 120 W	1	120	120
Weatherproof receptacles @ 120 W	1	120	120
Outdoor porch bracket fixture	1	100	100
Outdoor garage bracket fixtures @ 100 W each	2	200	200
Ceiling fixture	1	150	150
Closet recessed fixture	1	75	75
TOTALS	7	765	765

Table 11-1 Entry and porch: outlet count and estimated load. Circuit A15.

Figure 11-1 Cable layout of the entry-porch circuit A15. Note that cables from some of the other circuits are shown so that you can get a better idea of exactly how many wires will be found at the various locations.

Partially protected areas under roofs, open porches, or areas under canopies are defined by the Code as damp locations. Fixtures to be used in these locations must be marked "Suitable for Damp Locations." Many types of fixtures are available. Therefore, it is recommended that the electrician check the Underwriters Laboratories (U.L.) label on the fixture to determine the suitability of the fixture for a wet or damp location.

CIRCUIT A15

Circuit A15 is a rather simple circuit that has two sets of three-way switches: one set controlling the front entry ceiling fixture, the other set controlling the porch bracket fixture and the two bracket fixtures on the front of the garage. Also, one single-pole switch controls the weatherproof porch receptacle outlet, and another single-pole switch controls

Figure 11-2　Hall lighting fixtures: ceiling mount and chain mount. *Courtesy* Progress Lighting.

Figure 11-3　Outdoor porch and entrance lighting fixtures: wall bracket styles. *Courtesy* Progress Lighting.

the closet light. In the Review questions you will be asked to follow the suggested cable layout for making up all of the circuit connections.

Probably the most difficult part of this circuit is planning and making up the connections at the switch location just inside the front door. Here we find four toggle switches: (1) a three-way switch for ceiling fixture; (2) a three-way switch for porch and garage fixtures; (3) a single-pole switch for weatherproof receptacle on porch; and (4) a three-way switch for the fixture over the stairway leading to the basement (connected to the Recreation Room lighting circuit B12, not the entry-porch circuit A15).

This location is another challenge for determining the proper size wall box. Let's give it a try.

1. Add the circuit conductors
   $2 + 2 + 3 + 3 + 3 + 3 =$ 16

2. Add for equipment grounding wires (count 1 only)   1

3. Add eight for the four switches   8

4. Add for cable clamps (count 1 only)   1

   Total   26

To select a box that has adequate cubic-inch capacity for this example, refer to *Table 370-16(a)* in the *NEC*,® and Figures 2-19 and 2-34 in this text. One possibility would be to use four 3" × 2" × 3½" device boxes ganged together, providing 4 × 9 = 36 conductor count based on No. 14 AWG conductors.

When using nonmetallic boxes, refer to *Table 370-16(b)* and base the volume allowance on 2 cubic inches for each No. 14 AWG conductor and for the volume allowance for the wiring devices, cable clamps, and equipment grounding conductors. In the previous example, we are looking for a cubic-inch volume of 26 × 2 = 52 cubic inches.

An interesting possibility presents itself for the front entry closet. Although the plans show that the closet fixture is turned on and off by a standard single-pole switch to the left as you face the closet, a doorjamb switch could have been installed. A doorjamb switch is usually mounted about 6 feet (1.83 m) above the floor, on the inside of the 2 × 4 framing for the closet door. The electrician will run the cable to this point, and let it hang out until the carpenter can cut the proper size opening into the doorjamb for the box. After the finished woodwork is completed, the electrician then finishes installing the switch. Figure 11-4 illustrates a door switch. Note that the plunger on the switch is pushed inward when the edge of the door pushes on it as the door is closed, shutting off the light. The plunger can be adjusted in or out to make the switch work properly. These doorjamb switches come complete with their own special wall box.

The integral box furnished with door switches is generally suitable for one cable only. Wiring space is very limited. Do not plan on using this box for splices other than those necessary to make up the connections to the switch.

Figure 11-4   Door switch.

REVIEW

1. a. How many circuit wires enter the entry ceiling box? _____

   b. How many equipment grounding conductors enter the entry ceiling box? _____

2. Assuming that an outlet box for the entry ceiling has a fixture stud but has external cable clamps, what size box could be installed? Show calculations for both metal and nonmetallic boxes.

   _____

   _____

   _____

   _____

   _____

   _____

3. How many receptacle outlets and lighting outlets are supplied by Circuit A15?

   _____

4. Outdoor fixtures directly exposed to the weather must be marked as:

   a. suitable for damp locations

   b. suitable for dry locations only

   c. suitable for wet locations

   Circle correct answer.

5. Make a material list of all types of switches and receptacles connected to Circuit A15.

   _____

   _____

   _____

   _____

6. From left to right, facing the switches, what do the switches next to the front door control?

   _____

   _____

   _____

   _____

   _____

7. Who is to select the entry ceiling fixture? _____

8. The following layout is for Lighting Circuit A15, the entry, porch, and front garage lights. Using the cable layout shown in Figure 11-1, make a complete wiring diagram of this circuit. Use colored pencils to indicate the color of the conductors' insulation.

Lighting Branch-Circuit and Small Appliance Circuits for Kitchen

OBJECTIVES

After studying this Unit, the student will be able to

- discuss the features of an exhaust fan for the removal of kitchen cooking odors and humidity.
- explain the Code requirements for small appliance circuits in kitchens.
- understand the Code requirements for ground-fault protection for receptacles that serve countertops.
- discuss split-circuit receptacles.
- discuss multiwire circuits.
- discuss exhaust fan noise ratings.
- understand the overall general concept of grounding electrical equipment.
- know the difference between a receptacle outlet and a lighting outlet.
- discuss typical kitchen, dining room, undercabinet, and swag fixtures.

LIGHTING CIRCUIT B7

The lighting circuit supplying the kitchen and nook originates at Panel B. The circuit number is B7. The home run is very short, leading from Panel B in the corner of the recreation room to the switch box to the right of the kitchen sink. Here, the circuit continues upward through the junction box on the recessed fixture above the sink, which is approved for feed-through wiring (see Unit 7 for complete explanation of Code requirements for recessed fixtures) through the clock outlet box, and then across the kitchen ceiling connecting the ceiling light fixtures, the track light, the outdoor bracket fixture, and range hood exhaust fan/light unit, Figure 12-1.

Another way to run this circuit would be to run the branch-circuit home run to the clock outlet box, then over to the recessed fixture. This is a matter of considering how many conductors enter the boxes, the ease in making up the connections, and whether the junction box on the recessed fixture is "listed" for "through" wiring. Table 12-1 shows the outlets and estimated loads for the kitchen/nook area.

KITCHEN LIGHTING

The general lighting for the kitchen is provided by two four-lamp square fluorescent surface-mounted lighting fixtures shown in Figure 12-2. These fixtures

KITCHEN/NOOK LIGHTING

Figure 12-1   Cable layout for the kitchen lighting. This is Circuit B7. The appliance circuit receptacles are connected to 20-ampere circuits B13, B15, and B16 and are not shown on this lighting circuit layout.

Figure 12-2   Typical ceiling fixture of the type installed in the kitchen of this residence. This type of fixture might have four 20-watt fluorescent lamps, or it might have two 40-watt U-shaped lamps. *Courtesy* Progress Lighting.

DESCRIPTION	QUANTITY	WATTS	VOLT-AMPERES
Ceiling fixtures (eight 20-W FL lamps)	2	160	180
Recessed fixture over sink	1	100	100
Track light (nook)	1	100	100
Outdoor rear bracket fixture	1	100	100
Clock outlet	1	15	15
Range hood fan/light Two 60-W lamps		120	120
Fan motor (1⅓ A @ 120 V)		160	180
TOTALS	7	755	795

Table 12-1 Kitchen/Nook: outlet count and estimated load. Circuit B7.

are controlled by a single-pole switch located next to the doorway leading to the living room.

Above the breakfast nook table is a short section of track lighting controlled by two three-way switches: one located adjacent to the sliding door, the other at the doorway leading to the living room. Figure 12-3 illustrates just a few of the many types of lighting fixtures that might be secured to the track.

Figure 12-4 shows a variety of chain-suspended lighting fixtures of the type commonly used in dining rooms.

One goal of good lighting is to reduce shadows in work areas. This is particularly important in the kitchen. The two ceiling fixtures, the fixture above the sink, plus the fixture within the range hood will provide excellent lighting in the kitchen area. Of course, the track lighting in the nook area offers good lighting for that area.

Undercabinet Lighting

In some instances, particularly in homes with little outdoor exposure, the architect might specify strip fluorescent fixtures under the kitchen cabinets. Several methods may be used to install undercabinet lighting fixtures. They can be fastened to the wall just under the upper cabinets, Figure 12-5(A), installed in a recess that is part of the upper cabinets so that they are hidden from view, Figure 12-5(B), or fastened under and to the front of the upper cabinets, Figure 12-5(C). All three possibilities require close coordination with the cabinet installer to be sure that the wiring is brought out of the wall at the proper location to connect to the undercabinet fixtures.

Flat fluorescent and incandescent lighting fixtures are available that mount on the underside of upper cabinets. Some of these lighting fixtures are direct connected—others are cord- and plug-connected. These are pretty much out of eyesight, but provide excellent lighting for countertop work surfaces.

Lighting fixture manufacturers also offer shallow undercabinet track lighting systems in both 120 volts and 12 volts. These systems include track, track lighting heads for small halogen lamps, end fittings, and transformer with cord and plug. Some have two-level, high-low switching to vary the light intensity.

Visiting a lighting showroom, an electrical supply house, a home center, or a hardware store is an exciting experience. The choice of lighting fixtures is almost endless.

Figure 12-3 Fixtures that can be hung either single or on a track over dinette tables, such as in the nook area of this kitchen. *Courtesy* Progress Lighting.

Figure 12-4 Types of fixtures used over dining room tables. *Courtesy* Progress Lighting.

Figure 12-5 Methods of installing undercabinet lighting fixtures.

Lamp Type

Good color rendition in the kitchen area is achieved by using incandescent lamps, or by using warm white deluxe (WWX), warm white (WW), or deluxe cool white (CWX) 3000-K lamps wherever fluorescent lamps are used. See Unit 7 for details regarding energy saving ballasts and lamps.

FAN OUTLET

The fan outlet is connected to branch lighting circuit B7. Fans can be installed in a kitchen either to exhaust the air to the outside of the building or to filter and recirculate the air. Both types of fans can be used in a residence. The ductless hood fan does not exhaust air to the outside and does not remove humidity from the air.

Homes heated electrically with resistance-type units usually have excess humidity. The heating units neither add nor remove moisture from the air unless humidity control is provided. The proper use of weather stripping, storm doors, and storm windows, and a vapor barrier of polyethylene plastic film tends to retain humidity.

Vapor barriers are installed to keep moisture out of the insulation. This protection normally is installed on the warm side of ceilings, walls, and floors, under concrete slabs poured directly on the ground, and as a ground cover in crawl spaces. Insulation must be kept dry to maintain its effectiveness. For example, a 1% increase of moisture in insulating material can reduce its efficiency 5%. Humidity control using a humidistat on the attic exhaust fan is covered in Unit 22.

An exhaust fan simultaneously removes odors and moisture from the air and exhausts heated air, Figure 12-6. The choice of the type of fan to be installed is usually left to the owner. The electrician, builder, and/or architect may make suggestions to guide the owner in making the choice.

For the residence in the plans, a separate wall switch is not required because the speed switch, light, and light switch are integral parts of the fan.

Fan Noise

The fan motor and the air movement through an exhaust fan generate a certain amount of noise. The manufacturers of exhaust fans assign a sound-level rating so that it is possible to get some idea as to the noise level one might expect from the fan after it is installed.

The unit used to define fan noise is the *sone* (rhymes with tone). For simplicity, one sone is the noise level of an average refrigerator. The lower the sone rating, the quieter the fan. Technically, the sone is a unit of loudness equal to the loudness of a sound of one kilohertz at 40 decibels above the threshold of hearing of a given listener.

All manufacturers of exhaust fans provide this information in their descriptive literature.

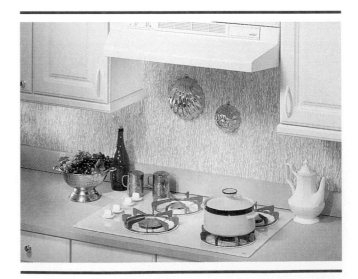

Figure 12-6 Typical range hood exhaust fan. Speed control and light are integral parts of the unit. *Courtesy* **NuTone.**

CLOCK OUTLETS

The clock outlet is connected to lighting circuit B7. The clock outlet in the face of the soffit may be installed in any sectional switch box or 4-inch square box with a single-gang raised plaster cover. A deep sectional box is recommended because a recessed clock outlet takes up considerable room in the box.

Some decorative clocks have the entire motor recessed so that only the hands and numbers are exposed on the surface of the soffit or wall. Some of these clocks require special outlet boxes, usually furnished with the clock. Other clocks require a standard 4-inch square outlet box. Accurate measurements must be taken to center the clock between the ceiling and the bottom edge of the soffit. If the clock is available while the electrician is roughing in the wiring, the dimensions of the clock can be checked to help in locating the clock outlet.

If an electrical clock outlet has not been provided, battery-operated clocks are available.

SMALL APPLIANCE BRANCH-CIRCUITS FOR RECEPTACLES IN KITCHEN

The Code requirements for small appliance circuits and the spacing of receptacles in kitchens are covered in *Sections 210-11(c)(1), 210-52(b),* and *220-16.* This was discussed in detail in Unit 3. Important points are:

- At least two 20-ampere small appliance circuits shall be installed, Figure 12-7.

- Either (or both) of the two circuits required in the kitchen is permitted to supply receptacle outlets in other rooms, such as a dining room, breakfast room, or pantry, Figure 12-7.

- These small appliance circuits shall be assigned a load of 1500 volt-amperes each when calculating feeders and service-entrance requirements.

- The receptacle for the refrigerator in the kitchen is *permitted* (not required) to be supplied by a separate 15- or 20-ampere branch circuit rather than connecting it to one of the two required 20-ampere small appliance branch circuits. This permission is found in *Section 210-52(b)(1), Exception No. 2.* No additional load has actually been added. The refrigerator load would have been there no matter what circuit it was connected to. The advantage of doing this is that it diverts the refrigerator load from the small appliance branch circuits in the kitchen. An additional 1,500 volt-amperes does not have to be added to the service-entrance calculations, *Section 220-16(a), Exception.*

- The clock outlet may be connected to an appliance circuit when this outlet is to supply and support the clock only (see Figure 3-3). In the residence in the plans, the clock outlet is connected to the lighting circuit.

- Countertop receptacles must be supplied by at least two 20-ampere small appliance circuits, Figure 12-7.

Figure 12-7 A receptacle outlet in a dining room may be connected to one of the two required small appliance circuits in the kitchen, *Section 210-52(b)(1).* **Countertop receptacles must be supplied by at least two 20-ampere small appliance circuits,** *Section 210-52(b)(3).*

- All 125-volt, single-phase, 15- and 20-ampere receptacles installed in kitchens to serve the countertops shall be GFCI protected, *Section 210-8(a)(6)*, Figure 12-8. A person can receive an electrical shock by touching a "live" wire and a grounded surface. The kitchen sink (plumbing, water) was always considered to be the hazard. The Code recognizes that there are other grounded "things" a person might touch, such as a range, oven, cooktop, refrigerator, microwave oven, or countertop food processor.

- A receptacle may be installed for plugging in a gas appliance to serve the ignition and/or clock devices on the appliance. See Figure 3-3.

- The small appliance circuits are provided to serve plug-in portable appliances. They are not permitted by the Code to serve lighting fixtures or appliances such as dishwashers, garbage disposals, or exhaust fans.

- When a receptacle is installed below the sink for easy plug-in connection of a food waste disposer, it is *not* to be connected to the required small appliance circuits, *Section 210-52(b)(2)*, and it is *not* required to be GFCI protected, *Section 210-8(a)(6)*. Refer to Unit 3 for spacing requirements for receptacles located in kitchens.

The 20-ampere small appliance circuits prevent

Figure 12-8 In kitchens, *all* 125-volt, single-phase, 15- and 20-ampere receptacles that serve countertop surfaces (walls, islands, or peninsulas) must be GFCI protected. The key words are "that serve countertop surfaces." GFCI personnel protection is *not* required for the refrigerator receptacle, a receptacle under the sink for plugging-in a food waste disposer, or a receptacle installed inside cabinets above the countertops for plugging-in a microwave oven attached to the underside of the cabinets. Where wet-bar sinks are installed outside of the kitchen area, *all* 125-volt, single-phase, 15- and 20-ampere receptacles that serve the countertops must be GFCI protected if they are within 6 feet (1.83 m) measured from the outer edge of the wet-bar sink. See *Sections 210-8(a)(6)* and *210-8(a)(7)*. Drawing is not to scale.

circuit overloading in the kitchen because of the heavy concentration and use of electrical appliances. For example, one type of cord-connected microwave oven is rated 1500 watts at 120 volts. The current required by this appliance alone is:

$$I = \frac{W}{E} = \frac{1500}{120} = 12.5 \text{ amperes}$$

Section 90-1(b) states that the Code requirements are considered necessary for safety. According to the Code, compliance with these rules will not necessarily result in an efficient, convenient, or adequate installation for good service or future expansion of electrical use.

Figure 12-9 shows four 20-ampere branch-circuits feeding the receptacles in the kitchen/nook area. These circuits are B13, B15, B16, and B22. The receptacle located behind the refrigerator is not required to be GFCI protected since it does not serve a counter surface area. By putting the refrigerator on a separate circuit, a ground-fault tripping of any of the GFCI-protected receptacles will not affect the refrigerator.

Figure 12-9 shows one way to arrange the 20-ampere branch-circuits in the kitchen/nook area. Obviously, there are other ways to arrange these branch-circuits. The intent is to arrange the circuits so as to provide the best availability of electrical power for appliances that will be used in the heavily concentrated work areas in the kitchen.

Section 210-8(a)(6) requires that all receptacles serving countertops in kitchens be GFCI protected.

Figure 12-9 Cable layout for kitchen/nook receptacles.

GFCI receptacles or GFCI circuit breakers can provide this protection.

In Figure 12-9, a feedthrough GFCI receptacle is installed to the right of the sink. Standard grounding-type receptacles are installed for the other two receptacles on Circuit B13. For Circuit B16, a feedthrough GFCI receptacle is installed for the first receptacle in the run, then a standard grounding-type receptacle is installed on the side of the range cooktop island.

Circuit B15 does not require any GFCI receptacles because none of these receptacles serve countertops.

Circuit B22 supplies only the non-GFCI receptacle behind the refrigerator.

Refer to Unit 6 for detailed discussion of GFCI protection.

Sliding Glass Doors

See Units 3 and 9 for a discussion on how to space receptacle outlets on walls where sliding glass doors are installed.

Small Appliance Branch-Circuit Load Calculations

The 20-ampere small appliance branch circuits are figured at 1,500 volt-amperes for each two-wire circuit when calculating feeders and services. The separate 20-ampere branch-circuit for the refrigerator is excluded from this requirement, *Section 220-16(a)*, *Exception*. See Unit 29 for complete coverage of load calculations.

SPLIT-CIRCUIT RECEPTACLES AND MULTIWIRE CIRCUITS

Split-circuit receptacles may be installed where a heavy concentration of plug-in load is anticipated, in which case each receptacle is connected to a separate circuit as in Figures 12-10 and 12-11.

A popular use of split-circuit receptacles is to control one receptacle with a switch and leave one receptacle "hot" at all times, as is done in the bedrooms and living room of this residence.

A PROPER WAY TO CONNECT GROUNDED NEUTRAL CONDUCTORS IN A MULTIWIRE BRANCH-CIRCUIT.

B IMPROPER WAY TO CONNECT GROUNDED NEUTRAL CONDUCTORS IN A MULTIWIRE BRANCH-CIRCUIT. NOT PERMITTED.

Figure 12-10 Connecting the grounded neutral conductors in a three-wire (multiwire) branch-circuit, *Section 300-13(b)*.

* SECTION 210-4(b) REQUIRES THAT A MEANS MUST BE PROVIDED TO SIMULTANEOUSLY DISCONNECT BOTH UNGROUNDED CONDUCTORS AT THE PANELBOARD WHERE THE BRANCH-CIRCUIT ORIGINATES. THIS COULD BE A 2-POLE SWITCH WITH FUSES, A 2-POLE CIRCUIT BREAKER, OR TWO 1-POLE CIRCUIT BREAKERS WITH A "LISTED" HANDLE TIE.

Figure 12-11   Split-circuit receptacle connected to multiwire circuit.

Split-circuit GFCI receptacles are not available. Refer to Unit 6, where GFCIs are discussed in great detail.

Therefore, the use of split-circuit receptacles is rather difficult to incorporate where GFCI receptacles are required. A multiwire branch-circuit could be run to a box where proper connections are made, with one circuit feeding one feedthrough GFCI receptacle and its downstream receptacles.

This would be used when it becomes more economical to run one three-wire circuit rather than installing two two-wire circuits. Again, it is a matter of knowing what to do, and when it makes sense to do it, Figure 12-12.

▶When multiwire circuits are installed, do not use the terminals of a receptacle or other wiring device to serve as the splice for the grounded neutral conductor. *Section 300-13(b)* states that: *"In multiwire branch circuits, the continuity of a grounded conductor shall not be dependent upon device."* Any splicing of the grounded conductor in a multiwire circuit must be made independently of the receptacle

Figure 12-12   This diagram shows how a multiwire circuit can be used to carry two circuits to a box, then split the three-wire circuit into two two-wire circuits. This circuitry does not require simultaneous disconnect of the ungrounded conductors as shown in Figure 12-11.

or lampholder, or other wiring device.◄ This requirement is illustrated in Figure 12-10. Note that the screw terminals of the receptacle *must* not be used to splice the neutral conductors. The hazards of an open neutral are discussed in Unit 17. Figure 12-13 shows two types of grounding receptacles commonly used for installations of this type.

When multiwire branch-circuits supply more than one receptacle on the same yoke, a means must be provided to disconnect simultaneously all of the "hot" conductors at the panelboard where the branch-circuit originates. Figure 12-11 shows that 240 volts are present on the wiring device; only 120 volts are connected to each receptacle on the wiring device. See *Sections 210-6(a)* and *(b)*.

The "simultaneous disconnect" rule applies to any receptacles, lampholders, or switches mounted on one yoke. See *Section 210-4(b)*.

RECEPTACLES AND OUTLETS

Article 100 defines a receptacle and a receptacle outlet, as illustrated in Figures 3-5, 12-14, and 12-15.

Receptacles are selected for circuits according to the following guidelines. A single receptacle connected to a circuit must have a rating not less than the rating of the circuit, *Section 210-21(b)*. In a residence, typical examples for this requirement for single receptacles are the clothes dryer outlet (30 amperes), or the range outlet (50 amperes), or the freezer outlet (15 amperes).

Circuits rated 15 amperes supplying two or more receptacles shall not contain receptacles rated over 15 amperes. For circuits rated 20 amperes supplying two or more receptacles, the receptacles connected to the circuit may be rated 15 or 20 amperes. See *Table 210-21(b)(3)*.

GENERAL GROUNDING CONSIDERATIONS

All electrical equipment located within reach of a grounded surface must be grounded. Metal water pipes, gas pipes, sinks, concrete floors in basements and garages, hot and cold air ducts, outdoor lighting, and similar scenarios leave little excuse for any electrical equipment not to be grounded. *Sections 250-110* and *112* provide the details for the above statement.

Section 250-114 tells us that cord- and plug-connected appliances must be grounded, unless they are designed with "double insulation." Appliances that are cord- and plug-connected with a 3-wire cord are grounded by the equipment grounding conductor in the cord. Appliances that are cord- and plug-connected with a 2-wire cord are "double-insulated." Double-insulated appliances provide two levels of protection against electric shock. Accepted levels of protection can be the insulation on a conductor, added insulating material, the plastic non-metallic enclosure of the appliance itself, or even a specified air-gap at an exposed terminal. It would

Figure 12-13 Grounding-type duplex receptacles, 15A–125V rating. Most duplex receptacles can be changed into a "split-circuit" receptacle by breaking off the small metal tab on the terminals. The above photo clearly shows this feature.

THESE ARE CONTACT DEVICES

THIS IS A RECEPTACLE

SINGLE RECEPTACLE "A" BY DEFINITION, "A" IS A SINGLE RECEPTACLE WIRING DEVICE.

MULTIPLE RECEPTACLE "B" BY DEFINITION, "B" IS A MULTIPLE RECEPTACLE WIRING DEVICE.

Figure 12-14 Receptacles: See *NEC*® definitions, *Article 100*.

THE BRANCH-CIRCUIT CONDUCTORS IN THE BOX ARE REFERRED TO AS AN *OUTLET.*

WHEN A RECEPTACLE IS CONNECTED TO THE CONDUCTORS, THE OUTLET IS REFERRED TO AS A *RECEPTACLE OUTLET.*

WHEN A LIGHTING FIXTURE IS CONNECTED TO THE CONDUCTORS, THE OUTLET IS REFERRED TO AS A *LIGHTING OUTLET.*

Figure 12-15   An *outlet* is defined in the *NEC*® as "A point on the wiring system at which current is taken to supply utilization equipment." When a receptacle is connected to the conductors, the outlet is referred to as a *receptacle outlet*. When a lighting fixture is connected to the conductors, the outlet is referred to as a *lighting outlet*.

take two failures to create a shock hazard to the user. U.L. Standard 1097 *Double-Insulated Systems for Use in Electrical Equipment* is the standard that electrical appliance manufacturers conform to.

The equipment grounding conductor found in nonmetallic-sheathed cables, the metal armor of Type AC armored cable, metal raceways such as EMT, and flexible metal conduit provide the required grounding means, *Section 250-118*.

In the case of nonmetallic-sheathed cable, never use the bare *equipment grounding conductor* as a *grounded circuit conductor* and never use the *grounded circuit conductor* as an *equipment grounding conductor*. These two conductors come together only at the main service panel, never in branch-circuit wiring. Refer to *Sections 250-140*, and *250-142*.

REVIEW

Note: Refer to the Code or plans where necessary.

1. If everything on Circuit B7 were turned on, what would be the total current draw?

   _____

   _____

2. From what panel does the kitchen lighting circuit originate? What size conductors are used? _____

   _____

3. How many lighting fixtures are connected to the kitchen lighting circuit? _____

_____

4. What color fluorescent lamps are recommended for residential installations?

_____

_____

5. a. What is the minimum number of 20-ampere small appliance circuits required for a kitchen according to the Code? _____

   b. How many are there in this kitchen? _____

6. How many receptacle outlets are provided in the kitchen? _____

7. What is meant by the term two-circuit (split-circuit) receptacles? _____

_____

_____

_____

_____

_____

8. Circle the correct response. Duplex receptacles connected to the 20-ampere small appliance branch-circuits in kitchens and dining rooms:

   a. may be rated 15 amperes.   (True) (False)

   b. must be rated 15 amperes.   (True) (False)

   c. may be rated 20 amperes.   (True) (False)

   d. must be rated 20 amperes.   (True) (False)

9. In kitchens, a receptacle must be installed at each counter space _____ inches or wider.

10. A fundamental rule regarding the grounding of metal boxes, fixtures, and so on, is that they must be grounded when "in reach of _____ ."

11. How many circuit conductors enter the box

   a. where the clock will be installed? _____

   b. in the ceiling box over which the track will be installed? _____

   c. at the switch location to the right of the sliding door? _____

12. How much space is there between the countertop and upper cabinets? _____

13. Where is the speed control for the range hood fan located? _____

14. Who is to furnish the range hood? _____

15. List the appliances in the kitchen that must be electrically connected. _____

_____

_____

_____

_____

_____

_____

16. Complete the wiring diagram, connecting feedthrough GFCI 2 to also protect receptacle 1, both to be supplied by Circuit A1. Connect feedthrough GFCI 3 to also protect
receptacle 4, both to be supplied by Circuit A2. Use colored pencils or markers to
show proper color. Assume that the wiring method is electrical metallic tubing, where
more freedom in the choice of insulation colors is possible.

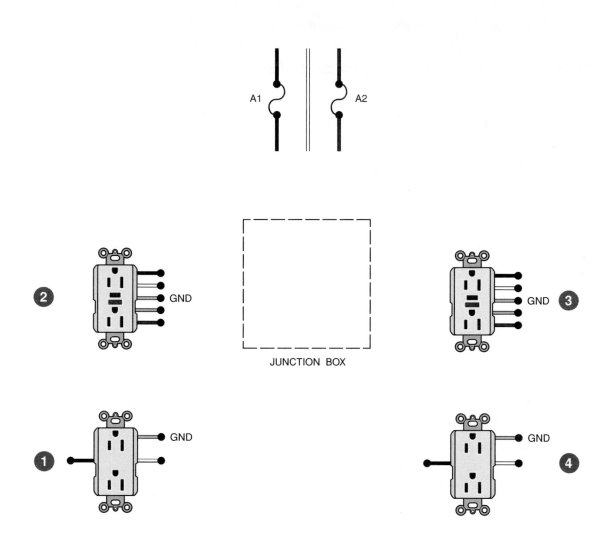

JUNCTION BOX

17. Each 20-ampere small appliance branch-circuit load demand shall be determined
at _____ . Choose one.

   a. 2400 volt-amperes   b. 1500 volt-amperes   c. 1920 volt-amperes

18. Is it permitted to connect an outlet supplying a clock receptacle to a 20-ampere small
appliance circuit? _____

_____

_____

_____

19. a. The Code requires a minimum of two small appliance circuits in a kitchen. Is it permitted to connect a receptacle in a dining room to one of the kitchen small appliance circuits? _____

What Code section applies to this situation? _____

b. May outdoor weatherproof receptacles be connected to a 20-ampere small appliance circuit? _____

c. Receptacles located above the countertops in the kitchen must be supplied by at least _____ 20-ampere small appliance circuits.

20. According to *Section 210-52*, no point along the floor line shall be more than _____ feet (_____ meters) from a receptacle outlet. A receptacle must be installed in any wall space _____ feet (_____ meters) wide or greater.

21. The Code states that in multiwire circuits, the screw terminals of a receptacle must *not* be used to splice the neutral conductors. Why?

_____

_____

_____

_____

_____

_____

22. Electric fans produce a certain amount of noise. It is possible to compare the noise levels of different fans prior to installation by comparing their _____ ratings.

23. Is it permitted to connect the white grounded circuit conductor to the equipment grounding terminal of a receptacle? _____

24. What is unique about 120-volt appliances that are two-wire cord- and plug-connected instead of being three-wire cord- and plug-connected? _____

_____

_____

25. a. All 125-volt, single-phase 15- and 20-ampere receptacles installed in a kitchen that serve countertop surfaces must be GFCI protected. (True) (False) Circle the correct answer.

b. All 125-volt, single-phase 15- and 20-ampere receptacles installed within 6 feet (1.83 m) of a wet-bar sink outside of the kitchen that serve countertop surfaces must be GFCI protected. (True) (False) Circle the correct answer.

26. The kitchen/nook features a sliding glass door. For the purpose of providing the proper receptacle outlets, answer the following statements *true* or *false*.

a. Sliding glass panels are considered to be wall space. _____

b. Fixed panels of glass doors are considered to be wall space. _____

27. The following is a layout for the lighting circuit for the Kitchen/Nook. Complete the wiring diagram using colored pens or pencils to show the conductors' insulation color.

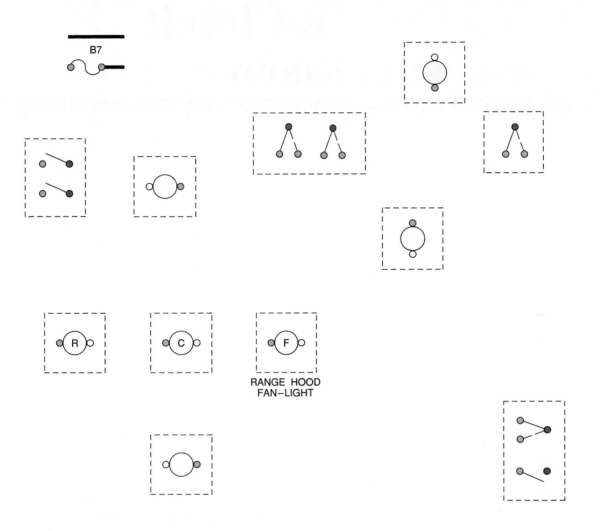

RANGE HOOD
FAN–LIGHT

KITCHEN–NOOK

Lighting Branch-Circuit for the Living Room

OBJECTIVES

After studying this Unit, the student will be able to

- understand various types of dimmer controls.
- connect dimmers to incandescent lamp loads and fluorescent lamp loads.
- understand the phenomenon of incandescent lamp inrush current.
- discuss Class P ballasts and ballast overcurrent protection.
- discuss the basics of track lighting.

LIGHTING CIRCUIT OVERVIEW

The feed for the living room lighting branch-circuit is connected to Circuit B17, a 15-ampere circuit. The home run is brought from Panel B to the weatherproof receptacle outside of the living room next to the sliding door. The circuit is then run to the split-circuit receptacle just inside the sliding door, almost back-to-back with the outdoor receptacle, Figure 13-1.

A three-wire cable is then carried around the living room feeding in and out of the eight receptacles. This three-wire cable carries the black and white circuit conductors plus the red wire, which is the switched conductor. Remember, because these receptacles will provide most of the general lighting for the living room through the use of floor lamps and table lamps, it is advantageous to have some of these lamps controlled by the three switches—two three-way switches and one four-way switch.

Note that the receptacle below the four-way switch is required according to the Code ruling that states that a receptacle outlet shall be installed in any wall space 2 feet (611 mm) or more in width, *Section 210-52(a)(2)(a)*. It is highly unlikely that a split-circuit receptacle connected for switch control,

as are the other receptacles in the living room, will ever be needed. Plus, the short distance to the study/bedroom receptacle makes it economically sensible to make up the connections as shown on the cable layout.

Accent lighting above the fireplace is in the form of two recessed fixtures controlled by a single-pole dimmer switch. See Unit 7 for installation data and Code requirements for recessed fixtures.

Table 13-1 summarizes the outlets and estimated load for the living room circuit.

The spacing requirements for receptacle outlets and location requirements for lighting outlets are covered in Unit 3, as is the discussion on the spacing of receptacle outlets on exterior walls where sliding glass doors are involved, *Section 210-52(a)(2)(b)*.

TRACK LIGHTING *(ARTICLE 410, PART R)*

The plans show that track lighting is mounted on the ceiling of the living room on the wall opposite the fireplace.

Track lighting provides accent lighting for a fireplace, a painting on the wall, some item (sculpture or collection) that the homeowner wishes to focus attention on, or it may be used to light work areas

Figure 13-1   Cable layout for the living room. Note that the receptacle on the short wall where the four-way switch is located is fed from the study/bedroom Circuit A21.

DESCRIPTION	QUANTITY	WATTS	VOLT-AMPERES
Receptacles @ 120 W Note: Receptacle under S₄ is connected to Study/Bedroom circuit.	8	960	960
Weatherproof receptacle	1	120	120
Track light five lamps @ 40 W each	1	200	200
Fireplace recessed fixtures @ 75 W each	2	150	150
TOTALS	12	1430	1430

Table 13-1   Living room outlet count and estimated load. Circuit B17.

such as counters, game tables, or tables in such areas as the kitchen nook.

Differing from recessed fixtures and individual ceiling fixtures that occupy a very definite space, track lighting offers flexibility because the actual lampholders can be moved and relocated on the track as desired.

Lampholders from track lighting selected from hundreds of styles are inserted into the extruded aluminum or PVC track at any point (the circuit conductors are in the track), and the plug-in connector on the lampholder completes the connection. In addition to being fastened to the outlet box, the track is generally fastened to the ceiling with toggle bolts

Figure 13-2   End-feed track light.

Figure 13-3   Center-feed track light.

or screws. Various track light installations are shown in Figures 13-2, 13-3, and 13-4.

Note on the plans for the residence that the living room ceiling has wood beams. This presents a challenge when installing track lighting that is longer than the space between the wood beams.

There are a number of things the electrician can do:

1. Install pendant kit assemblies that will allow the track to hang below the beam, Figure 13-5.

2. Install conduit conductor fittings on the ends of sections of the track, then drill a hole in the beam through which ½-inch electrical metallic

Figure 13-4   Plug-in track light.

tubing can be installed and connected to the conduit connector fittings, Figure 13-6. Running track lighting through walls or partitions is prohibited by *Section 410-101(c)(7)*.

3. Run the branch-circuit wiring concealed above the ceiling to outlet boxes located at the points where the track is to be fed by the branch-circuit wiring, Figure 13-7.

Figure 13-5   Track lighting mounted on the bottom of a wood beam. The supply wires to the track assembly are fed through the pendant.

Figure 13-6   Track lighting mounted on the ceiling between the wood beams. The two track assembly sections are connected together by means of a conduit that has been installed through a hole drilled in the beam.

Figure 13-7   Track lighting mounted on the ceiling between the wood beams. Each of the two track assembly sections are connected to outlet boxes in the ceiling in the same manner that a regular lighting fixture is hung.

Where to Mount Track Lighting

The ceiling height, the aiming angle, the type of lampholder, the type of lamp, and the type of lighting to be achieved are all important factors that must be considered in order to install the lighting track at the proper distance from the wall.

For example, consider an installation that will illuminate an oil painting that is to be centered 5'5" (1.67 m) (average eye level) from the floor. If the ceiling height is 9'0" (2.74 m), and the lampholder and lamp information tell us that the aiming angle should be 60°, we would find in the manufacturer's installation data that the track should be mounted 2 feet (610 mm) from the wall.

If, in this example, the ceiling height was 8'0" (2.44 m), then the distance to the track from the wall would be 18" inches (457 mm).

Because all of the factors are variables that affect the positioning of the track, it is absolutely necessary to consult the track lighting manufacturer's catalog for recommendations that will achieve the desired results.

Track lighting shall not be installed less than 5 feet (1.52 m) above the floor unless it is of the low-voltage type that has an open-circuit voltage of less than 30 volts RMS, or if the track is protected from physical damage. See *Section 410-101(c)(8)*.

▶ Track lighting is not permitted within 3 feet (914 mm) horizontal and 8 feet (2.44 m) vertical zone measured from the top of a bathtub rim, *Section 410-101(c)(9)*.◀ This restricted zone is explained in Unit 10.

The track itself must be grounded properly, *Section 410-105(b)*. The plug-in connector also has a ground contact to ensure that the track lighting fixture is safely grounded. Figure 13-8 shows tracks in cross-sectional views.

Figure 13-8   Tracks in cross-sectional views.

Typical fixtures that attach to track lighting systems are shown in Figure 13-9.

The track lighting in the living room is turned on and off by a single-pole dimmer switch.

Some sections of track systems are permitted to be cut to length in the field by the installer. Be sure to check the manufacturer's instructions on where and how to make the cut, and how to close off the cut with a proper end cap.

Portable Lighting Tracks

Portable lighting tracks per U.L. Standard 153 shall:

- not be longer than 8 feet (2.44 m).

- have a cord not less than 5 feet (1.52 m) permanently attached to the track by the manufacturer.

- have overcurrent protection (fuse or circuit breaker) built into the plug cap. Overcurrent protection in the plug cap is not required if the conductors in the cord are No. 12 AWG rated 20 amperes.

Figure 13-9   Typical fixtures that attach to track lighting systems. *Courtesy* Progress Lighting.

- not have its length altered in the field. To do so would void the U.L. listing.

▶ Remember, *Section 110-3(b)* of the *NEC*® states that "listed or labeled equipment shall be installed and used in accordance with any instructions included in the listing or labeling."◀

Supporting Track Lighting

Lighting tracks that are 4 feet (1.22 m) or less in length must be fastened at two places minimum. Then, if more track is needed, install at least one additional support for each individual extension that is 4 feet (1.22 m) long or less, *Section 410-104*.

Load Calculations for Track Lighting

▶ For track lighting installed in homes, it is not necessary to add additional loads for the purpose of branch-circuit, feeder, or service-entrance calculations. Track lighting is just another form of general lighting. This is confirmed in *Sections 220-3(b)(10)* and *220-12(b)*.◀

However, if the lighting track is to be installed in a commercial building, a loading factor of 150 VA for each 2 feet (610 mm) of lighting track must be added to the calculations for branch-circuit, feeder, and service-entrance computations, *Section 220-12(b)*.

For further information regarding track lighting, refer to *Part R* of *Article 410*.

DIMMER CONTROLS FOR HOMES

Dimmers are used in homes to lower the level of light. There are two basic types.

Electronic Dimmers

Figure 13-10 shows two types (slide and rotate) of electronic dimmers. These are very popular. They "fit" in a standard switch (device) box. They can replace standard single-pole or three-way toggle switches. See Figure 13-11 for the typical hookup of a single-pole and a three-way dimmer. Because dimmer and fan speed controls are physically larger than most wiring devices (plugs and switches), make sure there is enough room in the wall box to accommodate the dimmer, taking into consideration the number of conductors that are in the switch box. To "jam" the dimmer in may cause trouble, such as short-circuits and ground faults. The most common type of electronic dimmer for residential use is rated 125 volts, 60-hertz alternating current, 600 watts. Also available are 1000-watt, 1500-watt, and 2000-watt dimmer switches. The 1500-watt and 2000-watt dimmers fit into a single-gang wall box, but require a two-gang plate because of the need to cover the heat sinking

Figure 13-10 Some typical electronic dimmers. Both perform the same function of varying the voltage to the connected incandescent lamp load. (A) rotating knob. (B) slider knob. *Courtesy* Pass & Seymour/Legrand.

CAUTION: THE POWER MUST BE OFF BEFORE THE DIMMER IS CONNECTED TO THE CIRCUIT.

A LIGHT CONTROLLED BY A SINGLE-POLE DIMMER SWITCH.

B LIGHT CONTROLLED BY A THREE-WAY DIMMER SWITCH AND A STANDARD THREE-WAY SWITCH.

Figure 13-11   Use of single-pole and three-way electronic solid-state dimmer controls in circuits.

"fins." To prevent an overload, check the total wattage to be controlled with the rated capacity of the dimmer. The maximum wattage that can safely be connected is shown on the nameplate of the dimmer.

Dimmers that install in flush device boxes are intended only to control permanently installed incandescent. The instructions will state "Do not use to control receptacle outlets." The reasons for this are:

1. Serious overloading could result, because it may not be possible to determine or limit what other loads might be plugged in the receptacle.

2. Dimmers reduce voltage. Reduced voltage supplying a television, radio, stereo components, personal computer, vacuum cleaner, etc., can result in costly damage to the appliance.

Electronic dimmers for use *only* to control incandescent lamps are marked with a *wattage* rating. Serious overheating and damage to the transformer and to the electronic dimmer will result if an "incandescent only" dimmer is used to control inductive loads.

For example, a dimmer marked 600 watts @ 120 volts is capable of safely carrying 5 amperes.

$$\text{Amperes} = \frac{\text{Watts}}{\text{Volts}} = \frac{600}{120} = 5 \text{ amperes}$$

Dimmers that are suitable for control of "inductive" loads, such as lighting systems that incorporate a transformer as in low voltage outdoor lighting and low voltage track lighting, will be marked in *volt-amperes*. The instructions will clearly indicate that the dimmer can be used on inductive loads.

See Unit 7 for discussion on watts versus volt-amperes.

Do *not* hook up a dimmer with the circuit "hot." Not only is it dangerous because of the possibility of receiving an electrical shock, but the electronic solid-state dimmer can be destroyed when the splices are being made as a result of a number of "make and breaks" before the actual final connection is completed. The momentary inrush of current each time there is a "make and break" can cause the dimmer's internal electronic circuitry to heat up beyond its thermal capability. As one lead is being connected, the other lead could come in contact with "ground." Unless the dimmer has some sort of "built-in" short-circuit protection, the dimmer is destroyed. The manufacturer's warranty will probably be null and void if the dimmer is worked "hot." It is well worth the time to take a few minutes to de-energize the circuit.

Electronic dimmers use triacs to "chop" the ac sine wave, so some incandescent lamps might "hum" because the lamp filament vibrates. This hum can usually be eliminated or greatly reduced by turning the dimmer setting to a different brightness level, changing the lamp, installing a "rough service lamp," or as a last resort, changing the dimmer.

Autotransformer Dimmers

Figure 13-12 shows an autotransformer (one winding)-type dimmer. These are physically larger than the electronic dimmer, and require a special wall box. They are not ordinarily used for residential applications unless the load to be controlled is very large. They are generally used in commercial applications for the control of large loads. The lamp intensity is controlled by varying the voltage to the lamp load. When the knob is rotated, a brush contact moves over a bared portion of the transformer winding. As with electronic dimmers, they will not work on direct current, but that is not a problem because all homes are supplied with alternating current. The maximum wattage to be controlled is marked on the nameplate of the dimmer. Overcurrent protection is built into the dimmer control to prevent burnout due to an overload.

Figure 13-13 illustrates the connections for an autotransformer dimmer that controls incandescent lamps.

Dimming Fluorescent Lamps

To dim incandescent lamps, the voltage to the lamp filament is varied (raised or lowered) by the dimmer control. The lower the voltage, the less intense is the light. The dimmer is simply connected "in series" with the incandescent lamp load.

Fluorescent lamps are a different story. The connection is not a simple "series" circuit. To dim fluorescent lamps, special dimming ballasts and special dimmer switches are required. The dimmer control allows the dimming ballast to maintain a voltage to the cathodes that will maintain the cathodes' proper

Figure 13-12   Dimmer control, autotransformer type.

Figure 13-13   Dimmer control (autotransformer) wiring diagram for incandescent lamps.

operating temperature, and also allows the dimming ballast to vary the current flowing in the arc. This in turn varies the intensity of light coming from the fluorescent lamp.

There is an exception to the need to use special fluorescent dimming ballasts and special dimmer controls. Some compact fluorescent lamps are available that have an integrated circuit built right into the lamp, allowing the lamp to be controlled by a standard incandescent wall dimmer. These dimmable compact fluorescent lamps can be used to replace standard incandescent lamps of the medium screw shell base type. No additional wiring is neces-

sary. They are designed for use with dimmers, photocells, occupancy sensors, and electronic timers that are marked "Incandescent Only."

Figure 13-14 illustrates the connections for an autotransformer-type dimmer for the control of fluorescent lamps using a special dimming ballast. The fluorescent lamps are F40T12 rapid start lamps. With the newer electronic dimming ballasts, 32-watt, rapid start T8 lamps are generally used.

Figure 13-15 shows an electronic dimmer for the control of fluorescent lamps using a special dimming ballast.

Fluorescent dimming ballasts and controls are

Figure 13-14   Dimmer control wiring diagram for fluorescent lamps.

Figure 13-15   This wiring diagram shows the connections for an electronic dimmer designed for use with a special dimming ballast for rapid start lamps. Always check the dimmer and ballast manufacturers' wiring diagrams, as the color coding of the leads and the electrical connections may be different than shown in this diagram. The dimmer shown is a "slider" type, and offers wide-range (20% to 100%) adjustment of light output for the fluorescent lamp.

available in many types such as: simple "on-off," 100% to 1% variable light output, 100% to 10% variable light output, and two-level 100%/50% light output. Wirless remote control is also an option.

Here are some hints to ensure proper dimming performance of flourescent lamps:

- Make sure that the lamp reflector and the ballast case are solidly grounded.

- Do not mix ballasts or lamps from different manufacturers.

- Do not mix single- and two-lamp ballasts.

- "Age" (sometimes referred to as "seasoning") the lamps for approximately 100 hours before dimming. This allows the gas cathodes in the lamps to stabilize. Without "aging," it is possible to have unstable dimming, lamp striation (stripes), and/or unbalanced dimming characteristics between lamps.

- Because incandescent lamps and flourescent lamps have different characteristics, they cannot be controlled simultaneously with a single dimmer control.

- Do not control electronic ballasts and magnetic ballasts simultaneously with a single dimmer control.

Always read and follow the manufacturer's instructions furnished with dimmers and ballasts.

Without special dimming flourescent ballasts and controls, a simple way to have two light levels is to use 4-lamp flourescent fixtures such as those installed in the Recreation Room. Then, control the ballast that supplies the center two lamps with one wall switch, and control the ballast that supplies the outer two lamps with a second wall switch. This is sometimes referred to as an "inboard/outboard" connection.

REVIEW

Note: Refer to the Code or the plans where necessary.

1. To what circuit is the living room connected? _____

2. How many receptacles are connected to the living room circuit? _____

3. a. How many wires enter the switch box at the four-way switch location? _____

   b. What type and size of box may be installed at this location? _____

   _____

   _____

   _____

4. How many wires must be run between an incandescent lamp and its dimmer control?

   _____

5. Complete the wiring diagram for the dimmer and lamp.

120-VOLT SUPPLY

6. Is it possible to dim standard fluorescent ballasts? _____

7. a. How many wires must be run between a dimming-type fluorescent ballast and the dimmer control? _____

   b. Is a switch needed in addition to the dimmer control? _____

   _____

   _____

   _____

   _____

8. Explain why fluorescent lamps having the same wattage draw different current values.

   _____

   _____

   _____

   _____

   _____

9. What is the total current consumption of the track lighting and recessed fixtures above the fireplace? Show your calculations.

10. How many television outlets are provided in the living room? _____

11. Where is the telephone outlet located in the living room? _____

    _____

12. Solid-state electronic dimmers of the type sold for residential use (shall) (shall not) be used to control fluorescent fixtures. These dimmers (are) (are not) intended for speed control of small motors. Circle the correct answers.

13. When calculating a dwelling lighting load on the "volt-ampere per square foot" basis, the receptacle load used for floor lamps, table lamps, and so forth,
    a. must be added to the general lighting load at 1½ amperes per receptacle.
    b. must be added to the general lighting load at 1,500 volt-amperes for every 10 receptacles.
    c. is already included in the calculations as part of the volt-amperes per square foot.
    Circle the correct letter.

14. A layout of the outlets, switches, dimmers, track lighting, and recessed fixtures is shown in the following diagram. Using the cable layout of Figure 13-1, make a complete wiring diagram of this circuit. Use colored pencils or marking pens to indicate conductors.

15. Prepare a list of television outlets, telephone outlets, and wiring devices shown on the plans and cable layout for the living room area. Include the number of each type present.

_____

_____

_____

_____

_____

_____

16. The living room features a sliding glass door. For the purpose of providing the proper receptacle outlets, answer the following statements *true* or *false*.

   a. Sliding glass panels are considered wall space. _____

   b. Fixed panels of glass doors are considered wall space. _____

17. a. May lengths of track lighting be added when the track is permanently connected?

   _____

   b. May lengths of track lighting be added when the track is cord-connected?

   _____

18. Must track lighting always be fed (connected) at one end of the track? _____

   _____

19. May a standard electronic dimmer be used to control low-voltage lighting that incorporates a transformer, or for controlling the speed of a ceiling fan motor?

   Yes _____   No _____ .

20. Does the Code permit cutting lengths of portable track lighting to change their length?

   Yes _____   No _____ .

21. Why should the power be turned off when hooking up a dimmer? _____

   _____

   _____

Lighting Branch-Circuit for the Study/Bedroom

OBJECTIVES

After studying this Unit, the student will be able to

- discuss valance lighting.
- make all connections in the study/bedroom for the receptacles, switches, fan, and lighting.

CIRCUIT OVERVIEW

The study/bedroom circuit is so named because it can be used as a study for the present, providing excellent space for home office, personal computer, or den. Should it become necessary to have a third bedroom, the change is easily made.

Circuit A21 originates at Main Panel A. This circuit feeds the study/bedroom at the receptacle in the living room just outside of the study. From this point, the circuit feeds the split-circuit receptacle just inside the door of the study. A three-wire cable runs around the room to feed the other four split-circuit receptacles. Figure 14-1 shows the cable layout for Circuit A21.

The black and white conductors carry the "live" circuit, and the red conductor is the "switch return" for the control of one portion of the split-circuit receptacles.

There are four switched split-circuit receptacles in the study/bedroom controlled by two three-way switches. A fifth receptacle on the short wall leading to the bedroom hallway is not switched. The ready access to this receptacle makes it ideal for plugging in vacuum cleaners or similar appliances.

A ceiling fan/light fixture is mounted on the ceiling. This installation is covered in Unit 9. Table 14-1 summarizes the outlets and estimated load for the study/bedroom circuit.

VALANCE LIGHTING

Above the windows, indirect fluorescent lighting is provided. Figure 14-2 illustrates one method of installing fluorescent valance lighting.

If the fluorescent valance lighting is to have a dimmer control, then special dimming ballasts

DESCRIPTION	QUANTITY	WATTS	VOLT-AMPERES
Receptacles @ 120 W each	6	720	720
Closet fixture	1	75	75
Paddle fan/light	1		
Three 50-W lamps		150	150
Fan motor (0.75 A 120 V)		80	90
Valance lighting	1		
Two 40-W fluorescent lamps		80	90
TOTALS	9	1105	1125

Table 14-1 Study/bedroom outlet count and estimated load. Circuit A21.

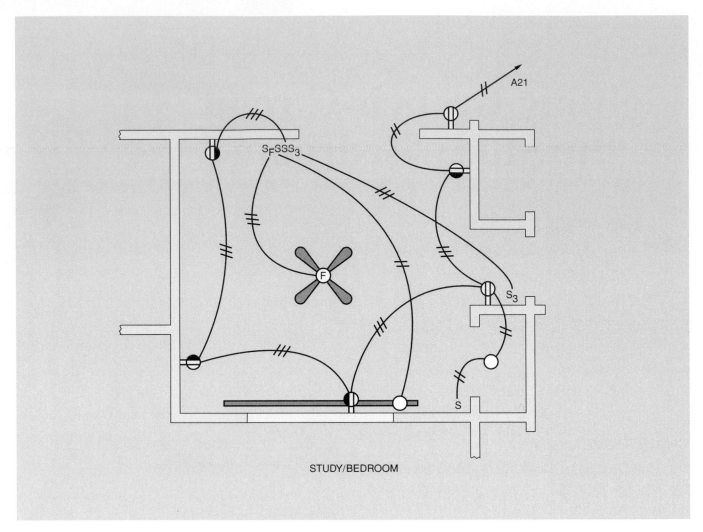

STUDY/BEDROOM

Figure 14-1   Cable layout for the study/bedroom.

would be required. See the discussion on fluorescent lamp dimming in Unit 13.

The study/bedroom wiring is rather simple. Most of the concepts have been covered in previous units in this text.

Figure 14-3 suggests one way to arrange the wiring for the ceiling fan/light control, the valance lighting control, and the receptacle outlet control.

SURGE SUPPRESSORS

Because this room is intended initially to be utilized as a study (home office), it is very likely that personal computers (PCs) will be located here. Transient voltage surge suppressors, covered in Unit 6 minimize the possibility that voltage surges might damage PCs.

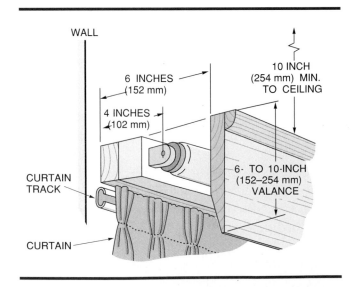

Figure 14-2   Fluorescent valance with draperies.

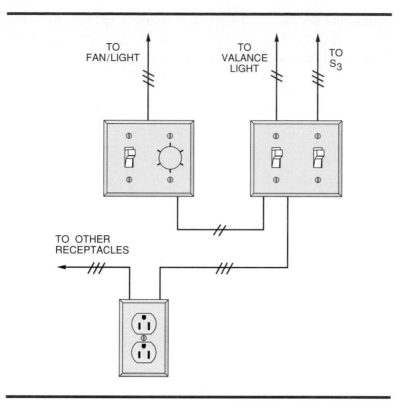

Figure 14-3   Conceptual view of how the study/bedroom switch arrangement is to be accomplished, Circuit A21.

REVIEW

Note: Refer to the Code or plans where necessary.

1. Based upon the total estimated load calculations, what is the current draw on the study/bedroom circuit? _____

2. The study/bedroom is connected to circuit _____

3. The conductor size for this circuit is _____

4. Why is it necessary to install a receptacle in the wall space leading to the bedroom hallway? _____

   _____

   _____

5. Show the calculations needed to select a properly sized box for the receptacle outlet mentioned in question 4. Nonmetallic-sheathed cable is the wiring method.

   _____

   _____

   _____

   _____

   _____

6. Show the calculations necessary to select a properly sized box for the receptacle outlet mentioned in question 4. The wiring method is electrical metallic tubing (EMT), usually referred to as "thin-wall conduit."

_____

_____

_____

_____

_____

7. Prepare a list of *all* wiring devices used in Circuit A21.

_____

_____

_____

_____

_____

_____

8. Using the cable layout shown in Figure 14-1, make a complete wiring diagram of Circuit A21. Use colored pencils or pens.

STUDY/BEDROOM

VALANCE
LIGHT

Dryer Outlet and Lighting Circuit for the Laundry, Powder Room, Rear Entry Hall, and Attic

OBJECTIVES

After studying this Unit, the student will be able to

- understand the Code requirements for receptacles installed in bathrooms.
- make the proper wiring and grounding connections for large appliances, based on the type of wiring method used.
- make load calculations for electric dryers.
- understand how to connect clothes dryers with cord- and plug-connections.
- understand the Code requirements governing the receptacle outlet(s) for laundry areas.
- calculate proper conduit sizing for conductors of the same size and for conductors of different sizes in one conduit.
- discuss the subject of "reduced size neutrals."
- discuss the Code rules pertaining to wiring methods in attics.
- demonstrate the proper way to connect pilot lights and pilot light switches.

LIGHTING CIRCUIT B10

The estimated loads for the various receptacles and fixtures in the laundry, rear entry, powder room, and attic are shown in Table 15-1.

RECEPTACLE CIRCUIT B21

Here is a brief recap of the Code requirements for 120-volt receptacles in bathrooms. This subject was also discussed in Unit 10.

- A bathroom is defined as an area that has a basin, plus one or more of the following: a toilet, a tub, or a shower. Although the plans refer to this particular area as a *powder room*, for Code purposes, it is a bathroom.

DESCRIPTION	QUANTITY	WATTS	VOLT-AMPERES
Rear entry hall receptacle	1	120	120
Powder room vanity fixture	1	200	200
Rear entrance hall ceiling fixtures @ 100 W each	2	200	200
Laundry room fluorescent fixture Four 40-W lamps	1	160	180
Attic lights @ 75 W each	4	300	300
Laundry exhaust fan	1	80	90
Powder room exhaust fan	1	80	90
TOTALS	11	1140	1180

Table 15-1 Laundry-rear entrance-powder room-attic. Circuit B10.

- At least one receptacle must be installed in a bathroom, *Section 210-52(d)*.

▶• The receptacle must be installed on the wall adjacent to the basin, and not more than 36 inches (914 mm) from the outside edge of the basin, *Section 210-52(d)*.◀ In large bathrooms, additional receptacles may be installed elsewhere in the bathroom.

- Receptacle(s) shall not be installed face up in the countertop, *Section 210-52(d)*.

- Receptacles in bathrooms must be GFCI protected, *Section 210-8(a)(1)*.

- The receptacle(s) in the bathroom(s) must be connected to a separate 20-ampere branch-circuit, *Section 210-11(c)(3)*. More than one bathroom may be served by the separate 20-ampere branch-circuit. In this residence, a separate 20-ampere branch-circuit has been run to each bathroom. Circuit B21 supplies the receptacle in the powder room.

- If the required 20-ampere branch-circuit for the receptacle(s) supplies more than one bathroom, then the circuit shall not supply other electrical equipment.

▶• If a separate 20-ampere branch-circuit supplies the receptacle(s) in a single bathroom, then the circuit is permitted to supply other equipment in that bathroom, *Section 210-11(c)(3)*.◀

CLOTHES DRYER CIRCUIT ▲ᴅ

The electric clothes dryer requires a separate 120/240-volt, single-phase, three-wire branch-circuit. In this residence, the electric clothes dryer is indicated on the Electrical Plans in the laundry by the symbol ▲ᴅ. It is supplied by Circuit B(1-3) located in Panel B.

Operation

Clothes dryers are available in many models. Electric clothes dryers have an electric heating element and a motor-operated drum that tumbles the clothes as the heat evaporates and expels the moisture. Dryers are generally vented to the outdoors to expel moisture-laden air. In this residence, a ceiling exhaust fan has also been installed in the laundry room.

A thermostat regulates the temperature and a timer regulates the length of time for the various drying cycles. Upscale electric clothes dryers have a

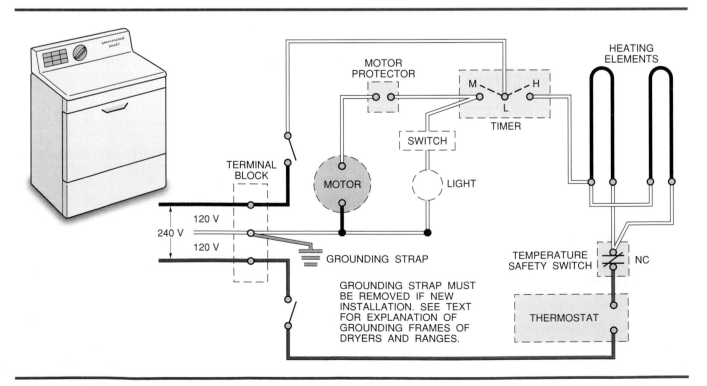

Figure 15-1   Clothes dryer: wiring and components.

moisture sensing feature that stops the drying process at a preselected level of dryness for various types of fabrics. Figure 15-1 shows the internal components and wiring of a typical electric clothes dryer.

The electrician is really interested in the dryer's wattage, ampere rating, and voltage rating so the proper size circuit can be determined and installed.

Dryer Connection Methods

Electric clothes dryers can be connected in several ways.

One method is to run a three-wire armored cable directly to a junction box on the dryer provided by the manufacturer, Figure 15-2. This flexible connection allows the dryer to be moved for servicing without disconnecting the wiring.

Another method is to run a conduit (EMT) to a point just behind the dryer. A combination coupling makes the transition from the EMT to a length of flexible metal conduit 24 inches (610 mm) to 36 inches (914 mm) long, Figure 15-3. This flexible connection also allows the dryer to be moved for servicing without disconnecting the wiring.

Figure 15-2 Dryer connected by armored cable.

Figure 15-3 Dryer connected by EMT and flexible metal conduit.

Probably the most common method is to install a receptacle on the wall behind the dryer. This arrangement is referred to as a cord- and plug-connection, Figure 15-4. The load rating of an appliance that is cord- and plug-connected shall not exceed 80% of the ampere rating of the branch-circuit, *Section 210-23(a)* and *(b)*.

Disconnecting Means

All electrical equipment is required to have a disconnecting means, *Section 422-30.*

For permanently connected appliances, *Section 422-31(b)* states that a disconnect switch or a circuit breaker can serve as the disconnecting means, provided the switch or circuit breaker is within sight of the appliance, or is capable of being locked in the "OFF" position. A circuit breaker in a load-center (panel) could serve as the required disconnecting means if it is within sight of the dryer, or if it can be locked in the "OFF" position. Breaker manufacturers supply "lock-off" devices that fit over the circuit breaker. Disconnect switches have provisions for locking the switch "OFF."

For cord- and plug-connected appliances, *Section 422-32(a)* accepts the cord- and plug arrangement as the disconnecting means.

Section 422-33 accepts the "unit switch" of an appliance as a disconnecting means, provided the unit switch has a marked "OFF" position. This certainly turns off the power within the appliance on the load side of the unit switch, but does not shut off the power to the appliance. To consider the unit switch as the appliance's disconnect, another disconnecting means must be provided, and this can be the service disconnect, *Section 422-33(c).*

Figure 15-4 shows a 30-ampere, 4-wire receptacle configuration for an electric clothes dryer installation

Figure 15-4 Dryer connection using a cord set.

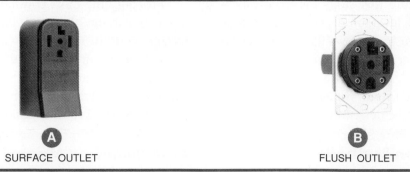

SURFACE OUTLET

FLUSH OUTLET

Figure 15-5   (A) shows a surface mount 4-wire receptacle. (B) shows a flush mount 4-wire receptacle. These are the types required for hooking up electric dryers and ranges. Prior to the adoption of the 1996 *NEC*, 3-wire receptacles and cords were permitted. This is no longer permitted. It is mandatory that a separate equipment grounding conductor be installed to ground the frames of electric clothes dryers and electric ranges. Also see Figure 20-2.

in conformance to the latest *NEC*.® Figure 15-5 shows a surface mount 4-wire receptacle (A) and a flush mount 4-wire receptacle (B).

For most residential electric clothes dryers, receptacles and cord sets are rated 30 amperes, 125/250 volts. Some large electric clothes dryers might require a 50-ampere, 125/250-volt receptacle and cord set.

Receptacles for dryers can be surface mounted or flush mounted. Flush-mounted receptacles are quite large, and require a large wall box. A 4-inch square, 1½-inch or 2⅛-inch deep outlet box with a suitable single-gang or two-gang plaster ring generally works fine. Correct box sizing is determined by referring to *Tables 370-16(a)* or *(b)*, or to Figures 2-19 and 2-34.

Dryer and range receptacles and cord sets are identical. To avoid duplication of text, refer to Unit 20 for a complete description of ratings and blade and slot configurations for three-wire and four-wire dryer and range receptacles, and their associated cord sets.

Load Calculations

In this residence, the dryer is installed in the laundry room. The Schedule of Special Purpose Outlets in the specifications shows that the dryer is rated 5,700 watts, 120/240 volts. The schematic wiring diagram in Figure 15-1 shows that the heating element is connected across 240 volts. The motor and lamp are connected to the 120-volt terminals.

It is impossible for the electrician to know the ampere rating of the motor, the heating elements, timers, lights, relays, etc., and in what sequence they actually operate. This dilemma is answered in the *NEC*® and U.L. "listing" requirements for all appliances.

The nameplate must show volts and amperes, or volts and watts, *Section 422-60(a)*. The nameplate must also show the minimum supply circuit conductor's ampacity and the maximum overcurrent protection, *Section 422-62*.

A minimum load of 5,000 watts (volt-amperes) or the nameplate rating of the dryer, whichever is larger is used when calculating feeder or service-entrance requirements, *Section 220-18*.

The rating of an appliance branch-circuit must not be less than the marked rating on the appliance, *Section 422-10(a)*.

The electric clothes dryer in this residence has a nameplate rating of 5,700 watts. This calculates out to be:

$$I = \frac{W}{E} = \frac{5,700}{240} = 23.75 \text{ amperes}$$

Rounded up, we have a minimum 30-ampere branch-circuit rating requirement.

Conductor Sizing

The minimum branch-circuit rating for the electric clothes dryer was found to be 30 amperes, and that is the minimum rating of the overcurrent protective device—fuses or circuit breaker. We need to find a conductor that has an ampacity of 30 amperes that will serve as the branch-circuit for the dryer. The conductors will be protected by the 30-ampere overcurrent device.

The appliance must be marked with the minimum supply circuit conductor ampacity, *Section 422-62*. The instruction may also indicate the minimum conductor size.

Conductor size is based on the "minimum supply circuit conductor ampacity" by referring to *Table 310-16*.

According to *Table 310-16*, the allowable ampacity for a No. 10 AWG copper conductor is 35 amperes. The maximum overcurrent protection for these conductors must not exceed 30 amperes, *Section 240-3(d)*.

In this residence, Circuit B(1-3) is a 3-wire, 30-ampere, 240-volt branch-circuit.

Nonmetallic-sheathed cable is run concealed in the walls from Panel B to a flush mounted 4-inch square box behind the dryer location in the laundry room. The nonmetallic-sheathed cable contains three No. 10 AWG conductors plus a No. 10 AWG equipment grounding conductor.

Where electrical metallic tubing is used as the wiring method, *Table C2, Appendix C* of the *NEC®* shows that three No. 10 AWG THHN conductors require a ½-inch EMT. The EMT serves as the equipment ground.

Overcurrent Protection for the Dryer

The motor of the dryer has integral thermal overload protection as required by *Section 422-11(g)*. Note the motor protector in Figure 15-1. This protection is required by U.L. standards, and prevents the motor from reaching dangerous temperatures as the result of an overload, bearing failure, or failure to start. Also note the temperature safety switch. This is a high-temperature cut-off that shuts off the heating element should the appliance's thermostat fail to operate. This high temperature cut-off is a U.L. requirement.

Overcurrent Protection for Branch-Circuit

The rating of the branch-circuit must not be less than the rating marked on the appliance, *Section 422-10(a)*.

Section 422-11(a) states that the overcurrent protection for the branch-circuit conductors must be sized according to *Section 240-3*, which refers us right back to *Article 422*—Appliances.

Section 422-62 requires the appliance to be marked with the minimum supply circuit conductor ampacity and the maximum size overcurrent device.

Section 240-3(d) indicates that the maximum overcurrent protection for No. 10 AWG copper conductors is 30 amperes.

Grounding Frames of Electric Dryers and Electric Ranges

For protection against electrical shock hazard, frames of electric clothes dryers and electric ranges must be grounded. Some key Code references are *Sections 250-134, 250-138, 250-140*, and *250-142(b)*.

Because of recent changes in the *NEC®* there are now two methods for grounding the frames of electric dryers and electric ranges, namely *New Installations* and *Existing Installations*.

New Installations. One accepted grounding means is to connect the appliance with a metal raceway such as EMT, flexible metal conduit (Greenfield), armored cable (BX), or other means listed in Section 250-118. There are limitations on the use of flexible metal conduit as an equipment grounding conductor. See Sections 250-118 and 350-5. This topic is covered in Unit 4.

The separate equipment grounding conductor found in nonmetallic-sheathed cable is also an acceptable grounding means. The minimum size of an equipment grounding conductor is found in *Table 250-122*. The equipment grounding conductor in nonmetallic-sheathed cable is sized according to *Table 250-122*. See Figure 15-6.

Type SE service-entrance cable may also be used provided all of the circuit conductors are insulated. The equipment grounding conductor may be insulated (green) or bare, *Section 338-3*. Type SE cable installed as interior wiring must be installed according to the installation requirements found in *Article 336* for nonmetallic-sheathed cable, *Section 338-4*.

When electric clothes dryers and electric ranges are cord- and plug-connected, the receptacle and the cord set must be four-wire, the fourth wire being the separate equipment grounding conductor.

Figure 15-6 New installations of branch-circuit wiring for electric ranges, wall-mounted ovens, surface cooking units, and clothes dryers require that all equipment grounding be done according to *Sections 250-134, 250-138, 250-140,* and *250-142(b)*. Receptacles and cord sets must be 4-wire (three circuit conductors *plus* equipment grounding conductor).

The small copper bonding strap (link) that is furnished with an electric clothes dryer or electric range *must not* be connected between the neutral terminal and the metal frame of the appliance.

Existing Installations.   Prior to the 1996 edition of the *NEC*, *Section 250-60* permitted the frames of electric ranges, wall-mounted ovens, surface cooking units, and clothes dryers to be grounded to the neutral conductor. By way of *Tentative Interim Amendment No. 53*, this special permission was put into effect in July of 1942, and was supposedly an effort to conserve raw materials during World War II. In effect, this special permission allowed the neutral conductor to serve a dual purpose: (1) the neutral conductor and (2) the equipment grounding conductor. This special permission remained in effect until the 1996 *NEC*.

Section 250-140 (*Section 250-60* in the 1996 *NEC* and previous editions of the *NEC*) now applies *only* to existing branch-circuit installations. For those installations wired according to the rules of *Section 250-60* prior to the 1996 *NEC*, the required appliance grounding was accomplished by the small copper bonding strap furnished by the appliance manufacturer that connected between the neutral terminal and the metal frame of the appliance.

For existing installations, the requirements were:

- the supply circuit had to be 120/240 volt, single-phase, 3-wire or 208Y/120 volt derived from a 3-phase, 4-wire wye-connected system.

- the neutral had to be no smaller than No. 10 AWG copper or No. 8 AWG aluminum.

- the neutral conductor had to be insulated. The exception to this was for Type SE cable where the circuit originated at the main service, in which case the neutral conductor did not have to be insulated but only covered by the outer jacket of the cable.

- a metal outlet box used to hold a receptacle was permitted to be grounded to the neutral, Figure 15-7.

If an electric clothes dryer or electric range is to be installed in a residence where the dryer branch-circuit wiring had been installed according to pre-1996 *NEC* rules, the wiring *does not* have to be changed. Merely use the grounding strap furnished by the dryer manufacturer to make a connection between the neutral conductor and the frame of the dryer.

THREE NO. 10 AWG
NONMETALLIC-SHEATHED
CABLE—NO EQ. GROUND

OUTLET
BOX

COVER FOR
OUTLET BOX

DRYER
RECEPTACLE
OUTLET

Figure 15-7   Prior to the 1996 *National Electrical Code,*® it was permitted to ground the junction box to the neutral only when the box was part of the circuit for electric ranges, wall-mounted ovens, surface cooking units, and clothes dryers, *Section 250-140.* The receptacle was permitted to be a 3-wire type because the appliance grounding was accomplished by making a connection between the neutral conductor and the metal frame of the appliance. This is no longer permitted.

RECEPTACLE OUTLETS—LAUNDRY

The Code in *Section 210-11(c)(2)* requires that a separate 20-ampere branch-circuit be provided for the laundry equipment and that this circuit "shall have no other outlets."

Section 210-50(c) requires that a receptacle outlet must be installed within 6 feet (1.83 m) of the intended location of the washer.

Section 210-52(f) states that *at least* one receptacle outlet must be installed for the laundry equipment.

Section 220-16(b) requires that 1500 volt-amperes must be included for each two-wire branch-circuit serving laundry equipment. This is for the purpose of computing the size of service-entrance and/or feeders. See the complete calculations in Unit 28.

In the laundry room in this residence, the 20-ampere Circuit B18 supplies the receptacle for the washer. Circuit B18 serves no other outlets. The receptacle is not required to be GFCI protected. Figure 15-8 shows the cable layout for the laundry room.

B1–3

B20

GFCI
WP

B18

LAUNDRY RECEPTACLES

Figure 15-8   Cable layout for laundry room receptacles.

Circuit B20 is a 20-ampere branch-circuit. It supplies two receptacles in the laundry room and will serve an iron, sewing machine, or other small portable appliance. This circuit also serves the weatherproof receptacle on the outside wall of the laundry area. This branch-circuit is *in addition* to the required separate 20-ampere circuit (B18) for the laundry equipment.

Circuits B18 and B20 are included in the calculations for service-entrance and feeder conductor sizing at a computed load of 1500 volt-amperes per circuit, *Section 220-16(b)*. This load is included with the general lighting load for the purpose of computing the service-entrance and feeder conductor sizing, and is subject to the demand factors that are applicable to these calculations. See the complete calculations in Unit 28.

COMBINATION WASHER/DRYERS

Combination washer/dryers are often found in multi-family dwellings where space is an issue. Combination washer/dryers take up half the floor space of individual conventional washers and dryers. If you are involved in an electrical installation, verify and resolve the electrical requirements before you start the project. These combination appliances might require a 30-ampere branch-circuit. Installation manuals will provide this information. The electrical inspector might *not* require the traditional 20-ampere branch-circuit and receptacle that is required by *Sections 210-52(f)* and *210-11(c)(2)*.

LIGHTING CIRCUIT

The general lighting circuit for the laundry, powder room, and hall area is supplied by Circuit B10. Note on the cable layout, Figure 15-9, that in order to help balance loads evenly, the attic lights are also connected to Circuit B10.

The types of vanity lights, ceiling fixtures, and wall receptacle outlets, as well as the circuitry and switching arrangements, are similar to types already discussed in other units of this text and will not be repeated.

Figure 15-9 Cable layout for laundry, rear entrance, powder room, and attic, Circuit B10.

The receptacle in the powder room is supplied by Circuit B21, a separate 20-ampere branch-circuit as required by *Section 210-52(d)*, discussed in Unit 10.

Exhaust Fans

Ceiling exhaust fans are installed in the laundry and in the powder room. These exhaust fans are connected to Circuit B10.

The exhaust fan in the laundry will remove excess moisture resulting from the use of the clothes washer.

Exhaust fans may be installed in walls or ceilings, Figure 15-10. The wall-mounted fan can be adjusted to fit the thickness of the wall. If a ceiling-mounted fan is used, sheet-metal duct must be installed between the fan unit and the outside of the house. The fan unit terminates in a metal hood or grille on the exterior of the house. The fan has a shutter that opens as the fan starts up and closes as the fan stops. The fan may have an integral pull-chain switch for starting and stopping, or it may be used with a separate wall switch. In either case, single-speed or multispeed control is available. The fan in use has a very small power demand, 90 volt-amperes.

To provide better humidity control, both ceiling-mounted and wall-mounted fans may be controlled with a humidistat. This device starts the fan when the humidity reaches a certain value. When the humidity drops to a preset level, the humidistat turns the fan off.

Some of the more expensive exhaust fans have a built-in sensor that will automatically turn on the exhaust fan at a predetermined humidity level. The fan will automatically turn off when the humidity has been lowered to a preset level. This eliminates the need for a separate wall-mounted humidity control (humidistat).

ATTIC LIGHTING AND PILOT LIGHT SWITCHES

Section 210-70(c) requires that at least one lighting outlet must be installed in an attic, and that this lighting outlet contain a switch such as a pull-chain lampholder, or controlled by a switch located near the entry to the attic. This section of the Code addresses attics that are used for storage or attics that contain equipment that might need servicing, such as air-conditioning equipment. *Section 210-70(c)* requires that the lighting outlet(s) be installed near this equipment.

Section 210-63 requires that a 125-volt, single-phase, 15- or 20-ampere receptacle outlet be installed in an accessible location within 25 feet (7.62 m) of air-conditioning or heating equipment in attics, in crawl spaces, or on the roof. Connecting this receptacle outlet to the load side of the equipment's disconnecting means is not permitted because this would mean that if you turned off the equipment, the power to the receptacle would also be off and thus would be useless for servicing the equipment. The *Exception* to *Section 210-63* exempts one- and two-family dwellings from having a receptacle outlet on the roof, although installing a receptacle outlet would certainly be a good idea.

The residence discussed in this text does not have air-conditioning, heating, or refrigeration equipment in the attic or on the roof.

WALL TYPE   CEILING TYPE

Figure 15-10   Exhaust fans. *Courtesy* NuTone

Figure 15-11   Pilot lamp connections.

The porcelain lampholders installed in the attic of the residence are available with a receptacle outlet for convenience in plugging in an extension cord. However, most electrical inspectors (authority having jurisdiction) would not accept the porcelain lampholder's receptacle in lieu of the required receptacle outlet as stated in *Section 210-63*.

The four porcelain lampholders in the attic are turned on and off by a single-pole switch on the garage wall close to the attic storable ladder. Associated with this single-pole switch is a pilot light. The pilot light may be located in the handle of the switch or it may be separately mounted. Figure 15-11 shows how pilot lamps are connected in circuits containing either single-pole or three-way switches.

Because the attic in this residence is served by a folding storable ladder, switch control in the attic is not required. Where a permanent stairway of six or more steps is installed, *Section 210-70(a)(2)* requires switch control at both levels.

If a neon pilot lamp in the handle (toggle) of a switch does not have a separate grounded conductor connection, then it will glow only when the switch is in the OFF position, as the neon lamp will then be in series with the lamp load, Figure 15-12.

The voltage across the load lamp is virtually zero, so it does not burn, and the voltage across the neon lamp is 120 volts, allowing it to glow. When the switch is turned on, the neon lamp is bypassed (shunted), causing it to turn off and the lamp load to turn on.

Use this type of switch when it is desirable to have a switch glow in the dark to make it easy to locate.

Installation of Cable in Attics

When nonmetallic-sheathed cable is installed in accessible attics, the installation must conform to the requirements of *Section 336-6(d)*. This section refers the reader directly to *Section 333-12*,

Figure 15-12   Example of neon pilot lamp in handle of toggle switch.

which describes how the cable is to be protected, Figure 15-13.

In accessible attics, Figure 15-13(A), cables must be protected by guard strips, Figure 15-14, when:

- they are run across the top of floor joists ①.

- they are run across the face of studs ② or rafters ③ within 7 feet (2.13 m) of the floor or floor joists.

Guard strips are *not* required if the cable is run along the sides of rafters, studs, or floor joists ④.

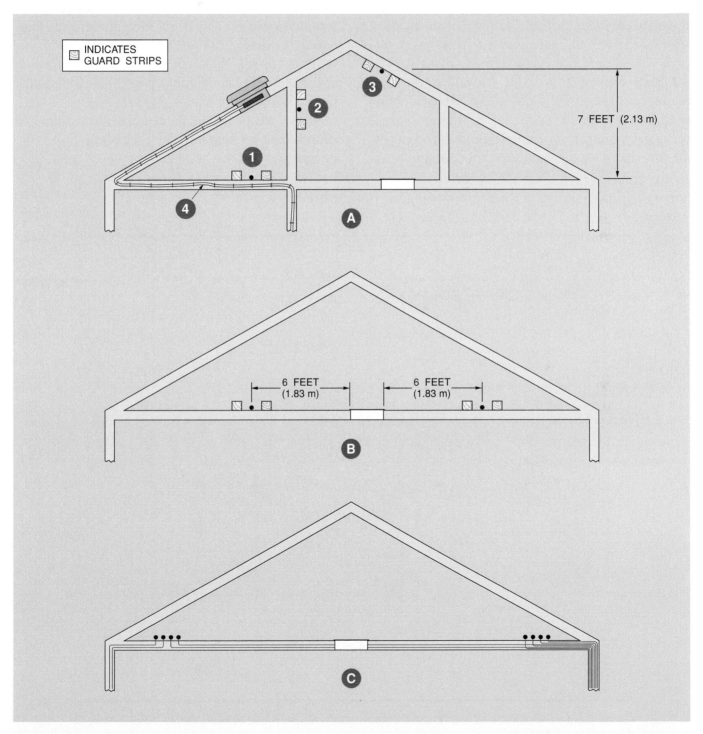

Figure 15-13 **Protection of nonmetallic-sheathed cable or armored cable in an attic. Refer to** *Sections 336-6(d)* **and** *333-12.*

In attics not accessible by permanent stairs or ladders, Figure 15-13(B), guard strips are required only within 6 feet (1.83 m) of the nearest edge of the scuttle hole or entrance.

Figure 15-13(C) illustrates a cable installation that most electrical inspectors consider to be safe. Because the cables are installed close to the point where the ceiling joists and the roof rafters meet, they are protected from physical damage. It would be very difficult for a person to crawl into this space, or store cartons in an area with such a low clearance. Although the plans for this residence show a 2-foot-wide catwalk in the attic, the owner may decide to install additional flooring in the attic to obtain more storage space. Because of the large number of cables required to complete the circuits, it would interfere with flooring to in-stall guard strips wherever the cables run across the tops of the joists, Figure 15-14. However, the cables can be run through holes bored in the joists and along the sides of the joists and rafters. In this way, the cables do not interfere with the flooring.

When running cables parallel to framing members, be careful to maintain at least 1¼ inches (31.8 mm) between the cable and the edge of the framing member. This is a Code requirement referenced in *Section 300-4(d)* to minimize the possibility of driving nails into the cable. All of *Section 300-4* is devoted to the subject of "protection against physical damage."

See Unit 4 for more detailed discussion and illustrations relating to the mechanical protection of cables.

GUARD STRIPS REQUIRED WHEN CABLES ARE RUN ACROSS THE TOP OF JOISTS.

GUARD STRIPS

CABLES RUN THROUGH BORED HOLES IN JOISTS ARE CONSIDERED PROTECTED.

Figure 15-14   Methods of protecting cable installations in accessible attics. See *Sections 333-12* and *336-6(d)*.

REVIEW

Note: Refer to the Code or plans where necessary.

1. List the switches, receptacles, and other wiring devices that are connected to Circuit B10. _____

_____

_____

_____

_____

_____

_____

_____

2. a. Fill in the blank spaces. At least _____ 120-volt receptacle must be installed in a bathroom within _____ inches of the basin, and it must be on the wall _____ to the basin. The receptacle must not be installed _____ on the countertop. The receptacle must be _____ protected.

   b. Circle the correct answers. A separate (15) (20)-ampere branch-circuit must supply the receptacle(s) in bathrooms and (shall) (shall not) serve other loads. If a separate 20-ampere branch-circuit serves a single bathroom, then that circuit (is permitted) (is not permitted) to supply other loads in that bathroom.

3. What special type of switch is controlling the attic lights? _____

_____

4. When installing cables in an attic along the top of the floor joists _____ _____ must be installed to protect the cables.

5. The total estimated volt-amperes for Circuit B10 has been calculated to be 1300 volt-amperes. How many amperes is this at 120 volts? _____

6. If an attic is accessible through a scuttle hole, guard strips are installed to protect cables run across the top of the joists only within 6 feet (1.83 m), 7 feet (2.13 m), 12 feet (3.66 m) of the scuttle hole. Circle correct answer.

7. What section(s) of the Code refers (refer) to the receptacle required for the laundry equipment? State briefly the requirements.

_____

_____

_____

_____

_____

_____

_____

8. What is the current draw of the exhaust fan in the laundry? _____

9. The following is a layout of the lighting circuit for the laundry, powder room, rear entry hall, and attic. Complete the wiring diagram using colored pens or pencils to indicate conductor insulation color. The GFCI receptacle in the powder room is not shown in the diagram because it is connected to a separate Circuit B21.

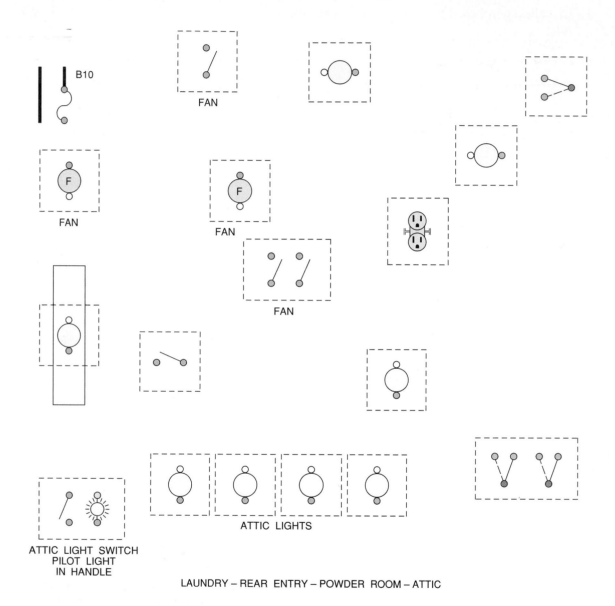

LAUNDRY – REAR ENTRY – POWDER ROOM – ATTIC

10. Laundry receptacle outlets are included in the residential load calculations at a value of _____ volt-amperes per circuit. Choose one: 1000, 1500, 2000 volt-amperes.

11. a. What regulates the temperature in the dryer? _____

   _____

   b. What regulates drying time? _____

   _____

12. List the various methods of connecting an electric clothes dryer. _____

   _____

13. a. What is the minimum power demand allowed by the Code for an electric dryer if no actual rating is available for the purpose of calculating feeder and service-entrance conductor sizing? _____

    b. What is the current draw? _____

14. What is the maximum permitted current rating of a portable appliance on a 30-ampere branch-circuit? _____

15. What provides motor running overcurrent protection for the dryer? _____

_____

16. a. Must the metal frame of an electric clothes dryer be grounded? _____

    b. May the metal frame of an electric clothes dryer be grounded to the neutral conductor? _____

17. An electric dryer is rated at 7.5 kW and 120/240 volts, three-wire, single-phase. The terminals on the dryer and panelboard are marked 75°C.

    a. What is the wattage rating? _____

    b. What is the current rating? _____

    c. What minimum size type THHN copper conductors are required? _____

    _____

    _____

    _____

    _____

    _____

    d. If electrical metallic tubing is used, what minimum size is required? _____

    _____

    _____

    _____

    _____

    _____

    _____

    _____

18. When a metal junction box is installed as part of the cable wiring to a clothes dryer or electric range, may this box be grounded to the circuit neutral? _____

19. A residential air-conditioning unit is installed in the attic.

    a. Is a lighting outlet required? _____

    b. What Code section? _____

    c. If a lighting outlet is required, how shall it be controlled? _____

    _____

    d. What Code section? _____

    e. Is a receptacle outlet required? _____

    f. What Code section? _____

20. *Section 422-30* of the Code states that a means must be provided to disconnect an appliance. In the case of an electric clothes dryer, the disconnecting means can be: Circle the correct statement(s).

a. a separate disconnect switch located within sight of the appliance.

b. a separate disconnect switch located out of sight from the appliance. The switch is capable of being locked in the OFF position.

c. a cord- and plug-connected arrangement.

d. a circuit breaker in a panel that is within sight of the dryer or capable of being locked in the OFF position.

Lighting Branch-Circuit for the Garage

OBJECTIVES

After studying this Unit, the student will be able to

- understand the fundamentals of providing proper lighting in residential garages.
- understand the application of GFCI protection for receptacles in residential garages.
- understand the Code requirements for underground wiring, both conduit and underground cable.
- complete the garage circuit wiring diagram.
- discuss typical outdoor lighting and related Code requirements.
- describe how conduits and cables are brought through cement foundations to serve loads outside of the building structure.
- understand the application of GFCI protection for loads fed by underground wiring.
- make a proper installation for a residential overhead garage door opener.
- select the proper overload protective devices based on the ampere rating of the connected motor load.

LIGHTING BRANCH-CIRCUIT

Circuit B14 originates at Panel B in the recreation room, and is brought into the wall receptacle box on the front wall of the garage. Figure 16-1 shows the cable layout for Circuit B14. From this point, the circuit is then carried to the switch box adjacent to the side door of the garage. From there the circuit conductors and the switch leg are carried upward to the ceiling fixtures, and then to the receptacle outlets on the right-hand wall of the garage.

The garage ceiling's porcelain lampholders are controlled from switches located at all three entrances to the garage.

The post light in the front of the house is fed from the receptacle located just inside the overhead door. The post light turns on and off by a photocell, an integral part of the post light.

Note that the fixtures on the outside next to the overhead garage door are controlled by two three-way switches, one just inside the overhead door and the other in the front entry. These fixtures are not connected to the garage lighting circuit.

All of the wiring in the garage will be concealed.

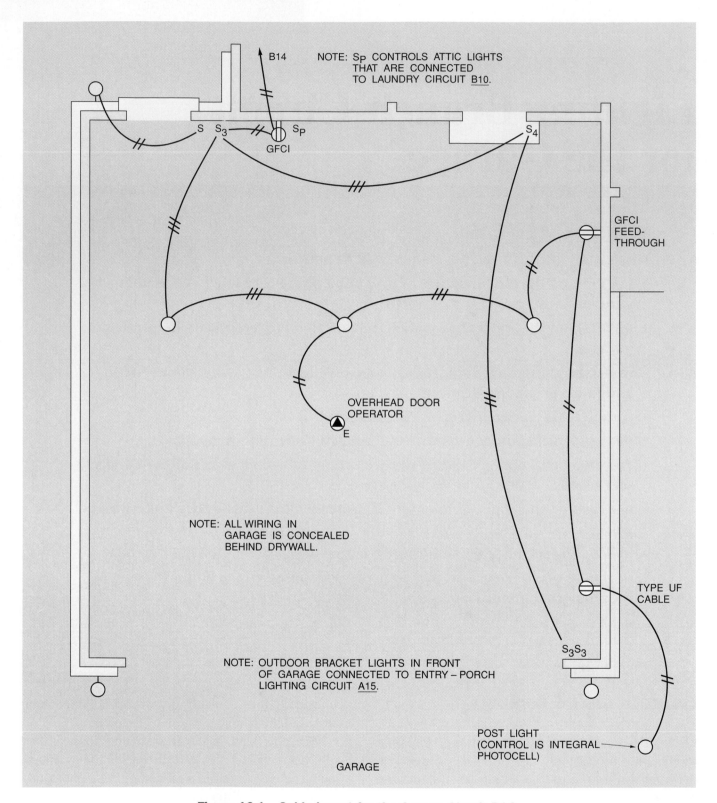

Figure 16-1   Cable layout for the Garage Circuit B14.

Building codes for one- and two-family dwellings require that the walls separating habitable areas and a garage have a fire resistance rating. In this residence, the walls and ceilings have a fire resistance rating of 1 hour. Fire resistance rating requirements were discussed in Unit 2.

Table 16-1 summarizes the outlets and the estimated load for the garage circuit.

DESCRIPTION	QUANTITY	WATTS	VOLT-AMPERES
Receptacles @ 180 W each	3	540	540
Ceiling fixtures @ 100 W each	3	300	300
Outdoor garage bracket fixture (rear)	1	100	100
Post light	1	100	100
Overhead door opener			
Two 40-W lamps	1	80	80
Motor (½ A @ 120 V)		180	200
TOTALS	9	1300	1320

Table 16-1 Garage outlet count and estimated load. Circuit B14.

LIGHTING A TYPICAL RESIDENTIAL GARAGE

The recommended approach to provide adequate lighting in a residential garage is to install one 100-watt lamp on the ceiling above each side of an automobile, Figure 16-2. Figure 16-3 shows typical lampholders commonly mounted in garages.

- For a one-car garage, a minimum of two fixtures is recommended.

- For a two-car garage, a minimum of three fixtures is recommended.

- For a three-car garage, a minimum of four fixtures is recommended.

Lighting fixtures arranged in this manner eliminate shadows between automobiles. Shadows are a hazard because they hide objects that can cause a person to trip or fall.

It is highly recommended that these ceiling fixtures be mounted toward the front end of the automobile as it is normally parked in the garage. This arrangement will provide better lighting where it is most needed when the owner is working under the hood. Do not place lights where they will be covered by the open overhead door.

Fluorescent fixtures are quite often used in the warmer parts (south and southwest) of the country. For example, instead of the three porcelain lampholders indicated on the electrical plans on the ceiling of the garage for the residence discussed in this text, three 8-foot fluorescent strip lights could be installed in the direction of front to back of the

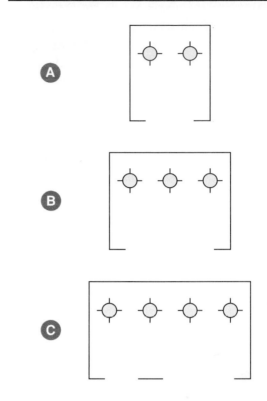

Figure 16-2 Positioning of lights in (A) a one-car garage, (B) a two-car garage, and (C) a three-car garage.

Figure 16-3 Typical porcelain lampholders that would be installed in garages, attics, basements, crawl spaces, and similar locations. Note that one has a pull-chain and one does not.

garage. These 8-foot fixtures each have two 40-watt fluorescent lamps, for a total of six 40-watt fluorescent lamps. Fluorescent lamps installed in a cold garage (northern parts of the country) would be difficult if not impossible to start, thus the need to use incandescent lamps.

RECEPTACLE OUTLETS IN A GARAGE

At least one receptacle must be installed in an attached residential garage, *Section 210-52(g)*. The garage in this residence has three receptacles. If a detached garage has no electricity provided, then obviously no rules are applicable, but just as soon as a detached garage is wired, then all pertinent Code rules become effective and must be followed. These would include rules relating to grounding, lighting outlets, receptacle outlets, GFCI protection, and so on.

Section 210-8(a)(2) requires that "all 125-volt, single-phase, 15- or 20-ampere receptacles installed in garages shall have ground-fault circuit-interrupter protection". GFCIs are discussed in detail in Unit 6.

There are a few exceptions to *Section 210-8(a)(2)* requiring GFCI protection of receptacles in residential garages. They are:

1. GFCI protection is *not* required if the receptacle is not readily accessible, such as the receptacle installed for the overhead door opener.

2. GFCI protection is *not* required for single or duplex receptacles intended for cord- and plug-connected appliances that are not easily moved, and are located in dedicated spaces. Examples include a freezer, refrigerator, or central vacuum equipment. The intent is that these dedicated receptacles will have the appliance cord plugged into it, and therefore would not be used to plug in portable electric tools and extension cords. This exception is difficult to enforce, because the electrician may not know during the rough-in that a certain receptacle is for the sole purpose of plugging in a specific appliance, say a large freezer. Such appliances are not at the construction site during rough-in.

Note that one GFCI receptacle is installed on the front wall of the garage, in the same box as the single-pole switch for the attic lighting. A second feedthrough GFCI receptacle is installed on the right-hand garage wall near the entry to the house. This GFCI receptacle also provides the required GFCI protection for the other receptacle on the same wall and for the protection of the underground wiring to the post light.

If we were to install one GFCI feedthrough receptacle at the point where Circuit B14 feeds the garage, then the entire circuit would be GFCI protected as required. But nuisance tripping of the GFCI might occur because of:

- the overall length of the circuit wiring.
- possible problems in the overhead door opener.
- possible problems in the underground run to the post light.
- problems in the post lamp itself.

All of these would result in a complete outage on Circuit B14. The electrician must consider the cost of GFCI receptacles and compare it with the cost of additional cable or conduit and wire needed. The possibility of a power outage must also be kept in mind when trying to get by with fewer GFCI receptacles.

All receptacles and switches in the garage are to be mounted 46 inches (1.22 m) to center according to a notation on the First Floor Electrical Plan. The Uniform Building Code requires that receptacles be mounted not less than 18 inches (457 mm) above the garage floor.

LANDSCAPE LIGHTING

One of the nice things about residential wiring is the availability of a virtually unlimited variety of sizes, types, and designs of outdoor lighting fixtures for gardens, decks, patios, and for accent lighting. Fixture showrooms at electrical distributors and the electrical departments of home centers offer a wide range of outdoor lighting fixtures, both 120-volt and low-voltage types. Figures 16-4 and 16-5 illustrate landscape lighting mounted on the ground. Figure 16-10 is frequently referred to as "treescape" lighting.

U.L. Standard 1838 covers *Low-Voltage Landscape Lighting Systems* generally used outdoors in wet locations.

U.L. Standard 2108 covers *Low-Voltage Lighting Systems* for use in dry locations.

Both of these U.L. standards require the manufacturer to include installation instructions with the product. It is extremely important to read these instructions to be sure the lighting fixtures and associated wiring are installed and used in a safe manner.

Low-voltage lighting products will bear markings such as *Outdoor Use Only, Indoor Use Only,* or

Indoor/Outdoor Use. If marked for use in damp or wet locations, the maximum voltage rating is 15 volts unless live parts are made inaccessible to contact during normal use. Low-voltage lighting fixtures are intended for specific applications, such as surface mounting, suspended or recessed installa-tions and are so marked. If of the recessed type, they must be installed according to the provisions of *Section 410-66* as discussed in Unit 7.

Low-voltage lighting systems are covered in *Article 411.* The basic requirements for these sys-tems are that they shall:

Figure 16-4 Weatherproof receptacle outlets stubbed out of the ground; either low-voltage decorative landscape lighting or 120-volt PAR lighting fixtures may be plugged into these outlets.

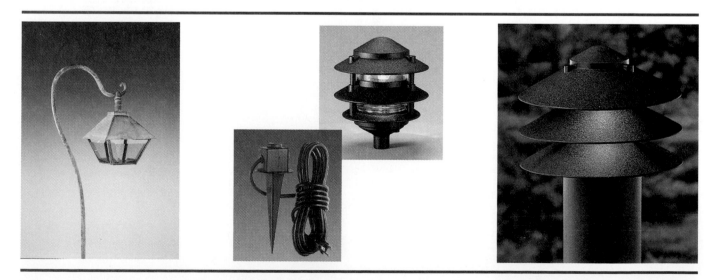

Figure 16-5 Fixtures used for decorative purposes outdoors, under shrubs and trees, in gardens, and for lighting paths and driveways. *Courtesy* Progress Lighting.

- operate at 30 volts rms (42.4 volts peak) or less.

- have one or more secondary circuits, each limited to 25 amperes maximum.

- be supplied through an isolating transformer. A standard step-down transformer does not meet this criteria. Look carefully at the nameplate on the transformer.

- be "listed" for the purpose.

- not have the wiring concealed or extended through a building unless the wiring meets the requirements of *Chapter 3, NEC.®* An example of this is running Type NMC nonmetallic-sheathed cable for the low-voltage portion of the wiring that is concealed or extended through the structure of the building.

- not be installed within 10 feet (3.05 m) of pools, spas, or water fountains unless specifically permitted in *Article 680.*

- not have the secondary of the isolating transformer grounded.

- not be supplied by a branch-circuit that exceeds 20 amperes.

OUTDOOR WIRING

If the homeowner requests that 120-volt receptacles be installed away from the building structure, the electrician can provide weatherproof receptacle outlets as illustrated in Figure 16-4.

RIGID METAL, RIGID NONMETALLIC, INTER-MEDIATE METAL CONDUIT, OR ELECTRICAL METALLIC TUBING OK TO SUPPORT BOXES THAT DO NOT HAVE DEVICES OR FIXTURES. RIGID METAL OR INTERMEDIATE METAL CONDUIT OK TO SUPPORT BOXES THAT DO HAVE DEVICES OR FIXTURES. RIGID NONMETALLIC CONDUIT NOT PERMITTED TO SUPPORT LIGHTING FIXTURES OR OTHER EQUIPMENT, *SECTION 347-3(b).*

THIS BODY IS CONSIDERED TO BE ADEQUATELY SUPPORTED.

A WHEN TWO OR MORE CONDUITS ARE TIGHTLY THREADED INTO THE HUBS OF A CONDUIT BODY, THE CONDUIT BODY IS CONSIDERED TO BE ADEQUATELY SUPPORTED, *SECTION 370-23(e).* FOR ENCLOSURES THAT SUPPORT FIXTURES OR CONTAIN WIRING DEVICES, THE ENCLOSURE MUST BE SUPPORTED WITHIN 18 INCHES (457 mm). CONDUITS COMING OUT OF THE GROUND ARE ACCEPTABLE AS MEETING THIS SUPPORT REQUIREMENT, *SECTION 370-23(f).*

B THIS CONDUIT BODY IS NOT ADEQUATELY SUPPORTED BY THE ONE CONDUIT THREADED INTO THE HUB. THIS CONDUIT BODY COULD TWIST VERY EASILY, RESULTING IN DAMAGED INSULATION ON THE CONDUCTORS AND A POOR GROUND CONNECTION BETWEEN THE CONDUIT AND THE CONDUIT BODY. SEE *SECTIONS 370-23(e)* AND *(f).*

THIS BODY IS NOT CONSIDERED TO BE ADEQUATELY SUPPORTED.

C CONDUCTORS MAY BE SPLICED IN THESE CONDUIT BODIES ONLY IF THE CONDUIT BODY IS MARKED WITH ITS CUBIC-INCH CAPACITY SO THAT THE PERMISSIBLE CONDUCTOR FILL MAY BE DETERMINED USING THE CONDUCTOR VOLUME FOUND IN *TABLE 370-16(b).*

SPLICES MAY BE MADE IN CONDUIT BODIES WHEN MARKED WITH THEIR CUBIC-INCH CAPACITY, *SECTION 370-16(c)(2).*

Figure 16-6   Supporting threaded conduit bodies, *Section 370-23.*

Outdoor receptacles must have GFCI protection. See Unit 6 for complete discussion of GFCI protection.

These types of boxes have ½-inch female threaded openings in which conduit fittings secure the conduit "stub-ups" to the box. Any unused openings are closed with ½-inch plugs that are screwed tightly into the threads. Refer to Figure 16-6.

Section 410-57(b) requires that receptacles installed outdoors maintain their weatherproof integrity when the receptacle is in use. This requires a self-closing cover that is deep enough to shelter the attachment plug cap of the cord. See Figure 16-4. This rule would apply to receptacles installed for decorative lighting, gutter heating cords, etc., where the cord will be plugged in for long periods of time.

Where flush-mounted weatherproof receptacles are desired for appearance purposes, a recessed weatherproof receptacle, as illustrated in Figure 16-7, may be used. This type of installation takes up more depth than a standard receptacle, and requires special mounting as indicated in the "exploded" view. Careful thought must be given during the "roughing-in" stages to ensure that the proper depth "roughing-in" box is installed.

An exception to this requirement is for weatherproof receptacles where the equipment plugged into the receptacle is "attended." This would apply to receptacles installed outside of a home that are used periodically for extension cords on tools, hedge trimmers, grills, and similar appliances. As soon as the homeowner is finished, the appliance is unplugged, and the self-closing cover returns to a closed position. Figures 5-7 and 16-6 show typical covers.

Wiring with Type UF Cable
(Article 339)

The plans show that Type UF underground cable, Figure 16-8, is used to connect the post light. The current-carrying capacity (ampacity) of Type UF cable is found in *Table 310-16*, using the 60°C column. According to *Article 339*, Type UF cable:

• is marked underground feeder cable.

Figure 16-7   A recessed type of weatherproof receptacle that meets the requirements of *Section 410-57(b)*, which requires that the enclosure be weatherproof while in use if unattended, where the cord will be plugged in for long periods of time. The "exploded" view illustrates one method of "roughing-in" this type of weatherproof receptacle. *Photos courtesy* of TayMac Corporation.

Figure 16-8   Type UF underground cable. *Courtesy* Southwire Company.

- is available in sizes from No. 14 AWG through No. 4/0 (for copper conductors) and from No. 12 AWG through No. 4/0 (for aluminum conductors).

- may be used in direct exposure to the sun if the cable is marked "sunlight resistant."

- may be used with nonmetallic-sheathed cable fittings.

- is flame-retardant.

- is moisture-, fungus-, and corrosion-resistant.

- may be buried directly in the earth.

- may be used for branch-circuit and feeder wiring.

- may be used in interior wiring for wet, dry, or corrosive installations.

- is installed by the same methods as non-metallic-sheathed cable *(Article 336)*.

- must not be used as service-entrance cable.

- must not be embedded in concrete, cement, or aggregate.

- must be buried in the same trench where single-conductor cables are installed.

- shall be installed according to *Section 300-5*.

Equipment fed by Type UF cable is grounded by properly connecting the bare equipment grounding conductor found in the UF cable to the equipment to be grounded, Figure 16-9.

For underground wiring, it is significant to note that whereas the Code does permit running up the side of a tree with conduit or cable (when protected from physical damage), Figure 16-10, the Code does *not* permit supporting conductors between trees, Figure 16-11. Figure 16-12 shows one type of lighting fixture that is permitted for the installation shown in Figure 16-10. Splices are permitted to be made directly in the ground, but only if the connectors are "listed" by a recognized testing laboratory for such use, *Sections 110-14(b)* and *300-5(e)*.

Figure 16-10 **This installation conforms to the Code.** *Section 225-26* **does not permit spans of conductors to be run between live or dead trees, but it does not prohibit installing decorative lighting as shown in this figure.** *Section 410-16(h)* **permits outdoor lighting fixtures and associated equipment to be supported by trees. Recognized wiring methods to carry the conductors up the tree must be used.**

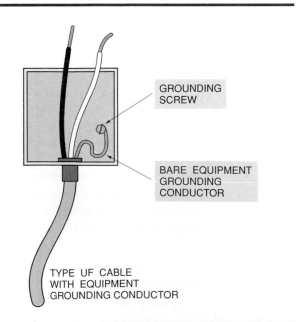

GROUNDING SCREW

BARE EQUIPMENT GROUNDING CONDUCTOR

TYPE UF CABLE WITH EQUIPMENT GROUNDING CONDUCTOR

Figure 16-9 **Illustration showing how the bare equipment grounding conductor of a Type UF cable is used to ground the metal box.**

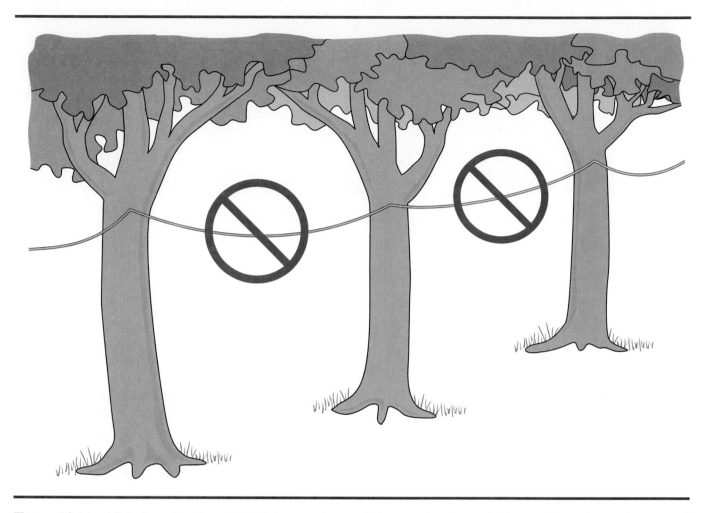

Figure 16-11 *Violation: Section 225-26* does not permit live or dead vegetation such as trees to support overhead conductor spans. Lumber and treated poles are not considered to be vegetation for this requirement.

Figure 16-12 Outdoor fixture with lamps. *Courtesy Progress Lighting.*

UNDERGROUND WIRING

Underground wiring is common in residential applications. Examples include decorative landscape lighting and wiring to post lamps and detached buildings such as garages or tool sheds.

Table 300-5 shows the minimum depths required for the various types of wiring methods. Refer to Figures 16-13, 16-14, and 16-15.

Across the top of *Table 300-5* are listed wiring methods such as direct burial cables or conductors, rigid metal conduit, intermediate metal conduit, rigid nonmetallic conduit that is approved for direct burial, special considerations for residential branch-circuits rated 120 volts or less having GFCI protection and overcurrent protection not over 20 amperes, and low-voltage landscape lighting supplied with Type UF cable.

Down the left-hand side of the table we find the location of the wiring methods that are listed across the top of the table.

The notes to *Table 300-5* are very important. Note 4 allows us to select the shallower of two depths when the wiring methods in Columns 1, 2, or 3 are combined with Columns 4 or 5. For example, the basic rule for direct buried cable is that it be covered by a minimum of 24 inches (610 mm). However, this minimum depth is reduced to 12 inches (305 mm) for residential installations when the branch-circuit is not over 120-volts, is GFCI protected, and the overcurrent protection is not over 20 amperes. This is for residential installations only.

TYPICAL POST LIGHT

"SOME" CURRENT FLOWS THROUGH THE PERSON.

TRANSFORMER

IMPROPER CONNECTION OF METAL FIXTURE TO GROUNDED CIRCUIT CONDUCTOR.

Ⓐ

THE CIRCUIT IS COMPLETED THROUGH THE LAMP... TO THE METAL FIXTURE... DOWNWARD ON METAL POST INTO THE GROUND.

Ⓑ

BUT NOTE AT POINTS Ⓐ & Ⓑ THE CURRENT ALSO FLOWS THROUGH THE PERSON... TO THE EARTH... WHERE THE CURRENTS JOIN TOGETHER.

MAIN BONDING JUMPER

UF CABLE

GROUNDING ELECTRODE

X

GROUNDED CIRCUIT CONDUCTOR "OPEN"

EARTH RETURN PATH BACK TO GROUNDING ELECTRODE... TO PANEL GROUNDED NEUTRAL BUS... THROUGH SERVICE ENTRANCE NEUTRAL... THROUGH TRANSFORMER... THEN TO SERVICE PANEL... THEN THROUGH "HOT" CONDUCTOR TO POST FIXTURE.

Figure 16-13 This diagram clearly illustrates the electric shock hazard associated with the code violation practice of grounding metal objects to the grounded circuit conductor. This is a violation of *Section 250-142(b)*.

The measurement for the depth requirement is from the top of the raceway or cable to the top of the finished grade, concrete, or other similar cover.

Do not back-fill a trench with rocks, debris, or similar coarse material, *Section 300-5(f)*.

Installation of Conduit Underground

Some local electrical codes require the installation of conduit for all underground wiring, using conductors identified for wet locations, such as Type TW, THW, or THWN.

All conduit installed underground must be protected against corrosion. The manufacturer of the conduit and U.L. Standards will furnish information as to whether the conduit is suitable for direct burial.

Additional supplemental protection of metal conduit can be accomplished by "painting" the conduit with a nonmetallic coating. See Table 16-2.

When metal conduit is used as the wiring method, the metal boxes, post lights, and so forth that are properly connected to the metal raceways are considered to be grounded because the conduit serves as the equipment ground, *Section 250-118*, Figure 16-16. See *Article 410, Part E* for details on grounding requirements for fixtures.

When nonmetallic raceways are used, then a separate equipment grounding conductor, either green or bare, must be installed to accomplish the adequate grounding of the equipment served. See Figure 16-17. This equipment grounding conductor must be sized according to *Table 250-122, NEC.®*

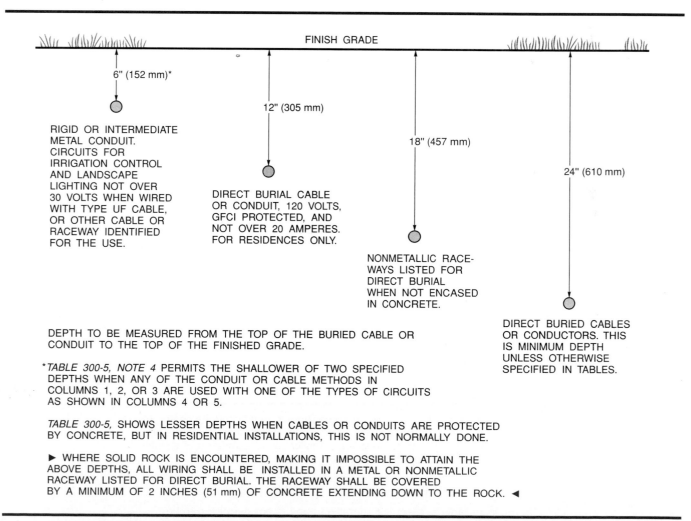

FINISH GRADE

6" (152 mm)*

RIGID OR INTERMEDIATE METAL CONDUIT. CIRCUITS FOR IRRIGATION CONTROL AND LANDSCAPE LIGHTING NOT OVER 30 VOLTS WHEN WIRED WITH TYPE UF CABLE, OR OTHER CABLE OR RACEWAY IDENTIFIED FOR THE USE.

12" (305 mm)

DIRECT BURIAL CABLE OR CONDUIT, 120 VOLTS, GFCI PROTECTED, AND NOT OVER 20 AMPERES. FOR RESIDENCES ONLY.

18" (457 mm)

NONMETALLIC RACE-WAYS LISTED FOR DIRECT BURIAL WHEN NOT ENCASED IN CONCRETE.

24" (610 mm)

DIRECT BURIED CABLES OR CONDUCTORS. THIS IS MINIMUM DEPTH UNLESS OTHERWISE SPECIFIED IN TABLES.

DEPTH TO BE MEASURED FROM THE TOP OF THE BURIED CABLE OR CONDUIT TO THE TOP OF THE FINISHED GRADE.

*TABLE 300-5, NOTE 4 PERMITS THE SHALLOWER OF TWO SPECIFIED DEPTHS WHEN ANY OF THE CONDUIT OR CABLE METHODS IN COLUMNS 1, 2, OR 3 ARE USED WITH ONE OF THE TYPES OF CIRCUITS AS SHOWN IN COLUMNS 4 OR 5.

TABLE 300-5, SHOWS LESSER DEPTHS WHEN CABLES OR CONDUITS ARE PROTECTED BY CONCRETE, BUT IN RESIDENTIAL INSTALLATIONS, THIS IS NOT NORMALLY DONE.

► WHERE SOLID ROCK IS ENCOUNTERED, MAKING IT IMPOSSIBLE TO ATTAIN THE ABOVE DEPTHS, ALL WIRING SHALL BE INSTALLED IN A METAL OR NONMETALLIC RACEWAY LISTED FOR DIRECT BURIAL. THE RACEWAY SHALL BE COVERED BY A MINIMUM OF 2 INCHES (51 mm) OF CONCRETE EXTENDING DOWN TO THE ROCK. ◄

Figure 16-14   Minimum depths for cables and conduits installed underground. See *Table 300-5* for depths for other conditions.

Table 300-5. Minimum Cover Requirements, 0 to 600 Volts, Nominal, Burial in Inches (Cover is defined as the shortest distance in inches measured between a point on the top surface of any direct-buried conductor, cable, conduit, or other raceway and the top surface of finished grade, concrete, or similar cover.)

	Type of Wiring Method or Circuit				
Location of Wiring Method or Circuit	Column 1 Direct Burial Cables or Conductors	Column 2 Rigid Metal Conduit or Intermediate Metal Conduit	Column 3 Nonmetallic Raceways Listed for Direct Burial Without Concrete Encasement or Other Approved Raceways	Column 4 Residential Branch Circuits Rated 120 Volts or Less with GFCI Protection and Maximum Overcurrent Protection of 20 Amperes	Column 5 Circuits for Control of Irrigation and Landscape Lighting Limited to Not More than 30 Volts and Installed with Type UF or in Other Identified Cable or Raceway
All locations not specified below	24	6	18	12	6
In trench below 2-in. thick concrete or equivalent	18	6	12	6	6
Under a building	0 (in raceway only)	0	0	0 (in raceway only)	0 (in raceway only)
Under minimum of 4-in. thick concrete exterior slab with no vehicular traffic and the slab extending not less than 6 in. beyond the underground installation	18	4	4	6 (direct burial) 4 (in raceway)	6 (direct burial) 4 (in raceway)
Under streets, highways, roads, alleys, driveways, and parking lots	24	24	24	24	24
One- and two-family dwelling driveways and outdoor parking areas, and used only for dwelling-related purposes	18	18	18	12	18
In or under airport runways, including adjacent areas where trespassing prohibited	18	18	18	18	18

Notes:

1. For SI units, 1 in. = 25.4 mm.

2. Raceways approved for burial only where concrete encased shall require concrete envelope not less than 2 in. thick.

3. Lesser depths shall be permitted where cables and conductors rise for terminations or splices or where access is otherwise required.

4. Where one of the wiring method types listed in Columns 1–3 is used for one of the circuit types in Columns 4 and 5, the shallower depth of burial shall be permitted.

5. Where solid rock prevents compliance with the cover depths specified in this table, the wiring shall be installed in metal or nonmetallic raceway permitted for direct burial. The raceways shall be covered by a minimum of 2 in. of concrete extending down to rock.

Reprinted with permission from NFPA 70-1999.

IS SUPPLEMENTAL CORROSION PROTECTION REQUIRED?

	In Concrete Above Grade?	In Concrete Below Grade?	In Direct Contact with Soil?
Rigid Conduit[1]	No	No	No[2]
Intermediate Conduit[1]	No	No	No[2]
Electrical Metallic Tubing[1]	No	Yes	Yes[3]

[1] Severe corrosion can be expected where ferrous metal conduits come out of concrete and enter the soil. Here again, some electrical inspectors and consulting engineers might specify the application of some sort of supplemental nonmetallic corrosion protection.

[2] Unless subject to severe corrosive effects. Different soils have different corrosive characteristics.

[3] In most instances, electrical metallic tubing is not permitted to be installed underground in direct contact with the soil because of corrosion problems.

Table 16-2

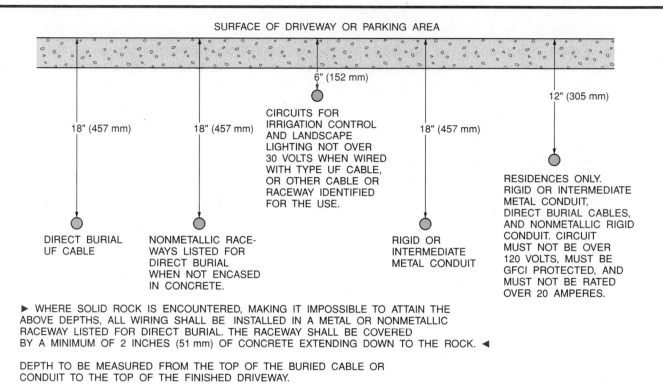

SURFACE OF DRIVEWAY OR PARKING AREA

6" (152 mm)

12" (305 mm)

18" (457 mm)

18" (457 mm)

CIRCUITS FOR
IRRIGATION CONTROL
AND LANDSCAPE
LIGHTING NOT OVER
30 VOLTS WHEN WIRED
WITH TYPE UF CABLE,
OR OTHER CABLE OR
RACEWAY IDENTIFIED
FOR THE USE.

18" (457 mm)

RESIDENCES ONLY.
RIGID OR INTERMEDIATE
METAL CONDUIT,
DIRECT BURIAL CABLES,
AND NONMETALLIC RIGID
CONDUIT. CIRCUIT
MUST NOT BE OVER
120 VOLTS, MUST BE
GFCI PROTECTED, AND
MUST NOT BE RATED
OVER 20 AMPERES.

DIRECT BURIAL
UF CABLE

NONMETALLIC RACE-
WAYS LISTED FOR
DIRECT BURIAL
WHEN NOT ENCASED
IN CONCRETE.

RIGID OR
INTERMEDIATE
METAL CONDUIT

► WHERE SOLID ROCK IS ENCOUNTERED, MAKING IT IMPOSSIBLE TO ATTAIN THE
ABOVE DEPTHS, ALL WIRING SHALL BE INSTALLED IN A METAL OR NONMETALLIC
RACEWAY LISTED FOR DIRECT BURIAL. THE RACEWAY SHALL BE COVERED
BY A MINIMUM OF 2 INCHES (51 mm) OF CONCRETE EXTENDING DOWN TO THE ROCK. ◄

DEPTH TO BE MEASURED FROM THE TOP OF THE BURIED CABLE OR
CONDUIT TO THE TOP OF THE FINISHED DRIVEWAY.

Figure 16-15   Depth requirements under residential driveways and parking areas. These depths are permitted for one- and two-family residences. See *Table 300-5*.

THIS METAL BOX IS CONSIDERED
TO BE PROPERLY GROUNDED
WHEN FASTENED PROPERLY TO
THE GROUNDED METAL RACEWAY.

GROUNDED METAL
CONDUIT

PROPER LOCKNUTS
AND BUSHINGS
TIGHTLY FASTENED

NONMETALLIC
RACEWAY

SEPARATE
EQUIPMENT
GROUND

Figure 16-17   Grounding a metal box with a separate equipment grounding conductor.

Figure 16-16   This illustration shows that a box is grounded when properly fastened to the grounded metal raceway. This is acceptable in *Section 250-118*. More and more Code-enforcing authorities and con-sulting engineers are requiring that a separate equipment grounding conductor be installed in all raceways. This ensures effective grounding of the installation.

Depending upon the type of soil and moisture content of the soil, the earth may or may not be a good ground. The Code recognizes that the earth by itself is not considered to be an adequate ground. ►*Section 250-2(d)*, second paragraph states that

"The earth shall not be used as the sole equipment grounding conductor or fault-return path." See Figure 16-18. ◄

Figure 16-19 shows two methods of bringing the conduit into the basement: (1) it can be run below ground level and then can be brought through the basement wall, or (2) it can be run up the side of the building and through the basement wall at the ceiling joist level. When the conduit is run through the basement wall, the opening must be sealed to prevent moisture from seeping into the basement. The electrician must decide which of the two methods is more suitable for each installation.

According to the Code, any location exposed to the weather is considered a wet location.

Section 410-4(a) states that outdoor fixtures must be constructed so that water cannot enter or accumulate in lampholders, wiring compartments, or other electrical parts. These fixtures must be marked "Suitable for Wet Locations."

Partially protected areas such as roofed open porches or areas under canopies are defined by the Code as damp locations. Fixtures to be used in these locations must be marked "Suitable for Damp Locations." Many types of fixtures are available. Therefore, it is recommended that the electrician check the U.L. label on the fixture to determine the suitability of the fixture for a wet or damp location.

The post in Figure 16-19 may or may not be embedded in concrete, depending on the consistency of the soil, the height of the post, and the size of the lighting fixture. Most electricians prefer to embed the base of the post in concrete to prevent rotting of wood posts and rusting of metal posts.

Figure 16-20 illustrates some typical post lights.

A

VIOLATION!!!
POST LAMP CONNECTED WITH 2-WIRE UF CABLE. THE EARTH IS THE ONLY GROUND FOR THE POST LAMP.

B

MEETS CODE
POST LAMP CONNECTED WITH 2-WIRE UF CABLE WITH GROUND WIRE. THE EQUIPMENT GROUND WIRE SERVES AS THE GROUNDING MEANS FOR THE POST LAMP.

C

VIOLATION!!!
POST LAMP WIRED WITH 2-WIRE UF CABLE. A GROUND ROD INSTALLED TO PROVIDE THE GROUND FOR THE POST LAMP.

►**Figure 16-18   These drawings illustrate the intent of *Section 250-2(d)*, second paragraph, which states that "The earth shall not serve as the sole equipment grounding conductor or fault-return path." ◄**

WEATHERPROOF
LAMP UNIT
SECTION 410-4(a)

X IF CONDUITS OR CABLES ARE RUN UNDER DRIVEWAYS, THEY MUST BE AT LEAST 18 INCHES (457 mm) DEEP. *TABLE 300-5.*

WHERE THERE IS LIKELIHOOD THAT THE EARTH WILL SETTLE OR BE SUBJECT TO FROST HEAVES, AN "S" LOOP MAY BE MADE IN THE UF CABLE WHERE IT ENTERS THE RACEWAY OR WALL. THIS WILL GENERALLY PROVIDE THE NEEDED FLEXIBILITY TO ALLOW FOR THE EARTH'S MOVEMENT. *SECTION 300-5(j).*

SEAL INSIDE OF RACEWAY SO MOISTURE AND COLD AIR CANNOT PASS THROUGH, *SECTION 300-5(g).*

METAL FIXTURE AND POST MUST BE GROUNDED. *ARTICLE 410, PART F.*

LB CONDUIT FITTING

CONDUIT TO PROTECT CABLE *SECTION 300-5(d).*

BOX ON CEILING JOIST IN BASEMENT

SOIL

UF CABLE

X BACKFILL *SECTION 300-5(f).*

BUSHING TO PROTECT CONDUCTORS *SECTION 300-5(h).*

SEE FIGURES 16-14, 16-15, AND *NEC® TABLE 300-5* FOR MINIMUM DEPTH REQUIREMENTS.

PENETRATING FOUNDATION WALL BELOW GRADE LEVEL NOT RECOMMENDED BECAUSE OF POSSIBLE SEEPAGE AND LEAKAGE OF WATER.

POST MAY OR MAY NOT BE EMBEDDED IN CONCRETE

Figure 16-19 Methods of bringing cable and/or conduit through a concrete wall and/or upward from a concrete wall into the hollow space within a framed wall.

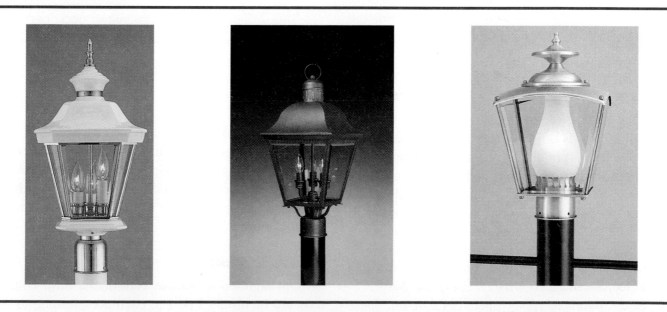

Figure 16-20 Post-type lights (lanterns). *Courtesy* Progress Lighting.

OVERHEAD GARAGE DOOR OPERATOR ⒶE

The overhead garage door operator in this residence is plugged into the receptacle provided for this purpose on the garage ceiling. This particular overhead door operator is rated 5.8 amperes at 120 volts. The actual installation of the overhead door will usually be done by a specially trained overhead door installer.

Principles of Operation

The overhead door operator contains a motor, gear reduction unit, motor reversing switch, and electric clutch. This unit is preassembled and wired by the manufacturer.

Almost all residential overhead door operators use split-phase, capacitor-start motors in sizes from ¼ to ½ horsepower. The motor size selected depends upon the size (weight) of the door to be raised and lowered. By using springs in addition to the gear reduction unit to counterbalance the weight of the door, a small motor can lift a fairly heavy door.

There are two ways to change the direction of a split-phase motor: (1) reverse the starting winding with respect to the running winding; (2) reverse the running winding with respect to the starting winding. The motor can be run in either direction by properly connecting the starting and running winding leads of the motor to the reversing switch. When the motor shaft is connected to a chain drive or screw drive, the door can be raised or lowered according to the direction in which the motor is rotating.

Wiring of Garage Door Operators

The wiring of overhead door operators for residential use is quite simple, because these units are completely prewired by the manufacturer. All residential-type overhead door operators come equipped with a cord and plug. This is shown in Figure 16-21. A 120-volt receptacle outlet is installed on the ceiling near the operator. Because a receptacle on a ceiling is not *readily accessible* as defined by the *NEC*® it does not have to be GFCI protected, *Section 210-8(a)(2), Exception No. 1.*

Some overhead door operators can be "hard-wired" as shown in Figure 16-22. This is not often done in residential installations.

Figure 16-21 Overhead door operator with three-wire cord connector.

Figure 16-22 Connections for overhead door operator.

The current draw of a typical residential overhead door operator is usually so small that it could be connected to the garage lighting branch-circuit. In the residence discussed in this text, the overhead door operator is served by a separate 15-ampere branch-circuit.

Push Buttons

In addition to remote controllers, push buttons should be installed next to each door leading to the garage, including the overhead door. Key-operated weatherproof switches can also be installed outside next to the overhead door. The wiring to each push button or key-operated switch can be a low-voltage type similar to that used for chime wiring. Figure 16-23 shows how push buttons are connected. Because the operating relay in an overhead door operator works on the momentary contact principle, a two-wire cable is all that is needed between the operator and the push buttons or key-operated switch.

Quite often, electricians will install the push-button wiring during the "rough-in" stage. An easy way to do this is to nail a single-gang, 4" square plaster "mud" ring to the stud at the desired locations, then run a 2-wire, No. 18 or No. 16 AWG low-voltage cable to a point where the overhead door operator will be mounted. When the house is "trimmed out," a single-gang wallplate with the push button is installed. See Figure 16-24.

Box-Cover Unit

The overhead door operator unit has overcurrent protection built into it according to U.L. standards. Additional "back-up" protection can be installed in the field. Here's how.

Instead of a standard receptacle on the ceiling near the operator, a box-cover unit can be installed. Box-cover units are available with *switch/fuse* or *receptacle/fuse* combinations, Figures 16-21, 16-22, 16-25, and 16-26. Box-cover units provide a conve-

Figure 16-24   A sketch of how a 4-inch square cover can be used to provide an opening for low-voltage wiring.

nient disconnecting means in addition to having a fuse holder in which to install a dual-element, time-delay Type S fuse.

Size the fuse at approximately 125% of the overhead door operator's ampere rating.

Figure 16-23   Push-button wiring for overhead door operator.

Figure 16-25   Box-cover unit that provides disconnecting means plus individual overcurrent protection. *Courtesy* Bussmann Division, Cooper Industries.

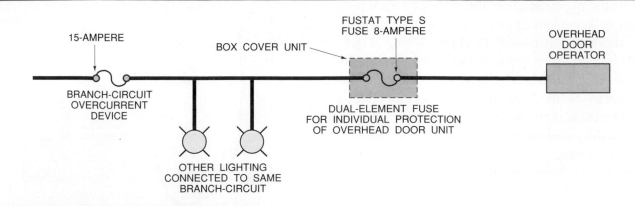

Figure 16-26 This diagram shows the benefits of installing a small-ampere-rated dual element time-delay fuse that is sized to provide overload protection for the motor. If an electrical problem occurs at the overhead door operator, only the 8-ampere fuse will open. The 15-ampere branch-circuit is not affected. All other loads connected to the 15-ampere circuit remain energized.

For example, let us say the overhead door operator is marked 5.8 amperes. Then:

$$8 \times 1.25 = 7.25 \text{ amperes}$$

The next standard size dual-element time-delay fuse is 8 amperes. An 8-ampere fuse will open faster than the 15- or 20-ampere branch-circuit breaker or fuse should a short-circuit occur within the operator. This isolates the door operator, and does not shut off the power to the rest of the branch-circuit.

VOLTAGE DROP PROBLEMS WITH LOW-VOLTAGE LIGHTING

There seems to be a mind-set that other than the fact that low-voltage lighting systems operate at a lower voltage than standard 120-volt circuits, no consideration is given to the much greater current draw for a given wattage at the lower voltage. This is a critical issue. When the length of the low-voltage wiring is long, larger conductors are necessary to compensate for voltage drop. If ignored, the lamps will not get enough voltage to burn properly (the lamps will burn dim), and the small conductors can reach dangerous temperatures that might cause a fire!

EXAMPLE:

A 480-watt load at 120 volts draws:
$$480 \div 120 = 4 \text{ amperes}$$

A 480-watt load at 12 volts draws:
$$480 \div 120 = 40 \text{ amperes}$$

You can readily see that for this 480-watt example, the conductors must be considerably larger for the 12-volt application to safely carry the current.

Prewired low-voltage lighting assemblies that bear the U.L. label have the correct size and length of conductors. Adding conductors in the field must be done according to the manufacturers' instructions.

Voltage drop calculations are covered in Unit 4.

REVIEW

Note: Refer to the Code or plans where necessary.

1. What circuit supplies the garage? _____

2. What is the circuit rating? _____

3. How many GFCI-protected receptacle outlets are connected to the garage circuit?

_____

4. a. How many cables enter the wall switch box located next to the side garage door?

   _____

   b. How many circuit conductors enter this box? _____

   c. How many equipment grounding conductors enter this box? _____

5. Show calculations on how to select a proper size box for question 4. What kind of box would you use? _____

   _____

   _____

   _____

   _____

   _____

6. a. How many lights are recommended for a one-car garage? _____ for a two-car garage? _____ for a three-car garage? _____

   b. Where are these lights to be located? _____

   _____

7. From how many points in the garage of this residence are the ceiling lights controlled? _____

8. GFCI breakers or GFCI receptacles are relatively expensive. How would *you* arrange the wiring of the garage circuit to make as economical installation as possible that complies with the Code? _____

   _____

   _____

   _____

   _____

   _____

9. The total estimated volt-ampere load of the garage circuit draws how many amperes? Show your calculations. _____

   _____

   _____

   _____

10. How high from the floor are the switches and receptacles to be mounted? _____

11. What type of cable feeds the post light? _____

12. All raceways, cables, and direct burial-type conductors require a "cover." Explain the term "cover." _____

   _____

   _____

   _____

13. In the spaces provided, fill in the cover (depth) for the following residential underground installations.

    a. Type UF. No other protection. _____ inches

    b. Type UF below driveway. _____ inches

    c. Rigid metal conduit under lawn. _____ inches

    d. Rigid metal conduit under driveway. _____ inches

    e. Electrical metallic tubing between house _____ inches
       and detached garage.

    f. UF cable under lawn. Circuit is 120 volts, _____ inches
       20 amperes, GFCI protected.

    g. Rigid nonmetallic conduit approved for direct
       burial. No other protection. _____ inches

    h. Rigid nonmetallic conduit. Circuit is 120 volts, _____ inches
       20 amperes, GFCI protected.

    i. Rigid conduit passing over solid rock and covered by _____ inches
       2 inches (51 mm) of concrete extending down to rock.

14. What section of the Code prohibits embedding Type UF cable in concrete? _____

_____

15. The following is a layout of the garage circuit. Using the suggested cable layout, make a complete wiring diagram of this circuit, using colored pens or pencils to indicate conductor insulation colors.

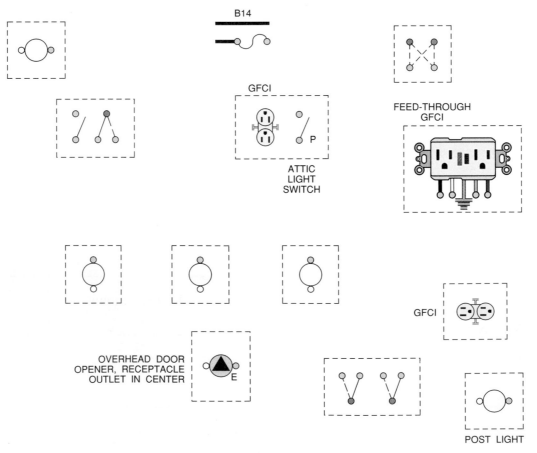

GARAGE

16. What type of motors are generally used for garage door openers? _____

17. How is the direction of a split-phase motor reversed? _____

_____

_____

18. When a box-cover unit is installed for an overhead door operator, a time-delay dual-element fuse ampere rating would generally be sized at _____% of the unit's name-plate current rating.

19. When outdoor receptacles are installed where a cord will be plugged in permanently, such as in the case of decorative lighting, the cover must provide weatherproof integrity when the cord is plugged in. True or false? _____ *Section* _____ .

20. A homeowner purchased some single-conductor cable that was marked as suitable for direct burial in the ground to wire a post light. A separate equipment grounding conductor to the post light was not carried. Instead, the bottom of the metal post buried two feet into the ground was used as a ground for the post light. The inspector "red tagged" (turned down) the installation. Who is right? The homeowner or the electrical inspector? Explain.

_____

_____

_____

_____

Recreation Room

OBJECTIVES

After studying this Unit, the student will be able to

- understand three-wire (multiwire) branch-circuits.
- understand how to install lay-in fixtures.
- calculate watts loss and voltage drop in two-wire and three-wire circuits.
- understand the term *fixture whips*.
- understand the advantages of installing multiwire branch-circuits.
- understand problems that can be encountered on multiwire branch-circuits as a result of open neutrals.

RECREATION ROOM LIGHTING

The recreation room is well-lighted through the use of six lay-in 2' × 4' fluorescent fixtures. These fixtures are exactly the same size as two 2' × 2' ceiling tiles. The fixtures rest on top of the ceiling tee bars, Figure 17-1.

Junction boxes are mounted above the dropped ceiling usually within 1 or 2 feet (305 to 610 mm) of the intended fixture location. Four to 6 feet (1.2 to 1.83 m) of flexible conduit are installed between this junction box and the fixture. These are commonly called *fixture whips*. They contain the correct type and size of conductor suitable for the temperature ratings and load required by the Code for recessed fixtures. See Unit 7 for a complete explanation of Code regulations for the installation of recessed fixtures, Figure 17-2.

Fluorescent fixtures of the type shown in Figure 17-1 might bear a label stating "Recessed Fluorescent Fixture." The label might also state "Suspended Ceiling Fluorescent Fixture." Although they might look the same, there is a difference.

Recessed fluorescent fixtures are intended for installation in cavities in ceilings and walls and are to be wired according to *Section 410-64*. These fixtures may also be installed in suspended ceilings if they have the necessary mounting hardware.

Suspended fluorescent fixtures are intended only for installation in suspended ceilings where the acoustical tiles, lay-in panels, and suspended grid are not part of the actual building structure.

Underwriters Laboratories (U.L.) Standard 1570 covers recessed and suspended ceiling fixtures in detail.

The six recessed "lay-in" fluorescent fixtures in the recreation room are four-lamp fixtures with warm white energy-saving F40SPX30/RS lamps installed in them. The ballasts in these fixtures are high power factor, energy-saving ballasts. Five of the fixtures are controlled by a single-pole switch located at the bottom of the stairs. One fixture is controlled by the three-way switches that control the stairwell light fixture hung from the ceiling above the stair landing midway up the stairs.

IMPORTANT: TO PREVENT THE FIXTURE FROM INADVERTENTLY FALLING, *SECTION 410-16(c)* OF THE CODE REQUIRES THAT 1) SUSPENDED CEILING FRAMING MEMBERS THAT SUPPORT RECESSED FIXTURES MUST BE SECURELY FASTENED TO EACH OTHER, AND MUST BE SECURELY ATTACHED TO THE BUILDING STRUCTURE AT APPPROPRIATE INTERVALS, AND 2) RECESSED FIXTURES MUST BE SECURELY FASTENED TO THE SUSPENDED CEILING FRAMING MEMBERS BY BOLTS, SCREWS, RIVETS, OR SPECIAL ►"LISTED"◄ CLIPS PROVIDED BY THE MANUFACTURER OF THE FIXTURE FOR THE PURPOSE OF ATTACHING THE FIXTURE TO THE FRAMING MEMBER.

"LISTED" STEEL CLIP

Figure 17-1   Typical lay-in fluorescent fixture commonly used in conjunction with dropped suspended ceilings.

JUNCTION BOX

FIXTURE WHIP

RECESSED FIXTURE

Figure 17-2   Typical recessed "lay-in" fluorescent fixture showing flexible fixture whip connection. Conductors in whip must have temperature rating as required on label in fixture. For example, "For Supply Use 90°C conductors."

This was done so that the entire recreation room lighting would not be on continually whenever someone came down the stairs to go into the workshop. This is a practical energy-saving feature.

Circuit B12 feeds the fluorescent fixtures in the recreation room. See cable layout, Figure 17-3. Table 17-1 summarizes the outlets and estimated load for the recreation room.

RECREATION ROOM

Figure 17-3 Cable layout for recreation room. The six fluorescent fixtures are connected to Circuit B12. The receptacles located around the room are connected to Circuit B11. The four recessed fixtures over the bar plus the three receptacles located on the bar wall plus the receptacles located to the right of the steps as you come down the steps are connected to Circuit B9. Circuits B9 and B11 are three-wire circuits. Junction boxes are located above the dropped ceiling, and to one side of each fluorescent fixture. The electrical connections are made in these junction boxes. A flexible fixture "whip" connects between each junction box and each fixture. See Figure 17-2.

How to Connect Recessed and Lay-In Lighting Fixtures

Figures 17-1 and 17-2 show typical lay-in fluorescent lighting fixtures. The installation and wiring connections for recessed lighting fixtures and lay-in suspended-ceiling fixtures was covered in Unit 7.

RECEPTACLES AND WET BAR

The circuitry for the recreation room wall receptacles and the wet-bar lighting area introduces a new type of circuit. This is termed a *multiwire branch-circuit* or a *three-wire branch-circuit*.

In many cases the use of a multiwire branch-

DESCRIPTION	QUANTITY	WATTS	VOLT-AMPERES
Circuit Number B11			
Receptacles @ 120 W each	7	840	840
Closet fixtures @ 75 W each	2	150	150
Exhaust fan (2.5 A × 120 V)	1	160	300
TOTALS	10	1150	1290
Circuit Number B12			
Fixture on stairway	1	100	100
Recessed fluorescent fixtures with four 40-W lamps each	6	960	1008
TOTALS	7	1060	1108
Circuit Number B9			
Receptacles @ 120 W each	4	480	480
Wet-bar recessed fixtures @ 75 W each	4	300	300
TOTALS	8	780	780

Table 17-1 Recreation room outlets and estimated load (three circuits).

circuit can save money in that one three-wire branch-circuit will do the job of two two-wire branch-circuits.

Also, if the loads are nearly balanced in a three-wire branch-circuit, the neutral conductor carries only the unbalanced current. This results in less voltage drop and watts loss for a three-wire branch-circuit as compared to similar loads connected to separate two-wire branch-circuits.

Figures 17-4 and 17-5 illustrate the benefits of a multiwire branch-circuit relative to watts loss and voltage drop. The example, for simplicity, is a purely resistive circuit, and loads are exactly equal.

The distance from the load to the source is 50 feet. The conductor size is No. 14 AWG solid, uncoated copper. From *Table 8* in *Chapter 9* of the *NEC*, we find that the resistance of 1,000 feet is 3.07 ohms. The resistance of 50 feet is:

$$\frac{3.07}{1,000} \times 50 = 0.1535 \text{ ohms (round off to 0.154)}$$

EXAMPLE 1:
TWO 2-WIRE BRANCH-CIRCUITS

Watts loss in each current-carrying conductor is

$$\text{Watts} = I^2R = 10 \times 10 \times 0.154 = 15.4 \text{ watts}$$

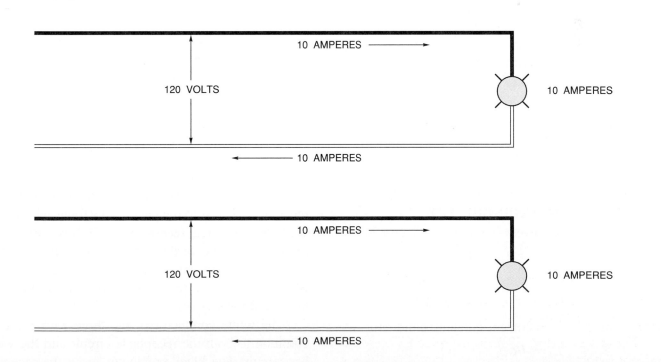

Figure 17-4 Two 2-wire branch-circuits.

Figure 17-5   A 3-wire branch-circuit.

Watts loss in all four current-carrying conductors is

$$15.4 \times 4 = 61.6 \text{ watts}$$

Voltage drop in each current-carrying conductor is

$$E_d = IR$$
$$= 10 \times 0.154$$
$$= 1.54 \text{ volts}$$

E_d for both current-carrying conductors in each circuit:

$$2 \times 1.54 = 3.08 \text{ volts}$$

Voltage available at load:

$$120 - 3.08 = 116.92 \text{ volts}$$

EXAMPLE 2:
ONE 3-WIRE BRANCH-CIRCUIT

Watts loss in each current-carrying conductor is

$$\text{Watts} = I^2R = 10 \times 10 \times 0.154 = 15.4 \text{ watts}$$

Watts loss in both current-carrying conductors is

$$15.4 \times 2 = 30.8 \text{ watts}$$

Voltage drop in each current-carrying conductor is

$$E_d = IR$$
$$= 10 \times 0.154$$
$$= 1.54 \text{ volts}$$

E_d for both current-carrying conductors is:

$$2 \times 1.54 = 3.08 \text{ volts}$$

Voltage available at loads:

$$240 - 3.08 = 236.92 \text{ volts}$$

Voltage available at each load:

$$\frac{236.92}{2} = 118.46 \text{ volts}$$

It is readily apparent that a multiwire circuit can greatly reduce watts loss and voltage drop. This is the same reasoning that is used for the three-wire service-entrance conductors and the three-wire feeder to Panel B in this residence.

In the recreation room, a three-wire cable carrying Circuits B9 and B11 feeds from Panel B to the receptacle closest to the panel. This three-wire cable continues to each receptacle wall box along the same wall. At the third receptacle along the wall, the multiwire circuit splits, where the wet-bar Circuit B9 continues on to feed into the receptacle outlet to the left of the wet bar.

The white conductor of the three-wire cable is common to both the receptacle circuit and the wet-bar circuit. The black conductor feeds the receptacles. The red conductor is spliced straight through the boxes and feeds the wet bar area.

Care must be used when connecting a three-wire circuit to the panel. The black and red conductors *must* be connected to the opposite phases in the panel to prevent heavy overloading of the neutral grounded (white) conductor.

The neutral (grounded) conductor of the three-wire cable carries the unbalanced current. This current is the difference between the current in the black wire and the current in the red wire. For example, if one load is 12 amperes and the other load is 10 amperes, the neutral current is the difference between these loads (2 amperes), Figure 17-6.

If the black and red conductors of the three-wire cable are connected to the same phase in Panel A, Figure 17-7, the neutral conductor must carry the total current of both the red and black conductors

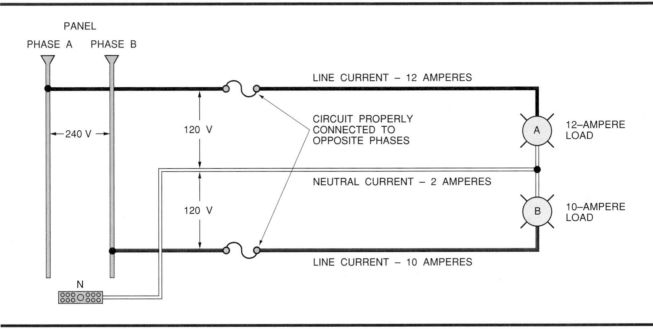

Figure 17-6   Correct wiring connections for three-wire (multiwire) branch-circuit.

Figure 17-7   Improperly connected three-wire (multiwire) branch-circuit.

rather than the unbalanced current. As a result, the neutral conductor will be overloaded. All single-phase, 120/240-volt panels are clearly marked to help prevent an error in phase wiring. The electrician must check all panels for the proper wiring diagrams.

Figure 17-7 shows how an improperly connected three-wire branch-circuit results in an overloaded neutral conductor.

If an open neutral occurs on a three-wire branch-circuit, some of the electrical appliances in operation may experience voltages higher than the rated voltage at the instant the neutral opens.

For example, Figure 17-8 shows that for an open neutral condition, the voltage across load A decreases and the voltage across load B increases. If the load on each circuit changes, the voltage on each circuit also changes. According to Ohm's law, the voltage drop across any device in a series circuit is directly proportional to the resistance of that device. In other words, if load B has twice the resistance of load A, then load B will be subjected to twice the voltage of load A for an open neutral condition. To ensure the proper connection, care must be used when splicing the conductors.

An example of what can occur should the neutral of a three-wire multiwire branch-circuit open is shown in Figure 17-9. Trace the flow of current from

phase A through the television set, then through the toaster, then back to phase B, thus completing the circuit. The following simple calculations show why the television set (or stereo or home computer) can be expected to burn up.

$$R_t = 8.45 + 80 = 88.45 \text{ ohms}$$

$$I = \frac{E}{R} = \frac{240}{88.45} = 2.71 \text{ amperes}$$

Voltage appearing across the toaster:

$$IR = 2.71 \times 8.45 = 22.9 \text{ volts}$$

Voltage appearing across the television:

$$IR = 2.71 \times 80 = 216.8 \text{ volts}$$

This example illustrates the problems that can arise with an open neutral on a three-wire, 120/240-volt multiwire branch-circuit.

The same problem can arise when the neutral of the utility company's incoming service-entrance conductors (underground or overhead) opens. The problem is minimized because the neutral of the service is solidly grounded to the metal water piping system within the building. However, there are cases on record where poor service grounding has resulted in serious and expensive damage to appliances within the home because of an open neutral in the incoming service-entrance conductors.

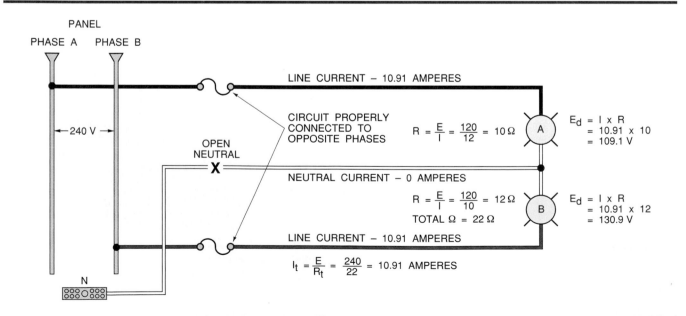

Figure 17-8   Example of an open neutral conductor.

For example, poor service grounding results when relying on a driven iron pipe (that rusts in a short period of time) as the only means of obtaining the service equipment ground. This was quite often done years ago when the electrical code was not as stringent as today's *NEC*. When working on older homes, check the service ground. It could very well be inadequate, unsafe, and in need of updating to the present *NEC*.

The integrity of using only ground rods is highly questionable, whether the rod is an iron pipe, a galvanized pipe, or a copper-clad rod. The Code prohibits the use of ground rods as the only means of obtaining a good service-entrance equipment ground.

See Unit 28 for details on the grounding and bonding of service-entrance equipment.

In summary, three-wire branch-circuits are used occasionally in residential installations. The two "hot" conductors of three-wire cable are connected to opposite phases. One white grounded (neutral) conductor is common to both "hot" conductors.

There are two receptacle outlets above the wet bar in the recreation room. ▶Receptacles installed to serve countertop surfaces that are within 6 feet (1.83 m) of the outer edge of a wet bar sink must be GFCI protected, and must not be installed face-up on the countertop surfaces or work surfaces, *Section 210-8(a)(7).*◀

A single-pole switch to the right of the wet bar controls the four recessed fixtures above it. See Unit 7 for details pertaining to recessed fixtures.

An exhaust fan of a type similar to the one installed in the laundry is installed in the ceiling of the recreation room to exhaust stale, stagnant, and smoky air from the room.

The basic switch, receptacle, and fixture connections are repetitive of most wiring situations discussed in previous units.

The wall boxes for the receptacles are selected based upon the measurements of the furring strips on the walls. Two-by-two furring strips will require the use of 4-inch square boxes with suitable raised covers. If the walls are furred with 2 × 4s, then possibly sectional device boxes could be used. Select a box that can contain the number of conductors, devices, and clamps to "meet Code." After all, the installation must be safe.

Probably the biggest factor in deciding what wiring method (cable or electrical metallic tubing) will be used to wire the recreation room is "what size and type of wall furring will be used?" See Unit 4 for the in-depth discussion on the mechanical protection required for nonmetallic sheathed cables where the cables will be less than $1\frac{1}{4}$ inch (31.8 mm) from the edge of the framing members.

Figure 17-9 Problems that can occur with an open neutral on a three-wire (multiwire) branch-circuit.

REVIEW

Note: Refer to the Code or plans where necessary.

1. What is the total current draw when all six fluorescent fixtures are turned on?

   _____

   _____

2. The junction box that will be installed above the dropped ceiling near the fluorescent fixtures closest to the stairway will have _____ (number of) No. 14 AWG conductors.

3. Why is it important that the "hot" conductor in a three-wire branch-circuit be properly connected to opposite phases in a panel? _____

   _____

4. In the diagram, Load A is rated at 10 amperes, 120 volts. Load B is rated at 5 amperes, 120 volts.

   a. When connected to the three-wire branch-circuit as indicated, how much current will flow in the neutral? _____

   b. If the neutral should open, to what voltage would each load be subjected, assuming both loads were operating at the time the neutral opened? Show all calculations.

5. Calculate the watts loss and voltage drop in each conductor in the following circuit. Show calculations.

```
50 FT  NO. 12  AWG  SOLID  COPPER


                                              15  AMPERE    ☀


50 FT  NO. 12  AWG  SOLID  COPPER
```

6. Unless specifically designed, all recessed incandescent fixtures must be provided with factory-installed protection (see Unit 7). _____

7. Where in the *NEC*® would you look for the installation requirements regarding flush and recessed lighting fixtures?

   _____

8. What is the current draw of the recessed fixtures above the bar? _____

   _____

   _____

9. a. VA load for Circuit B9 is _____ volt-amperes.

   b. VA load for Circuit B11 is _____ volt-amperes.

   c. VA load for Circuit B12 is _____ volt-amperes.

10. Calculate the total current draw for Circuits B9, B11, and B12.

11. Complete the wiring diagram for the recreation room. Follow the suggested cable layout. Use colored pencils or pens to identify the various colors of the conductor insulation.

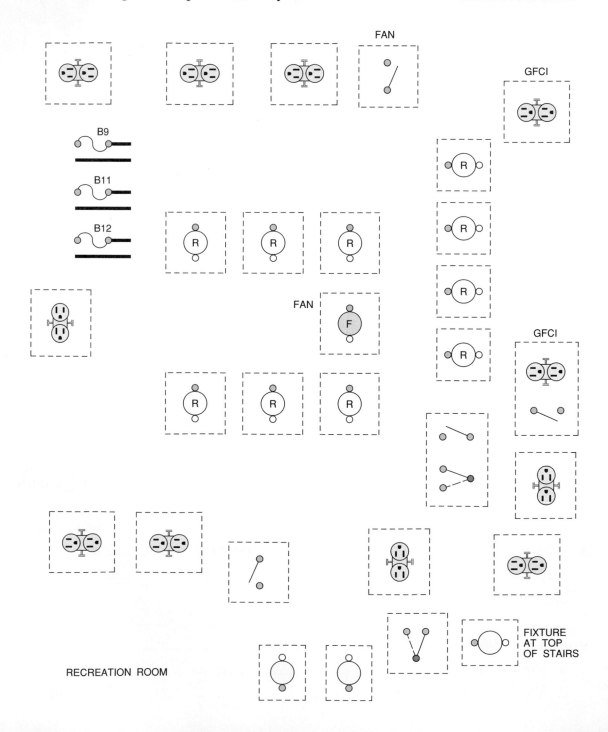

RECREATION ROOM

12. a. May a fluorescent fixture that is marked "Recessed Fluorescent Fixture" be installed in a suspended ceiling? _____

    b. May a fluorescent fixture that is marked "Suspended Ceiling Fluorescent Fixture" be installed in a recessed cavity of a ceiling? _____

13. The *NEC* ® requires GFCI protection for receptacles that serve countertop areas within 6 feet (1.83 m) of the edge of a wet-bar sink, as in the case of the two recreation room receptacles. This requirement is found in Section _____ .

14. Lay-in fluorescent fixtures that rest on the metal framing members of a suspended ceiling cannot just lay on the framing members. *Section 410-16(c)* requires that (1) framing members used to support recessed lighting fixtures must be securely fastened to each other and to the building structure at appropriate intervals, and (2) the fixtures must _____

_____ .

Lighting Branch-Circuit, Receptacle Circuits for Workshop

OBJECTIVES

After studying this Unit, the student will be able to

- understand the meaning, use, and installation of multioutlet assemblies.
- understand where GFCI protection is required in basements.
- understand the Code requirements for conduit installations.
- select the proper outlet boxes to use for surface mounting.
- make conduit fill calculations based upon the number of conductors in the conduit.
- make use of derating and correction factors for determining conductor current-carrying capacity.

The workshop area is supplied by more than one circuit:

A13 Separate circuit for freezer

A17 Lighting

A18 Plug-in strip (multioutlet assembly)

A20 Two receptacles on window wall

The wiring method in the workshop is electrical metallic tubing (EMT). Figure 18-1 shows the conduit layout for the workshop.

A single-pole switch at the entry controls all five ceiling porcelain lampholders. However, note on the plans that three of these lampholders have pull-chains, which would allow the homeowner to turn these lampholders on and off as needed, Figure 18-2. They would not have to be on all of the time, which would save energy.

In addition to supplying the lighting, Circuit A17 also feeds the smoke detectors, chime transformer, and ceiling exhaust fan. The current draw of the

smoke detectors and chime transformer is extremely small. Refer to Table 18-1. Refer to Table 18-2 for estimated loads on Circuits A13, A18, and A20.

Smoke detectors are discussed in detail in Unit 26. Smoke detectors are *not* to be connected to ground-fault circuit interrupter (GFCI)-protected circuits.

DESCRIPTION	QUANTITY	WATTS	VOLT-AMPERES
Ceiling lights @ 100 W each	5	500	500
Fluorescent fixtures Two 40-W lamps each	2	160	200
Chime transformer	1	8	10
Exhaust fan	1	80	90
Smoke detectors @ ½ W each	4	2	2
TOTALS	13	750	802

Table 18-1 Workshop: outlet count and estimated load. Lighting load. Circuit A17.

Figure 18-1 Conduit layout for workshop. Lighting is Circuit A17. Freezer single receptacle is Circuit A13. The two receptacles on the window wall are connected to Circuit A20. Plug-in strip receptacles and receptacle to the right of Main Panel are connected to Circuit A18. All receptacles are GFCI protected except the single receptacle outlet serving the freezer. A junction box mounted on the ceiling near the fluorescent fixtures above the workbench will provide a convenient place to make up the necessary electrical connections.

WORKBENCH LIGHTING

Two two-lamp, 40-watt fluorescent fixtures are mounted above the workbench to reduce shadows over the work area. The electrical plans show that these fixtures are controlled by a single-pole wall switch. A junction box is mounted immediately above or adjacent to the fluorescent fixture so that the connections can be made readily.

When armored cable or flexible metal conduit is used as the wiring method between the fixtures and the junction box or boxes on the ceiling, either of these wiring methods provide the necessary

Figure 18-2 Porcelain lampholders of both the keyless and pull-chain types. These lampholders are also available with a receptacle outlet, which must be GFCI protected when installed in locations where GFCI protection is required by the Code.

equipment grounding for the fluorescent fixtures, provided the metal outlet box is properly grounded. See *Section 250-118* for the many acceptable means to accomplish equipment grounding.

Many workshop fluorescent lighting fixtures are furnished with a three-wire (black, white, green) flexible cord suitable for the purpose that has the attachment plug cap factory attached to the cord. These fixtures can be plugged into a grounding-type receptacle mounted on the ceiling immediately above the fluorescent fixtures. This provides proper grounding of the lighting fixture.

RECEPTACLE OUTLETS

Sections 210-8(a)(4) and *(5)* of the *NEC®* require that all 125-volt, single-phase, 15- or 20-ampere receptacles installed in unfinished basements—such as workshop and storage areas that are not considered habitable, and crawl spaces that are at or below ground level—must be GFCI protected.

Exceptions to this requirement are:

- Receptacles that are not readily accessible.

- A single receptacle or a duplex receptacle for two appliances that occupy a dedicated space for each appliance, are cord- and plug-connected, and are not easily moved. Examples include freezers, refrigerators, water softeners, and central vacuum equipment.

- For the receptacle(s) installed for the laundry equipment as required by *Section 210-52(f)* and *Section 210-11(c)(2)*.

DESCRIPTION	QUANTITY	WATTS	VOLT-AMPERES
20-A Circuit, Number A18			
Receptacles next to main panel	1	180	180
A six-receptacle plug-in multioutlet assembly at 1½ A per outlet (180 VA)	1	1080	1080
TOTALS	02	1260	1260
20-A Circuit, Number A20			
Receptacles on window wall @ 180 W each	2	360	360
TOTALS	2	360	360
20-A Circuit, Number A13			
Single receptacle for freezer (*not* GFCI)	1	—	696
TOTALS	1	1620	2316

Table 18-2 Workshop: outlet count and estimated load for three 20-A circuits A13, A18, A20.

See Unit 6 for a full discussion of ground-fault circuit interrupters.

▶ *Section 210-52(g)* makes it mandatory to install at least one receptacle outlet in each unfinished basement area. ◀ If the laundry area is in the basement, then a receptacle outlet must *also* be installed for the laundry equipment, *Section 210-52(f)*. This additional receptacle must be within 6 feet (1.83 m) of the intended location of the washer, *Section 210-50(c)*, and must be a 20-ampere circuit.

The circuit feeding the required laundry receptacle shall have no other outlets connected to it, *Section 210-11(c)(2)*.

When a single receptacle is installed on an individual branch-circuit, the receptacle rating must not be less than the rating of the branch-circuit, *Section 210-21(b)(1)*.

CABLE INSTALLATION IN BASEMENTS

Unit 4 covers the installation of nonmetallic-sheathed cable (Romex) and armored cable (BX).

Some additional rules apply when nonmetallic-sheathed cable and armored cable are run exposed because of possible physical damage to the cables.

In unfinished basements, nonmetallic-sheathed cables made up of conductors smaller than two No. 6 or three No. 8 must be run through holes bored in joists, or on running boards, *Section 336-6(c)*. In unfinished basements, exposed runs of armored cable must closely follow the building surface or running boards to which it is fastened, *Section 333-11*. *Section 333-11* also permits armored cable (BX) to be run on the undersides of joists in basements, attics, or crawl spaces, where supported on each joist and so located as not to be subject to physical damage. This exception is not permitted when using nonmetallic-sheathed cable. The local inspection authority usually interprets the meaning of "subject to physical damage."

Review Units 4 and 15 for additional discussion regarding protection of cables installed in exposed areas of basements and attics.

CONDUIT INSTALLATION IN BASEMENTS

Where metal raceways are used in house wiring, in most cases the raceways will be electrical metallic tubing, *Article 348*, Figure 18-3. In some instances, rigid metal conduit *(Article 347)* will be used for a mast service; refer to Unit 28. The following discussion only briefly touches upon metal raceways, because most house wiring across the country is done with nonmetallic-sheathed cable, which was covered in detail in Unit 4.

Many local electrical codes do not permit cable wiring in unfinished basements other than where absolutely necessary. For example, a cable may be dropped into a basement from a switch at the head of the basement stairs. Transitions between cable and conduit also must be made. Usually this will be a junction box. The electrician must check the local codes before selecting conduit or cable for the installation, to prevent costly wiring errors.

The electrician should never assume that the *NEC*® is the recognized standard everywhere. Any city or state may pass electrical installation and licensing laws. In many cases these laws are more stringent than the *NEC*.®

Outlet boxes may be fastened to masonry walls with lead or plastic anchors, shields, concrete nails, or power-actuated studs.

Conduit may be fastened to masonry or wood surfaces using straps, such as the one-hole strap illustrated in Figure 18-4.

The conduits on the workshop walls and on the ceiling are exposed to view. The electrician must install the exposed wiring in a neat and skillful manner while complying with the following practices:

- The conduit runs must be straight.
- Bends and offsets must be true.
- Vertical runs down the surfaces of the walls must be plumb.
- To ensure that the conductor insulation will not be injured, be sure to ream the cut ends of rigid conduit and electrical metallic tubing to remove rough edges and burrs.

All conduits and boxes on the workshop ceiling are fastened to the underside of the wood joists, Figure 18-5. Raceways shall be installed as a complete

RIGID STEEL CONDUIT   ELECTRICAL METALLIC TUBING

Figure 18-3 Types of electrical raceways. *Courtesy Allied Tube & Conduit.*

Figure 18-4 A one-hole conduit strap.

system between pull points before pulling in the wires, Figure 18-6.

If the electrician can get to the residence before the basement concrete floor is poured, he might find it advantageous to install conduit runs in or under the concrete floor from the main panel to: the freezer outlet, the window wall receptacles, and the water pump disconnect switch location. This can be a cost-saving (labor and material) benefit.

Metal raceways and metal cables (BX) are acceptable equipment grounding conductors, *Section 250-118*. The runs would have to be continuous between other effectively grounded boxes, panels, and/or equipment. When a nonmetallic wiring method (NM, NMC, UF, etc.) is installed where short sections of metal raceway are provided for mechanical protection—or for appearance—these short sections of raceway need not be grounded, *Section 250-86, Exception No. 2*.

Outlet Boxes for Use in Exposed Installations

Figure 18-7 illustrates outlet boxes and plaster rings that are used for concealed wiring, such as for the recreation room and first floor wiring in the residence. For surface wiring such as the workshop, various types and sizes of outlet boxes may be used. For example, a handy box, Figure 18-8, could be used for the freezer receptacle. The remaining workshop boxes could be 4-inch square outlet boxes with raised covers, Figure 18-9. The box size and type is determined by the maximum number of conductors contained in the box, *Article 370*.

▶ *Section 410-56(f)(3)* prohibits mounting and supporting a receptacle to a cover (flat or raised) by only one screw unless the device assembly or box cover is "listed" and "identified" for securing by a single screw. ◀

The listing and identifying for the purpose is a decision that is made by a recognized testing laboratory, such as Underwriters Laboratories, Inc. In the past, securing a receptacle with a single screw often resulted in a loose receptacle when used over a long period of time. Many short-circuits and ground-faults have been reported because of this.

For the typical raised cover, as illustrated in Figure 18-9, the duplex receptacle would be fastened to the raised cover by two small No. 6-32 screws and nuts, usually furnished with the raised cover. The raised cover has two small "knock-out" holes for these screws. The center No. 6-32 screw could also be used.

Figure 18-5 Raceways, boxes, fittings, and cabinets must be securely fastened in place (points **X**), *Section 300-11(a)*. It is not permissible to hang other raceways, cables, or nonelectrical equipment from an electrical raceway (point **XX**), *Section 300-11(b)*. Refer to Figure 24-12 for exception that allows low-voltage thermostat cables to be fastened to the conduit that feeds a furnace or other similar equipment. See Unit 17 for discussion of the use of suspended-ceiling support wire for supporting equipment.

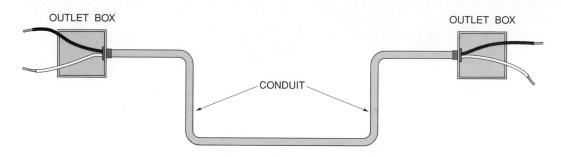

INSTALL CONDUIT COMPLETELY BETWEEN BOXES BEFORE
PULLING IN THE CONDUCTORS, *SECTION 300-18(a)*. NO MORE
THAN THE EQUIVALENT OF FOUR QUARTER BENDS (360° TOTAL)
PERMITTED BETWEEN PULL POINTS, *SECTIONS 331-10, 345-11,
346-11, 347-14, 348-10, 350-16, 351-10,* AND *351-30*.

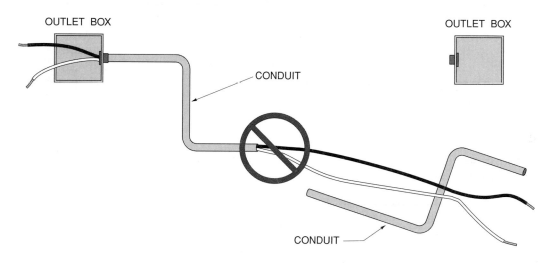

DO NOT PULL CONDUCTORS INTO A PARTIALLY INSTALLED
CONDUIT SYSTEM, AND THEN ATTEMPT TO SLIDE THE
REMAINING SECTION OF RACEWAY OVER THE CONDUCTORS.
THIS IS A VIOLATION OF *SECTION 300-18(a)*.

Figure 18-6   Do not pull wires in raceway until the conduit system is completely installed between pull points.

A SQUARE BOXES AND SINGLE-GANG
RAISED PLASTER COVERS OR RINGS

B THREE-GANG BOX WITH RAISED
PLASTER COVER OR RING

Figure 18-7   Typical outlet boxes and raised covers. *Courtesy* of Appleton Electric Company.

Figure 18-8 Handy box and covers. *Courtesy* of Appleton Electric Company.

Figure 18-9 Four-inch square boxes and raised covers. *Courtesy* of Appleton Electric Company.

Switch and outlet boxes are available with knockouts of the following sizes: ½ inch, ¾ inch, and 1 inch. The knockout size selected depends upon the size of the conduits entering the box. The conduit size is determined by the maximum percentage of fill of the conduit's total cross-sectional area by the conductors pulled into the conduit.

Supporting equipment and/or raceways to the joists, walls, and so on, is accomplished with proper nails, screws, anchors, and straps. Be careful when trying to support other equipment from raceways as this is not always permitted. Refer to Figure 18-5.

Conduit Fill Calculations

An electrician must be able to select the proper size raceways based upon the number, size, and type of conductors to be installed in a particular type of raceway. To do this, the electrician must become familiar with the use of certain tables found in *Chapter 9* and *Appendix C* of the *NEC*®

Percent Fill. The *NEC*® has established the following "percent fill" valves for conduit and tubing.

- one conductor . . . . 53% fill ⊙
- two conductors . . . . 31% fill ⊙⊙
- three or more conductors . . . . . 40% fill ⊛

These percentages are taken from *Chapter 9, Table 1* of the *NEC*.®

All Conductors the Same Size and Type When all conductors are the same size and type:

- Determine how many conductors of one size and type are to be installed in the raceway.
- Refer to *Tables C1* through *C12A* in *Appendix C* of the *NEC*® for the type of raceway, and for the specific size and type of conductors.
- Select the proper size raceway directly from the table.

These tables are not difficult to use. Here are a few examples of conductor fill for EMT. Look these up in the tables to confirm the accuracy of the examples.

- Not over five No. 10 THHN conductors may be pulled into a ½-inch EMT.

- Not over eight No. 12 THW conductors may be pulled into a ¾-inch EMT.

- Not over three No. 8 THHN conductors may be pulled into a ½-inch EMT.

- Not over six No. 8 THHW conductors may be pulled into a 1-inch EMT.

Conductors of Different Sizes and/or Types.
When conductors are different sizes or types:

- Refer to *Chapter 9, Table 5* of the *NEC*®

- Find the size and type of conductors used to determine the approximate cross-sectional area in square inches of the conductors.

- Refer to *Chapter 9, Table 8* for the square inch area of bare conductors.

- Total the square inch areas of all conductors.

- Refer to *Chapter 9, Table 4* for the type and size of raceway used.

- Determine the minimum size raceway based on the total square inch area of the conductors and the allowable square inch area fill for the particular type of raceway used.

TYPE	SIZE	APPROX. DIAM. INCHES	APPROX. AREA SQ. IN.
THHN	14	0.111	0.0097
THHN	12	0.130	0.0133
THHN	10	0.164	0.0211
THHN	8	0.216	0.0366
THHN	6	0.254	0.0507
THHN	3	0.352	0.0973
THW	6	0.304	0.0726
THW	8	0.266	0.0556

Figure 18-10 This sampling of typical conductors and their dimensions is taken from *Table 5, Chapter 9, NEC.*®

- The sum must not exceed the allowable percentage fill of the cross-sectional square inch area of the raceway used.

For illustrative purposes, only the data for a few types of wires from *Table 5* are shown in Figure 18-10. The dimensions and areas for some electrical metallic tubing from *Table 4* are shown in Figure 18-11. You will need to refer to *Chapter 9* and *Appendix C* of the *NEC*® for the many other sizes and types of conductors and raceways.

EXAMPLE: If three No. 6 THW conductors and two No. 8 THW conductors are to be installed in one EMT, determine the proper size EMT.

		ELECTRICAL METALLIC TUBING			
Trade Size in Inches	Internal Diameter in Inches	Total Area 100% in Sq. In.	2 Wires 31% in Sq. In.	Over 2 Wires 40% in Sq. In.	1 Wire 53% in Sq. In.
½	0.622	0.304	0.094	0.122	0.161
¾	0.824	0.533	0.165	0.213	0.283
1	1.049	0.864	0.268	0.346	0.458
1¼	1.380	1.496	0.464	0.598	0.793
1½	1.610	2.036	0.631	0.814	1.079
2	2.067	3.356	1.040	1.342	1.778
2½	2.731	5.858	1.816	2.343	3.105
3	3.356	8.846	2.742	3.538	4.688
3½	3.834	11.545	3.579	4.618	6.119
4	4.334	14.753	4.573	5.901	7.819

Figure 18-11 Dimensional and percent fill data for electrical metallic tubing. This is one of the twelve charts found in *Chapter 9, Table 4, NEC,*® for various types of raceways.

SOLUTION:

1. Find the square inch area of the conductors in *Chapter 9, Table 5.*

Three No. 6 THW conductors	0.0726 in²
	0.0726 in²
	0.0726 in²
Two No. 8 THW conductors	0.0556 in²
	0.0556 in²
Total Area	0.3290 in²

2. In *Chapter 9, Table 4, NEC,* look at the fifth column (over 2 wires, 40% fill) for an EMT that has an area of 0.3290 square inch or more. Note that a ¾-inch EMT holds up to 0.213 square inch of conductor fill, and a 1-inch EMT holds up to 0.346 square inch of conductor fill. Therefore, 1-inch EMT is the minimum size for the combination of conductors in the example.

Many electrical ordinances require that a separate equipment grounding conductor be installed in all raceways, whether rigid or flexible, whether metallic or nonmetallic. This requirement resulted from inspectors finding loose or missing set screws on set-screw type connectors and couplings. Corrosion of fittings also contributes to poor grounding. A separate equipment grounding conductor does take up space, and therefore must be counted when considering conduit fill.

EXAMPLE: A feeder to an electric furnace is fed with two No. 3 THHN conductors plus one equipment grounding conductor. This feeder is installed in EMT. The feeder overcurrent protection is 100 amperes. What is the minimum size EMT to be installed?

SOLUTION: According to *Table 250-122*, the minimum size equipment grounding conductor for a 100-ampere feeder is No. 8 AWG copper. Find the square inch area of the conductors in *Table 5, Chapter 9, NEC®*:

Two No. 3 THHN	0.0973 in²
	0.0973 in²
One No. 8 THHN	0.0366 in²
Total Area	0.2312 in²

From *Table 4, Chapter 9, NEC®*:

40% fill for ¾-inch EMT = 0.213 in²

40% fill for 1-inch EMT = 0.346 in²

Therefore, install 1-inch electrical metallic tubing. Remember, *Section 300-18(a)* states that a conduit system must be completely installed before pulling in the conductors. Refer to Figure 18-6.

Conduit Bodies (Section 370-16(c))

Conduit bodies are used with conduit installations to provide an easy means to turn corners, terminate conduits, and mount switches and receptacles. They are also used to provide access to conductors, to provide space for splicing (when permitted), and to provide a means for pulling conductors, Figure 18-12.

A conduit body:

- must have a cross-sectional area not less than twice the cross-sectional area of the largest conduit to which it is attached, Figure 18-13(A). This is a requirement only when the conduit body contains No. 6 AWG conductors or smaller.

- may contain the same maximum number of conductors permitted for the size raceway attached to the conduit body.

Figure 18-12 A variety of common conduit bodies and covers.

CROSS-SECTIONAL AREA
OF CONDUIT BODY HERE
MUST BE AT LEAST TWICE THE
CROSS-SECTIONAL AREA OF CONDUIT HERE.
FOR INSTANCE, A 1/2-INCH CONDUIT HAS A CROSS-
SECTIONAL AREA OF APPROXIMATELY 0.30 SQUARE INCHES.
THEREFORE, THE CROSS-SECTIONAL AREA OF THE
CONDUIT BODY MUST BE AT LEAST 0.60 SQUARE INCHES

A

8 CU. IN.

B

Figure 18-13   Requirements for conduit bodies.

NUMBER OF CURRENT-CARRYING CONDUCTORS	PERCENT OF VALUES IN TABLES. IF HIGH AMBIENT TEMPERATURES ARE PRESENT, ALSO APPLY CORRECTION FACTORS FOUND AT THE BOTTOM OF *TABLE 310-16.*
4 through 6	80
7 through 9	70
10 through 20	50
21 through 30	45
31 through 40	40
41 and above	35

Figure 18-14   Derating factors necessary to determine conductor ampacity when there are more than three current-carrying conductors in a raceway or cable. See *Section 310-15(b)(2)(a).*

NIPPLE
24" (610 mm)
OR LESS

Figure 18-15   The derating in Figure 18-14 for more than three current-carrying conductors in a raceway is *not* required when the conduit nipple does not exceed 24 inches (610 mm), *Section 310-15(b)(2)(a), Exception No. 3.* Derating is *not* required when cables are bundled in lengths not exceeding 24 inches (610 mm), *Section 310-15(b)(2)(a).*

• must not contain splices, taps, or devices unless the conduit body is marked with its cubic-inch capacity, Figure 18-13(B), and the conduit body must be supported "in a rigid and secure manner."

• the conductor fill volume must be properly calculated according to *Table 370-16(b)*, which lists the free space that must be provided for each conductor within the conduit body. If the conduit body is to contain splices, taps, or devices, it must be supported rigidly and securely.

DERATING FACTORS

For more than three current-carrying wires in conduit or cable, refer to Figure 18-14.

Section 310-15(b)(2)(a) states that for the conductors listed in the tables, the maximum allowable ampacity must be reduced when more than three current-carrying conductors are installed in a raceway. The reduction factors based on the number of conductors are noted in Figure 18-14. When the conduit nipple does not exceed 24 inches (610 mm), the reduction factor does not apply, Figure 18-15.

According to *Section 310-15(b)(4)(a)*, a neutral conductor of a multiwire branch-circuit carrying only unbalanced currents shall not be counted in determining current-carrying capacities, Figure 18-16.

EXAMPLE: What is the correct ampacity for No. 6 THHN when there are six current-carrying wires in the conduit?

$$75 \times 0.80 = 60 \text{ amperes}$$

CORRECTION FACTORS (DUE TO HIGH TEMPERATURES)

When conductors are installed in locations where the temperature is higher than 30°C (86°F), the correction factors noted below *Table 310-16* must apply.

Figure 18-16 A neutral conductor carrying the unbalanced currents from the other conductors need *not* be counted when determining the ampacity of the conductor, according to *Section 310-15(b)(4)(a)*. The Code would recognize this example as *two current-carrying conductors*. Thus, in the three multiwire circuits shown, the actual total is nine conductors, but only six current-carrying conductors are considered when the derating factor of Figure 18-14 is applied.

For example, consider that four current-carrying No. 3 THWN copper conductors are to be installed in one raceway in an ambient temperature of approximately 90°F. "Ambient" temperature means "surrounding" temperature. This temperature is frequently exceeded in attics where the sun shines directly onto the roof.

The maximum ampacity for these conductors is determined as follows:

- The allowable ampacity of No. 3 THWN copper conductors from *Table 310-16* = 100 amperes.

- Apply the correction factor for 90°F found below *Table 310-16*. The first column shows Celsius; the last column shows Fahrenheit.

$$100 \times 0.94 = 94 \text{ amperes}$$

- Apply the derating factor for four current-carrying wires in one raceway.

$$94 \times 0.80 = 75.2 \text{ amperes}$$

- Therefore, 75.2 amperes is the new corrected ampacity for the example given.

Some authorities having jurisdiction in extremely hot areas of the country, such as the southwestern desert climates, will require that service-entrance conductors running up the side of a building or above a roof exposed to direct sunlight be "corrected" per the correction factors below *Table 310-16*. Check this out before proceeding with an installation where this situation might be encountered.

OVERCURRENT PROTECTION FOR BRANCH-CIRCUIT CONDUCTORS

The basic rule is that overcurrent protection for conductors shall not be more than the allowable ampacity of the conductor, *Section 240-3*. Allowable ampacities for conductors are found in *Table 310-16*.

If the allowable ampacity of a conductor does not match a standard ampere rating of a fuse or circuit breaker, *Section 240-3(b)* permits the use of the next higher rated fuse or circuit breaker. For example, a conductor having an ampacity of 55 amperes could be protected by a 60-ampere overcurrent device. This permission does not apply if the conductors of the branch-circuit supply a number of receptacles for cord- and plug connection of portable loads because it would be impossible to know just how much load might be plugged in.

Small size branch-circuit conductors must be protected according to the values shown in Figure 18-17. This table is based on *Section 240-3(d)*.

An exception to this basic maximum size overcurrent protection rule is the protection of motor branch-circuit conductors. For motor branch-circuits, the maximum size overcurrent protection values are found in *Table 430-152*.

If the allowable ampacity of a given conductor is *derated* and/or *corrected*, the overcurrent protection must be based on the new *derated* or *corrected* ampacity.

CONDUCTOR SIZE (Copper)	MAXIMUM AMPERE RATING OF OVERCURRENT DEVICE
No. 14	15 amperes
No. 12	20 amperes
No. 10	30 amperes

Figure 18-17 Maximum size overcurrent protection for small size conductors, *Section 240-3(d)*.

SOME BASIC CODE CONSIDERATIONS FOR CONDUCTOR SIZING AND OVERCURRENT PROTECTION

The following is a quick review of some key Code requirements relating to conductor sizing and over-current protection.

►• *Section 210-19(a):* Branch-circuit conductors shall have an ampacity not less than the maximum load to be served. ◄

• *Section 210-20(b):* Overcurrent protection for branch-circuit conductors shall not exceed the ampacity of a conductor. There are exceptions to this as previously discussed, such as "next standard size permitted" if the OCD and the conductor ampacity do not match, and for motor branch-circuit conductors.

• *Section 210-23:* In no case shall the load exceed the branch-circuit ampere rating.

• *Section 215-2(a):* Feeder conductors shall have an ampacity not less than required to supply the load.

• *Section 215-3:* Overcurrent protection for feeder conductors shall not exceed the ampacity of a conductor. There are exceptions to this as previously discussed, such as "next standard size permitted" if the OCD and the conductor ampacity do not match, and for motor branch-circuit conductors.

• *Section 220-4:* The total load shall not exceed the rating of the branch-circuit.

• *Section 384-16(d):* The total load on any over-current device located in a panelboard shall not exceed 80 percent of its ampere rating where in normal operation, the load could be on for three hours or more.

• *Section 424-3(b):* For fixed electric space heating equipment such as an electric furnace, the branch-circuit conductors and the over-current protective devices shall not be less than 125% of the total load of the motor(s) and heater(s).

• *Section 426-4:* For fixed outdoor electric deicing and snow melting equipment such as heating cables buried in the concrete of a driveway, the branch-circuit conductors and the overcurrent protective devices shall not be less than 125% of the total load of the heaters.

• *Section 430-22:* Branch-circuit conductors for motors shall not be less than 125% of the motor's full-load current. Motor branch-circuit short-circuit and ground-fault protection is sized according to the percentages listed in *Table 430-152*.

A few of the above Code sections make reference to *continuous loads*. The Code defines a continuous load as: *"where the maximum current is expected to continue for three hours or more."*

Loads in homes are generally not considered to be continuous. However an electric furnace, or snow melting heating cables buried in the concrete of a driveway can be classified as a continuous load, because they could be on for periods of three hours or more. The 125% factor for these has already been included in *Sections 424-3(b)* and *426-4*.

Many loads in commercial and industrial installations can be considered to be continuous loads. For example, an electric range in a restaurant might be used continuously 24 hours a day, seven days a week. An electric range in a home would not be used in this manner.

For continuous loads, the overcurrent protective device and the conductors must not be less than 125% of the continuous load. For example, a continuous load is 40 amperes. The minimum overcurrent device rating and the minimum conductor ampacity is $40 \times 1.25 = 50$ amperes.

For combination continuous and noncontinuous loads, the overcurrent protective device and the conductors must not be less than 125% of the continuous load plus the noncontinuous load. For example, a feeder supplies a continuous load of 40 amperes and a noncontinuous load of 30 amperes. The minimum overcurrent device rating and the minimum conductor ampacity is $(40 \times 1.25) + 30 = 80$ amperes.

EXAMPLES OF DERATING, OVERCURRENT PROTECTION, AND CONDUCTOR SIZING

To illustrate what has been learned so far about derating and maximum overcurrent protection, suppose that eight THW copper conductors are installed in one raceway. All are considered to be current-carrying. The load on each conductor is 18 amperes. Let us consider the possibilities.

- No. 14 THW copper wire
  a. The maximum overcurrent protection from Figure 18-17 is 15 amperes.
  b. The ampacity of No. 14 THW copper wire from *Table 310-16* is 20 amperes.
  c. Apply the derating factor (8 current-carrying wires in the raceway) from Figure 18-14.

$$20 \times 0.70 = 14 \text{ amperes}$$
$$\text{(derated ampacity)}$$

  Thus, No. 14 THW wire is *too small* to supply the 18-ampere load.

- No. 12 THW copper wire
  a. The maximum overcurrent protection from Figure 18-17 is 20 amperes.
  b. The ampacity of No. 12 THW copper wire from *Table 310-16* is 25 amperes.
  c. Apply the derating factor (8 current-carrying wires in the raceway) from Figure 18-14.

$$25 \times 0.70 = 17.5 \text{ amperes}$$
$$\text{(derated ampacity)}$$

  Thus, No. 12 THW is *too small* to supply the 18-ampere load.

- No. 10 THW copper wire
  a. The maximum overcurrent protection from Figure 18-17 is 30 amperes.
  b. The ampacity of No. 10 THW copper wire from *Table 310-16* is 35 amperes.

  c. Apply the derating factor (8 current-carrying wires in the raceway) from Figure 18-14.

$$35 \times 0.70 = 24.5 \text{ amperes}$$
$$\text{(derated ampacity)}$$

  Thus, No. 10 THW copper wire is adequate to supply the 18-ampere load.

MULTIOUTLET ASSEMBLY

A multioutlet assembly has been installed above the workbench. For safety reasons, as well as for adequate wiring, it is recommended that any workshop receptacles be connected to a separate circuit. In the event of a malfunction in any power tool commonly used in the home workshop (saw, planer, lathe, or drill), only receptacle Circuit A18 is affected. The lighting in the workshop is not affected by a power outage on the receptacle circuit.

The GFCI feed-through receptacle to the left of the workbench provides GFCI protection for the entire plug-in strip, *Section 210-8(a)(5)*, Figure 18-18.

The installation of multioutlet assemblies must conform to the requirements of *Article 353* of the Code.

Load Considerations for Multioutlet Assemblies

Table 220-3(a) gives the general lighting loads in volt-amperes per square foot (0.093 m²) for types

Figure 18-18   Detail of how the workbench receptacle plug-in strip is connected by feeding through a GFCI feedthrough receptacle located below the Main Panel. This is Circuit A18.

of occupancies. *Section 220-3(b)(10)* indicates that *all* general-use receptacle outlets in one, two, and multifamily dwellings are to be considered outlets for general lighting, and as such have been included in the volt-amperes per square foot calculations. Therefore, no additional load need be added.

Because the multioutlet assembly, Figure 18-19, has been provided in the workshop for the purpose of plugging in portable tools, a separate Circuit A18 is provided. This circuit has been shown in the service-entrance calculations at 1500 volt-amperes, similar to the load requirements for small appliance circuits.

Many types of receptacles are available for various sizes of multioutlet assemblies. For example, duplex, single-circuit grounding, split-circuit grounding, and duplex split-circuit receptacles can be used with multioutlet assemblies.

Wiring a Multioutlet Assembly

A multioutlet assembly can be connected by bringing the supply into one end as in Figure 18-18, or into the back of the assembly as in Figure 18-21.

Receptacles for multioutlet assemblies can be spaced and wired on the job, in which case the covers are cut to the desired length. Receptacles can also be factory prewired, with six receptacles in a 36-inch (915 mm) long strip, and ten receptacles in a 60-inch (1.52 m) long strip.

For other strip lengths, standard spacing of receptacles for precut assemblies are 6 inches (152 mm), 9 inches (229 mm), 12 inches (305 mm), or 18 inches (457 mm). These measurements are center to center. Prewired multioutlet assemblies are generally wired with No. 12 AWG THHN conductors, and have either 15- or 20-ampere receptacles, depending upon how they are specified and ordered.

Multioutlet assemblies are not difficult to install. There is a wide variety of fittings, such as connectors, couplings, ground clamps, blank and end fittings, elbows for turning corners (flat, inside, outside, twist), flush plate adapters, starter boxes, raceway adapters, mounting clips, etc., to fit just about any application.

Figure 18-20 shows a cord-connected multioutlet assembly of the type used in workshops or other locations where there is a need for the

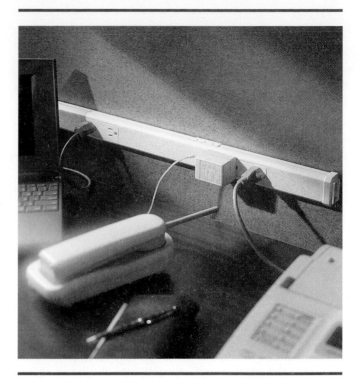

Figure 18-19 A typical metal multioutlet assembly. Multioutlet assemblies are permitted to be run through (not within) a dry partition when no outlets are located within the partition, and all exposed portions of the assembly can have the cover removed. Nonmetallic mulitoutlet assemblies are also available. *Courtesy* The Wiremold Company.

Figure 18-20 A cord-connected, prewired multioutlet assembly installed above a workbench. *Courtesy* The Wiremold Company.

Figure 18-21 A multioutlet assembly back plate with the armored cable supply feeding into the back of the assembly.

Figure 18-22 A multioutlet assembly installed above the backsplash of a kitchen countertop. *Courtesy* The Wiremold Company.

convenience of having a number of receptacles in close proximity. These prewired assemblies are available with six receptacles and circuit breaker, 40 inches long, or ten receptacles and circuit breaker, 52 inches long. These can also be used under kitchen cabinets. Figure 18-21 shows a multioutlet assembly back plate with the electrical supply feeding into the back of the assembly. Figure 18-22 shows a multioutlet assembly above the backsplash of a kitchen countertop.

EMPTY CONDUITS

Although not truly part of the workshop wiring, Note 8 to the first-floor electrical plans indicates that two empty ½-inch EMT raceways are to be installed, running from the workshop to the attic. This installation is unique, but certainly welcomed if at some later date additional wiring (telephone, television, etc.) might be required. The empty raceways will provide the necessary route from the basement to the attic, thus eliminating the need to fish through the wall partitions.

REVIEW

1. a. What circuit supplies the workshop lighting? _____

   b. What circuit supplies the plug-in strip over the workbench? _____

   c. What circuit supplies the freezer receptacle? _____

2. Approximately how much ½-inch EMT is used for connecting Circuits A13, A17, A18, and A20? _____

3. a. How is EMT fastened to masonry? _____

    _____

   b. The cut ends of rigid and electrical metallic tubing must be _____
       to prevent the insulation on conductors from being damaged.

4. What type of lighting fixtures are installed on the workshop ceiling? _____

    _____

5. A check of the plans indicates that two empty ½-inch thinwall conduits are installed
   between the basement and the attic. In your opinion do you feel this is a good idea?
   Briefly explain your thoughts. _____

    _____

    _____

    _____

    _____

    _____

    _____

6. To what circuit are the smoke detectors connected? _____

7. Is it a Code requirement to connect smoke detectors to a GFCI-protected circuit?

    _____

8. When a freezer is plugged into a receptacle, be sure that the receptacle is protected by
   a GFCI. Is this statement true or false? Explain. _____

    _____

    _____

    _____

    _____

    _____

    _____

9. A sump pump is plugged into a single receptacle outlet. This receptacle shall

   a. be GFCI protected

   b. not be GFCI protected

   Circle the correct answer.

10. When a 120-volt receptacle outlet is provided for the laundry equipment in an unfin-
    ished basement, what is the *minimum* number of additional receptacle outlets required
    in that basement? What Code reference? _____

    _____

11. Derating factors for conductor ampacities must always be taken into consideration
    when (a) _____ are contained in one raceway.
    Correction factors are applied when (b) _____ .

12. What current is flowing in the neutral conductor at:

   a. _____ amperes   b. _____ amperes   c. _____ amperes   d. _____ amperes

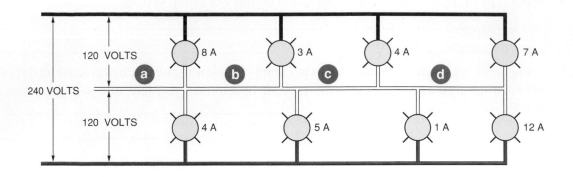

13. When a single receptacle is installed on an individual branch-circuit, the receptacle must have a rating _____ than the rating of the branch-circuit.

14. Calculate the total current draw of Circuit A17 if all lighting fixtures and the exhaust fan were turned on.

15. A conduit body must have a cross-sectional area not less than (two) (three) (four) times the cross-sectional area of the (largest) (smallest) conduit to which it is attached. Circle correct answers.

16. A conduit body may contain splices if marked with:

   a. the U.L. logo.

   b. its cubic-inch area.

   c. the size of conduit entries.

   Circle the correct answer.

17. What size box would you use where Circuit A17 enters the junction box on the ceiling? _____

   _____

18. List the proper size of electrical metallic tubing for the following. Assume that it is new work and that THHN insulated conductors are used. See *Chapter 9, Table C1, NEC.®*

   a. 3 No. 14 AWG _____   f. 4 No. 12 AWG _____

   b. 4 No. 14 AWG _____   g. 5 No. 12 AWG _____

   c. 5 No. 14 AWG _____   h. 6 No. 12 AWG _____

   d. 6 No. 14 AWG _____   i. 6 No. 10 AWG _____

   e. 3 No. 12 AWG _____   j. 4 No. 8 AWG _____

19. According to the Code, what size electrical metallic tubing is required for each of the following combinations of conductors? Assume that these are new installations. Show all calculations. Refer to *Chapter 9*, *NEC*, *Tables 4*, *5*, and *8*.

a. Three No. 14 AWG THHN, four No. 12 AWG THHN

b. Two No. 12 AWG TW, three No. 8 AWG TW

c. Three No. 3 THW, one No. 4 Bare

d. Three No. 3/0 THHN, two No. 8 THHN

20. a. When more than three current-carrying conductors are installed in one raceway, their allowable ampacities must be reduced according to *Section* _____

   b. Four No. 10 THW current-carrying conductors are installed in one raceway. The room temperature will not exceed 86°F. The connected load is nonmotor, noncontinuous. (1) What is the ampacity of these conductors before derating? (2) What is the maximum overcurrent protection permitted to protect these conductors? (3) What is the ampacity of these conductors after derating? _____

   _____

   _____

   _____

   _____

   _____

   c. Why is it not necessary to count equipment grounding conductors and/or all of the "travelers" when installing 3-way and 4-way switches for the purposes of derating?

   _____

   _____

   _____

   _____

21. What is the minimum number of receptacles required by the Code for basements?

   _____

22. For laundry equipment in basements, what sort of electrical circuitry is required by the Code? _____

   _____

   _____

   _____

   _____

23. How many receptacles are required to complete the multioutlet assembly above the workbench? _____

24. A Type NM cable carries a branch-circuit to the outside of a building, where it terminates in a metal weatherproof junction box. From this box, a short length of EMT is installed upward to another metal weatherproof junction box on which is attached a cluster of outdoor flood lights. The first box is located approximately 2 feet (610 mm) above the ground. To "meet Code," do the metal boxes and metal raceway have to be grounded?   Yes _____   No _____   *Sections* _____ .

25. When raised covers are used for receptacles, does the Code permit fastening the receptacle to the cover with the one center 6/32 screw? Yes _____   No _____ *Section* _____ .

26. The following is a layout of the lighting circuits for the workshop. Using the conduit layout in Figure 18-1, make a complete wiring diagram. Use colored pencils or pens to indicate conductors.

27. *Section 210-8(a)(5)* requires that GFCI protection be provided for all receptacles in unfinished basements. There are some exceptions. What are these exceptions?

   _____

   _____

   _____

28. Where loads are likely to be on continuously, the *computed* load for branch-circuits and feeders must be figured at (100%) (125%) (150%) of the continuous load, (plus) (minus) the noncontinuous load. This requirement generally (does) (does not) apply to residential wiring. Circle the correct answers.

Special-Purpose Outlets– Water Pump, Water Heater

OBJECTIVES

After studying this Unit, the student will be able to

- list the requirements for deep-well jet pump and submersible pump installations.
- calculate (given the rating of the motor) the conductor size, conduit size, and overcurrent protection required for pump circuits.
- list some of the electrical circuits used for connecting and metering electric water heaters.
- describe the basic operation of water heaters.
- understand the meaning of the term "Time-of-Use."
- list the functions of the tank, the heating elements, and the thermostat for water heaters.
- understand "recovery" time for electric water heaters.
- make the proper electrical and grounding connections for water heaters, water pumps, and metal well casings.
- understand the hazards of possible scalding.
- discuss heat pump water heaters.

WATER PUMP CIRCUIT (▲)ᴮ

All dwellings need a good water supply. City dwellings generally are connected to the city water system. In rural areas, where there is no public water supply, each dwelling has its own water supply system in the form of a well. The residence plans show that a deep-well jet pump is used to pump water from the well to the various plumbing outlets. The outlet for this jet pump is shown by the symbol (▲)ᴮ. An electric motor drives the jet pump. A circuit must be installed for this system.

JET PUMP OPERATION

Figure 19-1 shows the major parts of a typical jet pump. The pump impeller wheel ① forces water down a drive pipe ② at a high velocity and pressure to a point just above the water level in the well casing. Some well casings may be driven to a depth of more than 100 feet (30.5 m) before striking water. Just above the water level, the drive pipe curves sharply upward and enters a larger vertical suction pipe ③. The drive pipe terminates in a small nozzle or jet ④ in the suction pipe. The water emerges from

Figure 19-1   Components of a jet pump.

the jet with great force and flows upward through the suction pipe.

Water rises in the suction pipe, drawn up through the tailpipe ⑤ by the action of the jet. The water rises to the pump inlet and passes through the impeller wheel of the pump. Part of the water is forced down through the drive pipe again. The remaining water passes through a check valve and enters the storage tank ⑥.

The tailpipe is submerged in the well water to a depth of about 10 feet (3.05 m). A foot valve ⑦ and strainer ⑧ are provided at the end of the tailpipe. The foot valve prevents water in the pumping equipment from draining back into the well when the pump is not operating.

When the pump is operating, the lower part of the storage tank fills with water and air is trapped in the upper part. As the water rises in the tank, the air is compressed. When the air pressure is 40 pounds per square inch, a pressure switch ⑨ dis-

connects the pump motor. The pressure switch is adjusted so that as the water is used and the air pressure falls to 20 pounds per square inch, the pump restarts and fills the tank again.

The Pump Motor

The jet pump motor in this residence is a 3,600-rpm, 1-horsepower (hp) single-phase, capacitor start, dual-voltage (115/230) volt motor. A capacitor start motor is designed to produce a starting torque adequate for pumps with comparatively low-starting current. In pumping operations, a high-starting torque helps to overcome the back pressure within the tank and the weight of the water being lifted from the well.

A dual-voltage motor can be connected to either voltage by changing the connections in the junction box on the motor. When connected for the higher voltage, the current draw will be half than when connected for the lower voltage. For example, in *Table 430-148*, the full-load current rating of a 230-volt, single-phase motor is 8 amperes. The same motor connected for 115 volts has a current draw of 16 amperes. Current is a big factor in determining conductor size, watts loss in the conductors, voltage drop, disconnect switch size, and motor controller size.

Note that the current values shown in *Table 430-148* for 115-volt motors covers the voltage range of 110 to 120 volts. For 230-volt motors, the voltage range is 220 to 240 volts. Many single-phase motors are marked 115/208-230 volt. The 120/208-volt wye systems are popular in commercial buildings, where the line-to-line voltage is 208 volts, and the line-to-neutral voltage is 120 volts.

The NEMA standard for conventional electric motors is that they are designed to operate satisfactorily at ±10% of the motor's rated voltage.

The Pump Motor Circuit

Figure 19-2 shows the electrical circuit that operates the jet pump motor and pressure switch. This circuit originates at Main Service Panel A, Circuit A5-7. *Table 430-148* indicates that a 1-hp single-phase motor has a rating of 8 amperes at 230 volts. Using this information, the conductor size, conduit size, branch-circuit overcurrent protection, and running overcurrent protection can be determined for the motor.

A 230-VOLT BUS IN PANEL

B 20-AMPERE, TWO-POLE CIRCUIT BREAKER

C NO. 12 AWG CONDUCTORS

D TWO-POLE DISCONNECT SWITCH AND MOTOR OVERLOAD PROTECTION

E TWO-POLE PRESSURE SWITCH

F 1-HP, 230-VOLT, SINGLE-PHASE MOTOR

Figure 19-2   The pump circuit.

Conductor Size *(Section 430-22)*

The motor has a rating of 8 amperes. To calculate conductor size, use 125% of 8 amperes.

$$1.25 \times 8 = 10 \text{ amperes}$$

Table 310-16 of the Code indicates that the minimum conductor size permissible is No. 14 AWG. However, the Schedule of Special Purpose Outlets found in this text, which is part of the Specifications for the residence, indicates that No. 12 AWG conductors are to be used for the pump circuit. Installing larger size conductors can minimize voltage drop, and might be suitable at a later date should the need arise to install a larger capacity pump.

Conduit Size

Two No. 12 Type THHN conductors require ½-inch EMT, *Appendix C, Table C1, NEC.*

Motor Branch-Circuit Short-Circuit and Ground-Fault Protection
(Section 430-52, Table 430-152)

Fuses and circuit breakers are the most commonly used forms of motor branch-circuit short-circuit and ground-fault protection. Should the motor windings or branch-circuit conductors short circuit or go to ground, the motor's branch-circuit short-circuit and ground-fault overcurrent device, either fuses or circuit breakers, will open.

For the jet pump motor in this residence connected for 230 volts, the various choices for the motor branch-circuit short-circuit and ground-fault protection are found in *Section 430-52* and in *Table 430-152*. They are:

- Non-Time-Delay Fuses
  $8 \times 3 = 24$ amperes (use 25-ampere fuses)
  Maximum size: 400% (4.0)

- Dual-Element, Time-Delay Fuses
  $8 \times 1.75 = 14$ amperes
  (use 15-ampere fuses)
  Maximum size: 225% (2.25)

- Instant Trip Breakers
  $8 \times 8 = 64$ amperes setting
  Maximum setting: 1300% (13)

- Inverse Time Breakers
  $8 \times 2.5 = 20$-ampere setting
  Maximum setting:
     400% for F.L.A. of 100 or less.
     300% for F.L.A. of over 100.

The motor branch-circuit overcurrent device must be capable of allowing the motor to start, *Section 430-52(b)*.

Where the values for branch-circuit protective devices (fuses or breakers) as determined by *Table 430-152* do not correspond to the standard sizes, ratings, or settings as listed in *Section 240-6*, the next higher size, rating, or setting is permitted, *Section 430-52(c), Exception No. 1*.

If, after applying the values found in *Table 430-152* and after selecting the next higher standard size, rating, or setting of the branch-circuit overcurrent device, the motor still will not start, *Section 430-52(c), Exception No. 2* permits using an even larger size, rating, or setting. The maximum values (percentages) are shown in *Section 430-52(c)*.

A nuisance opening of a fuse, or tripping of a breaker, can easily be predicted by referring to time/current curves for the intended fuse or breaker. This topic is covered briefly in Unit 28.

Motor Overload Protection

(Sections 430-31 through 430-44)

Overload protection is needed to keep the motor from dangerous overheating due to overloads or failure of the motor to start. *Sections 430-31* through *430-44* cover virtually every type and size of motor, starting characteristics, and duty (continuous or non-continuous) in use today. The overload protection for a motor might be thermal overloads, electronic sensing overload devices, built-in (inherent) thermal protection, or time-delay fuses.

Although there are exceptions, most of the common installations require motor overload protection not to exceed 125% of the motor's full load current draw. For the jet pump motor, this would be:

$$1.25 \times 8 = 10 \text{ amperes.}$$

Instructions furnished with a particular motor or motor operated appliance will probably indicate the maximum size overload protection and maximum size branch-circuit protection. These values must be followed.

Wiring for the Pump Motor Circuit

The water pump circuit is run in ½-inch electrical metallic tubing (EMT) from Main Panel A to a disconnect switch and controller that are mounted on the wall next to the pump. The EMT may be run across the ceiling of the workshop or be embedded in the concrete floor. Thermal overload devices (sometimes called *heaters*) are installed in the controller to provide running overcurrent protection. Two No. 12 THHN conductors are connected to the two-pole, 20-ampere branch-circuit breaker in Panel A, Circuit A(5-7). Dual-element, time-delay fuses are installed in the disconnect switch located near the pump. We have the following choices for sizing these fuses:

Section 430-32(a)(1):	$8 \times 1.25 = 10$ amperes
Table 430-152:	$8 \times 1.75 = 14$ amperes
Section 430-52(c)(1), Exception No. 1:	Next standard size is 15 amperes.
Section 430-52(c)(1), Exception No. 2:	If 15-ampere fuses do not allow the motor to start, it is permitted to size at not to exceed 225%.

$8 \times 2.25 = 18$ amperes maximum

This puts us back to the 15-ampere rating. However, 17½-ampere, dual-element, time-delay fuses are available.

Any of these ampere ratings of dual-element, time-delay fuses are suitable, and they all fit into a 30-ampere disconnect switch.

A short length of flexible metal conduit is connected between the load side of the thermal overload switch and the pressure switch on the pump. This pressure switch is usually connected to the motor at the factory. In Figure 19-2, a double-pole pressure switch is shown at Ⓔ. When the contacts of the pressure switch are open, all conductors feeding the motor are disconnected.

SUBMERSIBLE PUMP

A submersible pump, Figure 19-3, consists of a centrifugal pump driven by an electric motor. The pump and the motor are contained in one housing, submersed below the permanent water level within the well casing. When running, the pump raises the water upward through the piping to the water tank. Proper pressure is maintained in the system by a pressure switch. The disconnect switch, pressure and limit switches, and controller are installed in a logical and convenient location near the water tank.

Most water pressure tanks have a precharged air chamber in an elastic bag (bladder) that separates the air from the water. This ensures that air is not absorbed by the water. Water pressure tanks compress air to maintain delivery pressure in a usable range without cycling the pump at too fast a rate. When in direct contact with the water, air is gradually absorbed into the water, and the cycling rate of the pump will become too rapid for good pump life, unless the air is periodically replaced. With the elastic bag, the initial air charge is always maintained, so recharging is unnecessary.

Most pumps commonly called "two-wire pumps" contain all required starting and protection components and connect directly to the pressure switch—with no aboveground controller. Generally, there are no moving electrical parts within the pump, such as

CONTROLLER
USED ON
"3-WIRE"
PUMPS
ONLY

SWITCH

SPECIAL CABLE
SUITABLE FOR USE
WITH SUBMERSIBLE
PUMPS

PRESSURE
SWITCH

PRECHARGED
WATER TANK

DEEP-WELL
PIPE

HIGH-PRESSURE
SAFETY CUTOFF
SWITCH

CENTERING
GUIDE

CABLE STRAPPED TO
WATER PIPE (DROP PIPE)

CENTERING GUIDE

PERMANENT WATER LEVEL

IMPELLER TO DRIVE
WATER UPWARD

WATER INTAKE SCREEN

SQUIRREL-CAGE
INDUCTION MOTOR

WELL CASING DRIVEN
TO DEPTH NECESSARY TO
REACH BELOW PERMANENT
WATER LEVEL

Figure 19-3 Submersible pump.

the centrifugal starting switch found in a typical single-phase, split-phase induction motor.

Other pumps, referred to as "three-wire pumps," require an aboveground controller that contains any required components not in the motor, such as a starting relay, overload protection, starting and running capacitors, lightning arrester, and terminals for making the necessary electrical connections.

Rating data are on the pumps themselves, on the "three-wire" controllers, and on a separate nameplate furnished with each "two-wire" pump for aboveground identification.

The calculations for sizing the conductors, and the requirements for the disconnect switch, the motors' branch-circuit fuses, and the grounding connections are the same as those for the jet pump motor. The nameplate data and instructions fur-

nished with the pump must be followed. The manufacturer's recommendations for conductor size, type, and length are extremely important if the pump is to operate properly.

See Unit 16 for information about underground wiring.

Submersible pumps are covered by Underwriters Laboratories (U.L.) Standard 778.

Submersible Pump Cable

Power to the motor is supplied by a cable especially designed for use with submersible pumps. This cable is marked "submersible pump cable," and it is generally supplied with the pump and the pump's various other components. The cable is cut to the proper length to reach between the pump and its controller, Figure 19-3, or between the pump and the well-cap, Figure 19-4. When needed, the cable may be spliced according to the manufacturer's specifications.

Submersible water pump cable is "tag-marked" for use within the well casing for wiring deep-well water pumps where the cable is not subject to repetitive handling caused by frequent servicing of the pump units.

Submersible pump cable is *not* designed for direct burial in the ground unless it is marked Type USE or Type UF. Should it be required to run the pump's circuit underground for any distance, it is necessary to install underground feeder cable (UF), Type USE, or a raceway with suitable conductors, and then make up the necessary splices in a "listed" weatherproof junction box, Figure 19-4.

GROUNDING

Figure 19-4 illustrates how to provide proper grounding of a submersible water pump and the well casing. Proper grounding of the well casing and submersible pump motor will minimize or eliminate stray voltage problems that could occur if the pump motor is not grounded.

Key code sections that relate to grounding water pumping equipment are:

- *250-86:*   Metal enclosures for circuit conductors shall be grounded, other than short sections of metal enclosures that provide support or protection.

SPLICES MADE IN SPACE
PROVIDED IN WELL CAP
OR JUNCTION BOX.

WELL
CAP

EQUIPMENT GROUNDING
CONDUCTOR CONNECTED
TO CASING AND TO PUMP

PUMP CONTROLLER AND
DISCONNECT SWITCH

METAL WELL
CASING MUST
BE BONDED
TO PUMP
EGC OR TO
EQUIPMENT
GROUND
BUS AT THE
PANELBOARD
WHERE PUMP
CIRCUIT
ORIGINATES.
*SECTION
250-112(m).*

PUMP CABLE CONTAINS WIRES PLUS
EQUIPMENT GROUNDING CONDUCTOR.

TYPE UF OR TYPE USE CABLE
OR RACEWAY WITH PROPERLY
SIZED CONDUCTORS.
MUST CARRY EQUIPMENT
GROUNDING CONDUCTOR
SIZED PER *TABLE 250-122.*

PUMP AND METAL WELL
CASING SHALL BE GROUNDED
SECTION 250-112(l)
AND *SECTION 250-112(m).*

PUMP

INLET

MOTOR

• DISCONNECT MUST BE IN
SIGHT OF CONTROLLER
AND MUST DISCONNECT
THE CONTROLLER,
SECTION 430-102(a).

• DISCONNECT MUST BE IN
SIGHT OF MOTOR OR BE
CAPABLE OF BEING LOCKED
IN THE "OFF" POSITION
SECTION 430-102(b).

• DISCONNECT MUST OPEN
ALL UNDERGROUND
(HOT) CONDUCTORS
SECTION 430-103.

• DISCONNECT MAY BE
IN SAME ENCLOSURE
AS CONTROLLER
SECTION 430-103.

• DISCONNECT MUST BE
HORSEPOWER RATED
SECTION 430-109.

Figure 19-4   Grounding and disconnecting means requirements for submersible water pumps.

- *250-112(l):* Motor-operated water pumps, including submersible pumps, must be grounded.

- *250-112(m):* Bond metal well casing to the pump's equipment grounding conductor, or run a separate equipment grounding conductor all the way back to the ground bus at the same panel that supplies the submersible pump circuit.

- *250-134(b):* Equipment fastened in place or connected by permanent wiring must have the equipment grounding conductor run with the circuit conductors.

- ▶• *250-2(d):* The earth shall not be used as the sole equipment grounding conductor or fault-return path. ◀

- *430-141:* Grounding of motors—general.

- *430-142:* Grounding of motors—specific.

- *430-144:* Grounding of controllers.

- *430-145:* Acceptable methods of grounding.

Because so much nonmetallic (PVC) water piping is used today, some electrical codes require that a grounding electrode conductor be installed between the neutral bar of the main service disconnecting means and the metal well casing. This would satisfy the requirements of *Sections 250-104* (bonding), *250-50* (grounding electrode system), *250-52* (made and other electrodes), and *250-56* (resistance of made electrodes). This then is the service ground and is sized per *Table 250-66* of the Code. At the metal casing, the grounding electrode conductor is attached to a lug termination, or by means of exothermic (Cadweld) welding.

WATER HEATER CIRCUIT Ⓐc

Electrical contractors many times are also in the plumbing, heating, and appliance business. They need to know more than how to do the electrical hookup. The following text discusses electrical as well as other data about electric water heaters that will prove useful.

All homes require a supply of hot water. To meet this need, one or more automatic water heaters are generally installed as close as practical to the areas having the greatest need for hot water. Water piping carries the heated water from the water heater to the various plumbing fixtures and to appliances such as dishwashers and clothes washers.

For safety reasons, in addition to the regular temperature control thermostat that can be adjusted by the homeowner, the *NEC*® and U.L. Standards require that electric water heaters be equipped with a high temperature cut-off. This high temperature control limits the maximum water temperature to 190°F. This is 20°F below the 210°F temperature that causes the "temperature/pressure" relief valve to open. The high temperature cutoff is factory preset, and should never be tampered with or modified in the field. In conformance to *Section 422-47* of the *NEC*,® the high temperature limiting control must disconnect all ungrounded conductors.

In most residential electric water heaters, the high temperature cutoff and upper thermostat are combined into one device, Figure 19-5.

Figure 19-5 Typical electric water heater showing location of heating elements and thermostats.

Combination Pressure/Temperature Safety Relief Valves

Look at Figure 19-5. A pressure/temperature relief valve is an important safety device installed into an opening in the water heater tank within 6 inches (152.4 mm) of the top. The opening is provided for and clearly marked by the manufacturer. Many water heater pressure/temperature relief valves are factory installed, although some models still require installation of the valve by the installing contractor. Proper discharge piping must be installed from the pressure/temperature relief valve downward to within 6 inches (152 mm) of the floor, preferably near a floor drain. Terminating the discharge pipe to within 6 inches (152 mm) of the floor is important to protect people from injury from scalding water or steam discharge should the relief valve activate for any reason. This also protects electrical equipment in and around the water heater from becoming soaked should the relief valve operate.

Pressure/temperature relief valves are installed for two reasons. Their pressure/temperature sensing features are interrelated.

- *Pressure:* Water expands when heated. Failure of the normal temperature thermostat and the high temperature limiting device to operate could result in pressures that exceed the rated working pressure of the water heater tank and the plumbing system components. The relief valve, properly installed and properly rated to local codes, will open to relieve the excess pressure in the tank to the predetermined setting marked on the relief valve. The rating is stamped on the valve's identification plate.

- *Temperature:* Should runaway water heating conditions occur within the water heater, temperatures far exceeding the atmospheric boiling point of water (212°F) can take place. Under high pressures, water will stay liquid at extremely high temperatures. For a typical municipal water supply pressure of 45 psi, water will not boil until it reaches approximately 290°F. At 290°F (143°C), opening a faucet, a broken pipe, or a ruptured water heater tank would allow the pressure to instantly drop to normal atmospheric pressure. The superheated water would immediately flash into steam, having

about 1,600 times more volume than the liquid water occupied. This blast of steam can be catastrophic, and could result in severe burns or death.

Thus, the function of the temperature portion of the pressure/temperature relief valve is to open when a temperature of 210°F (98.7°C) occurs within the water heater tank. Properly rated and maintained, the pressure/temperature relief valve will allow the water to flow from the water heater through the discharge pipe at a rate faster than the heating process, thereby keeping the water temperature below the boiling point. Should the cold water supply be shut off, the relief valve will allow the steam to escape without undue buildup of temperature and pressure in the tank.

Sizes

Residential electric water heaters are available in many sizes, such as 6, 12, 15, 20, 30, 32, 40, 42, 50, 52, 66, 75, 80, 82, 100, and 120 gallons.

Corrosion of Tank

To reduce corrosion of the steel tank, most water heater tanks are glass lined. Glass lining is a combination of silica and other minerals (rasorite, rutile, zircon, cobalt, nickel oxide, and fluxes). This special mix is heated, melted, cooled, crushed, then sprayed onto the interior steel surfaces and baked on at about 1,600°F. To further reduce corrosion, aluminum or magnesium anode rods are installed in special openings, or as part of the hot water outlet fitting, to allow for the flow of a protective current *from* the rod(s) *to* any exposed steel of the tank. This is commonly referred to as *cathodic protection*. Corrosion problems within the water heater tank are minimized as long as the aluminum or magnesium rod(s) remain in an active state. Anodes should be inspected at regular intervals to determine when replacement is necessary, usually when the rod(s) is reduced to about $\frac{1}{3}$ of its original diameter, or when the rod's core wire is exposed. Further information can be obtained by contacting the manufacturer, and consulting the installation manual.

Some water heater tanks are fiberglass or plastic lined.

Heating Elements

The wattage ratings of electric water heaters can vary greatly, depending upon the size of the heater in gallons, the speed of recovery desired, local electric utility regulations, and codes. Typical wattage ratings are 1,500, 2,000, 2,500, 3,000, 3,800, 4,500, and 5,500 watts.

A resistance heating element might be dual rated. For example, the element might be marked 5,500 watts at 250 volts, and 3,800 watts at 208 volts.

Most residential-type water heaters are connected to 240 volts except the smaller 2-, 4-, and 6-gallon *point of use* sizes generally rated 1,500 watts at 120 volts. Commercial electric water heaters can be rated single-phase or three-phase, 208, 240, 277, or 480 volts.

U.L. requires that the power (wattage) input must not exceed 105% of the water heater's nameplate rating. All testing is done with a supply voltage equal to the heating element's *rated* voltage. Most heating elements may burn out prematurely if operated at voltages 5% higher than for which they are rated.

To help reduce the premature burnout of heating elements, some manufacturers will supply 250-volt heating elements, yet will mark the nameplate 240 volts with its corresponding wattage at 240 volts. This allows a safety factor if slightly higher than normal voltages are experienced.

Heating Element Construction

Heating elements generally contain a nickel-chrome (nichrome) resistance wire embedded in compacted powdered magnesium oxide to ensure that no grounds occur between the wire and the sheath. The magnesium oxide is an excellent insulator of electricity, yet it effectively conducts the heat from the nichrome wire to the sheath.

The sheath can be made of tin-coated copper, stainless steel, or incolloy. The stainless steel and incolloy types can withstand higher operating temperatures than the tin-coated copper elements. They are used in premium water heaters because of their resistance to lime scaling and dry firing burnout.

High watt density elements generally have a U-shaped tube, and give off a great amount of heat per square inch of surface. Figure 19-6(A) illustrates a flange-mounted high-density heating element. Figure 19-6(B) illustrates a high-density heating element that screws into a threaded opening in the tank. High-density elements are very susceptible to lime-scale burnout.

Low watt density elements, Figure 19-6(C), generally have a double loop, like a U tube bent in half. They give off less heat per square inch, have a longer life, and are less noisy than high-density elements.

Figure 19-6   Heating elements.

Figure 19-7 A dual-wattage heating element. The two wires of the incoming 240-volt circuit are connected to L1 and L2. These leads connect to the 3800-watt element. To connect the 1700-watt element, the jumper is connected between L2 and L3. With both heating elements connected, the total wattage is 5500 watts.

An option offered by at least one manufacturer of electric water heaters is a dual-wattage heating element. A dual-wattage heating element actually contains two heating elements in the one jacket. One element is rated 3800 watts and one element is rated 1700 watts. Three leads are brought out to the terminal block. The 3800-watt heating element is connected initially, with the option of connecting the second 1700-watt heating element if the wiring to the water heater is capable of handling the higher total combined wattage of 5500 watts. To attain the higher wattage, a jumper (provided by the manufacturer) is connected between two terminals on the terminal block, Figure 19-7.

Any electric heating element coated with lime scale may be noisy during the heating cycle. Limed up heating elements should be cleaned or replaced, following the manufacturer's instructions.

Do not energize the electrical supply to a water heater unless you are sure it is full of water. Most water heater heating elements are designed to operate only when submersed in water.

Residential electric water heaters are available with one or two heating elements. Figure 19-5 shows two heating elements. Single-element water heaters will have the heating element located near the bottom of the tank. If the water heater has two heating elements, the upper element will be located about two-thirds of the way up the tank.

Speed of Recovery

Speed of recovery is the time required to bring the water temperature to satisfactory levels in a given length of time. The current accepted industry standard is based on a 90°F rise. For example, referring to Figure 19-8, we find that a 3,500-watt element can raise the water temperature 90°F of a little more than 16 gallons an hour. A 5,500-watt element

HEATING ELEMENT WATTAGE	GALLONS PER HOUR RECOVERY FOR INDICATED TEMPERATURE RISE					
	60°F	70°F	80°F	90°F	100°F	Equivalent NET Btu output
750	5.2	4.4	3.9	3.5	3.1	2,559
1,000	6.9	5.9	5.2	4.6	4.1	3,412
1,250	8.6	7.4	6.5	5.7	5.2	4,266
1,500	10.3	8.9	7.8	6.9	6.2	5,119
2,000	13.8	11.8	10.3	9.2	8.3	6,825
2,250	15.5	13.3	11.6	10.3	9.3	7,678
2,500	17.2	14.8	12.9	11.5	10.3	8,531
3,000	20.7	17.7	15.5	13.8	12.4	10,238
3,500	24.1	20.7	18.1	16.1	14.5	11,944
4,000	27.6	23.6	20.7	18.4	16.6	13,650
4,500	31.0	26.6	23.3	20.7	18.6	15,356
5,000	34.5	29.6	25.9	23.0	20.7	17,063
5,500	37.9	32.5	28.4	25.3	22.5	18,769
6,000	41.4	35.5	31.0	27.6	24.6	20,475

NOTES: If the incoming water temperature is 40°F (Northern states), use the 80° column to raise the water temperature to 120°F.

If the incoming water temperature is 60°F (Southern states), use the 60° column to raise the water temperature to 120°F.

For example, approximately how long would it initially take to raise the water temperature in a 42-gallon electric water heater to 120°F where the incoming water temperature is 40°F? The water heater has a 2,500-watt heating element.

Answer: 42 ÷ 12.9 = 3.26 hours (approximately 3 hours and 16 minutes)

Figure 19-8 Electric water heater recovery rate per gallon of water for various wattages.

can raise water temperature 90°F of about 25 gallons an hour. Speed of recovery is affected by the type and amount of insulation surrounding the tank, the supply voltage, and the temperature of the incoming cold water.

Figure 19-9 shows the approximate time it takes to run out of hot water using typical shower heads with various capacity electric water heaters. Regular residential-type shower heads deliver 5 to 6 gallons per minute. Water saver shower heads deliver 1 to 2 gallons per minute. This varies with water pressure. An easy way to determine a shower head's flow rate is to turn on the water, hold a large bucket of known capacity under it, and time how long it takes to fill the bucket.

Scalding from Hot Water

To avoid scalding, many states have laws regulating the maximum temperature settings for water heaters. This maximum setting is usually 120°F. Where such laws do not exist, the Consumer Product Safety Commission suggests a maximum temperature setting of 130°F. Manufacturers post caution labels on their water heaters that read something like this:

SCALD HAZARD: WATER TEMPERATURE OVER 125°F CAN CAUSE SEVERE BURNS INSTANTLY OR DEATH FROM SCALDS. SEE INSTRUCTION MANUAL BEFORE CHANGING TEMPERATURE SETTING.

LENGTH OF CONTINUOUS SHOWER TIME—IN MINUTES—FOR TYPICAL SIZES OF RESIDENTIAL ELECTRIC WATER HEATERS. CHART BASED ON 70% DRAW EFFICIENCY. RECOVERY RATE IS NOT FIGURED IN BECAUSE IT IS NORMALLY NOT A FACTOR IN CONTINUOUS DRAW OF WATER SITUATIONS.

Shower head(s) rate in gallons per minute	Hot water flow rate in gallons per minute based on 120°F. Stored hot water to provide 105°F. mixed water out of shower head.	Hot water flow rate in gallons per minute based on 120°F. Stored hot water to provide 105°F. mixed water out of shower head.	100 GALLON		80 GALLON		66 GALLON		50 GALLON		40 GALLON		30 GALLON	
	40°F. INCOMING WATER	60°F. INCOMING WATER	40°F. INCOMING WATER	60°F. INCOMING WATER	40°F. INCOMING WATER	60°F. INCOMING WATER	40°F. INCOMING WATER	60°F. INCOMING WATER	40°F. INCOMING WATER	60°F. INCOMING WATER	40°F. INCOMING WATER	60°F. INCOMING WATER	40°F. INCOMING WATER	60°F. INCOMING WATER
2.0	1.6	1.5	43	47	34	37	28	31	22	23	17	19	13	14
2.5	2.0	1.9	34	38	28	30	23	25	17	19	14	15	10	11
3.0	2.4	2.3	29	31	23	25	19	21	14	16	11	12	9	9
4.0	3.3	3.0	22	23	17	19	14	15	11	12	9	9	6	7
5.0	4.0	3.8	17	19	14	15	11	12	9	9	7	7	5	6
6.0	4.9	4.5	14	16	11	12	9	10	7	8	6	6	4	5
8.0	6.5	6.0	11	12	9	9	7	8	5	6	4	5	3	4
12.0	9.8	9.0	7	8	6	6	5	5	4	4	3	3	2	2

Example #1: A 40-gallon electric water heater has a 40°F incoming water supply. How many minutes can a 3 GPM shower head be used to obtain 105°F hot water?

Answer: 11 minutes

Example #2: A residence has an 80-gallon electric water heater. The incoming water supply is approximately 60°F. How many minutes can two 2 GPM shower heads be used at the same time to obtain 105°F hot water? Hint: Two 2 GPM shower heads are the same as one 4 GPM shower head.

Answer: 19 minutes

Figure 19-9   This chart shows the approximate time in minutes it takes to run out of hot water when using various shower heads, different incoming water temperatures, and different capacity water heaters.

Certain states do not accept the water heater thermostat as an acceptable water temperature control device, and require that showers and shower/bath combinations be provided with an automatic safety mixing device to prevent sudden unanticipated changes in water temperatures. Some manufacturers use nonadjustable upper thermostats, factory set at 120°F. Because 120°F is not hot enough to dissolve grease and to activate power detergents, most newer dishwashers boost the water temperature in the dishwasher from 120°F to 140°F–145°F using the same heating element that provides the heat for the drying cycle.

Figure 19-10 shows time/temperature relationships relating to scalding.

Thermostats/High Temperature Limit Controls

All electric water heaters have a thermostat(s) and a high temperature limit control(s).

Thermostats and high temperature limit controls are available in many configurations. They are available as separate controls and as combination controls.

Figure 19-11 illustrates a combination thermostat/high temperature limit control. This particular control is an interlocking type sometimes referred to as a snap-over type. It is used on two-element water heaters and controls the upper heating element as well as provides the safety feature of limiting the water temperature to a factory preset value. Another

thermostat is used to control the lower heating element. When all of the water in the water heater is cold, the upper element heats the upper third of the tank. The lower thermostats are closed, calling for heat, but the interlocking characteristic of the upper thermostat/high limit will not allow the lower heating element to become energized. With this connection, both heating elements cannot be on at the same time. This is referred to as *nonsimultaneous* or *limited demand* operation.

When the water in the upper third of the tank reaches the temperature setting of the upper thermostat, the upper thermostat "snaps over" completing the circuit to the lower element. The lower element then begins to heat the lower two-thirds of the tank. Most of the time, it is the lower heating element that keeps the water at the desired temperature. Should a large amount of water be used in a short time, the upper element comes on. This provides fast recovery of the water in the upper portion of the tank, which is where the hot water is drawn from the tank.

Safety Note: Most residential water heater thermostats are of the single-pole type. They do not open both ungrounded conductors of the 240-volt supply. They open one conductor only, which is all that is needed to open the 240-volt circuit to the

Figure 19-11   Typical electric water heater controls. The seven terminal control combines water temperature control plus high temperature limit control. This is the type commonly used to control the upper heating element as well as providing a high temperature limiting feature. The two terminal control is the type used to control water temperature for the lower heating element. *Courtesy* THERM-O-DISC, Inc., subsidiary of Emerson Electric.

TEMPERATURE	TIME TO PRODUCE SERIOUS BURN
120°F	Approx. 9½ minutes
125°F	Approx. 2 minutes
130°F	Approx. 30 seconds
135°F	Approx. 15 seconds
140°F	Approx. 5 seconds
145°F	Approx. 2½ seconds
150°F	Approx. 1⁸⁄₁₀ seconds
155°F	Approx. 1 second
160°F	Approx. ½ second

Figure 19-10   Time/temperature relationships related to scalding of an adult subjected to moving water. For children, the time to produce a serious burn is less than for adults. *Courtesy* of Shriners Burn Institute, Cincinnati Unit.

heating element. What this means is that a reading of 120 volts to ground will always be present at the terminals of the heating element on a 240-volt, single-phase system such as found in residential wiring. If the heating element becomes grounded to the metal sheath, the element can continue to heat even when the thermostat is in the OFF position because of the presence of the 120 volts.

The high temperature limit control opens both ungrounded conductors.

Various Types of Electrical Connections

There are many variations for electric water heater hookups. Every electric utility has its own unique time-of-use programs and specific electrical connection requirements for electric water heaters. It is absolutely essential that you contact your local electric utility for this information.

Because of the cumulative large power consumption of electric water heater loads, electric utilities across the country have been innovative in creating special lower rate structures (programs) for electric water heaters, air conditioners, and heat pumps. Electric utilities want to be able to shed some of this load during their peak periods. They will offer the homeowner choices of how the water heater, air conditioner, or heat pump is to be connected, and in some cases, will provide financial incentives (lower rates) to those who are willing to use these restricted "time-of-day" connections. "Time-of-day," "time-of use," and "off-peak" are terms used by different utilities. If homeowners have a large storage capacity electric water heater, they will have an adequate supply of hot water during those periods of time the power to the water heater is off. If they cannot live with this, they have other options. Check with the local power company for details on their particular programs.

The following text discusses some of the methods in use around the country for connecting electric water heaters, and in some cases, air-conditioning and heat pump equipment.

Utility Controlled

The world of electronics has opened up many new ways for an electric utility to meter and control residential electric water heater, air-conditioning,

and heat pump loads. Electronic watt-hour meters can be read remotely, programmed remotely, and can detect errors remotely. The options are endless.

The utility can install watt-hour meters that combine a watt-hour meter and a set of switching contacts. The contacts are switched on and off by the utility during peak hours using a power line carrier signal. Older style versions of this type of watt-hour meter incorporated a watt-hour meter, time clock, and switching contacts. The time clock would be set by the utility for the desired times of the on-off cycle. Today, the metering, and the programming for timing and control of the on-off cycling is accomplished through electronic solid-state devices. Electronic meters contain a microprocessor that records, and an optical assembly that scans the rotating disc to record the number of turns and speed of the disc. These electronic digital meters can record the energy consumption during normal hours and during peak hours, allowing the electric utility to bill the homeowner at different rates for different periods of time. One type of residential single-phase watt-hour meter can register four different "time-of-use" rates. This type of meter is shown in Figure 29-2.

One major utility programs their residential meters for "peak periods" from 9 A.M. to 10 P.M., Monday through Friday—except for holidays. Their "off-peak" periods are weekend days and all other hours. Programming of electronic meters can extend out for more than five years. The possibilities for programming electronic meters are endless. Older style mechanical meters used gears to record this data, and did not have the programming capabilities that electronic meters do.

The switching contacts in the meter are connected in *series* with the water heater, air-conditioner, or heat pump load that is to be controlled for the special "time-of-day" rates, Figures 19-12, and 19-13. The utility determines when they want to have the electric water heater on or off.

Figure 19-12 shows a separate combination meter/time clock for the electric water heater load. The utility sets the clock to be "off" during their peak hours of high power consumption. Thus the term *off-peak metering*. Off-peak metering is oftentimes called "time-of-use" or "time-of-day" metering. In exchange for lower electric rates, the homeowner runs the risk of being without hot water.

A 240/120-VOLT SERVICE

B MAIN WATT-HOUR METER

C WATER HEATER WATT-HOUR METER

D SERVICE CONDUCTORS TO MAIN PANEL

E TWO-POLE DISCONNECT SWITCH
OR CIRCUIT BREAKER

F COMBINATION HIGH TEMPERATURE LIMIT
AND UPPER INTERLOCKING THERMOSTAT

G UPPER HEATING ELEMENT

H LOWER THERMOSTAT

I LOWER HEATING ELEMENT

Notations used in Figures 19-12, 19-13, 19-15, and 19-16.

Should homeowners run out of hot water during these off periods, they have no recourse but to wait until the power to the water heater comes on again.

For this type of installation, large storage capacity water heaters are needed so as to be able to have hot water during these periods.

Figure 19-13 illustrates a type of meter that contains an integral set of contacts, controlled by the utility via signals sent over the power lines. These meters are available with either 30-ampere or 50-ampere contacts, the contact rating selected to switch the load to be controlled. The electric utility determines when to shed the load. The homeowner must be willing to live with the fact that the power to the water heater, air-conditioner, or heat pump might be turned off during the peak periods.

In Figure 19-14, the utility mounts a radio-controlled relay to the side of the meter socket. One major utility using this method makes use of ten different radio frequencies. When the utility needs to shed load, it can signal one or all of the relays to turn the water heater, air conditioner, or heat pump load off and on. This might only be a few times each year, as opposed to time-clock programs that operate daily. This method is a simple one. One watt-hour meter registers the total power con-

Figure 19-12 Wiring for a typical "off-peak" electric water heater circuit. A separate watt-hour meter/time clock controls the circuit to the owater heater. The utility sets the time clock to turn the power off during certain peak periods of the day. This type of circuitry is called "time-of-day" programming.

Figure 19-13   The meter contains a set of contacts that can be controlled by the utility. The water heater circuit is fed from the main panel, through the contacts, then to the water heater.

Figure 19-14   In this diagram, the utility sends a radio signal to the relay, which in turn can "shed" the load of a residential water heater, air-conditioner, or heat pump. If the relay has two sets of contacts, four conductors would be required. Low-voltage relays are also used, in which case low-voltage conductors are run between the radio control unit and a relay in the appliance. For the control of an air-conditioner or heat pump, the radio control unit is connected "in series" with the room thermostat. This arrangement can be used with one watt-hour meter using a sliding rate schedule, with a "time-of-day" control. Contact the electric utility for technical data regarding how they want the electrical connections to be made for their particular programs.

sumption. The homeowner signs up for this program. The program offers a financial incentive, and the homeowner knows up front that the utility might occasionally turn off the load during crucial peak periods.

Customer Controlled

There are some rate schedules where customers can somewhat control their energy costs by doing their utmost to use electrical energy during off-peak hours. For example, they would not wash clothes, use the electric clothes dryer, take showers, or use the dishwasher during peak (premium) hours. The choice is up to homeowners. To accomplish this, watt-hour meters can be installed that have two sets of dials: one set showing the *total kilowatt hours*, and the second set showing *premium time kilowatt hours*. See Figure 29-2. With this type of meter, the utility will have the total kilowatt hours used as well as the premium time kilowatt hours used. With this information, the utility bills customers at one rate for the energy used during peak hours, and at another rate for the difference between

total kilowatt hours and premium time kilowatt hours. The electric utility will determine and define *premium time*.

Some homeowners have installed time-clocks on their water-heater circuit to shut-off during peak premium time hours to be sure they are not using energy during the high rate per kilowatt-hour periods.

Uncontrolled

The simplest uncontrolled method is shown in Figure 19-15. There are no extra meters, switching contacts in the meter, extra disconnect switches, or controllable relays. The water heater circuit is connected to a two-pole branch-circuit breaker in the

KILOWATT-HOUR POWER CONSUMPTION OF WATER HEATER IS METERED BY SAME METER USED FOR NORMAL HOUSEHOLD LOADS. ONLY ONE METER IS REQUIRED. POWER RATES ARE USUALLY ON A SLIDING SCALE; THAT IS, THE MORE POWER IN KILOWATT-HOURS USED, THE LOWER RATE PER KILOWATT-HOUR.

Figure 19-15 Water heater connected to two-pole circuit in Main Panel.

main distribution panel. The electrical power to the water heater is on 24 hours a day. This is the connection for the water heater in the residence in this text. Instead of separate "low-rate" metering that requires an additional watt-hour meter, the utility offers a sliding scale.

EXAMPLE:

METER READING 06-01-99	00532	
METER READING 05-01-99	00007	
TOTAL KILOWATT-HOURS USED	00525	
BASIC MONTHLY ENERGY CHARGE		$11.24
1ST. 400 kWh × 0.10206		40.82
NEXT 125 kWh × 0.06690		8.36
AMOUNT DUE		$60.42

State, local, and applicable regulatory taxes would be added to the amount due.

Certain utilities will use the higher energy charge during the June, July, August, and September summer months.

Another uncontrolled method is to use two watt-hour meters, Figure 19-16. The water heater has power 24 hours per day. One meter will register the normal lighting load at the regular residential energy charge. The second meter will register the energy

Figure 19-16 Twenty-four-hour water heater installation. The total energy demand for the water heater is fed through a separate watt-hour meter at a lower kWh rate than the regular meter.

consumed by the water heater, air conditioner, or heat pump at some lower energy rate.

Are Water Heater Control Conductors Permitted in the Same Raceway as Service-Entrance Conductors?

Conductors other than service-entrance conductors are not permitted in the same raceway as the service-entrance conductors, *Section 230-7*. However, *Exception No. 2* to this section does permit load management control conductors that have overcurrent protection to be installed in the service raceway.

ELECTRIC WATER HEATER BRANCH-CIRCUITS

Branch-Circuit Rating

Section 422-13 applies to all fixed storage electric water heaters having a capacity of 120 gallons (454.2 L) or less. This would include most typical residential electric water heaters. The branch-circuit rating must not be less than 125% of the nameplate rating of the water heater.

Disconnecting Means

Section 422-30 requires that the appliance must have a means of disconnect. This would normally be located in the panel (load center), or it could be provided by a separate disconnect switch. Although not normally done, *Section 422-33* does allow the unit switch that is an integral part of an appliance having an OFF position that disconnects all ungrounded conductors from the source to be considered the required **disconnecting means**. In many local codes, this is not acceptable. *Section 422-33* further requires that there be another means of disconnect for the appliance, which in the case of a one-family dwelling is the main service disconnect.

Section 422-34 requires that switches and circuit breakers used as the disconnecting means must be of the indicating type, meaning they must clearly show that they are in the ON or OFF position.

Section 422-31(b) requires that for permanently connected appliances rated greater than 300 volt-amperes or ⅛ horsepower, the disconnecting means must be within sight of the appliance, or be capable of being locked in the OPEN position. The electric water heater in the residence provides a good example of this Code requirement.

Branch-Circuit Overcurrent Protection

Section 422-11(e) states that for single non-motor-operated electrical appliances, the overcurrent protective device shall not exceed the protective device rating marked on the appliance. If the appliance has no such marking, then the overcurrent device is to be sized as follows:

- If appliance does not exceed 13.3 amperes— 20 amperes
- If appliance draws more than 13.3 amperes— 150% of the appliance rating

For a single non-motor-operated appliance, if the 150% sizing does not result in a standard size overcurrent device rating as listed in *Section 240-6*, then it is permitted to go to the next standard size.

EXAMPLE: A water heater nameplate indicates 4,500 watts, 240 volts. What is the maximum size fuse permitted by the Code?

SOLUTION:

$$I = \frac{W}{E} = \frac{4,500}{240} = 18.75 \text{ amperes}$$

Section 422-13 requires that the branch-circuit rating must not be less than 125% of the water heater's nameplate rating. The minimum branch-circuit rating is:

$$18.75 \times 1.25 = 23.44 \text{ amperes}$$

This would require a minimum 25-ampere fuse or circuit breaker, which is the next standard size larger than the calculated 23.44 amperes. See *Section 240-6(a)* for standard ampere ratings for fuses and circuit breakers.

▶If the water heater or any other nonmotor-operated appliance is not marked with the required overcurrent protective device rating, *Section 422-11(e)* permits us to use a fuse or circuit breaker as high as 150% of the appliance's ampere rating for a nonmotor-operated appliance rated over 13.3 amperes. If the 150% sizing does not result in a standard size fuse or breaker, then we could even go to the next higher rating. ◀

For nonmotor-operated appliances rated 13.3 amperes or less that are not marked with their required overcurrent protective device, the maximum overcurrent protection is 20 amperes. The maximum branch-circuit overcurrent device could be:

$$18.75 \times 1.5 = 28.125 \text{ amperes}$$

This would be a 30-ampere fuse or circuit breaker, which is the next standard size larger than the calculated 28.125 amperes. But there is no reason to use a 30-ampere branch circuit fuse or circuit breaker because we have already determined that a 25-ampere fuse or circuit breaker is suitable for the 4,500-watt load.

Conductor Size

The conductors supplying the 4,500-watt water heater will be protected by the 25-ampere or breaker as determined above.

Checking *Table 310-16*, we find the allowable ampacity of conductors in the 60°C column as required by *Section 110-14(c)*. We also find the maximum overcurrent protection for small conductors listed in *Section 240-3(d)*. Putting this all together, we now have the following data:

CONDUCTOR SIZE (THHN Copper)	ALLOWABLE AMPACITY FROM 60°C COLUMN OF *TABLE 310-16*	MAXIMUM AMPERE RATING OF OVERCURRENT DEVICE
No. 14	20 amperes	15 amperes
No. 12	25 amperes	20 amperes
No. 10	30 amperes	30 amperes

By trial and error, we can see that although the No. 12 THHN conductors have an ampacity of 25 amperes, which would be suitable for the minimum branch-circuit rating, the maximum OCD is 20 amperes. Therefore, since we selected a 25-ampere OCD, the conductor would have to be No. 10 THHN, which has an allowable ampacity of 30 amperes, and can be protected by a OCD rated not over 30 amperes.

You might want to review Unit 4 for a refresher regarding why we use the 60°C column of *Table 310-16* in selecting the conductor size.

The water heater for the residence is connected to Circuit A6-8. This is a 20-ampere, two-pole circuit breaker located in the main panel in the workshop. This circuit is straight 240 volts and does not require a neutral conductor.

Checking the Schedule of Special-purpose Outlets, we find that the electric water heater has two heating elements: a 2000-watt element and a 3000-watt element. Because of the thermostats on the water heater, both elements cannot be energized

at the same time. Therefore, the maximum load demand is

$$\text{amperes} = \frac{\text{watts}}{\text{volts}} = \frac{3000}{240} = 12.5 \text{ amperes}$$

For water heaters having two heating elements connected for nonsimultaneous operation, the nameplate on the water heater will be marked with the largest element's wattage at rated voltage. For water heaters having two heating elements connected for simultaneous operation, the nameplate on the water heater will be marked with the total wattage of both elements at rated voltage.

Conductor Size and Conduit Size for the Water Heater in This Residence

The specifications for the residence call for conductors to be No. 12 Type THHN. The branch-circuit overcurrent protection is 20 amperes.

Table C1, Appendix C of the *NEC®* indicates that ½-inch EMT is required for two No. 12 THHN conductors. To simplify the installation, a short length (18 to 24 inches, or 457 to 610 mm) of ½-inch flexible metal conduit is attached to the ½-inch EMT and is connected to the knockout provided for the purpose on the water heater.

Cord/Plug Connections Not Permitted for Water Heaters

Sections 400-7(a)(8) and *422-16(a)* state the uses where flexible cords are permitted. A key code requirement, often violated, is that flexible cords shall be used only "where the fastening means and mechanical connections are specifically designed to permit ready removal for maintenance and repair, and the appliance is intended or identified for flexible cord connection." Certainly the plumbing does not allow for the "ready removal" of the water heater. The conductors in flexible cords cannot handle the high temperatures encountered on the water heater terminals. Check the instruction manual for proper installation methods.

Equipment Grounding

The water heater is grounded through the ½-inch conduit and the flexible metal conduit. Code references are *Sections 250-110, 250-118, 350-14*, and *422-16*. The subject of equipment grounding is covered in Unit 4.

EFFECT OF VOLTAGE VARIATION

As previously stated, the heating elements in the heater in this residence are rated 240 volts. They will operate at a lower voltage with a reduction in wattage. If connected to voltages above their rating, they will have a very short life.

Ohm's law and the wattage formula show how the wattage and current depend upon the applied voltage.

$$R = \frac{E^2}{W} = \frac{240 \times 240}{3,000} = 19.2 \text{ ohms}$$

$$I = \frac{E}{R} = \frac{240}{19.2} = 12.5 \text{ amperes}$$

If a different voltage is substituted, the wattage and current values change accordingly.

At 220 volts:

$$W = \frac{E^2}{R} = \frac{220 \times 220}{19.2} = 2,521 \text{ watts}$$

$$I = \frac{W}{E} = \frac{2521}{220} = 11.46 \text{ amperes}$$

Another way to calculate the effect of voltage variance is:

$$\text{Correction Factor} = \frac{\text{Applied Voltage Squared}}{\text{Rated Voltage Squared}}$$

Using the previous example:

$$\text{Correction Factor} = \frac{220 \times 220}{240 \times 240} = \frac{48,400}{57,600} = 0.84$$

then, $3,000 \times 0.84 = 2,521$ watts

In resistive circuits, current is directly proportional to voltage and can be simply calculated using a ratio and proportion formula. For example, if a 240-volt heating element draws 12.7 amperes at rated voltage, the current draw can be determined at any other applied voltage, say 208 volts.

$$\frac{208}{240} = \frac{X}{12.7}$$

$$\frac{208 \times 12.7}{240} = 11.006 \text{ amperes}$$

To calculate this example if the applied voltage is 120 volts:

$$\frac{120}{240} = \frac{X}{12.7}$$

$$\frac{120 \times 12.7}{240} = 6.35 \text{ amperes}$$

Also, in a resistive circuit, wattage varies as the square of the current. Therefore, when the voltage on a heating element is doubled, the current also doubles and the wattage increases four times. When the voltage is reduced to one-half, the current is halved and the wattage is reduced to one-fourth.

Using the above formulae, it will be found that a 240-volt heating element when connected to 208 volts will have approximately three-quarters of the wattage rating than at rated voltage.

HEAT PUMP WATER HEATERS

With so much attention given to energy savings, heat pump water heaters have entered the scene.

Residential heat pump water heaters can save 50 to 60% over the energy consumption of resistance-type electric water heaters, depending upon energy rates. The heat energy transferred to the water is three to four times greater than the electrical energy used to operate the compressor, fan, and pump.

Heat pump water heaters work in conjunction with the normal resistance-type electric water heater. Located indoors near the electric water heater and properly interconnected, the heat pump water heater becomes the primary source for hot water. The resistance electric heating elements in the electric water heater are the back-up source should it be needed.

Heat pump water heaters remove low-grade heat from the surrounding air, but instead of transferring the heat outdoors as does a regular air-conditioning unit, it transfers the heat to the water. The unit shuts off when the water reaches 130°F. Moisture is removed from the air at approximately one pint per hour, reducing the need to run a dehumidifier. The exhausted cool air from the heat pump water heater is about 10°–15°F below room temperature. The cool air can be used for limited cooling purposes, or it can be ducted outdoors.

A typical residential heat pump water heater rating is 240 volts, 60 Hz, single-phase, 500 watts.

Heat pump water heaters operate with a hermetic refrigerant motor compressor. The electric hookup is similar to a typical 240-volt, single-phase, residential air-conditioning unit. *Article 440* of the *NEC®* applies. This is covered in Unit 23.

Electric utilities are a good source to learn more about the economics of installing heat pump water heaters to save energy.

REVIEW

Note: Refer to the Code or the plans where necessary.

WATER PUMP CIRCUIT Ⓐ B

1. Does a jet pump have any electrical moving parts below the ground level? _____

2. Which is larger, the drive pipe or the suction pipe? _____

3. Where is the jet of the pump located? _____

4. What does the impeller wheel move? _____

5. Where does the water flow after leaving the impeller wheel?

   a. _____ b. _____

6. What prevents water from draining back into the pump from the tank? _____

7. What prevents water from draining back into the well from the equipment? _____

   _____

8. What is compressed in the water storage tank? _____

9. Explain the difference between a "two-wire" submersible pump and a "three-wire" submersible pump. _____

   _____

   _____

   _____

10. What is a common speed for jet pump motors? _____

11. Why is a 240-volt motor preferable to a 120-volt motor for use in this residence?

    _____

    _____

12. How many amperes does a 1-hp, 240-volt, single-phase motor draw? (See *Table 430-148.*) _____

13. What size are the conductors used for this circuit? _____

14. What is the branch-circuit protective device? _____

15. What provides the running overload protection for the pump motor? _____

    _____

    _____

16. What is the maximum ampere setting permitted for running overload protection of the 1-hp, 240-volt pump motor? _____

17. Submersible water pumps operate with the electrical motor and actual pump located (Circle correct answer.)
    a. above permanent water level
    b. below permanent water level
    c. half above and half below permanent water level

18. Because the controller contains the motor starting relay and the running and starting capacitors, the motor itself contains _____

19. What type of pump moves the water upward inside of the deep-well pipe? _____

    _____

20. Proper pressure of the submersible pump system is maintained by a _____

    _____

21. Fill in the data for a 16-ampere electric motor, single-phase, no Code letters.

    a. Branch-circuit protection, non-time-delay fuses.

       Normal size _____ A   Switch size _____ A

       Maximum size _____ A   Switch size _____ A

    b. Branch-circuit protection dual-element, time-delay fuses.

       Normal size _____ A   Switch size _____ A

       Maximum size _____ A   Switch size _____ A

    c. Branch-circuit protection instant-trip breaker.

       Normal setting _____ A

       Maximum setting _____ A

    d. Branch-circuit protection—inverse time breaker.

       Normal rating _____ A

       Maximum rating _____ A

    e. Branch-circuit conductor size, Type THHN _____

       Ampacity _____ A

    f. Motor overload protection using dual-element time-delay fuses.

       Maximum size _____ A

22. The *NEC®* is very specific in its requirement that submersible electric water pump motors be grounded. Where is this specific requirement found in the Code?

    _____

    _____

    _____

23. Does the *NEC®* allow submersible pump cable to be buried directly in the ground?

    _____

    _____

24. Must the disconnect switch for a submersible pump be located next to the well?

    _____

    _____

25. A metal well casing (shall) (shall not) be bonded to the pump's equipment grounding conductor or by grounding it with a separate equipment grounding conductor run all the way back to the same ground bus in the panel that supplies the pump circuit, *Section* _____ .

WATER HEATER CIRCUIT Ⓐc

1. According to *Section 422-47(b)* the high temperature limiting control must disconnect _____ of the ungrounded conductors. The high temperature control limits the maximum water temperature to _____°C.

2. A major hazard involved with water heaters is that they operate under whatever pressure the serving water utility supplies. Water stays liquid at temperatures higher than the normal boiling point of 212°F. If the high temperature limit control failed to operate for whatever reason, should a pipe burst or a faucet be opened, the pressure would instantly drop to normal atmospheric pressure, causing the superheated water to turn into _____ that could result in _____ . To prevent this from happening, water heaters are equipped with pressure/temperature relief valves.

3. Magnesium rods are installed inside the water tank to reduce _____ .

4. The heating elements in electric water heaters are generally classified into two categories. These are _____ density and _____ density elements.

5. An 80-gallon electric water heater is energized for the first time. Approximately how many hours would it take to raise the water temperature 80°F (from 40°F to 120°F)? The water heater has a 3,000-watt heating element.

   _____

   _____

   _____

6. What term is used by utilities when the water heater power consumption is measured using different rates during different periods of the day? _____

7. Explain how the electric utility in your area meters residential electric water heater loads.

   _____

   _____

   _____

8. For residential water heaters, the Consumer Product Safety Commission suggests a maximum temperature setting of _____ °F. Some states have laws stating a maximum temperature setting of _____ °F.

9. An 80-gallon electric water heater has 60°F incoming water. How many minutes can a 3 gallon/minute showerhead be used to draw 105°F water? The water heater thermostat had been set at 120°F. _____

10. Approximately how long would it take to produce serious burns to an adult with 140°F water? _____

11. Two thermostats are generally used in an electric water heater.

    a. What is the location of each thermostat? _____

       _____

    b. What type of thermostat is used at each location? _____

       _____

12. a. How many heating elements are provided in the heater in the residence discussed in this text? _____

    b. Are these heating elements allowed to operate at the same time? _____

13. When does the lower heating element operate? _____

_____

14. The Code states that water heaters having a capacity of 120 gallons (454.2 L) or less shall be considered _____ duty and, as such, the circuit must have a rating of not less than _____ % of the rating of the water heater.

15. Why does the storage tank hold the heat so long? _____

16. The electric water heater in this residence is connected for "limited demand" so that only one heating element can be on at one time.

    a. What size wire is used to connect the water heater? _____

    b. What size overcurrent device is used? _____

17. a. If both elements of the water heater in this residence are energized at the same time, how much current will they draw? (Assume the elements are rated at 240 volts.)

    b. What size and Type THHN wire is required for the load of both elements? Show calculations.

18. a. How much power in watts would the two elements in question 17 use if connected to 220 volts? Show calculations.

    b. What is the current draw at 220 volts? Show calculations.

19. A 240-volt heater is rated at 1,500 watts.
    a. What is its current rating at rated voltage? _____
    b. What current draw would occur if connected to a 120-volt supply?

    c. What is the wattage output at 120 volts?

20. For a single, non-motor-operated electrical appliance rated greater than 13.3 amperes that has no marking that would indicate the size of branch-circuit overcurrent device, what percentage of the nameplate rating is used to determine the maximum size branch-circuit overcurrent device? _____

21. A 7,000-watt resistance-type heating appliance is rated 240 volts. What is the maximum size fuse permitted to protect the branch-circuit supplying this appliance?

    _____

22. The Code requirements for standard resistance electric water heaters are found in *Article 422* of the Code. Because heat pump water heaters contain a hermetic refrigerant motor compressor, the Code requirements are found in *Article* _____ of the Code.

Special-Purpose Outlets for Ranges, Counter-Mounted Cooking Unit ⏏G, and Wall-Mounted Oven ⏏F

OBJECTIVES

After studying this Unit, the student will be able to

- interpret electrical plans to determine special installation requirements for counter-mounted cooking units, wall-mounted ovens, and freestanding ranges.

- understand the latest *NEC®* connection requirements for electric ranges, ovens, and counter-mounted cooking units.

- identify and understand the purpose of the National Electrical Manufacturers Association (NEMA) blade and slot configurations on 30-ampere and 50-ampere receptacles and cords.

- identify the terminals of 30-ampere and 50-ampere receptacles.

- understand when to use 3-wire and 4-wire receptacles and cords for electric range hookups.

- compute loads and select proper conductor sizes for electric ranges, wall-mounted ovens, and counter-mounted cooking units.

- properly ground electric ranges, wall-mounted ovens, and counter-mounted cooking units in conformance to the *NEC®* for both new installations and existing installations.

- describe a seven-heat control and a three-heat control for a dual-element heating unit for a cooktop.

- understand how infinite heat temperature controls operate.

- supply counter-mounted cooking units and wall-mounted ovens by connecting them to one feeder using the "tap" rule.

- understand how to install a feeder to a load center, then divide the feeder to individual circuits to supply the appliances.

BASIC CIRCUIT REQUIREMENTS FOR ELECTRIC RANGES, COUNTER-MOUNTED COOKING UNITS, AND WALL-MOUNTED OVENS

Connection Methods

Any recognized wiring method can be used to hook up appliances, such as nonmetallic-sheathed cable, armored cable, EMT, flexible metal or nonmetallic conduit, or service-entrance cable. Local electrical codes can contain information as to the restriction on the use of any of the wiring methods listed above.

Here are some of the more common methods for connecting electric ranges, ovens, and cooktops.

Direct Connection. One method is to run the branch-circuit wiring directly to a junction box on the appliance provided by the manufacturer of the appliance. When "roughing-in" the branch-circuit wiring, sufficient length must be provided so as to allow flexibility to make up the connections, and to place the appliance into position, and for servicing.

Junction Box. Another method is to run the branch-circuit wiring to a junction box (an outlet box) behind, under, or adjacent to the range, oven, or cooktop location. By definition, this junction box is accessible because should the need arise, the junction box can be reached by removing the appliance. The "whip" on the appliance is then connected to the

junction box where the splicing of the tap conductors in the "whip" and the branch-circuit conductors are made. According to U.L. Standard No. 858, the flexible metal "whip" must not be less than 3 feet and no longer than 6 feet. The conductors in the "whip" are rated for the high temperatures encountered in the appliance.

Depending on the location of the appliances, it might be possible to connect more than one cooking appliance to a single branch-circuit from a single junction box or from two junction boxes. This is illustrated in Figure 20-1.

Cord- and Plug-Connection. Still another very common method is to use a cord- and plug-connection, *Section 422-32(a)*. This is particularly true for freestanding ranges. A cord- and plug connection also serves as the disconnecting means for household cooking equipment. The range receptacle is permitted to be at the rear base of an electric range. Removing the bottom drawer makes this connection accessible, *Section 422-32(b)*. The receptacle must have an ampere rating not less than the rating of the appliance, *Section 422-32(c)*. For household cooking equipment, the demand factors of *Table 220-19* are also applicable to the receptacle, *Section 422-32(c)*. Range and dryer receptacles are identical, and are discussed later in this unit.

Figure 20-1 A counter-mounted cooktop unit and a wall-mounted oven connected to one branch-circuit. Depending on the location of the appliances, it might be possible to make the installation with one junction box.

The load of an appliance that is cord- and plug-connected shall not exceed 80% of the ampere rating of the branch-circuit, *Section 210-23(a)* and *(b)*.

Cord sets rated 30 amperes are referred to as *dryer cords*. Cord sets rated 40, 45, or 50 amperes are referred to as *range cords*.

Receptacles rated 30 amperes are referred to as *dryer receptacles*. Receptacles rated 50 amperes are referred to as *range receptacles*.

Nameplate on Appliance

Section 422-60(a) requires that the nameplate must show the appliance's rating in volts and amperes, or in volts and watts.

Load Calculations

In *Section 210-19(c)* and *Exceptions*, we find that:

• the branch-circuit conductors must have an ampacity not less than the branch-circuit rating.

►• the branch-circuit conductors' ampacity must not be less than the maximum load being served. ◄

• the branch-circuit rating must not be less than 40 amperes for ranges of 8¾ kW or more.

• for ranges 8¾ kW or more, the neutral of a three-wire branch-circuit may be 70% of the branch-circuit rating, but not smaller than No. 10.

• tap conductors connecting ranges, ovens, and cooking unit to a 50-ampere branch-circuit must be rated at least 20 amperes, be adequate for the load, and must not be any longer than necessary for servicing the appliance.

• *Section 220-19* permits the use of the values found in *Table 220-19* and its footnotes for the load computation.

• *Table 220-19* shows the kW-demand for different kW rated household cooking appliances that are over 1¾ kW. Column A is used in most situations. Note that the maximum demand is considerably less than the actual appliance's nameplate rating. This is because rarely, if ever, will all of the heating elements be used at the same time. *Note 3* to the table tells us when we are permitted to use Columns B and C. This is discussed later in this unit.

►• *Section 220-22* states that for calculating a feeder or service, the neutral load can be figured at 70% of the load on the ungrounded conductors for ranges, ovens, and counter-mounted cooktops using *Table 220-19* as the basis for the load calculation. ◄

In *Section 422-10(a)* we find that the branch-circuit rating shall not be less than the marked rating of the appliance. However, there is an exception for household cooking appliances. The last paragraph of this section states that *"Branch circuits for household cooking appliances shall be permitted to be in accordance with Table 220-19."*

The demand load for individual wall-mounted ovens or individual counter-mounted cooking units must be calculated at the appliance's full nameplate rating, *Table 220-19, Note 4*.

Wire Size

After making the load calculations to determine the appliance's computed load in amperes, we then refer to *Table 310-16* to select a conductor that has an allowable ampacity equal to or greater than the computed load. As mentioned many times in this text, *Section 110-14(c)* as well as the U.L. Standards require that we use the 60°C column, regardless of the conductor's insulation rating.

An exception to this basic rule is when the equipment is marked for 75°C. This was discussed in detail in Unit 4 in the section entitled *"The Weakest Link of The Chain."*

Overcurrent Protection

Section 422-11(a) states that the branch-circuit overcurrent protection shall not exceed the rating marked on the appliance.

For appliances that have no marked maximum overcurrent protection, then the branch-circuit overcurrent protection is sized at the ampacity of the branch-circuit conductors supplying the appliance. See *Section 240-3(b)*, *(d)*, and *(e)*.

Some large commercial electric ranges might have integral overcurrent protection that subdivides the range into two or more circuits, each overcurrent device rated not over 50 amperes, *Section 422-11(b)*.

Disconnecting Means

All appliances require a means of disconnecting it from the power source, *Section 422-30*.

Section 422-32(a) accepts a cord- and plug arrangement as the required disconnecting means.

A branch-circuit switch or circuit breaker is acceptable as the disconnecting means if it is within sight of the appliance or is capable of being locked in the "OFF" position, *Section 422-31(b)*. Breaker manufacturers supply "lock-off" devices that fit over the circuit breaker. Individual disconnect switches also have a "lock-off" provision.

Section 422-33 accepts the "unit switches" that have a marked "OFF" position to be an acceptable disconnect for the appliance because they disconnect all of the ungrounded conductors. This enables a person to work on the appliance knowing that the power is off to the various electrical components of the range. However, the line side terminals of the appliance would still be "hot." The branch-circuit switch or circuit breaker serves as the disconnect that will turn off the total power to the appliance. This branch-circuit breaker disconnecting means is very important when appliances are "hard wired" as opposed to being cord- and plug-connected.

GROUNDING FRAMES OF ELECTRIC RANGES, WALL-MOUNTED OVENS, AND COUNTER-MOUNTED COOKING UNITS

As with most electrical equipment, the frames of electric ranges, wall-mounted ovens, and counter-mounted cooking units must be grounded. Grounding is accomplished through the equipment grounding conductor in nonmetallic-sheathed cable, the metal armor of BX, or with metal raceways such as EMT and flexible metal conduit. See *Section 250-118* for a list of acceptable equipment grounding means.

Electric ranges, wall-mounted ovens, and counter-mounted cooking units are not permitted to be grounded to the neutral branch-circuit conductor, *Section 250-140*. The small copper grounding strap (link)(jumper) furnished with these appliances *must not* be connected between the neutral terminal and the metal frame of the appliance. It is best to remove and discard this grounding jumper.

However, unusual as it may seem, grounding the frames of these appliances to the neutral was permitted prior to the 1996 *NEC.* Beginning with the 1996 edition of the *NEC,* the Code recognized this previously permitted practice with restrictions. Read the next sentence carefully! When installing a new electric range, wall-mounted oven, or counter-mounted cooking unit on an *existing* installation branch-circuit where the wiring method did not contain a separate equipment grounding conductor, it is permitted to ground their frames to the neutral circuit conductor. In this situation, the small copper grounding strap (link)(jumper) furnished with these appliances *must* be used. Note that this practice is permitted only for existing installations.

This subject was covered in detail in Unit 15, and will not be repeated here.

Receptacles for Electric Ranges and Dryers

Because of their similarities, the following text applies to both range and dryer receptacles.

Residential electric ranges, electric dryers, wall-mounted ovens, and surface-mounted cooking units may be connected with cords designed specifically for this use. Depending upon the appliance's wattage rating, the receptacles are usually rated 30 amperes or 50 amperes, 125/250 volts. Cord sets rated 40 amperes or 45 amperes contain 40- or 45-ampere conductors, yet the plug cap is rated 50 amperes. It becomes a cost issue to use a full 50-ampere rated cord set in all cases, or to match the cord ampere rating as close as possible to that of the appliance being connected.

Range and dryer receptacles are available in both surface mount and flush mount.

Blade and Slot Configuration for Plug Caps and Receptacles

Figure 20-2 shows the different configurations for receptacles commonly used for electric range and dryer hookups. The National Electrical Manufacturers Association (NEMA) uses the letter "R" to indicate a receptacle, and the letter "P" for plug cap.

- **30-ampere, 3-wire:** The L-shaped slot on the receptacle (NEMA 10-30R) and the L-shaped blade on the plug cap (NEMA 10-30P) are for the white neutral wire. Frames of electric ranges, ovens, and clothes dryers installed prior to the adoption of the 1996 *National Electrical Code* were permitted to be grounded to the neutral conductor as stated in *Sections 250-140* and *250-142* of the 1996 edition and previous

50-AMPERE
3-POLE, 3-WIRE
125/250-VOLT
NEMA 10-50R
PERMITTED
PRIOR TO 1996 *NEC*®

50-AMPERE
4-POLE, 4-WIRE
125/250-VOLT
NEMA 14-50R
REQUIRED BY 1996 *NEC*®

30-AMPERE
3-POLE, 3-WIRE
125/250-VOLT
NEMA 10-30R
PERMITTED
PRIOR TO 1996 *NEC*®

30-AMPERE
4-POLE, 4-WIRE
125/250-VOLT
NEMA 14-30R
REQUIRED BY 1996 *NEC*®

Figure 20-2   Illustrations of 30-ampere and 50-ampere receptacles used for cord- and plug-connection of electric ranges, ovens, counter-mounted cooking units, and electric clothes dryers. Four-wire receptacles and four-wire cord sets are required for installations according to the 1999 *NEC*.® Three-wire receptacles and three-wire cord sets were permitted prior to adoption of the 1996 *NEC*.® See *Sections 250-140* and *250-142*.

editions of the *NEC*.® The neutral conductor served a dual-purpose, that of the neutral and that of the appliance's required equipment ground. Today, this permission is for existing branch-circuits only.

- **30-ampere, 4-wire:** The L-shaped slot on the receptacle (NEMA 14-30R) and the L-shaped blade on the plug cap (NEMA 14-30P) are for the white neutral wire. The horseshoe-shaped slot on the receptacle and the round or horseshoe-shaped blade on the plug cap are for the equipment ground.

- **50-ampere, 3-wire:** The wide flat slot on the receptacle (NEMA 10-50R) and the matching wide blade on the plug cap (NEMA 10-50P) are for the white neutral wire. Frames of electric ranges, ovens, and clothes dryers installed prior to adoption of the 1996 *National Electrical Code*® were permitted to be grounded to the neutral conductor as stated in *Sections 250-140* and *250-142*. This permission is for existing branch-circuits only. The neutral connection served a dual-purpose: that of the neutral and that of the appliance's required equipment ground.

- **50-ampere, 4-wire:** The wide flat slot on the receptacle (NEMA 14-50R) and the matching wide blade on the plug cap (NEMA 14-50P) are for the white neutral wire. The horseshoe-shaped slot on the receptacle and the round or horseshoe-shaped blade on the plug cap are for the equipment ground.

Cord Sets for Electric Ranges and Dryers

A 30-ampere cord set contains:

- **3-wire:** three No. 10 AWG conductors. The attachment plug cap (NEMA 10-30P) is rated 30 amperes.

- **4-wire:** four No. 10 AWG conductors. The attachment plug cap (NEMA 14-30P) is rated 30 amperes.

A 40-ampere cord set contains:

- **3-wire:** three No. 8 AWG conductors and one No. 10 AWG conductor. The attachment plug cap (NEMA 10-50P) is rated 50 amperes.

- **4-wire:** three No. 8 AWG conductors and one No. 10 AWG conductor. The attachment plug cap (NEMA 14-50P) is rated 50 amperes.

A 45-ampere cord set contains:

- **3-wire:** two No. 6 AWG conductors and one No. 8 AWG conductor. The attachment plug cap (NEMA 10-50P) is rated 50 amperes.

- **4-wire:** three No. 6 AWG conductors and one No. 8 AWG conductor. The attachment plug cap (NEMA 14-50P) is rated 50 amperes.

A 50-ampere cord set contains:

- **3-wire:** two No. 6 AWG conductors and one No. 8 AWG conductor. The attachment plug cap (NEMA 10-50P) is rated 50 amperes.

- **4-wire:** three No. 6 AWG conductors and one No. 8 AWG conductor. The attachment plug cap (NEMA 14-50P) is rated 50 amperes.

Terminal Identification for Receptacles and Cords

Receptacle and cord terminals are marked:

- "X" and "Y" for the ungrounded conductors.

- "W" for the white grounded conductor. The "W" terminals will generally be whitish or silver (tinned) in color, in accordance with *Section 200-9* and *200-10* of the *NEC*.®

- "G" for the equipment grounding conductor. This terminal is green colored, hexagon shaped, and is marked "G," "GR," "GRN," or "GRND" in accordance with *Sections 250-126* and *410-58(b)* of the *NEC*.® The grounding blade on the plug cap must be longer than the other blades so that its connection is made before the ungrounded conductors make connection in accordance with *Section 410-58(d)* of the *NEC*.®

WALL-MOUNTED OVEN CIRCUIT Ⓐꜰ

The wall-mounted oven in this residence is located to the right of the kitchen sink.

The branch-circuit wiring, using nonmetallic-sheathed cable is run from Panel B to the oven location, coming into the cabinet space where the oven will be installed from behind or from the bottom. Earlier in this unit we discussed the various methods for hooking up appliances. The choices are direct connection, junction box, or cord- and plug connection. In most cases, a built-in wall-mounted oven will have a flexible metal conduit attached to it, which is run to a junction box where the electrical connections are completed.

Circuit Number. The circuit number is B (6-8). This information is found in the Schedule of Special Purpose Outlets contained in the Specifications at the rear of this text, and on the Directory for Panel B found in Unit 28.

Nameplate Data. This appliance is rated 6.6 kW (6,000 watts) @ 120/240 volts.

Load Calculations. (Use nameplate rating per *Section 220-19, Note 4.*)

$$I = \frac{W}{E} = \frac{6,600}{240} = 27.5 \text{ amperes}$$

Wire Size. The conductors in nonmetallic-sheathed cable are rated for 90°C. However, in conformance to *Section 336-26*, the allowable ampacity is determined using the 60°C column of *Table 310-16*. Here we find that a No. 10 conductor has an allowable ampacity of 30 amperes—more than adequate for the appliance's calculated load of 27.5 amperes. This will be a No. 10 three-conductor plus equipment grounding conductor nonmetallic-sheathed cable.

Overcurrent Protection. A two-pole, 30-ampere circuit breaker located in Panel B.

Disconnecting Means. The disconnecting means is the 30-ampere, two-pole circuit breaker in Panel B. If a cord- and plug connection were to be made, that also would serve as a disconnecting means.

Grounding. The equipment grounding conductor contained in the nonmetallic-sheathed cable provides the means for grounding the frame of the oven. According to *Table 250-122*, the equipment grounding conductor for a circuit protected by a 30-ampere overcurrent device is a No. 10 copper conductor. That is the size EGC contained in No. 10 nonmetallic-sheathed cable.

Figure 20-3 shows a typical wiring diagram for a standard oven unit.

COUNTER-MOUNTED COOKING UNIT CIRCUIT Ⓐɢ

The cooktop in this residence is located in the island.

Counter-mounted cooking units are available in many styles. They may have two, three, or four surface heating elements. Instead of individual exposed surface heating elements, some have a smooth ceramic surface where the electric heating elements are in direct contact with the underside of the ceramic top. These units have the same controls as conventional surface heating elements and are wired according to the general *NEC*® guidelines previously discussed.

Most of these ceramic-top cooking units have a faint design to indicate the location of the heating elements. The ceramic is rugged, attractive, and easy to clean.

Figure 20-3   Wiring diagram for an oven unit.

The branch-circuit wiring is run from Panel B to the island. The nonmetallic-sheathed cable is concealed above the recreation room ceiling, upward into the island cabinetry, ending in a junction box below the cooking unit. Earlier in this unit we discussed the various methods for hooking up appliances. The choices are direct connection, junction box, or cord- and plug connection. In most cases, the cooktop will have a flexible metal conduit attached to it, which also runs to the junction box where the electrical connections are completed.

Note on the plans that a receptacle outlet is to be installed on the side of the island. This receptacle outlet is supplied by small appliance branch-circuit B16.

Circuit Number.   The circuit number is B (2-4). This information is found in the Schedule of Special Purpose Outlets contained in the Specifications at the rear of this text, and on the Directory for Panel B found in Unit 28.

Nameplate Data.   This appliance is rated 7,450 watts @ 120/240 volts.

Load Calculations.   (Use nameplate rating per *Table 220-19, Note 4.*)

$$I = \frac{W}{E} = \frac{7,450}{240} = 31.04 \text{ amperes}$$

Wire Size.   The conductors in nonmetallic-sheathed cable are rated for 90°C. However, in conformance to *Section 336-26*, the allowable ampacity is determined using the 60°C column of *Table 310-16*. Here we find that a No. 8 conductor has an allowable ampacity of 40 amperes—more than adequate for the appliance's calculated load of 31.04 amperes. This will be a No. 8 three-conductor plus equipment grounding conductor nonmetallic-sheathed cable.

Overcurrent Protection.   A two-pole, 40-ampere circuit breaker located in Panel B.

Disconnecting Means.   The disconnecting means is the 40-ampere, two-pole circuit breaker in Panel B. A cord- and plug connection would also be a disconnecting means.

Grounding.   The equipment grounding conductor contained in the nonmetallic-sheathed cable provides the means for grounding the frame of the appliance. According to *Table 250-122*, the equipment grounding conductor for a circuit protected by a 40-ampere overcurrent device is a No. 10 copper conductor. That is the size EGC contained in No. 8 nonmetallic-sheathed cable.

FREESTANDING RANGE

Load Calculations

The load calculations for a freestanding electric range are simple.

According to *Table 220-19*, the demand is permitted to be figured at 8 kW for an electric range that is rated 12 kW or less. For example, if an electric range is marked 11.4 kW, the maximum demand can be figured at 8 kW because under normal situations, the surface heating elements, broiler element, baking element, and lights would not all be on at the same time.

Larger double-oven freestanding ranges are probably rated more than 12 kW. Let us consider a range that has a rating of 14,050 watts. This is the same wattage as the previously discussed wall-mounted oven and counter-mounted cooking unit.

The total connected load of this freestanding electric range is:

$$I = \frac{W}{E} = \frac{14,050}{240} = 58.5 \text{ amperes}$$

However, *Table 220-19* permits the computed load to be considerably less than the connected load.

Table 220-19 and the footnotes provide the information we need to make the load calculation.

1. 14,050 − 12,000 watts = 2,050 watts (2 kW)

2. According to *Note 1*, *Table 220-19*, a 5% increase for each kW and major fraction in excess of 12 kW must be added to the load demand in Column A. 2 kW (the kW over 12 kW) × 5% per kW = 10%.

3. 8 kW (from Column A) × 0.10 = 0.8 kW

4. The calculated load is: 8 kW + 0.8 kW = 8.8 kW

5. In amperes, the calculated load is:

$$I = \frac{W}{E} = \frac{8,800}{240} = 36.7 \text{ amperes}$$

Wire Size. The conductors in nonmetallic-sheathed cable are rated for 90°C. However, in conformance to *Section 336-26*, the allowable ampacity is determined using the 60°C column of *Table 310-16*. Here we find that a No. 8 conductor has an allowable ampacity of 40 amperes—more than adequate for the appliance's calculated load of 36.7 amperes. This will be a No. 8 three-conductor plus equipment grounding conductor nonmetallic-sheathed cable.

Section 210-19(b) states that for ranges of 8¾ kW or more, where the maximum demand has been computed using Column A of *Table 220-19*, the neutral may be reduced to not less than 70% of the branch-circuit rating, but never smaller than No. 10.

$$40 \times 0.70 = 28 \text{ amperes}$$

Checking *Table 310-16*, we find that a No. 10 conductor has an allowable ampacity of 30 amperes—more than adequate to carry the computed neutral load. If the wiring method were a raceway method, this electric range could be served by two No. 8 phase conductors and one No. 10 neutral conductor.

Using the 70% multiplier is a moot point when installing nonmetallic-sheathed cable or armored cable, because the insulated phase conductors in these cables are all the same size.

Overcurrent Protection. A two-pole 40-ampere circuit breaker.

Disconnecting Means. The disconnecting means would be the 40-ampere, two-pole circuit breaker. A cord- and plug connection behind the base of a freestanding electric range would also serve as a disconnecting means, *Section 422-32(b)*.

Grounding. The equipment grounding conductor contained in the nonmetallic-sheathed cable provides the means for grounding the frame of the appliance. According to *Table 250-122*, the equipment grounding conductor for a circuit protected by a 40-ampere overcurrent device is a No. 10 copper conductor. That is the size EGC contained in No. 8 nonmetallic-sheathed cable.

CALCULATIONS WHEN MORE THAN ONE WALL-MOUNTED OVEN AND COUNTER-MOUNTED COOKING UNIT ARE SUPPLIED BY ONE BRANCH-CIRCUIT

Load Calculations

Note 4 to *Table 220-19* states that when a single branch-circuit supplies a counter-mounted cooking unit, and not more than two wall-mounted ovens, all located in the same room, the calculation is made by adding up the nameplate ratings of the individual appliances, then treating this total as if it were one electric range.

Figure 20-1 shows a wall-mounted oven and a

counter-mounted cooking unit supplied by one 40-ampere branch-circuit.

The total connected load of these two appliances is:

$$I = \frac{W}{E} = \frac{14,050}{240} = 58.5 \text{ amperes}$$

However, since we are permitted to treat the two individual cooking appliances as one, we use the same calculations as we did for the free-standing range previously discussed.

The resulting conductor size, overcurrent protection, disconnecting means, and grounding are the same as for the equivalent freestanding range.

The cost of installing one 40-ampere branch-circuit might be higher than the cost of installing separate branch-circuits to each appliance. There will be additional junction boxes, different size cables, cable connectors, conduit fittings, and splices. In addition, there are restrictions in the *NEC®* on how to size the smaller "tap" conductors.

Factory-installed flexible metal conduit furnished and connected to these built-in cooking appliances will have the proper size and temperature-rated conductors for the particular appliance.

If you provide the flexible connections (whips) in the field, the minimum conductor sizing would be calculated as discussed above for each appliance. After the minimum conductor size is determined, we then must make sure that the conductors meet the "tap" rule. This is important because the "taps" will have overcurrent protection rated greater than the ampacity of the "tap" conductors.

Section 210-23(c) states that fixed cooking appliances that are fastened in place may be connected to 40- or 50-ampere branch-circuits. If "taps" are to be made to a 40- or 50-ampere branch-circuit, the taps must be able to carry the load, and, in no case, may the taps be less than 20 amperes, *Section 210-19(c)*.

For example, the 40-ampere circuit shown in Figure 20-1 has No. 10 THHN tap conductors for the oven, and No. 8 THHN tap conductors for the counter-mounted cooking unit. Each of these taps has an ampacity of greater than 20 amperes.

According to *Table 310-16*, a No. 8 THHN conductor has an allowable ampacity of 40 amperes, using the 60°C column of the table.

USING A LOAD CENTER

In this residence, the built-in oven and range are very close to Panel B, in which case it is economical to install a separate branch-circuit to each appliance. However, there are situations where the panel and the built-in cooking appliances are far apart.

The decision becomes one of installation cost (labor and material) as to whether to run separate branch-circuits, or to run one larger branch-circuit and use the "tap" rules.

We just finished discussing the issues involved when installing one branch-circuit for more than one built-in cooking appliance, then applying the rather restrictive "tap" rules.

The tap rules can be ignored if a smaller load center is installed as illustrated in Figure 20-4. There will be no "taps"—just branch-circuits to each appliance sized for that particular appliance.

The calculations for the feeder from the main panel to another smaller load center are based on the two cooking appliances being treated as one. The conductors from the smaller load center are calculated for each appliance as previously discussed.

In homes with a basement, the load center could be installed below the kitchen where the appliances are installed.

So now you have it! Run separate branch-circuits to each appliance, run one large branch-circuit and make "taps" to serve each appliance, or run one large branch-circuit and install a small load center. The choice is yours. Whichever method you choose, make sure the installation "Meets Code."

CALCULATIONS WHEN MORE THAN ONE ELECTRIC RANGE, WALL-MOUNTED OVEN, OR COUNTER-MOUNTED COOKING UNIT ARE SUPPLIED BY A FEEDER OR SERVICE

Although generally not an issue for wiring one- and two-family dwellings, the calculations for services and feeders of larger multifamily dwellings is more complex where a number of household cooking units are installed.

We need to refer to *Table 220-19*. This table has three columns. Column A is always used, with one exception. *Exception No. 3* shows that for household

Figure 20-4   Appliances connected to a load center.

cooking units having a nameplate rating of more than 1¾ kW through 8¾ kW, Columns B and C are permitted to be used.

Column B is for household cooking units rated more than 1¾ kW but less than 3½ kW. Column C is for household cooking units rated 3½ kW to 8¾ kW.

When using Columns B and C, apply the demand factors for the number of household cooking units whose kW ratings fall within that column. Then add the results of each column together.

EXAMPLE: Calculate the demand for a feeder that has four 3-kW wall-mounted ovens and four 6-kW counter-mounted cooktops connected to it.

Step 1: 4 ovens × 3 kW = 12 kW
12 kW × 66% = 7.92 kW

Step 2: 4 cooktops × 6 kW = 24 kW
24 kW × 50% = 12 kW

Step 3: 7.92 kW + 12 kW = 19.92 kW

Therefore, 19.92 kW is the demand load for the four ovens and four cooktops, even though the actual connected load is 36 kW. Diversity plays a large factor in arriving at the computed demand load.

Heating Elements

Electric heating elements generate a large amount of radiant heat. Heat is measured in British thermal units (Btu). A Btu is defined as the amount of heat required to raise the temperature of one pound of water by one degree Fahrenheit. A 1,000-watt heating element generates approximately 3,412 Btu of heat per hour.

HEATING CONTROL OPTIONS
Surface Heating Elements for Cooking Units

The heating elements used in surface-type cooking units are manufactured in several steps. First, spiral-wound nichrome resistance wire is impacted in magnesium oxide (a white, chalklike powder). The wire is then encased in a nickel-steel alloy sheath, flattened under very high pressure, and formed into coils. The flattened surface of the coil makes good contact with the bottoms of cooking utensils. Thus, efficient heat transfer takes place between the elements and the utensils.

Surface unit heating elements with one-coil (one heating element) construction are used in conjunction with infinite heat switches that "pulse" the circuit to the heating element. Two-coil (two heating elements) surface unit heating elements are used with fixed-heat 3-, 5-, and 7-position switches. Three-coil (three heating elements) surface unit heating elements are used with 3-position switches.

Figures 20-5 and 20-6 illustrate typical electric range surface units.

Figure 20-5 Typical 240-volt electric range surface heating element, used with infinite heat controls.

Figure 20-6 Typical electric range surface unit of the type used with 7-position controls.

TEMPERATURE CONTROLS

Infinite-Position Temperature Controls

Modern electric ranges are equipped with "infinite"-position temperature knobs (controls). Newer switches have a rotating cam that provides an infinite number of heat positions. Older style heat controls connect the heating elements of a surface unit to 120 volts or 240 volts in series or in parallel to attain a certain number of specific heat levels.

The contacts open and close as a result of a heater-bimetal device that is an integral part of the rotary control. When the control is turned on, a separate set of pilot light contacts closes and remains closed, while the contacts for the heating element open and close to attain and maintain the desired temperature of the surface heating unit, Figure 20-7.

Inside these infinite position temperature controls are contacts that open and close repetitively. The circuit to the surface heating element is

Figure 20-7 Typical internal wiring of an infinite heat surface element control.

regulated by the ratio of the time the contacts are in a closed position versus the open position. This is termed *input percentage*.

The contacts in a temperature control of this type have an expected life of over 250,000 automatic cycles. The knob rotation expected life exceeds 30,000 operations.

One-coil (single-element) surface units are used with infinite heat switches.

Fixed-Heat Temperature Controls

Still available are the older style rotary controls that have specific "indents" that lock the switch contacts into place as the knob is rotated to the various heat positions. Similar switching arrangements are found in push-button-type controls, where one button is pushed at a time. Pushing another button will raise and cancel the first button. Fixed-heat types of rotary and push-button switches, with or without pilot lights, are used in conjunction with two-coil surface unit heating elements, Figures 20-8 and 20-9. The different heat levels are obtained by switching between different voltages and coil circuitry. Their unique switching operation offers an excellent opportunity to make electrical calculations for series, parallel, and series/parallel circuits.

Seven-Heat Control for Two-Coil Heating Unit

One method of obtaining several levels of heat is to use a two-element unit, Figure 20-8. Figure 20-10 shows that seven levels of heat can be obtained from such a unit if a seven-point "fixed-heat" switch is used. Note that the two elements of the surface heating unit are both contained in the same steel jacket, and they are connected together at the inner end of the steel jacket. A return wire (not a resistance wire) is carried back to the terminal block. This arrangement means that the elements can be operated singly, together, or in series.

Figure 20-8 shows the heating element connected with element A as lead 2, element B as lead 1, and the return wire as the common (C). Resistance readings can be taken between these leads as follows:

Figure 20-8   Typical surface unit wiring.

Figure 20-9   Surface unit with three-heat series-parallel switch.

Figure 20-10   Surface heating element with seven levels of heat.

Test between C and 2   90 ohms (element A)

Test between C and 1
(single operation)   74 ohms (element B)

Test between 2 and 1   164 ohms (elements A
(elements in series)   and B in series)

Test between C and 1–2
connected together
(elements connected   40.6 ohms (elements A
in parallel)   and B in parallel)

Element B allows more current to flow than does element A because element B has 74 ohms of resistance as compared to 90 ohms of resistance for element A. As the current increases, the wattage also increases. The eight switch positions shown in Figure 20-10 result in the wattages shown in the following list. For each switch position, the formula $W = E^2/R$ is used to find the wattage value.

1. A and B parallel,
      240 volts   40.6 ohms   1418 watts
2. B only, 240 volts   74   ohms   778 watts
3. A only, 240 volts   90   ohms   640 watts
4. A and B parallel,
      120 volts   40.6 ohms   _____ watts*
5. B only, 120 volts   74   ohms   _____ watts*
6. A only, 120 volts   90   ohms   _____ watts*
7. A and B series,
      120 volts   164 ohms   _____ watts*
8. OFF position   Infinity   0 watts

---

*These values are omitted because the student is required to calculate the values in the Review section. A sample calculation is completed as follows:

$$W = \frac{E^2}{R} = \frac{240 \times 240}{40.6} = \frac{57,600}{40.6} = 1418 \text{ watts}$$

Three-Heat and Five-Heat Control for Two-Coil Heating Units

Three-heat and five-heat switches are for use with two-coil heating elements. These switches function in much the same manner as the seven-heat switch, except for the fewer heat levels available.

A simple, three-heat series-parallel switch is shown in Figure 20-9. A five-heat switch is not shown.

Coil Size Selector Switch

Coil size selector switches are used with three-coil heating elements. They control the heat by surface area rather than by varying the heat over the entire heating element. In the first switch position, the center 4-inch heating element comes on. In the second switch position, the center 4-inch element stays on, and the middle 6-inch element also comes on. In the third switch position, the center 4-inch, the middle 6-inch, and the outer 8-inch element are all on. These units supposedly save energy because the heating can be focused on the contact area directly beneath a cooking utensil.

Automatic Sensor Control

Infinite heat is attained using a one-coil surface heating element that has a sensor in the center of the surface unit. A special 12-volt control circuit is provided through a small transformer. The control knob is set to the desired temperature. A "responder" pulses the heating element off and on to maintain the set temperature. To operate properly, the cooking utensil must maintain good contact with the sensor. Use flat bottom utensils only.

High-Speed Control

Another type of heating element is the 120-volt flash or high-speed element. When connected to a special control switch, this type of element is connected briefly across 240 volts. In a few seconds, the element is returned automatically to the 120-volt supply for continued operation. A 1,250-watt flash unit connected to 240 volts produces 5,000 watts of heat in a matter of seconds. This brings the unit to maximum heat quickly. When the flash element is returned to the 120-volt supply, the unit continues to produce 1,250 watts, the rated value.

To calculate the maximum wattage at 240 volts, assume that a 1,250-watt heating element is rated at 120 volts. This unit has a resistance of:

$$R = \frac{E^2}{W} = \frac{120 \times 120}{1,250} = \frac{14,400}{1,250} = 11.52 \text{ ohms}$$

When this 11.52-ohm element is connected to a 240-volt supply, the wattage is:

$$W = \frac{E^2}{R} = \frac{240 \times 240}{11.52} = \frac{57,600}{11.52} = 5,000 \text{ watts}$$

Note that the wattage quadruples when the supply voltage is doubled. Conversely, the wattage is reduced to one-fourth its original value when the supply voltage is reduced to one-half its original value.

Heat Generation by Surface Heating Elements

The surface heating elements generate a large amount of radiant heat. For example, a 1,000-watt heating unit generates approximately 3,412 Btu of heat per hour. Heat is measured in British thermal units (Btu). A *Btu* is defined as the amount of heat required to raise the temperature of one pound of water by one degree Fahrenheit.

Caution When Making Connections

After determining the current rating ($I = W/E$), refer to *Table 310-16*. We find that for 27.5 amperes, a No. 10 AWG copper TW, THW, or THHN may be used to supply this wall-mounted oven. As previously discussed, for nonmetallic-sheathed cable, even though the conductor insulation is 90°C, the allowable ampacity must be determined according to the 60°C column of *Table 310-16*. This is stated in *Section 336-30(b)*.

When using armored cable, if the conductors are rated for 75°C as in Type ACTH, the ampacity is permitted to be based upon the 60°C or 75°C column of *Table 310-16*, depending upon the equipment and/or device terminal marking. If the conductors are rated for 90°C as in Type ACTHH, the ampacity would still have to be based upon the 60°C or 75°C column of *Table 310-16*, again depending upon the terminal marking. When armored cable is buried in insulation, it must have 90°C rated conductors, and its ampacity is based upon that of 60°C conductors, *Section 333-20*.

Recall from Unit 4 that, according to Underwriters Laboratories Standards (U.L.), and *Section 110-14(c)* of the *NEC*,® when conductor sizes No. 14 through No. 1 AWG are installed, their allowable current-carrying capacity (ampacity) is to be that of the 60°C column of *Table 310-16* unless the

terminals are marked with some other temperature rating, such as 75°C or 90°C.

That is why it is so important to carefully read the appliance manufacturer's installation data for recommended conductor size and type. Check particularly for the temperature rating.

Self-Cleaning Oven

Self-cleaning ovens are popular appliances because they automatically remove cooking spills from the inside of the oven. Lined with high temperature material, a self-cleaning oven can be set for a cleaning temperature that is much higher than baking and broiling temperatures. When the oven control is set to its "self-clean" position, a *sensor* located inside of the oven monitors the oven temperature to a preset value that approximates 800°F. The sensor is a resistor. The resistance of the sensor increases in proportion to the temperature inside the oven. This change in sensor resistance signals the electronic circuitry to provide the oven temperature needed for proper cleaning of the oven. At this temperature, all residue in the oven, such as grease and drippings, is burned until it is fine ash that can be removed easily. During the self-clean cycle, for safety reasons, a special latching mechanism on the oven door makes it impossible to open the oven door until a predetermined safe temperature is reached. On some self-cleaning electric ranges, the door locking lever also connects a *high temperature limit switch* into the circuit that protects against excessive high temperatures during the cleaning cycle.

Although the temperature and timing control features, and the internal associated circuitry found in modern freestanding and built-in electric ranges can be very complex, the actual branch-circuit sizing is the same for self-cleaning ranges and non-self-cleaning ranges.

Operation of the Oven

The oven heating elements are similar to the surface elements. The oven elements are mounted in removable frames. A thermostat controls the temperature of these heating units. Any oven temperature can be obtained by setting the thermostat knob to the proper point on the dial. The thermostat controls both the baking element and the broiling element. These elements can be used together to preheat the oven.

A combination clock and timer is used on most ovens. The timer can be preset so that the oven turns on automatically at a preset time, heats for a preset duration, and then turns off.

Microwave Oven

Modern microwave ovens, a result of electronic technology, have introduced many features never before possible. Such features include:

- control of time and temperature.

- programmable multiple-state settings.

- temperature probes for insertion into the food to ensure desired temperature.

- ability to "hold" food temperatures following cooking cycle.

- touch-sensitive controls.

- programmable defrost settings with lower temperatures than those required for actual cooking or baking so that premature cooking does not occur.

- computer calculations to automatically increase the time of operation when recipe quantities are doubled, tripled, or more.

- delayed starting of oven for up to 12 hours.

- preprogrammed temperature and time for recipes.

The electrician's primary concern is how to calculate the microwave oven load, and how to install the circuit for microwave appliances so as to conform to the Code. The same Code rules apply for both electrical and electronic microwave cooking appliances.

Ovens are insulated with fiberglass or polyurethane foam insulation placed within the oven walls to prevent excessive heat leakage.

REVIEW

Note: Refer to the Code or the plans where necessary.

COUNTER-MOUNTED COOKING UNIT CIRCUIT Ⓐ₉

1. a. What circuit supplies the counter-mounted cooking unit in this residence? _____

   b. What is the rating of this circuit? _____

2. What three methods may be used to connect counter-mounted cooking units? _____

   _____

   _____

   _____

   _____

   _____

   _____

3. Is it permissible to use standard 60°C insulated conductors to connect all counter-mounted cooking units? Why? _____

   _____

   _____

4. What is the maximum operating temperature (in degrees Celsius) for a

   a. Type TW conductor? _____

   b. Type THW conductor? _____

   c. Type THHN conductor? _____

5. The *NEC*® permits the following wiring methods to supply household built-in ovens and ranges. True or False (Circle the correct answer.)

   a. Nonmetallic-sheathed cable   (True) (False)

   b. Armored cable   (True) (False)

   c. Flexible metal conduit   (True) (False)

6. For electric ranges that have a computed demand of 8¾ kW or more, the neutral conductor can be reduced to _____% of the branch-circuit rating, but shall not be smaller than No. _____ AWG.

7. In the illustration of a typical seven-heat, eight-position switch, Figure 20-6A, the wattage values for positions 4, 5, 6, and 7 are omitted. Calculate these values. Show all calculations.

   Position 4

   Position 5

   Position 6

   Position 7

8. A 120-volt flash or high-speed unit rated at 1000 watts produces _____ watts when connected briefly to the 240-volt source. Show all calculations.

9. One kilowatt equals _____ Btu per hour.

10. Older style electric ranges had heat control knobs that had "indents" that could be felt when rotating the knob to adjust temperature to three, five, or seven different heat positions. Most new electric ranges have control knobs that do not have "indents." This type of heat control is known as an _____ heat control.

WALL-MOUNTED OVEN CIRCUIT Ⓐ_F

1. To what circuit is the wall-mounted oven connected? _____

2. An oven is rated at 7.5 kW. This is equal to

   a. _____ watts.

   b. _____ amperes at 240 volts.

3. a. What section of the Code governs the grounding of a wall-mounted oven? _____

   _____

   b. By what methods may wall-mounted ovens be grounded? _____

   _____

   _____

   _____

   _____

   _____

4. What is the type and ampere rating of the overcurrent device protecting the wall-mounted oven in this residence? Hint: Refer to the Schedule of Special Purpose Outlets in the Specifications. _____

   _____

   _____

   _____

5. Approximately how many feet (meters) of cable are required to connect the oven in the residence? _____

   _____

6. When connecting a wall-mounted oven and a counter-mounted cooking unit to one feeder, how long are the taps to the individual appliances? _____

_____

_____

_____

7. The branch-circuit load for a single wall-mounted self-cleaning oven or counter-mounted cooking unit shall be the _____ rating of the appliance.

8. A 6-kW counter-mounted cooking unit and a 4-kW wall-mounted oven are to be installed in a residence. Calculate the maximum demand according to *Column A, Table 220-19*. Show all calculations. Both appliances will be connected to the same branch-circuit.

9. The size of the neutral conductor supplying an electric range may be based on _____ % of the ampacity of the ungrounded conductors.

10. A freestanding electric range is rated 11.8 kW, 120/240 volts. The wiring method is nonmetallic-sheathed cable. Use the 60°C column of *Table 310-16* in conformance to *Section 110-14(c)*. Answer the following questions.

    a. According to Column A, *Table 220-19*, what is the maximum demand? _____kW.

    b. What is the minimum size for the ungrounded conductors?   No. _____ AWG

    c. What is the minimum size neutral conductor?   No. _____ AWG

d. What is the correct rating for the branch-circuit overcurrent protection?

_____ amperes

11. A double-oven electric range is rated at 18 kW, 120/240 volts. Calculate the maximum demand according to *Table 220-19*. Show all calculations.

12. For the range discussed in question 11:

a. What size 60°C ungrounded conductors are required? _____

b. What size 60°C neutral conductor is required? _____

c. What size 75°C ungrounded conductors are required? The terminal block and lugs on the range, the branch-circuit breaker, and the panelboard are marked as suitable for 75°C wire. _____

d. What size 75°C neutral conductor is required? Terminations the same as in c.

_____

13. For ranges of 8¾ kW or higher rating, the minimum branch-circuit rating is _____ amperes.

14. When a separate circuit supplies a counter-mounted cooking unit, what Code reference tells us that it is "against Code" to apply the demand factors of *Table 220-19*, and requires us to compute the load based upon the appliance's actual nameplate rating? _____

15. A nonmetallic-sheathed cable is used to connect a wall-mounted oven. The insulated conductors are No. 10. What is the size of the equipment grounding conductor in this cable? _____

16. Match the following statements with the correct letter. Some letters may be used more than once.

_____ A 30-ampere, 3-wire receptacle   a. for the equipment grounding conductor

_____ A 50-ampere, 4-wire receptacle   b. for the white neutral conductor

_____ The L-shaped blade on a   c. for the "hot" ungrounded conductors
30-ampere plug cap   d. NEMA 10-30R

_____ The horseshoe-shaped slot   e. NEMA 14-50R
on a 50-ampere receptacle   f. NEMA 14-50P

_____ A 50-ampere, 4-wire plug cap

_____ Terminals "X" and "Y"

_____ Terminal "W"

_____ Terminal "G"

Special-Purpose Outlets – Food Waste Disposer ▲H, Dishwasher ▲I

This Unit discusses in detail the circuit requirements for food waste disposers and dishwashers as installed in residences. However, because there are many other types of appliances found in homes, the electrician needs to read carefully *Article 422* of the Code. *Article 422* covers all types of appliances in general, branch-circuit requirements, and installation requirements. It also discusses specific types of appliances.

CAUTION: Keep in mind that these 120-volt appliances are *not* permitted to be connected to the small appliance circuits provided for in the kitchen. The small appliance circuits are intended to be used for the numerous types of portable appliances used in the kitchen area.

FOOD WASTE DISPOSER ▲H

The food waste disposer outlet is shown on the plans by the symbol ▲H. The food waste disposer, rated at 7.2 amperes, is connected to Circuit B19, a separate 20-ampere, 120-volt branch-circuit in Panel B. To meet the specifications, No. 12 THHN conductors are used. The conductors terminate in a junction box provided on the disposer.

Overload Protection

The food waste disposer is a motor-operated appliance. Food waste disposers normally are driven by a $\frac{1}{4}$- or $\frac{1}{3}$-hp, split-phase, 120-volt motor. Therefore, running overcurrent protection must be provided.

The manufacturers of food waste disposers provide the required motor overload protection with an integral thermal overload protector. Usually this is a small manual reset button on the unit. In the event of an overload caused by prolonged use of the appliance, or if something falls into the drain and jams the appliance, the thermal overload protector trips, turning off power to the appliance. After the problem has been taken care of, merely push the reset button.

For cord- and plug-connected food waste disposers, a conventional grounding-type receptacle can be installed under the sink. A fuse/receptacle

SINGLE-POLE OFF-ON SWITCH LOCATED ABOVE COUNTERTOP NEAR DISPOSER UNIT

FOOD WASTE DISPOSER

120-VOLT SUPPLY

TWO-WIRE BX, IF NMC, THEN EQUIPMENT GROUNDING CONDUCTOR REQUIRED

RUNNING OVERCURRENT PROTECTION BUILT INTO DISPOSER UNIT

Figure 21-1   Wiring for a food waste disposer operated by a separate switch located above countertop near the sink. Many local codes require this wall switch for all installations, to be sure that there is a safe means of easily disconnecting the appliance.

box cover unit could also be installed under the sink as illustrated in Figure 21-2. A dual-element, time-delay plug fuse sized at approximately 125% of the motor's full load current is inserted into the fuse holder of the box-cover unit. This fuse serves as additional back-up overload protection and short-circuit protection for the appliance.

Disconnecting Means

All electrical appliances must be provided with some means of disconnecting the appliance, *Article 422, Part (C)*. The disconnecting means may be a separate ON-OFF switch, Figure 21-1, a cord connection, Figure 21-2, or a branch-circuit switch or circuit breaker, within sight of the appliance or capable of being locked in the OFF position.

Some local codes require that food waste disposers, dishwashers, and trash compactors be cord connected, as illustrated in Figure 21-2, to make it easier to disconnect the unit, to replace it, to service it, and to reduce noise and vibration. The appliance shall be intended or identified for flexible cord connection, *Sections 400-7, 422-16(a)* and *(b)*.

U.L. lists food waste disposer with and without a cord- and plug-assembly.

FLUSH MOUNTING BOX-COVER UNIT, WITH THREE-WIRE GROUNDING RECEPTACLE AND DUAL-ELEMENT FUSE SIZED AT APPROXIMATELY 125% OF THE FULL-LOAD CURRENT RATING OF THE MOTOR. FUSE ADDS EXTRA OVERCURRENT PROTECTION TO THE OVERLOAD PROTECTION BUILT INTO THE DISPOSER. RECEPTACLE MUST BE ACCESSIBLE AND LOCATED TO AVOID DAMAGE TO CORD.

EGC

BRANCH-CIRCUIT WIRING CONCEALED IN WALL BEHIND CABINETS. WIRING FEEDS INTO FLUSH-MOUNTED SWITCH BOX UNDER COUNTERTOP IN SPACE NEAR FOOD WASTE DISPOSER.

RECEPTACLE MUST BE ACCESSIBLE.

FOOD WASTE DISPOSER

THREE-WIRE GROUNDING-TYPE ATTACHMENT PLUG CAP. PERMITTED TO SERVE AS THE DISCONNECTING MEANS *(SECTION 422-32.)* APPLIANCE MUST COME WITH CORD OR BE MARKED AS SUITABLE FOR CONNECTION.

FLEXIBLE CORD MUST BE IDENTIFIED FOR THE PURPOSE, SHALL HAVE A GROUNDING-TYPE ATTACHMENT PLUG CAP, AND MUST NOT BE LESS THAN 18 INCHES (457 mm) LONG—NOR LONGER THAN 36 INCHES (914 mm). EXAMPLES OF SUITABLE THREE-CONDUCTOR CORDS: SP-3, SPE-3, OR SPT-3. SEE *TABLE 400-4* FOR FLEXIBLE CORD TYPES AND PERMITTED USES.

Figure 21-2   Typical cord connection for a food waste disposer, *Section 422-16(a)* and *(b)*. The same rules apply to built-in dishwashers and trash compactors, except the cord length must be 3 to 4 feet (0.914 to 1.22 m).

Turning the Food Waste Disposer On and Off

Separate On-Off Switch. When a separate ON-OFF switch is used, a simple circuit arrangement can be made by running a two-wire supply cable to a switch box located next to the sink at a convenient location above the countertop. Not only is this convenient, but it positions the switch out of the reach of children, Figure 21-1. A second cable is run from the switch box to the junction box of the food waste disposer. A single-pole switch in the switch box provides ON-OFF control of the disposer.

In the kitchen of this residence, the "continuous feed" food waste disposer is controlled by a single-pole switch located to the right of the kitchen sink above the countertop, in combination with the light switch and receptacle.

Another possibility, but not as convenient, is to install the switch for the disposer inside the cabinet space directly under the sink near the food waste disposer.

Integral ON-OFF Switch. A "batch-feed" food waste disposer is equipped with an integral prewired control switch. This integral control starts and stops the disposer when the user twists the drain cover into place. An extra ON-OFF switch is not required with this type of control. To connect the disposer, the electrician runs the supply conductors directly to the junction box on the disposer and makes the proper connections, Figure 21-3.

Disposer with Flow Switch in Cold Water Line. Some manufacturers of food waste disposers recommend that a flow switch be installed in the cold water line under the sink, Figure 21-4. This switch is connected in series with the disposer motor. The

Figure 21-3 Wiring for a food waste disposer with an integral ON-OFF switch.

Figure 21-4 Disposer with a flow switch in the cold water line.

flow switch prevents the disposer from operating until a predetermined quantity of water is flowing through the disposer. The cold water helps prevent clogged drains by solidifying any grease in the disposer. Thus, the addition of a flow switch means that the disposer cannot be operated without water.

Figure 21-4 shows one method of installing a food waste disposer with a flow switch in the cold water line.

Grounding

- *Article 250, Part F* covers equipment grounding and equipment grounding conductors.

- *Article 250, Part G* covers methods of equipment grounding.

- *Section 250-118:* This section lists all of the acceptable methods of grounding equipment, such as the equipment grounding conductor in nonmetallic-sheathed cable, armored cable, or flexible metal conduit.

- *Section 250-134:* Refers back to *Section 250-118.*

- *Section 250-138:* Accepts the equipment grounding conductor in flexible cords for cord- and plug-connected appliances.

- *Section 250-142(b):* Never ground electrical appliances to the grounded branch-circuit conductor.

DISHWASHER ⒶI

The dishwasher is supplied by a separate 20-ampere circuit connected to Circuit B5. No. 12 THHN conductors are used to connect the appliance. The dishwasher has a ⅓-hp motor rated at 7.2 amperes and 120 volts. The outlet for the dishwasher is shown by the symbol Ⓐ₁. During the drying cycle of the dishwasher, a thermostatically controlled, 1000-watt electric heating element turns on.

Actual connected load		Circuit calculation	
Motor	7.20 A	$7.20 \times 1.25 = 9.00$ A	
Heater	8.33 A		8.33 A
Total	15.53 A		17.33 A

For most dishwashers, the motor does not run during the drying cycle. Thus, the actual maximum demand on the branch-circuit would only be the larger of the two loads, which in the previous example is the 1,000-watt heating element.

Water heater manufacturers ship their water heaters with the thermostat set as low as possible. When the water heater is installed, the thermostat must be set at recommended settings (such as 120°F). In fact, because of the problem of people being scalded, many municipalities and states have implemented code requirements that water heater thermostats not be set over 120°F. This temperature is acceptable for baths and showers, but 120°F is not hot enough to properly clean dishes and glasses. The manufacturers of the dishwashers provide a feature that turns on the heating element for the final rinse cycle. This raises the water temperature from 120°F to the desired 140° to 145°F.

See Unit 19 for a discussion on water heaters and the dangers of scalding with hot water of varying temperatures.

Wiring the Dishwasher

The dishwasher manufacturer supplies a terminal or junction box for the electrical hookup. The electrician must make sure the supply conductors (cable or flex conduit) are brought in at the proper location. Usually, the electrician brings the supply conductors into the space under the sink, but the supply conductors could also be brought up from underneath or from behind the dishwasher. The manufacturer's installation manual provides all of the necessary information.

The dishwasher is a motor-operated appliance and requires motor overloaded protection. Overload protection is an integral part of the dishwasher's motor, and prevents the motor from burning out if it becomes overloaded, stalled, or overheated for any reason. The branch-circuit protection for the appliance is provided by the branch-circuit fuse or circuit breaker. Supplemental overcurrent protection is often provided by installing a box-cover "switch/fuse unit" under the sink of a type similar to that shown in Figure 16-25. For a cord/plug connection, the box-cover unit would be a "receptacle/fuse" type, Figure 21-2. Time-delay, dual-element Type S fuses sized at approximately 125% of the nameplate rating of the dishwasher would be installed in the box-cover unit. This also serves as the disconnecting means required by *Section 422-30.*

Common Feed for Dishwasher and Food Waste Disposer

There may be instances where the distance from the kitchen sink area to the electrical load center is very long, in which case it may be more economical to run one branch-circuit to feed the dishwasher and food waste disposer, instead of two separate circuits. In this case, the electrician must split the circuit in a junction box mounted in the cabinet space below the food waste disposer. If a box-cover unit is used and has the proper fuse sizes, then separate overcurrent protection and separate disconnecting means are available. The connections for this circuit are similar to those for the overhead garage door opener circuit described in Figures 16-21, 16-25, and 16-26.

Both the food waste disposer and dishwasher in this residence are permitted to be supplied with one 20-ampere circuit because the circuit calculations are:

Food waste disposer
$$= 7.2 \times 1.25 \quad = \quad 9 \text{ amperes}$$

Heater in dishwasher
$$= 1000 \div 120 \quad = \quad \underline{8.33 \text{ amperes}}$$

Maximum calculated demand $= 17.33$ amperes

For higher current or wattage rated appliances, it may be necessary to install a 30-ampere branch-circuit. Do not connect the garbage disposer or dishwasher to any of the small appliance branch-circuits in the kitchen. To do so would be a violation of the Code. The small appliance circuits are intended to supply cord-connected *portable* appliances only. See Unit 12.

Grounding

The *NEC*® requirements for grounding food waste disposers and dishwashers are the same. The subject has been discussed previously in the food waste disposer text.

PORTABLE DISHWASHERS

In addition to built-in dishwashers, portable models are available. These portable units have one hose that is attached to the water faucet and a water drainage hose that hangs in the sink. The obvious location for a portable dishwasher is near the sink.

Thus, the dishwasher probably will be plugged into the receptacle nearest the sink.

Section 210-8(a)(6) requires that *all* 125-volt, single-phase, 15- and 20-ampere receptacles that serve countertop surfaces in kitchens be ground-fault circuit interrupter (GFCI) protected.

Portable dishwashers are supplied with a three-wire cord containing two circuit conductors, one equipment grounding conductor, and a three-wire grounding-type attachment plug cap. If the three-wire plug cap is plugged into the three-wire grounding-type receptacle, the dishwasher is adequately grounded.

In older homes that do not have grounding-type receptacles, the use of a portable cord- and plug-connected dishwasher will require the installation of a grounding-type receptacle(s). Assuming the appliance will be used in the kitchen, the receptacle will also have to be GFCI protected. This topic was covered in Unit 6.

CORD CONNECTION OF FIXED APPLIANCES

As previously discussed, food waste disposers, built-in dishwashers, and trash compactors in residential occupancies are permitted to be cord connected, Figure 21-2.

The Code electrical installation requirements for dishwashers and trash compactors are very similar to the requirements for food waste disposers, except the cord is required to be 3 to 4 feet (0.914 to 1.22 m) long. The receptacle outlet should be located in the under-cabinet space adjacent to the appliance, which generally would be under the sink, allowing for easy access if the appliance needs to be disconnected. It is *not* recommended that the receptacle outlet be installed behind the appliance.

Receptacle outlets installed beneath the sink for the purpose of cord connecting a dishwasher, trash compactor, or food waste disposer are not required to be of the GFCI type, because these receptacles are not intended to serve the countertop area. Repeating what was stated at the beginning of this Unit, these receptacle outlets must not be connected to the required small appliance 20-ampere branch-circuits that serve the kitchen and/or dining room areas.

Some electrical inspectors feel that using the

cord-connected method of connecting these appliances is better than direct connection (sometimes called "hard-wired") because servicing these appliances is generally done by appliance repair people.

Section 400-7 covers the uses permitted for flexible cords. Specifically, *Section 400-7(a)(8)* states that the appliance must be fastened in such a manner as to allow ready removal of the appliance for maintenance and repairs. The appliance must be intended or identified for flexible cord connection. Similar

wording on the use of flexible cord connections for appliances is covered in *Section 422-16(a)* and *(b)*.

Simply stated, these *NEC*® requirements as well as the U.L. standards mean that unless the appliance is clearly marked as "identified for flexible cord connection," a connection in the field would be a violation and would invalidate the appliance's "Listing." Some appliances are identified for both flexible and direct connection. As always, pay attention to what the labeling and instructions have to say.

REVIEW

Note: Refer to the Code or the plans where necessary.

FOOD WASTE DISPOSER CIRCUIT Ⓐₕ

1. How many amperes does the food waste disposer draw? _____

2. a. To what circuit is the food waste disposer connected? _____

   b. What size wire is used to connect the food waste disposer? _____

3. Means must be provided to disconnect the food waste disposer. The homeowner need not be involved in electrical connections when servicing the disposer if the disconnecting means is _____

   _____

4. How is running overcurrent protection provided in most food waste disposer units?

   _____

5. When running overcurrent protection is not provided by the manufacturer or if additional backup overcurrent protection is wanted, dual-element, time-delay fuses may be installed in a separate box-cover unit. These fuses are sized at not over _____ % of the full-load rating of the motor.

6. Why are flow switches sometimes installed on food waste disposers? _____

   _____

7. What Code sections relate to the grounding of appliances? _____

   _____

8. Do the plans show a wall switch for controlling the food waste disposer? _____

9. A separate circuit supplies the food waste disposer in this residence. Approximately how many feet (meters) of cable will be required to connect the disposer?

   _____

10. If a receptacle is installed underneath the sink for the purpose of plugging in a cord-connected food waste disposer, must the receptacle be GFCI protected? _____

DISHWASHER CIRCUIT ⚠,

1. a. To what circuit is the dishwasher in this residence connected? _____

   b. What size wire is used to connect the dishwasher? _____

2. The motor on the dishwasher is (circle one)

   a. ¼ hp   b. ⅓ hp   c. ½ hp

3. The heating element is rated at (circle one)

   a. 750 watts   b. 1000 watts   c. 1250 watts

4. How many amperes at 120 volts do the following heating elements draw?

   a. 750 watts _____

   b. 1000 watts _____

   c. 1250 watts _____

5. How is the dishwasher in this residence grounded? _____

   _____

   _____

   _____

   _____

6. What type of cord is used on most portable dishwashers? _____

7. How is a portable dishwasher grounded? _____

   _____

   _____

8. Who is to furnish the dishwasher? _____

   _____

9. What article of the Code specifically addresses electrical appliances? _____

Special-Purpose Outlets for the Bathroom Ceiling Heat/Vent/Lights ▲K ▲J, the Attic Fan ▲L, and the Hydromassage Tub ▲A

OBJECTIVES

After studying this Unit, the student will be able to

- explain the operation and switching sequence of the heat/vent/light.
- describe the operation of a humidistat.
- install attic exhaust fans with humidistats, both with and without a relay.
- list the various methods of controlling exhaust fans.
- understand advantages of installing exhaust fans in residences.
- understand the electrical circuit and Code requirements for hydromassage bathtubs.
- understand grounding requirements for electrical equipment.

BATHROOM CEILING HEATER CIRCUITS ▲K ▲J

Both bathrooms contain a combination heater, light, and exhaust fan installed in the ceiling. The heat/vent/light is shown in Figure 22-1. The symbols ▲K and ▲J represent the outlets for these units.

Each heat/vent/light contains a heating element similar to the surface burners of electric ranges. The appliances also have a single-shaft motor with a blower wheel, a lamp with a diffusing lens, and a means of discharging air to the outside of the dwelling.

The heat/vent/lights specified for this residence are rated 1500 watts at 120 volts. The unit in the

master bedroom bathroom ▲J is connected to Circuit A12, a 20-ampere, 120-volt circuit. In the front bedroom bathroom, the unit is connected to Circuit A11, also a 20-ampere, 120-volt circuit.

The current rating of the 1500-watt unit is

$$I = \frac{W}{E} = \frac{1500}{120} = 12.5 \text{ amperes}$$

Article 424 of the *NEC*® covers fixed electric space heating. *Section 424-3(b)* requires that the branch-circuit conductors and the overcurrent protective device (fuses or circuit breakers) shall be not less than 125% of the appliances' ampere rating. Thus, 20-ampere, 120-volt branch-circuits rather than 15-ampere branch-circuits have been chosen to supply the electric ceiling heaters in the bathroom.

Figure 22-1 Heat/vent/light. *Photo courtesy of NuTone.*

Wiring

In Figure 22-2, we see the wiring diagram for the heat/vent/light. The 120-volt line is run from Panel A to the wall box. A raceway (flexible or electrical metallic tubing) is run between the wall box where the switches are located to the heat/vent/light

NOTE: HEATER CANNOT OPERATE WITHOUT FAN, BUT FAN CAN OPERATE WITHOUT HEATER.

120-VOLT LINE

FAN

HEATER

LAMP

NIGHT LIGHT

CONNECTIONS AS PER MANUFACTURER'S RECOMMENDATIONS

Figure 22-2 Wiring for heat/vent/light.

unit. In this raceway, five wires are needed—one common white grounded circuit conductor, and one conductor each for the heating element, fan motor, 100-watt lamp, and the 7-watt night light. The four ungrounded switch leg wires may be any color except white, natural gray, or green. Color coding of conductors was discussed in detail in Unit 5. Because the circuit rating is 20 amperes, No. 12 AWG conductors are installed. See the Schedule of Special Purpose Outlets contained in the Specifications for this text.

Nontrained electricians often violate the code by running one 3-wire and one 2-wire cable between the wall box and heat/vent/light. *Section 300-3(b)* states that "all conductors of the same circuit and, where used, the neutral and all equipment grounding conductors shall be contained within the same raceway, cable tray, trench, cable, or cord."

The switch location requires a two-gang opening, with the wall box being large enough to accommodate seven No. 12 AWG conductors, four switches, and cable clamps. Box sizing was covered in Units 2 and 8.

Operation of the Heat/Vent/Light

The heat/vent/light is controlled by four switches, as shown in Figure 22-2:

- one switch for the heater
- one switch for the light
- one switch for the exhaust fan
- one switch for the night light

When the heater switch is turned on, the heating element begins to give off heat. The heat activates a bimetallic coil attached to a damper section in the housing of the unit. The heat-sensitive coil expands until the damper closes the discharge opening of the exhaust fan. Air is taken in through the outer grille of the unit. The air is blown downward over the heating element. The blower wheel circulates the heated air back into the room area.

If the heater is turned off so that only the exhaust fan is on, the air is pulled into the unit and is exhausted to the outside of the house by the blower wheel.

Exhaust fans and fixtures are available with integral automatic moisture sensors and motion detectors. Step into a dark bathroom, and the light goes on. Step into and run the shower, and the exhaust fan turns on at a predetermined moisture level.

Grounding the Heat/Vent/Light

The heat/vent/light must be grounded!

The grounding of this appliance is accomplished by using any of the accepted equipment grounding methods found in *Section 250-118*. For residential wiring, the equipment grounding conductor will be in nonmetallic-sheathed cable, armored cable, or flexible metal conduit. Never attempt to ground an appliance to the white grounded circuit conductor, *Section 250-142(b)*.

ATTIC EXHAUST FAN CIRCUIT Ⓐ L

An exhaust fan is mounted in the hall ceiling between the master bedroom and the front bedroom, Figure 22-3(A). When running, this fan removes hot, stagnant, humid, or smoky air from the dwelling. It will draw in fresh air through open windows. Such an exhaust fan can be used to lower the indoor temperature of the house as much as 10° to 20°F.

The exhaust fan in this residence is connected to Circuit A10, a 120-volt, 15-ampere circuit in the main panel utilizing No. 14 THHN wire.

Exhaust fans can also be installed in a gable of a house, Figure 22-3(B), or in the roof.

On hot days, the air in the attic may reach a temperature of 150°F or more. Thus, it is desirable to provide a means of taking this attic air to the outside. The heat from the attic can radiate through the ceiling into the living areas. As a result, the temperature in the living area is raised and the air-conditioning load increases. Personal discomfort increases as well. These problems are minimized by properly vented exhaust fans.

Figure 22-4 shows a typical ceiling-mounted exhaust fan, sometimes referred to as whole-house ventilator fans.

Sizing an Exhaust Fan

Exhaust fans are rated in CFM, which means cubic feet of air per minute of ventilating capability.

To determine the minimum amount of CFM needed, multiply the square footage of the house by 3. For example, an 1,800-square foot house would need an exhaust fan rated: $1,800 \times 3 = 5,400$ CFM.

To determine the minimum amount of attic exhaust louver square footage area, divide the CFM by 750. In the above example, $5,400 \div 750 = 7.2$ square feet.

Exhaust fan manufacturers' literature contains sizing recommendations for their particular product.

Fan Operation

Many sizes and types of attic exhaust fans are available. The attic exhaust fan in this residence has a two-speed, ¼-horsepower direct-drive, permanent split capacitor motor. Direct-drive means that the fan blade is attached directly to the shaft of the motor. It is rated 120 volts, 60 Hz (60 cycles per second), 5.8 amperes, 696 VA, and operates at a maximum speed of 1050 rpm. The motor is protected against overload by

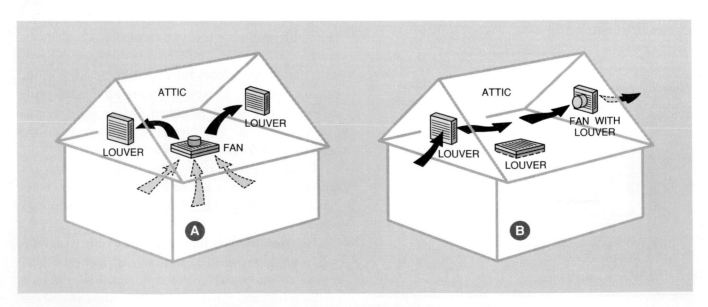

Figure 22-3 Exhaust fan installation in an attic.

Figure 22-4 Ceiling exhaust fan as viewed from the attic. *Courtesy* NuTone.

internal thermal protection. The motor is mounted on rubber cushions for quiet operation.

When the fan is not running, the louver remains in the CLOSED position. When the fan is turned on, the louver opens. When the fan is turned off, the louver automatically closes to prevent the escape of air from the living quarters.

Fan Control

Typical residential exhaust fans can be controlled by a variety of switches.

Some electricians, architects, and home designers prefer to have fan switches (controls), Figures 22-5 and 22-6, mounted 6 feet (1.83 m) above the floor so that they will not be confused with other

Figure 22-5 Speed control switches. *Courtesy* NuTone.

Figure 22-6 Timer switch. *Courtesy* NuTone.

wall switches. This is a matter of personal preference and should be verified with the owner.

Figure 22-7 shows some of the ways that an exhaust fan can be connected.

1. A simple ON-OFF switch, Figure 22-7(A).

2. A speed control switch that allows multiple and/or infinite number of speeds, Figure 22-7(B). (The exhaust fan in this residence is controlled by this type of control.) See also Figure 22-5.

3. A timer switch that allows the user to select how long the exhaust fan is to run, up to 12 hours, Figure 22-7(C). See also Figure 22-6.

4. A humidity control switch that senses moisture buildups, Figure 22-7(D). See "Humidity Control" later in this Unit for further details on this type of switch.

5. Exhaust fans mounted into end gables or roofs of residences are available with an adjustable temperature control on the frame of the fan. This thermostat generally has a start range of 70° to 130°F, and will automatically stop at a temperature 10° below the start setting, thus providing totally automatic ON-OFF control of the exhaust fan, Figure 22-7(E).

6. A high temperature automatic heat sensor that will shut off the fan motor when the temperature reaches 200°F. This is a safety feature so that the fan will not spread a fire. Connect in series with other control switches, Figure 22-7(F).

7. A combination of controls, Figure 22-7(G). This circuit shows an exhaust fan controlled by an infinite-speed switch, a humidity control, and a high temperature heat sensor. Note that the speed control and the humidity control are connected in parallel so that either can start the fan. The high temperature heat sensor is in series so that it will shut off the power to the fan when it senses 200°F, even if the speed control or humidity control are in the ON position.

Overload Protection for the Fan Motor

Several methods can be used to provide running overload protection for the ¼-hp attic fan motor. For

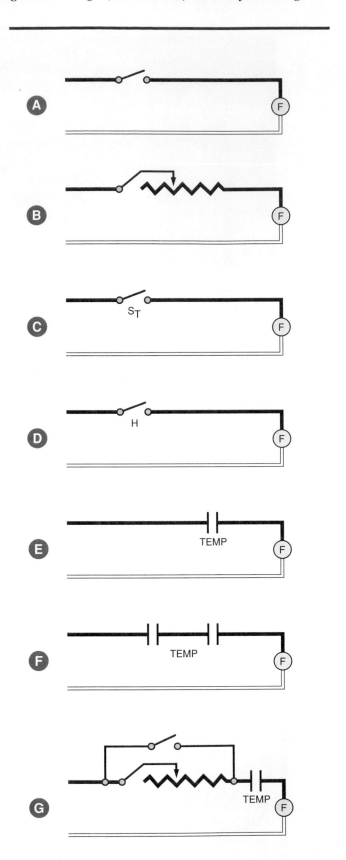

Figure 22-7 Different options for the control of exhaust fans.

example, an overload device can be built into the motor or a combination overload tripping device and thermal element can be added to the switch assembly. Several manufacturers of motor controls also make switches with overcurrent devices. Such a switch may be installed in any standard flush switch box opening (2" × 3"). A pilot light may be used to indicate that the fan is running. Thus a two-gang switch box or 1½" × 4" square outlet box with a two-gang raised plaster cover may be used for both the switch and the pilot light.

Section 430-32(c) states that for automatically started motors, the running overload device must not exceed 125% of the full-load running current of the motor. *Table 430-148* shows that the ¼-hp exhaust fan motor has a full-load current rating of 5.8 amperes. Thus, the overload device rating is 5.8 × 1.25 = 7.25 amperes.

Branch-Circuit Short-Circuit and Ground-Fault Protection for the Fan Motor

The attic exhaust fan is connected to Circuit A10, which is a 15-ampere, 120-volt circuit. This branch-circuit supplies the attic exhaust fan only. It supplies no other loads.

The fan motor is equipped with built-in thermal overload protection, and has a full-load current rating of 5.8 amperes at 115 volts. We have already discussed the subject of motor branch-circuit short-circuit and ground-fault protection in Unit 19. Here are the basics.

Section 430-32(a)(1): 5.8 × 1.25 = 7.25 amperes for overload protection.
(7½-ampere, dual-element, time-delay fuses are available.)

Table 430-152: 5.8 × 1.75 = 11+ ampere dual-element, time-delay fuses. (12 ampere dual-element, time-delay fuses are available.)

Section 430-52(c)(1), Exception No. 1: Next standard size is 15 amperes.
(12-ampere time-delay, dual-element fuses are available.)

Section 430-52(c)(1), Exception No. 2: The maximum size must not exceed 225%.

5.8 × 2.25 = 13+ amperes maximum. This puts us back to the 12-ampere, dual-element, time-delay fuses.

A switch/fuse box-cover unit can be mounted near the attic exhaust fan.

Since the motor has built-in thermal overload protection, installing either a 7½-, 8-, or 10-ampere, time-delay, dual-element fuse would be a good choice. Should the motor short-circuit or go to ground, the fuse would open without tripping the branch-circuit breaker back in the Panel A.

HUMIDITY CONTROL

An electrically heated dwelling can experience a problem with excess humidity due to the "tightness" of the house, because of the care taken in the installation of the insulation and vapor barriers. High humidity is uncomfortable. It promotes the growth of mold and the deterioration of fabrics and floor coverings. In addition, the framing members, wall panels, and plaster or drywall of a dwelling may deteriorate because of the humidity. Insulation must be kept dry or its efficiency decreases. A low humidity level can be maintained by automatically controlling the exhaust fan. One type of automatic control is the *humidistat*. This device starts the exhaust fan when the relative humidity reaches a high level. The fan exhausts air until the relative humidity drops to a comfortable level. The comfort level is about 50% relative humidity. Adjustable settings are from 0 to 90% relative humidity.

The electrician must check the maximum current and voltage ratings of the humidistat before the device is installed. Some humidistats are low-voltage devices and require a relay. Other humidistats are rated at line voltage and can be used to switch the motor directly (a relay is not required). However, a relay must be installed on the line-voltage humidistat if the connected load exceeds the maximum allowable current rating of the humidistat.

The switching mechanism of a humidistat is controlled by a nylon element, Figure 22-8. This element is very sensitive to changes in humidity. A bimetallic element cannot be used because it reacts to temperature changes only.

Figure 22-8 Details of a humidity control used with an exhaust fan.

Wiring

The humidistat and relay in the dwelling are installed using two-wire No. 14 AWG cable. The cable runs from Circuit A10 to a 4-inch square, 1½-inch-deep outlet box located in the attic near the fan, Figure 22-9(B). A box-cover unit is mounted on this outlet box. The box-cover unit serves as a disconnecting means within sight of the motor as required by *Sections 422-30* and *430-102*.

Mounted next to the box-cover unit is a relay that can carry the full-load current of the motor. Low-voltage thermostat wire runs from this relay to the humidistat. The humidistat is located in the hall between the bedrooms. The thermostat cable is a two-conductor cable because the humidistat has a

Figure 22-9 Wiring for the humidistat control.

single-pole, single-throw switching action. The line-voltage side of the relay provides single-pole switching to the motor. Normally the white grounded conductor is not broken.

Although the example in this residence utilizes a relay in order to use low-voltage wiring, it is certainly proper to install line-voltage wiring and use line-voltage thermostats and humidity controls. These are becoming more and more popular, with the switches actually being solid-state speed controllers permitting operation of the fan at an infinite number of speeds between high and low settings.

Grounding the Fan

The metal frame and motor must be grounded. It is possible that this appliance could be within reach of metal ducts in the attic.

Equipment grounding is accomplished by using any of the accepted equipment grounding methods found in *Section 250-118*. For residential wiring, the equipment grounding conductor will be in non-metallic-sheathed cable, armored cable, or flexible metal conduit. Never attempt to ground an appliance to the white grounded circuit conductor, *Section 250-142(b)*.

APPLIANCE DISCONNECTING MEANS

To clean, adjust, maintain, or repair an appliance, it must be disconnected to prevent personal injury. *Sections 422-30* through *422-35* outline the basic methods for disconnecting fixed, portable, and stationary electrical appliances. Note that each appliance in the dwelling conforms to one or more of the disconnecting methods listed. The more important Code rules for appliance disconnects are as follows:

- each appliance must have a disconnecting means.

- the disconnecting means may be a separate disconnect switch if the appliance is permanently connected.

- the disconnecting means may be the branch-circuit switch or circuit breaker if the appliance is permanently connected and not over ⅛ hp or 300 volt-amperes.

- the disconnecting means may be an attachment plug cap if the appliance is cord connected.

- the disconnecting means may be the unit switch on the appliance only if other means for disconnection are also provided. In a single-family residence, the service disconnect serves as the other means.

- the disconnect must have a positive ON-OFF position.

- the disconnect must be capable of being locked in the OFF position or be within sight of a motor-driven appliance where the motor is more than ⅛ hp.

Read *Sections 422-30* through *422-35* for the details of appliance disconnect requirements.

HYDROMASSAGE TUB CIRCUIT ▲A

The master bedroom is equipped with a hydromassage tub. A hydromassage bathtub is sometimes referred to as a whirlpool bath.

Section 680-4 defines a hydromassage bathtub as "a permanently installed bathtub equipped with a recirculating piping system, pump and associated equipment. It is designed so that it can accept, circulate, and discharge water upon each use."

In other words, fill—use—drain.

The significant difference between a hydromassage bathtub and a regular bathtub is the recirculating piping system and the electric pump that circulates the water. Both types of tubs are drained completely after each use.

Spas and hot tubs are intended to be filled, then used. They are *not* drained after each use because they have a filtering and heating system.

Another basic difference between a spa and a hot tub is that spas are constructed of manufactured material such as fiberglass, acrylics, plastics, or concrete, whereas hot tubs are made of wood, such as redwood, cypress, cedar, oak, or teak.

Electrical Connections

The hydromassage tub is fed with a separate Circuit A9, a 15-ampere, 120-volt circuit using No. 14 AWG THHN conductors.

The Schedule of Special-Purpose Outlets indicates that the hydromassage bathtub in this residence has a ½-hp motor that draws 10 amperes.

Section 680-70 requires ground fault circuit interrupter (GFCI) protection for hydromassage tubs and all their associated electrical components. *Section 680-70* also requires GFCI protection for all 125-volt, single-phase receptacles located within 5 feet (1.52 m) of the inside wall of a hydromassage tub.

Section 680-71 tells us that lighting fixtures, switches, receptacles, and other electrical equipment located in the same room, and not directly associated with the hydromassage bathtub, shall be installed in accordance with the requirements of *Chapters 1 through 4* of the Code. This section recognizes that a hydromassage tub is used in much the same manner as a regular bathtub, and as such, does not introduce any additional hazards other than those of a regular bathtub.

Spas, hot tubs, and swimming pools do introduce additional electrical shock hazards, and these issues are covered in Unit 30.

The hydromassage bathtub electrical control is prewired by the manufacturer. All the electrician needs to do is run the separate 15-ampere, 120-volt, GFCI-protected circuit to the end of the tub where the pump and control are located. Generally the manufacturer will supply a 3-foot (914 mm) length of watertight flexible conduit that contains one black, one white, and one green equipment grounding conductor, Figure 22-10. Make sure that the equipment grounding conductor of the circuit is properly connected to the green grounding conductor of the hydromassage tub.

Do not connect the green grounding conductor to the grounded (white) circuit conductor!

Proper electrical connections are made in the junction box where the branch-circuit wiring is to be connected to the hydromassage wires. This junction box, because it contains splices, must be accessible.

The pump and power panel may also need servicing. Access may be from underneath or the end, whichever is convenient for the installation, Figure 22-11.

Figure 22-12 is a photograph of a typical hydromassage tub.

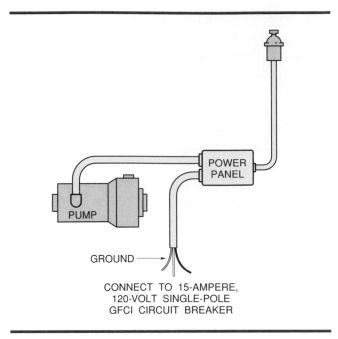

CONNECT TO 15-AMPERE, 120-VOLT SINGLE-POLE GFCI CIRCUIT BREAKER

Figure 22-10 Typical wiring diagram of a hydromassage tub showing the motor, power panel, and electrical supply leads.

Figure 22-11 The basic roughing-in of a hydromassage bathtub is similar to that of a regular bathtub. The electrician runs a separate 15-ampere, 120-volt GFCI-protected circuit to the area where the pump and control are located. Check manufacturer's specifications for this data. An access panel from the end or from below is necessary to service the wiring, the pump, and the power panel.

Figure 22-12 A typical hydromassage bathtub, sometimes referred to as a whirlpool. *Courtesy* **of the Kohler Company.**

REVIEW

Note: Refer to the Code or the plans where necessary.

BATHROOM CEILING HEATER CIRCUITS Ⓐₖ Ⓐⱼ

1. What is the wattage rating of the heat/vent/light? _____

2. To what circuits are the heat/vent/lights connected? _____

3. Why is a 4-inch square, 2⅛-inch deep box with a two-gang raised plaster ring or similar deep box used for the switch assembly for the heat/vent/light? _____

   _____

   _____

   _____

   _____

4. a. How many wires are required to connect the wall switches and the heat/vent/light?

   _____

   b. What size wires are used? _____

5. Can the heating element be energized when the fan is not operating? _____

6. Can the fan be turned on without the heating element? _____

7. What device can be used to provide automatic control of the heating element and the fan of the heat/vent/light? _____

8. Where does the air enter the heat/vent/light? _____

9. Where does the air leave this unit? _____

10. Who is to furnish the heat/vent/light? _____

   _____

11. For a ceiling heater rated 1,200 watts at 120 volts, what is the current draw? _____

   _____

12. An electrician wired up a heat/vent/light as described in this Unit. The wiring diagram showed that the electrician needed to run five conductors between the switches and the heat/vent/light. The electrician used one 12/3 nonmetallic-sheathed cable and one 12/2 nonmetallic-sheathed cable. Both cables had equipment grounding conductors. During the rough-in inspection, the electrical inspector cited "Noncompliance with *Section 300-3(b)*." What was the problem? _____

   _____

   _____

ATTIC EXHAUST FAN CIRCUIT ⓐ L

1. What is the purpose of the attic exhaust fan? _____

   _____

   _____

2. At what voltage does the fan operate? _____

3. What is the horsepower rating of the fan motor? _____

4. Is the fan direct or belt driven? _____

5. How is the fan controlled? _____

6. What is the setting of the running overcurrent device? _____

7. What is the rating of the running overcurrent protection if the motor is rated at 10 amperes? _____

8. What is the basic difference between a thermostat and a humidistat? _____

   _____

   _____

9. What size conductors are to be used for this circuit? _____

10. How many feet (meters) of cable are required to complete the wiring for the attic exhaust fan circuit? _____

11. May the metal frame of the fan be grounded to the grounded circuit conductor?

_____

12. What section of the Code prohibits grounding equipment to a grounding circuit conductor? _____

HYDROMASSAGE BATHTUB CIRCUIT Ⓐ A

1. What circuit supplies the hydromassage bathtub? _____

2. Which of the following statements "meets Code"?

   a. Hydromassage tubs and all of their electrical components shall be GFCI protected.

   b. Hydromassage tubs and all of their electrical components shall not be GFCI protected.

   Circle the correct statement.

3. What conductor size feeds the hydromassage tub? _____

4. What is the fundamental difference between a hydromassage bathtub and a spa?

   _____

   _____

5. What sections of the Code reference hydromassage bathtubs? _____

6. Must the metal parts of the pump and power panel of the hydromassage tub be grounded? _____

   _____

   _____

   _____

7. Is it permissible to connect the hydromassage tub's green equipment grounding conductor to the branch-circuit's grounded (white) conductor? _____

8. All 120-volt, single-phase receptacles within (5) (10) (15) feet of a hydromassage tub must be GFCI protected. Circle correct answer.

UNIT 23

Special-Purpose Outlets– Electric Heating ⒜M, Air Conditioning ⒜N

OBJECTIVES

After studying this Unit, the student will be able to

- list the advantages of electric heating.
- describe the components and operation of electric heating systems (baseboard, cable, furnace).
- describe thermostat control systems for electric heating units.
- install electric heaters with appropriate temperature control according to *NEC®* rules.
- discuss air conditioners, heat pumps, terminology, Code requirements, and electrical connections.
- explain how heating and cooling may be connected to the same circuit.
- understand where to mount the disconnect switch outdoors for air conditioners and heat pumps.

There are many types of electric heat available for heating homes (e.g., heating cable, unit heaters, boilers, electric furnaces, duct heaters, baseboard heaters, heat pumps, and radiant heating panels). This Unit discusses many of these types. The residence discussed in this text is heated by an electric furnace located in the workshop.

Detailed Code requirements are found in *Article 424*.

This text cannot cover in detail the methods used to calculate heat loss and the wattage required to provide a comfortable level of heat in the building. For this residence, the total estimated wattage is 13,000 watts. Depending upon the location of the residence (in the Northeast, Midwest, or South, for example), the heating load will vary.

Electric heating has gained wide acceptance when compared with other types of heating systems. It has a number of advantages. Electric heating is flexible when baseboard heating is used because each room can have its own thermostat. Thus, one room can be kept cool while an adjoining room is warm. This type of zone control for an electric, gas- or oil-fired central heating system is more complex and more expensive.

Electric heating is safer than heating with fuels. The system does not require storage space, tanks, or chimneys. Electric heating is quiet. Electric heat does not add or remove anything from the air. As a result, electric heat is cleaner. This type of heating is considered to be healthier than fuel heating systems that remove oxygen from the air. The only moving part of an electric baseboard heating system is the thermostat. This means that there is a minimum of maintenance.

If an electric heating system is used, adequate insulation must be provided. Proper insulation can keep electric bills to a minimum. Insulation also helps to keep the residence cool during the hot summer months. The cost of extra insulation is offset through the years by the decreased burden on the air-conditioning equipment. Energy conservation measures require the installation of proper and adequate insulation.

TYPES OF ELECTRIC HEATING SYSTEMS

Electric heating units are available in baseboard, wall-mounted, and floor-mounted styles. These units may or may not have built-in thermal overload protection. The type of unit to be installed depends on structural conditions and the purpose for which the room is to be used.

Another method of providing electric heating is to embed resistance-type cables in the plaster, or between two layers of drywall on the ceiling, *Sections 424-34* through *424-45*. Because premise wiring (nonmetallic-sheathed cable, etc.) will be located above these ceilings and will be subjected to the heat created by the heating cables, the Code requires the following:

- keep the wiring not less than 2 inches (50.8 mm) above the ceiling, and *also* reduce the ampacity of the wiring to the 50°C correction factors found at the bottom of *Table 310-16*, or

- keep the wiring above insulation that is at least 2 inches (50.8 mm) thick—in which case the correction factors do not have to be applied.

For ease in identification, *Section 424-35* requires that the leads on heating cables be color coded as follows:

- 120 volts   yellow
- 208 volts   blue
- 240 volts   red
- 277 volts   brown
- 480 volts   orange

Electric heat can also be supplied by an electric furnace and a duct system similar to the type used on conventional hot-air central heating systems. The heat is supplied by electric heating elements rather than the burning of fuel. Air conditioning, humidity control, air circulation, electronic air cleaning, and zone control can be provided on an electric furnace system.

CIRCUIT REQUIREMENTS FOR ELECTRIC FURNACES

In an electric furnace, the source of heat is the resistance heating element(s). The blower motor assembly, filter section, condensate coil, refrigerant line connections, fan motor speed control, high temperature limit controls, relays, and similar components are much the same as those found in gas furnaces.

You probably have noticed that there is a repetition of *NEC®* requirements for various appliances. As with all appliances, many of the *NEC®* requirements have already been met by virtue of the product being U.L. "Listed." We might look at this in this manner: do the premise wiring according to the *NEC®* requirements and let the U.L. requirements take care of the safety requirements of the appliance itself. Your big concern when hooking up an electric furnace is to size the conductors, the branch-circuit overcurrent protection, and the disconnecting means properly.

Rather than repeating the Code requirements in detail, here is a summary of the basic code rules:

- *Section 220-15:* When making the calculation for sizing a service or a feeder, include 100% of the rating of the total connected fixed space heating load. This is shown in Unit 29.

- *Article 250, Part F:* This part of the *NEC®* contains most of the requirements for which equipment and appliances are required to be grounded.

- *Article 250, Part G:* This part of the *NEC®* lists the acceptable methods for equipment and appliance grounding. This is generally the equipment grounding conductor in nonmetallic-sheathed cable, the metal armor and bonding strip in armored cable, or a metal raceway such as electrical metallic tubing or flexible metal conduit.

- *Section 422-10(a):* The branch-circuit rating must not be less than the marked rating. This governs the overcurrent protection and conductor sizing issues.

- *Section 422-11:* The branch-circuit overcurrent protection must be sized according to the value indicated on the nameplate. The electrician in the field need do nothing more than size the conductors and overcurrent protection according to the nameplate data and instructions furnished with the furnace. The manufacturer of the furnace has done all of the calculations according to U.L. standards to determine the actual current draw of the appliance.

  The motor has built-in thermal overload protection. The overcurrent protection for other internal components is provided by the manufacturer. This is why it is important to purchase a U.L. "Listed" electric furnace.

- *Section 422-12:* Central heating equipment must be supplied by a separate branch-circuit. This includes electric, gas, and oil furnaces. It also includes heat pumps. Associated equipment such as humidifiers, electrostatic air cleaners, and so on, are permitted to be connected to the same branch-circuit. The logic behind this "separate circuit" requirement is that if the heating system were to be connected to some other branch-circuit, such as a lighting branch-circuit, a fault on that circuit would shut off the power to the heating system. This could cause freezing of water pipes.

- *Section 422-30:* A disconnecting means must be provided.

- *Section 422-31(b):* The disconnecting means must be within sight of the furnace. For electric furnaces, the disconnect might be mounted on the side of the furnace, or on an adjacent wall. For gas furnaces, the disconnect is simply a snap switch or switch/fuse mounted in a handy box on the side of the furnace. In some instances, the manufacturer of the furnace incorporates the disconnecting means right into the equipment.

- *Section 422-35:* Another reminder is that the disconnect for motor-driven appliances having a motor of greater than $\frac{1}{8}$ horsepower must be within sight of the appliance. This section refers us to *Section 430-102.*

- *Section 422-62(a):* The nameplate shall specify the minimum size supply circuit conductor's ampacity and the maximum branch-circuit overcurrent protection.

- *Section 430-102(a):* The disconnect must be within sight of the controller and shall disconnect the controller.

- *Section 430-102(b):* The disconnect shall be within sight of the motor or be capable of being locked in the "OFF" position.

- *Section 422-60(a):* The nameplate must show volts and watts, volts and amperes, or volts and kilowatts.

Figure 23-1 illustrates the key *NEC®* requirements for hooking up an electric furnace.

Supply Voltage

Because proper supply voltage is critical to ensure full output of the furnace's heating elements, voltage must be maintained at not less than 98% of the furnace's rated voltage. The effect of voltage drop has already been discussed in Units 4 and 19. Mathematically, for every 1% drop in voltage, there will be a 2% drop in wattage output of a resistance heating element. Manufacturers' installation instructions usually have tables that show the wattage or kilowatt output at different voltages. Here is a typical table:

KILOWATT RATING	
@ 240 volts	@ 208 volts
12.0	9.0
15.0	11.3
20.0	15.0
25.0	18.8
30.0	22.5

Therefore, the actual voltage at the supply terminals of an electric furnace or electric baseboard heating unit will determine the true wattage output of the heating elements.

Cord- and Plug-Connection Not Permitted

Sections 400-7(a)(8) and *422-16(a)* describe the permitted uses for flexible cords. A key requirement, oftentimes violated, is that flexible cords shall be used only *"where the fastening means and mechanical connections are specifically designed to permit ready removal for maintenance and repair, and the*

THERMOSTAT

FIELD WIRING OF LOW-VOLTAGE CLASS 2 CONTROL CIRCUIT CONDUCTORS SHALL NOT BE PLACED IN SAME RACEWAY, BOX, OR ENCLOSURE WITH POWER CONDUCTORS EXCEPT WHERE INTRODUCED SOLELY TO CONNECT THE EQUIPMENT, *SECTION 725-54(a)(1)* AND *EXCEPTION NO. 2.*

TYPICAL ELECTRIC FURNACE

DISCONNECTING MEANS:
• SHALL HAVE AMPERE RATING *NOT LESS* THAN 125% OF THE TOTAL LOAD OF THE MOTOR AND HEATER. SEE NAMEPLATE ON FURNACE.
• MUST DISCONNECT THE EQUIPMENT FROM ALL UNGROUNDED CONDUCTORS.
• MUST BE WITHIN SIGHT OF FURNACE OR MUST BE CAPABLE OF BEING LOCKED IN OFF POSITION.
• SHALL INDICATE "ON-OFF."
• FUSES MUST BE INSTALLED UNLESS THE FURNACE NAME-PLATE STATES THAT AN HACR BREAKER IS PERMITTED.
• FUSES SHALL BE SIZED AT *NOT LESS* THAN 125% OF THE TOTAL LOAD OF THE MOTOR AND HEATERS. SEE NAMEPLATE ON FURNACE.

MUST BE SEPARATE CIRCUIT, *SECTION 422-12.*

BRANCH-CIRCUIT CONDUCTORS:
• SHALL BE SIZED AT *NOT LESS* THAN 125% OF THE TOTAL LOAD OF THE MOTORS AND HEATERS. SEE NAMEPLATE ON FURNACE.

NAMEPLATE WILL SHOW MANUFACTURER'S NAME, VOLTS AND AMPERES, OR VOLTS AND WATTS, OR VOLTS AND KILOWATTS.

FURNACE MAY ALSO CONTAIN SUPPLEMENTAL OVERCURRENT PROTECTION SO THAT NONE OF THE RESISTANCE-TYPE HEATING ELEMENTS ARE FUSED AT NOT OVER 60 AMPERES. THE FURNACE'S RESISTANCE-TYPE HEATING ELEMENTS MUST BE SUBDIVIDED INTO LOADS NOT TO EXCEED 48 AMPERES, *SECTION 422-11(f).* U.L. LISTED FURNACES CONFORM TO THIS REQUIREMENT.

EXAMPLE: WHAT SIZE COPPER CONDUCTORS (THHN), FUSES, AND DISCONNECT SWITCH ARE REQUIRED FOR A FURNACE MARKED 79 AMPERES, 240 VOLT, SINGLE PHASE, 60 CYCLES? TERMINALS ON FURNACE AND SWITCH MARKED 75°C. TO SELECT THE PROPER AMPACITY OF THE CONDUCTORS IN ACCORDANCE WITH *SECTION 110-14(c),* BE SURE TO USE THE 75°C AMPACITY COLUMN IN *TABLE 310-16.*

ANSWER: CONDUCTOR SIZE: 79 x 1.25 = 98.75 AMPERES
FROM *TABLE 310-16,* SELECT NO. 3 THHN (100 AMPERES AT 75°C.)

FUSE SIZE: 79 x 1.25 = 98.75 AMPERES
INSTALL 100-AMPERE FUSES.

SWITCH: 100-AMPERE SWITCH

Figure 23-1   Basic Code requirements for an electric furnace. These "package" units have all of the internal components, such as limit switches, relays, motors, and contactors prewired. The electrician generally need only provide the branch-circuit supply and disconnecting means. See *Article 424* for additional Code information.

appliance is intended or identified for flexible cord connection." Certainly the gas piping and the size of a gas or electric furnace do not allow for "ready removal" of the furnace. Furthermore, the conductors in flexible cords cannot withstand the high temperature encountered on the terminals of a furnace.

ANSI Standard Z21.47 does not allow cord- and plug-connection for gas furnaces.

CONTROL OF ELECTRIC BASEBOARD HEATING UNITS

Line-voltage and low-voltage thermostats can be used to control electric baseboard heating units, Figure 23-2. When the current rating or ampere rating of the load exceeds the rating of a line-voltage thermostat, then it is necessary to utilize a relay in conjunction with the low-voltage thermostat. This is

Figure 23-2 **Thermostats for electric heating systems. A thermostat or other switching device that is marked with an OFF position must open all ungrounded conductors of the circuit when turned to the OFF position. Typical mounting height is 52 inches (1.52 m) to center.** *Courtesy* **of Honeywell, Inc.**

illustrated in Figure 23-3. Color-coding requirements are explained later in this unit, and was covered in detail in Unit 5.

Common ratings for line-voltage thermostats are 2,500 watts, 3,000 watts, and 5,000 watts. Other ratings are available. You might be working with watts, with amperes, or both. For example, if a thermostat has a 2,500-watt rating at 120 volts and 5,000 watts at 240 volts, its current rating is:

$$I = \frac{W}{E} = \frac{2,500}{120} = 20.8 \text{ amperes}$$

—or—

$$I = \frac{5,000}{240} = 20.8 \text{ amperes}$$

The total connected load for this thermostat must not exceed 20.8 amperes.

CIRCUIT REQUIREMENTS FOR BASEBOARD ELECTRIC HEATING UNITS

Figure 23-3 shows an electric baseboard heater controlled by a low-voltage thermostat. The low-voltage contacts of the relay are connected to the thermostat. The line-voltage contacts of the relay are used to switch the actual heater load. Relays are used whenever the load exceeds the ampere or wattage limitations of a line-voltage thermostat. Color coding for conductors was explained in Unit 5 and is also discussed later in this Unit.

Figure 23-4 shows typical baseboard electric heating units. The heater to the right has an integral thermostat on it. Baseboard electric heating units are available in both 240-volt and 120-volt ratings and in a variety of wattage ratings.

The wiring for an individual baseboard heating

Figure 23-3 **Wiring for baseboard electric heating unit having a low-voltage thermostat and relay.** *Section 200-7* **requires that a white wire used as an ungrounded conductor must be permanently reidentified at points where connections are made or where it is visible and accessible. This becomes necessary when nonmetallic-sheathed cable or armored cable is the wiring method. Reidentification can be accomplished with black paint or black tape. Color coding and reidentification of conductors was discussed in Unit 5 and later in this Unit.**

unit or group of units is shown in Figures 23-5 and 23-6. A two-wire cable (armored cable or non-metallic-sheathed cable with ground) would be run from a 240-volt, two-pole circuit in the main panel to the outlet box or switch box installed at the thermostat location. A second two-wire cable runs from the thermostat to the junction box on the heater unit. The proper connections are made in this junction box. Most heating unit manufacturers provide knockouts at the rear and on the bottom of the junction box. The supply conductors can be run through

these knockouts. Most baseboard units also have a channel or wiring space running the full length of the unit, usually at the bottom. When two or more heating units are joined together, the conductors are run in this wiring channel. Most manufacturers indicate the type of wire required for these units because of conductor temperature limitations.

Supply conductors generally must be rated 90°C (194°F). This would be Type THHN conductors, or nonmetallic-sheathed cable or armored cable that has 90°C (194°F) conductors. The nameplate and/or

Figure 23-4   Baseboard electric heating systems.

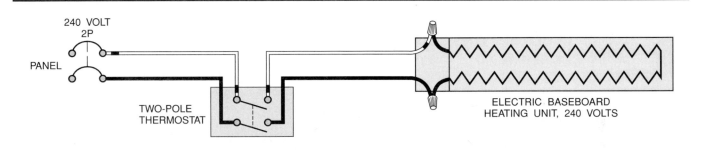

Figure 23-5   Wiring for a single baseboard electric heating unit. The thermostat is a line-voltage thermostat. See Figure 23-3 caption for color-coding requirements.

Figure 23-6   Wiring for two electric baseboard heating units. Note the circuit connections at the thermostat location. The thermostat is a line-voltage thermostat. See Figure 23-3 caption for color-coding requirements.

instructions furnished with the baseboard electric heating unit provides this information.

A variety of fittings such as internal and external elbows (for turning corners) and blank sections are available from the manufacturer.

Most wall and baseboard heating units are available with built-in thermostats, Figure 23-4. The supply cable for such a unit runs from the main panel to the junction box on the unit.

Some heaters have receptacle outlets. Underwriters Laboratories (U.L.) states that when receptacle sections are included with the other components of baseboard heating systems, they must be supplied separately using conventional wiring methods.

Manufacturers of electric baseboard heaters list "blank" sections ranging in length from 2 to 10 feet (0.6 to 3.05 m). These blank sections give the installer the flexibility needed to spread out the heater sections, and to install these blanks where wall receptacle outlets are encountered, Figure 23-7.

LOCATION OF ELECTRIC BASEBOARD HEATERS IN RELATION TO RECEPTACLE OUTLETS

Listed electric baseboard heaters shall not be installed below wall receptacle outlets unless the instructions furnished with the baseboard heaters indicate that they may be installed below receptacle outlets. See the *Fine Print Note* to *Section 210-52*. The reasoning for generally not allowing electric baseboard heaters to be installed below receptacle outlets is the possible fire and shock hazard resulting when a cord hangs over and touches the heated electric baseboard unit. The insulation on the cord can melt from the heat.

Electricians must pay close attention to the location of receptacle outlets and the electric baseboard heaters. They must study the plans and specifications carefully.

They may have to fine-tune the location of the receptacle outlets and the electric baseboard heaters, and/or install blank spacer sections, keeping in mind the Code requirements for spacing receptacle outlets and just how much (wattage and length) baseboard heating is required, Figures 23-8 and 23-9.

The Code does permit factory-installed or

Figure 23-7 Use of blank baseboard heating sections.

Figure 23-8 *Code violation.* Unless the instructions furnished with the electric baseboard heater specifically state that the unit is listed for installation below a receptacle outlet, then this installation is a violation of *Section 110-3(b)*. In this type of installation, cords attached to the receptacle outlet would hang over the heater, creating a fire hazard, and possible shock hazard should the insulation on the cord melt. See *Exception* and *Fine Print Note* to *Section 210-52*.

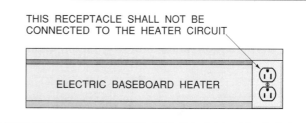

Figure 23-9 Factory-installed receptacle outlets or receptacle outlet assemblies provided by the manufacturer for use with its electric baseboard heaters may be counted as the required receptacle outlet for the space occupied by a permanently installed heater. See the first paragraph of *Section 210-52*.

factory-furnished receptacle outlet assemblies as part of permanently installed electric baseboard heaters. These receptacle outlets shall not be connected to the heater circuit, *Section 210-52*, Figure 23-10.

A POSITION ELECTRICAL BASEBOARD HEATING UNITS SO THEY WILL *NOT* BE DIRECTLY BELOW A WALL RECEPTACLE OUTLET.

B IF INSTALLED AS SHOWN, ELECTRICAL CORDS COULD COME IN CONTACT WITH THE BASEBOARD UNIT, SUBJECTING THIS CORD TO RUBBING (ABRASION) AND HEAT, WHICH MIGHT RESULT IN FAILURE OF THE INSULATION OF THE CORD, A POTENTIAL FIRE AND SHOCK HAZARD.

BASEBOARD HEATER

BASEBOARD HEATER

RECEPTACLE OUTLET

C AN EXAMPLE OF HOW THE RECEPTACLE OUTLETS IN THE WINDOW CORNER OF BEDROOM #1 MIGHT BE INSTALLED.

Figure 23-10   Position of electric baseboard heaters.

HEAT PUMPS

A heat pump is simply an air conditioner operating in reverse. A "reversing valve" inside the unit changes the direction of the flow of the system's refrigerant.

A "split" heat pump system consists of an outdoor unit with a hermetic motor compressor, a fan, and a coil. The indoor unit consists of a coil inside of the furnace. The movement of air inside the home is accomplished by the blower fan on the furnace. The "split system" is the most common type for residential use.

When all of the components are housed in one outdoor unit, the system is referred to as a "packaged system."

The hermetic motor compressor is the heart of the system. The refrigerant circulates between the indoor and outdoor units, through the coils, tubing, and compressor, absorbing and releasing heat as it travels through the system.

In the winter, the heat pump (the outdoor coil) extracts heat from the outdoor air and distributes it through the warm air ducts in the home. In extremely cold climates, a heat pump is supplemented by electric heating elements in the air handler.

In the summer, the process reverses. The heat pump (indoor coil) absorbs heat and condenses humidity from the indoor air. The heat is transferred to the outside by the refrigerant, and the humidity (condensation) is removed by piping it to a suitable drain.

Heat pump water heaters were discussed in Unit 19.

Because a heat pump contains a hermetic motor compressor, the Code requirements found in *Article 440* apply. These are discussed later in this unit.

GROUNDING

Grounding of electrical equipment has already been discussed and will not be repeated here. Grounding is covered in *Article 250* of the *NEC*. To repeat a caution, *never* ground electrical equipment to the grounded (white) circuit conductor of the circuit.

MARKING THE CONDUCTORS OF CABLES

Two-wire cable contains one white and one black conductor. Because 240-volt circuits can supply electric baseboard heaters, it would appear that the use of two-wire cable is in violation of the Code. However, *Section 200-7* shows that this is not a Code violation.

According to *Section 200-7*, two-wire cable may be used for the 240-volt heaters if the white conductor is permanently re-identified by paint, colored tape, or other effective means. This step is necessary because it may be assumed that an unmarked white conductor is a grounded conductor having no voltage to ground. Actually, the white wire is connected to a hot phase and has 120 volts to ground. *Thus, a person can be subject to a harmful shock by touching this wire and the grounded baseboard heater (or any other grounded object) at the same time.* Most electrical inspectors accept black paint or tape as a means of changing the color of the white conductor. The white conductor must be permanently re-identified at the electric heater terminals and at the panels where these cables originate. Re-read the section on "Color Coding" in Unit 5.

ROOM AIR CONDITIONERS

For homes that do not have central air conditioning, window or through-the-wall air conditioners may be installed. These types of room air conditioners are available in both 120-volt and 240-volt ratings. Because room air conditioners are plug-and-cord connected, the receptacle outlet and the circuit capacity must be selected and installed according to applicable Code regulations. The Code rules for air conditioning are found in *Article 440*. The Code requirements for room air-conditioning units are found in *Sections 440-60* through *440-64*.

The basic Code rules for installing these units and their receptacle outlets are as follows:

- the air conditioners must be grounded.

- the air conditioners must be connected using a cord and attachment plug.

- the air-conditioner rating may not exceed 40 amperes at 250 volts, single phase.

- the rating of the branch-circuit overcurrent device must not exceed the branch-circuit conductor rating or the receptacle rating, whichever is less.

- the air-conditioner load shall not exceed 80% of the branch-circuit ampacity if no other loads are served.

- the air-conditioner load shall not exceed 50% of the branch-circuit ampacity if other loads are served.

- the attachment plug cap may serve as the disconnecting means.

RECEPTACLES FOR AIR CONDITIONERS

Figure 23-11 shows two types of receptacles that can be used for 240-volt installations. A combination receptacle is shown in Figure 23-11(B). The lower portion of the receptacle is for 120-volt use and the upper portion is for 240-volt use. Note that the tandem and grounding slot arrangement for the 240-volt receptacle meets the requirements of *Section 210-7(f)*. This section states that receptacle outlets for different voltage levels must be noninterchangeable with each other.

Figure 23-11 Types of 240-volt receptacles.

CENTRAL HEATING AND AIR CONDITIONING

The residence discussed throughout this text has central electric heating and air conditioning consisting of an electric furnace and a central air-conditioning unit.

The wiring for central heating and cooling systems is shown in Figure 23-12. Basic circuit requirements are shown in Figure 23-13. Note that one branch-circuit runs to the electric furnace and another branch-circuit runs to the air conditioner or heat pump outside the dwelling. Low-voltage wiring is used between the inside and outside units to provide control of the systems. The low-voltage Class 2 circuit wiring must not be run in the same raceway as the power conductors, *Section 725-54(a)(1)*.

Refer to Unit 24 for more information about low-voltage circuitry.

"Time-of-Use" Lower Energy Rates

Many utilities offer lower energy rates for air-conditioner and/or heat pump loads through special programs, such as "time-of-use" usage and sliding scale energy rates. This is covered in Unit 19.

Figure 23-12 Connection diagram showing typical electric furnace and air-conditioner/heat pump installation.

BRANCH-CIRCUIT OVERCURRENT PROTECTION:
- MUST BE ABLE TO CARRY STARTING CURRENT.
- SIZE ACCORDING TO DATA ON THE EQUIPMENT LABEL.
- MUST BE FUSES UNLESS LABEL ON EQUIPMENT SHOWS THAT HACR BREAKERS ARE PERMITTED.

DISCONNECTING MEANS:
- SELECT SIZE BASED ON NAMEPLATE RATED-LOAD CURRENT OR BRANCH-CIRCUIT SELECTION CURRENT – WHICHEVER IS GREATER – AND LOCKED ROTOR CURRENT.
- AMPERE RATING OF SWITCH MUST BE AT LEAST 115% OF NAMEPLATE RATED-LOAD CURRENT OR BRANCH-CIRCUIT SELECTION CURRENT – WHICHEVER IS GREATER.
- MUST ALSO BE HORSEPOWER RATED. CHECK *TABLES 430-148, 149, 150,* AND *151* TO COMPARE THE RATED-LOAD CURRENT, BRANCH-CIRCUIT SELECTION CURRENT, AND HORSEPOWER RATING.
- MUST BE WITHIN SIGHT OF EQUIPMENT.

BRANCH-CIRCUIT CONDUCTORS:
- THE CONDUCTOR AMPACITY RATING REQUIRED FOR THE AIR-CONDITIONING UNIT IS FOUND ON THE LABEL. THIS HAS BEEN DETERMINED BY THE MANUFACTURER TAKING INTO CONSIDERATION THE MOTOR COMPRESSOR CURRENT, FAN MOTOR CURRENT, AND HEATER CURRENT. THIS IS GENERALLY 125% OF THE LARGEST MOTOR PLUS THE FULL-LOAD RATING OF THE REST OF THE EQUIPMENT'S LOADS, SUCH AS FANS AND HEATERS.

TYPICAL AIR-CONDITIONING UNIT

OVERLOAD PROTECTION:
- THIS IS USUALLY AN INTEGRAL PART OF THE EQUIPMENT, SUPPLIED BY THE MANUFACTURER.

LABEL:
- MANUFACTURER'S NAME
- VOLTAGE
- FREQUENCY
- PHASES
- MINIMUM CIRCUIT AMPACITY
- MAXIMUM RATING OF BRANCH-CIRCUITS, SHORT-CIRCUIT, AND GROUND-FAULT PROTECTIVE DEVICE
- WILL STATE "MAXIMUM SIZE FUSE" OR "MAXIMUM SIZE FUSE OR HACR BREAKER"

FAN MOTOR

HERMETICALLY SEALED MOTOR

Figure 23-13 Basic circuit requirements for a typical residential-type air conditioner or heat pump. Reading the label is important in that the manufacturer has determined the minimum conductor size and branch-circuit fuse size, *Article 440*.

UNDERSTANDING THE DATA FOUND ON AN HVAC NAMEPLATE

The letters HVAC stand for heating, ventilating, and air conditioning.

Article 440 applies when HVAC equipment employs a hermetic refrigerant motor-compressor(s). *Article 440* is supplemental to the other articles of the *NEC,* and is needed because of the unique characteristics of hermetic refrigerant motor-compressors.

The most common examples are air conditioners, heat pumps, and refrigeration equipment.

Hermetic refrigerant motor-compressors:

- combine the motor and compressor into one unit
- have no external shaft
- operate in the refrigerant
- do not have horsepower and full-load current (often referred to as full-load amperes or FLA)

ratings like standard electric motors. This is because as the compressor builds up pressure, the current increases, causing the windings to get hot. At the same time, the refrigerant gets colder, passes over the windings, and cools them. Because of this, a hermetically sealed motor can be "worked" much harder than a regular electric motor of the same size.

Special terminology is needed to understand how to properly make the electrical installation. These special terms are found in *Article 440* and in Underwriters Laboratories (U.L.) Standard 1995. An air-conditioning nameplate from Shiver Manufacturing Company will be used to help explain these special terms.

- **Rated-Load Current (RLC):** Rated-load current is established by the hermetic motor-compressor manufacturer under actual operation at rated refrigerant pressure and temperature, rated voltage, and rated frequency. RLC is marked on the nameplate. In most instances, the marked RLC is at least equal to 64.1% of the hermetic refrigerant motor-compressor's maximum continuous current (MCC). In our example, the RLC is 17.8 amperes. See *Section 440-2*.

- **Maximum Continuous Current (MCC):** This is the maximum current that the hermetic refrigerant motor-compressor can draw before damage occurs. Manufacturers know the damage capabilities of their equipment, and generally provide the necessary internal overcurrent protection that keeps the hermetic motor-compressor from burning out. This value is 156% of the marked RLC or branch-circuit selection current (BCSC). The overload protective system for the hermetic refrigerant motor-compressor must operate for currents in excess of the 156% value. Do not look for the MCC value on the nameplate of the end-use equipment. The MCC value might be marked on the motor-compressor. It is nice, but not necessary for the electrician to know the MCC value of the motor-compressor. See *Section 440-52*.

- **Branch-Circuit Selection Current (BCSC):** The manufacturer of the HVAC end-use equipment might design better cooling and better heat dissipation into the equipment, in which case the

hermetic refrigerant motor-compressor might be capable of being continuously "worked" harder than other equipment not so designed. "Working" the hermetic refrigerant motor-compressor harder will result in a higher current draw. This is safe insofar as the motor-compressor is concerned, but the conductors, disconnect switch, and branch-circuit overcurrent devices must also be capable of safely carrying this higher current draw. This higher value of current is marked on the nameplate using the term *Branch-Circuit Selection Current (BCSC)*.

Because the BCSC ampere value is always equal to or greater than the RLC ampere value, this BCSC value must be used instead of the RLC value when selecting conductors, disconnects, overcurrent devices, and other associated electrical equipment. In our example, the BCSC is 19.9 amperes. See *Sections 440-2* and *440-4(c)*.

SHIVER MANUFACTURING COMPANY						
MODEL NO. XYZ – ELECTRICAL RATINGS –						
	VAC	HZ	PH	RLC	LRC	FLA
Compressor	230	60	1	17.8	107	—
Outdoor fan motor ¼ hp	230	60	1	—	—	1.5
Branch circuit selection current			19.9 amperes			
Minimum circuit ampacity			26.4 amperes			
Maximum fuse or HACR type brkr.			45.0 amperes			
Operating voltage range:			197 min. 253 max.			

- **Minimum Circuit Ampacity (MCA):** This is one of two things (the other is MOP) the electrician needs to look for on the nameplate. MCA is the minimum circuit ampacity requirement to determine the conductor size and switch rating. MCA data is always marked on the nameplate of the end-use equipment, and is determined by the manufacturer of the end-use equipment as follows:

MCA = (RLC or BCSC × 1.25) + other loads

Other loads would include condensing fans, electric heaters, coils, etc., that operate concurrently (at the same time).

In our example, the marked MCA is 26.4 amperes, calculated by the manufacturer of the end-use equipment as follows:

$$(19.9 \times 1.25) + 1.5 = 26.375 \text{ amperes}$$
$$\text{(round off to 26.4)}$$

Referring to *Table 310-16*, using the 60°C column, a No. 10 AWG copper conductor would be adequate for this air-conditioning unit.

See *Part D* of *Article 440*.

A mistake that electricians often make is to multiply the MCA by 1.25 again. Certainly there is no hazard in doing this, but it might result in an unnecessary higher installation cost. For long runs, larger conductors would keep voltage drop to a minimum.

- *Type of Overcurrent Protection:* A typical nameplate will show the "maximum size fuse," or "maximum size fuse or **HACR**-type circuit breaker." If maximum size fuse is specified, then circuit breakers are not permitted. *Section 110-3(b)* of the Code states that "listed or labeled equipment shall be used or installed in accordance with any instructions included in the listing or labeling." Figure 23-14 illustrates

the importance of using the proper type of overcurrent protective devices. In our example, the nameplate indicates "maximum fuse or HACR-type breaker 40 amperes."

In recent years, most 240-volt circuit breakers rated 100 amperes or less of the types used in residential wiring bear the letters HACR. But be very careful when connecting an air conditioner or heat pump to older style breakers that were not tested to the HACR requirements and are not marked with the letters HACR. To make such a hookup would be in violation of *Section 110-3(b)*.

See Figures 23-14, 23-15, 23-16, and 23-17.

- *Disconnecting Means to Be* **in Sight**: *Section 440-14* of the Code requires that the disconnecting means must be within sight of the unit, and it must be readily accessible. See Figures 23-15, 23-16, and 23-17. Mounting the disconnect on the side of the house instead of on the air conditioner itself allows for easy replacement of the air-conditioning unit, should that become necessary. Mounting the disconnect behind the air conditioner or heat pump would not be considered "readily accessible" by most electrical

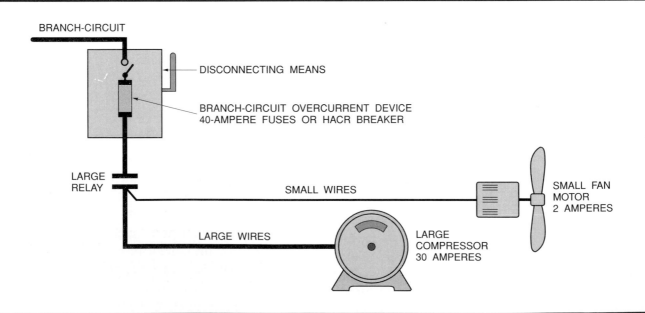

Figure 23-14 In a typical air-conditioner unit, the branch-circuit overcurrent device must protect the large components (large wires, large relay, hermetic motor compressor) as well as the small components (small wires, fan motor, crankcase heater) under short-circuit and/or ground fault situations. It is extremely important to install the proper *size* and *type* of overcurrent device. Read the nameplate on the equipment and the instructions furnished with the equipment!

Figure 23-15 This installation conforms to the Code, *Section 440-14.* The disconnect switch is within sight of the unit and contains the 40-ampere fuses called for on the air-conditioner nameplate as the branch-circuit protection.

Figure 23-17 This installation "meets Code" because the circuit breaker in the panel is marked "HACR." This is permitted when the label on the breaker and the label on the air-conditioner bear the letters "HACR."

Figure 23-16 This installation *violates* the Code, *Section 110-3(b).* Although the disconnect is within sight of the air conditioner, it does not contain fuses. Note that the branch-circuit protection is provided by the 40-ampere circuit breaker inside the building. Note also that the nameplate requires that the branch-circuit protection be 40-ampere fuses maximum. If fused branch-circuit protection were provided at the panel inside the building, the installation would meet Code requirements.

inspectors. Generally, there is just not enough space to work safely on the disconnect. Squeezing behind, or leaning across the top of the air conditioner to gain access to the disconnect is certainly not a safe practice. Mounting the disconnect on the outside wall of the house, to one side of the AC unit is the accepted practice.

• *Disconnecting Means Rating:* Figure 23-18 shows a fusible and a non-fusible "pull-out" disconnect commonly used for residential air-conditioning and/or heat pump installations. The horsepower rating of the disconnecting means must be at least equal to the sum of all of the individual loads within the end-use equipment, at rated load conditions, and at locked rotor conditions, *Section 440-12(b)(1).*

The ampere rating of the disconnecting means must be at least 115% of the sum of all of the individual loads within the end-use equipment, at rated load conditions, *Section 440-12(b)(2).* Installing a disconnect switch that equals or exceeds the MOP value will in almost all cases be the correct choice. The MOP in our example is 40 amperes. Thus, a 60-ampere disconnect switch is the correct size.

Figure 23-18 (A) is a fusible disconnect. (B) is a non-fusible disconnect. These are available in 30- and 60-ampere rating at 240 volts. *Courtesy* of Midwest Electric Products.

In cases where the MOP value is close to the ampere rating of the disconnect switch (i.e., 30, 60, 100, 200, etc.), *locked-rotor current* values must be considered.

- **Locked-Rotor Current** *(LRC)*: This is the maximum current draw when the motor is in a LOCKED position. When the rotor is locked, it is not turning. The disconnect switch and controller (if used) must be capable of safely interrupting locked-rotor current. See *Sections 440-12* and *440-41*. In our example, the nameplate indicates that the hermetic refrigerant motor-compressor has a LRC of 107 amperes. The fan motor has a FLA of 1.5 amperes, and would have a LRC of approximately 6 times. According to *Section 440-12(b)(1)(b)*, all locked-rotor currents and other loads are added together when combined loads are involved. Therefore, the total locked-rotor current that the disconnect switch in our example would be called upon to interrupt is:

$$(1.5 \times 6) + 107 = 116 \text{ locked-rotor amperes}$$

Checking *Table 430-151A* for the conversion of locked-rotor current to horsepower, we find that the locked-rotor current for a 230-volt, single-phase, 3-horsepower motor is 102 amperes, and the locked-rotor current for a 230-volt, single-phase 5-horsepower motor is 168 amperes. Selecting a disconnect switch on the basis of the locked-rotor current in our example, the air-conditioning unit is considered to be a 5-horsepower unit, because 116 falls between 102 and 168. Manufacturers' technical literature indicates that a 30-ampere, 240-volt heavy-duty, single-phase disconnect switch has a 3-horsepower rating, and a 60-ampere, 240-volt heavy duty, single-phase disconnect switch has a 10-horsepower rating. Thus, a 60-ampere disconnect is selected. Note that this is the same size disconnect selected previously on the basis of MOP.

- *Maximum Overcurrent Protection (MOP)*: The MOP value is the maximum overcurrent protection. The MOP value is marked on the nameplate, and is determined by the manufacturer of the end-use equipment as follows:

$$(\text{RLC or BCSC} \times 2.25) + \text{other loads}$$

Other loads would include condensing fans, electric heaters, coils, etc., that operate concurrently.

In our example, the marked MOP is 45 amperes, calculated by the manufacturer of the end-use equipment as follows:

$$(19.9 \times 2.25) + 1.5 = 46.275 \text{ amperes}$$
(round down to the next lower standard size: 45 amperes)

See *Section 440-22*.

ENERGY RATINGS

Energy ratings indicate the efficiency of heating and cooling equipment. Basically, these ratings are a comparison of output (heating or cooling) to input (electricity, gas, or oil).

EER (Energy Efficiency Rating): Cooling efficiency rating for room air conditioners. The ratio of the rated cooling capacity in Btu per hour divided by the amount of electrical power used in kilowatt hours. The higher the EER number, the greater the efficiency.

SEER (Seasonal Energy Efficiency Rating): Cooling efficiency rating for central air conditioners and heat pumps. SEER is determined by the total cooling of an air conditioner or heat pump in Btu during its normal usage period for cooling divided by the total electrical energy input in kilowatt-hours during the same period. The higher the SEER number, the greater the efficiency.

AFUE (Annual Fuel Utilization Efficiency): Tells you how efficient a furnace converts fuel (gas or oil) to heat. For example, an AFUE of 85% means that 85% of the fuel is used to heat your home, and the other 15% goes up the chimney. The higher the efficiency, the lower the operating cost. Old furnaces might have an AFUE rating as low as 60%. Mid-efficiency ratings are approximately 80%. High-efficiency ratings are 90% or higher. Maximum furnace efficiency available is approximately 96.6%. The higher the AFUE number, the greater the efficiency.

HSPF (Heating Seasonal Performance Factor): Heating efficiency of a heat pump. HSPF is determined by the total heating of a heat pump in Btu during its normal usage period for heating divided by the total electrical energy input in kilowatt-hours during the same period. The higher the HSPF rating, the greater the efficiency.

NONCOINCIDENT LOADS

▶ Loads such as heating and air-conditioning are not likely to operate at the same time. The *NEC®* recognizes this diversity in *Section 220-21*. Therefore, when calculating a feeder that supplies both types of loads, or when sizing a service-entrance, only the larger of the two loads need be considered. This is discussed in Unit 29 where service-entrance calculations are presented. Of course, the branch-circuit supplying the heating load is sized for that particular load, and the branch-circuit supplying the air-conditioning load is sized for that particular load. ◀

Gas Explosion Hazard

Although it might seem a bit out of place in an electrical book to talk about gas explosions, we must talk about it.

Often overlooked is a requirement in *NFPA Standard 54* (Fuel Gas Code), *Section 2.7.2(c)* that requires that gas meters be located at least 3 feet from sources of ignition. An electric meter or a disconnect switch are possible sources of ignition. Some utility regulations require a minimum of 3 feet clearance between electric metering equipment and gas meters and gas-regulating equipment. It is better to be safe than sorry! Check this issue out with the local electrical inspector and/or the local electric utility before installing the air conditioner or heat pump disconnect on the outside of the house. Also consider the location of the service watt-hour meter.

REVIEW

Note: Refer to the Code or the plans where necessary.

ELECTRIC HEAT

1. a. What is the allowance in watts made for electric heat in this residence? _____

   b. What is the value in amperes of this load? _____

   _____

2. What are some of the advantages of electric heating? _____

_____

_____

_____

3. List the different types of electric heating system installations. _____

_____

_____

_____

4. There are two basic voltage classifications for thermostats. What are they? _____

_____

5. What device is required when the total connected load exceeds the maximum rating of a thermostat? _____

6. The electric heat in this residence is provided by what type of equipment? _____

_____

7. At what voltage does the electric furnace operate? _____

8. A certain type of control connects electric heating units to a 120-volt supply or a 240-volt supply, depending upon the amount of the temperature drop in a room. These controls are supplied from a 120/240-volt, three-wire, single-phase source. Assuming that this type of device controls a 240-volt, 2000-watt heating unit, what is the wattage produced when the control supplies 120 volts to the heating unit? Show all calculations.

9. What advantages does a 240-volt heating unit have over a 120-volt heating unit?

_____

_____

_____

_____

_____

_____

10. The white wire of a cable may be used to connect to a "hot" ungrounded circuit conductor only if _____

_____

11. Receptacle outlets furnished as part of a permanently installed electric baseboard heater, when not connected to the heater's branch-circuit, (may) (may not) be counted as the required receptacle outlet for the space occupied by the baseboard heater. Electric baseboard heaters (shall) (shall not) be installed beneath wall receptacle outlets unless the instructions furnished with the heaters indicate that it is acceptable to install the heater below a receptacle outlet. Circle correct answers.

12. The branch-circuit supplying a fixed electric space heater must be sized to at least _____ % of the heater's rating according to *Section* _____ .

13. Compute the current draw of the following electric furnaces. The furnaces are all rated 240 volts.
    a. 7.5 kW
    b. 15 kW
    c. 20 kW

14. For "ballpark" calculations, the wattage output of a 240-volt electric furnace connected to a 208-volt supply will be approximately 75% of the wattage output had the furnace been connected to a 240-volt supply. In question 13, calculate the wattage output of (a), (b), and (c) if connected to 208 volts. _____

    _____

    _____

    _____

    _____

15. A central electric furnace heating system is installed in a home. The circuit supplying this furnace:
    a. may be connected to a circuit that supplies other loads.
    b. shall be connected to a circuit that supplies other loads.
    c. shall be connected to a separate circuit.

16. What section of the Code provides the correct answer to question 15? _____

17. Electric heating cable embedded in plaster, or "sandwiched" between layers of dry wall, creates heat. Therefore, any nonmetallic-sheathed cable run above these ceilings and buried in insulation must have its current-carrying capacity derated according to the (40°C) (50°C) (60°C) correction factors found at the bottom of *Table 310-16* in the *NEC.*® Circle the correct answer.

AIR CONDITIONING

1. a. When calculating air-conditioner load requirements and electric heating load requirements, is it necessary to add the two loads together to determine the combined load on the system? _____

   b. Explain the answer to part (a). _____

   _____

   _____

2. The total load of an air conditioner shall not exceed what percentage of a separate branch-circuit? (circle one)
   a. 75%   b. 80%   c. 125%

3. The total load of an air conditioner shall not exceed what percentage of a branch circuit that also supplies lighting? (circle one)

   a. 50%   b. 75%   c. 80%

4. a. Must an air conditioner installed in a window opening be grounded if a person on the ground outside the building can touch the air conditioner? _____

   b. What Code section governs the answer to part (a)? _____

5. A 120-volt air conditioner draws 13 amperes. What size is the circuit to which the air conditioner will be connected? _____

6. What is the Code requirement for receptacles connected to circuits of different voltages and installed in one building? _____

7. When a central air-conditioning unit is installed and the label states "Maximum size fuse 50-amperes," is it permissible to connect the unit to a 50-ampere circuit breaker?

   _____

8. When the nameplate on an air-conditioning unit states "Maximum Size Fuse or HACR Circuit Breaker," what type of circuit breaker must be used? _____

   _____

9. Briefly, what is the reasoning behind the U.L. requirement that air conditioners and other multimotor appliances be protected with fuses or special HACR circuit breakers?

   _____

   _____

   _____

   _____

   _____

   _____

   _____

10. What section of the *NEC*® prohibits the installing of Class 2 control circuit conductors in the same raceway as the power conductors? _____

    _____

11. Match the following terms with the statement that most closely defines the term. Enter the letters in the blank space.

    RLC _____   a) The maximum ampere rating of the overcurrent device.

    MCA _____   b) The current value to be used instead of the rated-load ampere.

    BCSC _____   c) The value used to determine the minimum ampere rating for the branch-circuit.

    MCC _____   d) A current value for the hermetic motor compressor that was determined by operating it at rated temperature, voltage, and frequency.

    MOP _____   e) The maximum current that the compressor can draw continuously without damage.

12. The disconnect for an air-conditioner or heat pump must be installed _____
    _____ of the unit.

Gas and Oil Central Heating Systems

OBJECTIVES

After studying this Unit, the student will be able to

- Understand the basics of typical home heating systems.
- Understand some of the more important components of typical home heating systems.
- Interpret basic schematic wiring diagrams.
- Understand how a thermocouple and thermopile operate.
- Understand and apply the Code requirements for control-circuit wiring, *Article 725, NEC.*®

In Unit 23, we discussed electrical branch-circuit requirements for electric heating (furnace, baseboard, ceiling cable). In this unit, typical residential gas and oil heating systems are discussed.

Every manufacturer of residential heating systems provides installation manuals that must be followed.

WARM AIR FURNACES

By far, the most popular method for heating a home is with a gas-fired, forced-warm air furnace. Oil-fired, forced-warm air furnaces are also available. Warm air furnaces are available with upflow, downflow, and horizontal air flow. A fan forces hot air through the hot air ducts to various rooms in the house. Cold air ducts, usually referred to as "cold air returns" bring the air back to the furnace where the air is again filtered, heated, and forced out through the hot air duct.

Humidity control is another feature commonly installed in home heating systems.

Some systems bring fresh air into the home. This ventilation equipment brings fresh air into the home, then filters and circulates this fresh air through the ductwork in the home.

For larger homes, by installing special dampers in the hot air ducts and the necessary associated controls, heating systems can provide zone control so that different areas of the home can be controlled by their own thermostats.

HOT WATER SYSTEM

Hot water is circulated through the piping system by a circulating pump. This is often referred to as a *hydronic* system. The heating unit is referred to as a boiler, and can be gas or oil fired. A hot water system is very adaptable to zone control by using a combination of one or more circulating pumps and/or valves for the different zones. Hot water baseboards look very much like electric baseboard heaters.

Hot water piping systems are not easily adaptable to central air conditioning. Additional ductwork is required.

Upscale homes might have special plastic tubing secured to the underside of the floors to heat the floors from below. Hot water is circulated through this tubing when the thermostat calls for heat. This type of system is used under tile floors in bathrooms, kitchens, and similar areas.

PRINCIPLE OF OPERATION

Most residential heating systems operate as follows. A room thermostat is connected to the proper electrical terminals on the furnace or boiler. Inside the equipment, the various controls and valves are interconnected in a manner that will provide safe and adequate operation of the furnace.

Most residential heating systems are "packaged units" in which the burner, hot surface igniter or hot spark igniter, safety controls, valves, fan controls, sensors, high temperature limit switches, fan blower motor, blower fan speed control, draft inducer motor, primary and secondary heat exchanger, and so on, are preassembled, prewired parts of the furnace or boiler. Many of these items are included in a solid-state integrated printed circuit board module, which is the "brain" of the system. This module provides proper sequencing and safety features.

Older models have standing gas pilot lights and thermocouples.

Similar electronic circuitry is used in oil burners where the ignition, the oil burner motor, blower fan motor, high temperature limit controls, hot water circulating pump in the case of hot water systems, and the other devices previously listed are all interconnected to provide a packaged oil burner unit.

Evaporator coils for air-conditioning purposes might be an integral part of the unit, or they might be mounted in the ductwork above the furnace.

Wiring a Residential Central Heating System

For the electrician, the "field wiring" for a typical residential furnace or boiler consists of:

1. Installing and connecting the low-voltage wiring between the furnace and the thermostat. This might be a cable consisting of two, three, four, or five small No. 18 or No. 20 AWG conductors. Small knockouts are provided on the furnace through which the low-voltage Class 2 wires are brought into the furnace's wiring compartment (see Figure 24-1). Class 2 wiring must not be run through the same raceway or cable as power wiring, *Section 725-54(a)(1).*

2. Installing and connecting the branch-circuit power supply to the furnace. Knockouts are

ALL INTERNAL COMPONENTS PREWIRED BY MANUFACTURER OF FURNACE

THERMOSTAT WIRING INSTALLED BY ELECTRICIAN

POWER SUPPLY INSTALLED BY ELECTRICIAN

DISCONNECT SWITCH MUST BE WITHIN SIGHT OF FURNACE, *SECTIONS 422-31, 424-19,* AND *430-102.*

Figure 24-1 Diagram showing thermostat wiring, power supply wiring, and disconnect switch. ANSI Standard Z21.47 does not allow cord- and plug connections for a gas furnace.

provided on the furnace through which the line voltage power supply is brought into the wiring compartment of the furnace. Depending on local codes, the line voltage power supply wiring might be in EMT, armored cable, or other accepted wiring methods (Figure 24-1).

The disconnecting means must be within sight of the furnace, *Sections 422-31, 424-19,* and *430-102.* Some furnaces come with a disconnect switch. If not, a toggle switch can be mounted on a box on the side of the furnace. Some electric furnaces come complete with circuit breakers that provide the overcurrent protection for the heating elements as required by U.L., and to serve as the disconnecting means required by the *NEC.*

Separate Branch-Circuit Required

Section 422-12 requires that central heating equipment must be supplied by a separate branch-circuit.

The size of the branch-circuit depends on the requirements of the particular furnace. A typical gas-forced warm air furnace might only require a

*DISCONNECTING MEANS AND OVERCURRENT PROTECTION AS REQUIRED. DISCONNECTING MEANS INSTALLED WITHIN SIGHT OF CONTROLLER, *SECTIONS 430-102* AND *422-31*.

Figure 24-2   Typical wiring diagram for a gas burner, forced hot water system.

120-volt, 15-ampere branch-circuit, whereas an electric furnace requires a 240-volt branch-circuit, and much larger conductors than a gas- or oil-fired furnace or boiler. Instructions furnished with a given furnace will indicate the branch-circuit requirements.

Some typical wiring diagrams are shown in Figures 24-2, 24-3, and 24-4. These wiring diagrams show the individual electrical components of a typical residential heating system. In older systems, many of these components were wired by the electrician. In newer systems, most components other than the thermostat are prewired as part of a packaged furnace or boiler. The manufacturer's literature includes detailed wiring diagrams for a given model.

*DISCONNECTING MEANS AND OVERCURRENT PROTECTION AS REQUIRED. DISCONNECTING MEANS INSTALLED WITHIN SIGHT OF CONTROLLER, *SECTION 430-102* AND *422-31*.

Figure 24-3   Typical wiring diagram for a gas burner, forced warm air system.

Figure 24-4 Typical wiring diagram of an oil burner heating installation.

MAJOR COMPONENTS

Gas- and oil-forced warm air systems, and hot water boilers contain many individual components. Here is a brief description of these components.

Fan Control

Fan control is used to control the blower fan on forced warm air systems. On newer furnaces, the fan control is actually an electronic timer whereby the fan will start running after so many seconds of time has passed after the main burner has come on. This is sufficient time to allow the burner to warm the air so that the blower does not blow cold air. Where adjustable, fan controls are usually set to turn on between 120 and 140°F, and to shut off at 90 to 100°F.

Fan Motor

Fan motor is the electric motor that forces the warm air through the ducts of a heating system. It may be single, multiple, or variable speed. The blower fan itself might be directly connected to the shaft or it may be belt driven. In many cases, a squirrel cage fan is connected to the shaft of the motor. The motor in most cases is provided with integral overload protection in conformance to the *NEC*®.

High Temperature Limit Control

Should the temperature in the plenum of the furnace reach dangerously high temperatures (approximately 200°F), the high temperature limit control will shut off the power to the burner. The fan will continue to run. When the temperature in the plenum returns to a safe level, the high temperature limit control will allow the burner to resume normal operation. In newer furnaces, the high temperature limit control is part of the integrated circuit board, and is communicated to by a sensor strategically located in the plenum. Older furnaces might have individual high temperature limit control and a separate fan control, or they might have a combination high temperature limit/fan controls.

Hot Surface Igniter

In furnaces that do not have a standing constant pilot light, the hot surface igniter heats to a cherry red/orange color. When it reaches the proper temperature, the main gas valve is allowed to open. If the hot surface igniter does not come on, the main gas valve will not open. Energy efficient furnaces may use this concept for ignition.

Induced Draft Blower

When the thermostat calls for heat, the induced draft blower starts first to expel any gases remaining in the combustion chamber from a previous burning cycle. It continues to run, pulling hot combustion gases through the heat exchangers, then vents the gases to the outdoors.

Integrated Control

Integrated control is a solid-state, printed circuit board containing many electronic components. When the thermostat calls for heat, the integrated circuit board takes over to manage the sequence of events that allow the burner to operate safely.

Liquid Immersion Control

Liquid immersion control controls high water and circulating water temperatures in boilers. When acting as a high limit control, it will shut off the burner if the water temperature in the boiler reaches dangerous levels. When acting as a circulating water temperature control, it makes sure that the water being circulated through the piping system is at the desired temperature, not cold. A liquid immersion control is mounted so that its sensing tube is either in direct contact with the water, or inserted into an immersion well that is in direct contact with the boiler water. Temperature settings must be in conformance to the manufacturers' instructions. In packaged boilers, the settings are usually set for the particular boiler. Field inspection to make sure the setting is correct is recommended.

Low Water Control

Sometimes referred to as a low water cutoff, this device senses low water levels in a boiler. When a low water situation occurs, the low water control shuts off the electrical power to the burner.

Main Gas Valve

The main gas valve is the valve that allows the gas to flow to the main burner of a gas-fired furnace. It may also function as a pressure regulator, a safety shutoff, and as the pilot and main gas valve. Should there be a "flame-out," or if the pilot light fails to operate, or in the event of a power failure, this valve will shut off the flow of gas to the main burner.

Primary Control

Primary control controls the oil burner motor, ignition, and oil valve in response to commands from a thermostat. If the oil fails to ignite, the controller shuts down the oil burner. A primary control is sometimes referred to as a "stack switch." Primary controllers are available for both constant and intermittent ignition burners. Both types may recycle once if the burner fails to ignite the oil, in which case the controller trips off to a safety mode.

Spark Ignition

In spark ignition models, the spark comes on while pilot gas flows to the pilot orifice. Once the pilot flame is "proven" through an electronic sensing circuit, the main gas valve opens. Once ignition takes place, the sensor will monitor and prove existence of a main flame. If for some reason the main flame goes off, the furnace shuts down. Energy-efficient furnaces may use this concept for ignition.

Switching Relay

A switching relay is used between the low-voltage wiring and line voltage wiring. When a low-voltage thermostat calls for heat, the relay "pulls in," closing the line voltage contacts on the relay, which turns on the main power to the furnace or boiler. Usually they have a built-in transformer that provides the low voltage for the low-voltage Class 2 control circuit.

Transformer

The transformer converts (transforms) the branch-circuit voltage to low voltage. For example, the furnace's branch-circuit might be 120 volts. The primary of the transformer is connected to the 120-volt supply. The transformer steps down the 120-volt primary voltage to 24 volts on the secondary side of the transformer. Class 2 circuits for the low-voltage wiring between the thermostat and a furnace or boiler are generally 24 volts.

Thermocouple

A thermocouple is used in systems having a standing pilot to hold open a gas valve pilot solenoid magnet. A thermocouple consists of two dissimilar metals connected together to form a circuit. The

Figure 24-5   Principle of a thermocouple.

metals might be iron or copper, copper and iron constantan, copper-nickel alloy and chrome-iron alloy, platinum and platinum-rhodium alloy, or chromel and alumel. The types of metals used depend on the temperatures involved. When one of the junctions is heated, an electrical current flows (Figure 24-5). To operate, there must be a temperature difference between the metal junctions. In a gas burner, the source of heat for the thermocouple is the pilot light. The cold junction of the thermocouple remains open and is connected to the pilot safety shutoff gas valve circuit.

A single thermocouple develops a voltage of about 25 to 30 dc millivolts. Because of this extremely low voltage, circuit resistance is kept to a very low value.

Thermopile

When thermocouples are connected in series, a thermopile is formed (Figure 24-6). The power output of a thermopile is greater than that of a single thermocouple—typically either 250 or 750 dc millivolts.

For example, 10 thermocouples (25 dc millivolts each) connected in series result in a 250 dc millivolt thermopile. Twenty-six thermocouples connected in series results in a 750 dc millivolt thermopile. For comparison, 750 millivolts is one-half the voltage of a 1½-volt "D" size flashlight battery.

Thermostat

Depending on the complexity of the heating/cooling system, a simple thermostat might merely turn the furnace on and off, or it might be a more sophisticated programmable combination thermostat.

A programmable combination thermostat might have features such as temperature control for both heating and air conditioning, day/night or multiple time period setbacks for comfort and energy savings, a fan switch for continuous operation of the air circulating fan motor, a clock, a thermometer,

Figure 24-6   Principle of a thermopile.

Figure 24-7   A digital electronic programmable thermostat that provides proper cycling of the heating system, resulting in comfort as well as energy savings. This type of thermostat, depending on the model, can be programmed to control temperature, time of operation, humidity, ventilation, filtration, circulation of air, as well as zone control. Installation and operational instructions are furnished with the thermostat. *Courtesy of Honeywell, Inc.*

outdoor temperature sensing, humidity control, ventilation and filtration control, and indoor air quality control. A programmable thermostat is shown in Figure 24-7. Quality thermostats generally are accurate to ±1 degree.

Most residential thermostats have some kind of adjustment feature to control proper cycling.

Some thermostats feature an integral heat anticipator and a small adjusting lever. The adjusting lever point to the numbers .10, .20, .25, .30, etc., which correspond to the current draw of the circuit controlled by the thermostat.

Electronic thermostats use a logic circuit to control the cycling rate instead of a manual adjusting lever.

Thermostats in residences should be mounted approximately 52 inches above the finished floor and not mounted where influenced by drafts, concealed hot or cold water pipes or ducts, on outside walls, lighting fixtures, direct rays from the sun, heat from fireplaces and heating appliances, or air currents from hot or cold air registers.

Some thermostats contain a small vial of mercury. When the vial "tips," the mercury rolls downward, opening and/or closing the circuit by bridging the gap between the electrical contacts inside the vial. To ensure accuracy, these types of thermostats must be mounted perfectly level.

Water Circulating Pump

In hydronic (wet) systems, this pump(s) circulates hot water through the piping system. Multizone systems may have a number of circulating pumps, each controlled by a thermostat for a particular zone. In multizone systems, a high limit control might be set to maintain the boiler water temperature to a specific temperature.

Some multizone systems have one circulating pump, and use electrically operated valves that open and close on command from the thermostat(s) located in a particular zone(s).

CONTROL CIRCUIT WIRING

Article 725 covers most of the control-circuit wiring requirements. *Article 430, Part F* covers motor control circuits.

Class 1, 2, and 3 circuits offer alternative wiring methods regarding wire size, derating factors, overcurrent protection, insulation, and materials that differentiate these types of circuits from that of regular light and power wiring as covered in *Chapters 1 through 4* of the *NEC.*®

Figure 24-8 defines remote control, signaling, and power-limited terminology. More discussion regarding this is found in Units 25, 26, and 27 of this text.

Article 725 discusses three classes of control circuits. Because electricians in their career will come across other types of control-circuit wiring, the following text provides a brief description of the three classes.

Class 1

- Class 1 remote-control and signaling circuits have their voltage and power limited to the values shown in *Section 725-21.*

- Class 1 remote-control and signaling circuits must not exceed 600 volts, *Section 725-21(b).*

- Class 1 conductor must be rated 600 volts, *Section 725-27(b).*

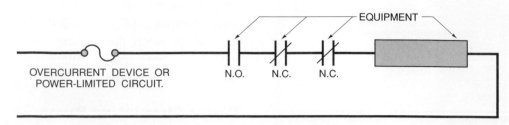

A A REMOTE-CONTROL, SIGNALING, OR POWER-LIMITED CIRCUIT IS THAT PORTION OF THE WIRING SYSTEM BETWEEN THE LOAD SIDE OF THE OVERCURRENT DEVICE OR THE POWER-LIMITED SUPPLY AND ALL CONNECTED EQUIPMENT, *SECTION 725-2.*

B A REMOTE-CONTROL CIRCUIT IS ANY ELECTRICAL CIRCUIT THAT CONTROLS ANY OTHER CIRCUIT THROUGH A RELAY OR EQUIVALENT DEVICE, *ARTICLE 100, DEFINITIONS.*

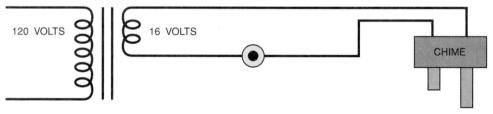

C A SIGNAL CIRCUIT IS ANY ELECTRICAL CIRCUIT THAT ENERGIZES SIGNALING EQUIPMENT, *ARTICLE 100, DEFINITIONS.*

Figure 24-8 **The above diagrams explain a remote-control circuit and a signaling circuit.**

- Class 1 circuits and power supply circuits may be run in the same raceway only if the interconnected equipment is *functionally associated, Section 725-26(b)* (Figure 24-9).

- Class 1 wiring for safety control equipment, such as high limit and low water controls, must be installed in rigid or intermediate metal conduit, rigid nonmetallic conduit, EMT, Type MI, MC, or LWA cable, or be otherwise suitably protected from physical damage. A failure of the equipment, or a short circuit or open circuit in the wiring would "introduce a direct fire or life hazard." See *Section 725-8(a)* and *(b)* (Figure 24-10).

- Class 1 circuits must have overcurrent protection sized to the ampacity of the conductors to be protected, *Section 725-23*. The overcurrent protection must be located where the conductors receive their supply, *Section 725-24*.

Class 1 circuits are rarely found in residential equipment, but is mentioned here as part of the overall understanding for control-circuit wiring.

UNIT 24 Gas and Oil Central Heating Systems **427**

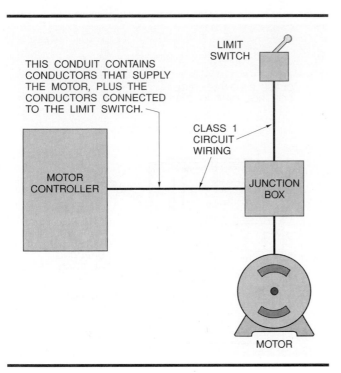

Figure 24-9 The power conductors supplying the motor and the Class 1 circuit remote-control conductors to the limit switch and controller may be run in the same conduit when the interconnected equipment is "functionally associated," *Section 725-26(b)*. The insulation on the Class 1 circuit conductors must be insulated for the maximum voltage of any other conductors in the raceway. The minimum voltage rating for Class 1 conductors is 600 volts, *Section 725-27(b)*.

Class 2

In residential wiring, most of the low-voltage wiring comes under the Class 2 circuit category. Some examples of Class 2 circuits are the thermostat wiring for heating, air conditioning, and heat pumps. The wiring for security systems, intercom systems, and chime wiring is Class 2.

• Class 2 circuits are "power-limited," and are considered to be safe from fire hazard and from electric shock hazard.

• Class 2 wiring does not have to be in a raceway or conduit because should a failure in the wiring occur, such as a short circuit or an open circuit in a thermostat cable, there would not be a direct fire or shock hazard. The operation of the safety controls in the heating equipment would prevent the occurrence of a fire or life hazard.

• Class 2 circuits shall not be run in the same raceway, cable, compartment, outlet box, or fitting with electric light and power conductors, *Section 725-54(a)(1)*. Even if the Class 2 conductors have the same 600-volt insulation as the power conductors, this is not permitted (Figure 24-11). A barrier inside an enclosure that keeps the Class 2 wiring and power conductors apart is permitted.

Figure 24-10 Any failure in Class 1 circuit wiring would introduce a direct fire or life hazard. A failure in the Class 2 circuit wiring would *not* introduce a direct fire or life hazard. See *Section 725-8(a)* and *(b)*.

AIR-CONDITIONER

BASEMENT

CAUTION: IF THE LOW-VOLTAGE WIRING IS CLASS 2, KEEP SEPARATE FROM THE POWER WIRING, SECTION *725-54(a)(1)*. IF THE LOW-VOLAGE WIRING IS CLASS 1, IT IS ACCEPTABLE TO RUN WITH THE POWER CON- DUCTORS, *SECTION 725-26(b)*.

Figure 24-11 Most residential air-conditioning units, furnaces, and heat-pumps have their low-voltage circuitry classified as Class 2. *Section 725-54(a)(1)* **prohibits installing low-voltage Class 2 conductors in the same raceway as the power conductors . . . even if the Class 2 conductors have 600-volt insulation. Therefore, the conduit running out of the basement wall in the above diagram would not be permitted to contain both the 240-volt power conductors and the 24-volt low-voltage Class 2 conductors. Always read the instructions furnished with these types of appliances to be sure your wiring is in compliance with these instructions, per** *Section 110-3(b)* **and the** *NEC.* **Review Unit 4 for grounding methods.**

Manufacturers of heating and cooling appliances provide separate openings for bringing in the Class 2 wiring, and separate knockouts for bringing in the power supply.

- Class 2 wiring may be fastened to the conduit that contains the power supply conductors to specific equipment, *Sections 725-54(d)* and *300-11(b)* (Figure 24-12).

- Transformers that are intended to supply Class 2 and Class 3 circuits are "listed" by U.L. and are marked "Class 2 Transformer" or "Class 3 Transformer." These transformers have "built-in" protection against the possibility of overheating should a short circuit occur somewhere in the secondary wiring. These transformers may be connected to circuits having overcurrent protection not over 20 amperes, *Section 725-51*.

CLASS 2 LOW-VOLTAGE CONTROL CIRCUIT CABLE

THIS CONDUIT CONTAINS THE POWER CONDUCTORS SUPPLYING THE FURNACE

DISCONNECT SWITCH

Figure 24-12 The Code in *Section 300-11(b)* **and** *725-54(d)*, **allows Class 2 control circuit conductors (cables) to be supported by the raceway that contains the power conductors supplying electrical equipment such as the furnace shown.**

Class 3

- Class 3 circuits are power limited and are considered to be safe from fire hazard, but could present an electrical shock hazard because of the higher voltage and current levels permitted, *Section 725-2*.

- Class 3 circuits are similar to Class 2 circuits, except that they operate at higher voltage and current levels.

- Class 3 conductors have a voltage rating of not less than 300 volts, *Section 725-71(f)*.

Class 2 and Class 3 Power Source Requirements

Class 2 and Class 3 power source voltage and current limitations, and overcurrent protection requirements are found in *Chapter 9*, *Tables 11(a)* and *11(b)* of the *NEC.* You should refer to the nameplate on Class 2 and Class 3 transformers, power supplies, and other equipment to confirm that they are U.L. "listed" for a particular class.

Cables for Class 2 and Class 3 Circuits

Low-voltage conductors for Class 2 and Class 3 wiring are rated in a "hierarchy" based on their ability to withstand fire. Their permitted uses and permitted substitutions between the different types are listed in *Table 725-61* and *725-71* of the *NEC*. These tables are presented in descending order relative to their fire resistance.

The low cost of Class 2 and Class 3 conductors generally installed in homes are Types CL2X and CL3X. When you ask for "bell wire," you are probably getting Type CL2X.

Securing Class 1, 2, and 3 Cables

Class 1, 2, and 3 cables must be installed in a neat and workmanship manner, and shall be adequately supported by the building structure in such a manner that they will not be damaged by normal building use, *Section 725-7.*

Section 300-11(b) states that "*raceways shall not be used as a means of support for other raceways, cables, or nonelectric equipment.*" Taping, strapping, tie wrapping, or hanging or securing Class 1, 2, or 3 cables to electrical raceways, piping, or ducts is generally prohibited.

However, the exception to *Section 725-54(d)* refers back to *Section 300-11(b)*, which allows Class 2 conductors to be supported by the raceway that contains the power supply conductors for the same equipment that the Class 2 conductors are connected to (see Figure 24-12).

Figure 24-13(A) shows the control circuit totally contained within the equipment. Factory internal wiring is not governed by *Article 725.* This wiring is covered in the U.L. Standards.

Figure 24-13(B) shows the control wiring leave the equipment. The external wiring in this illustration comes under the requirements of *Article 430, Part F.* If the control circuit is supplied from other than being connected to the load side of the motor's branch-circuit overcurrent device, the control-circuit wiring would be governed by the requirements of *Article 725.*

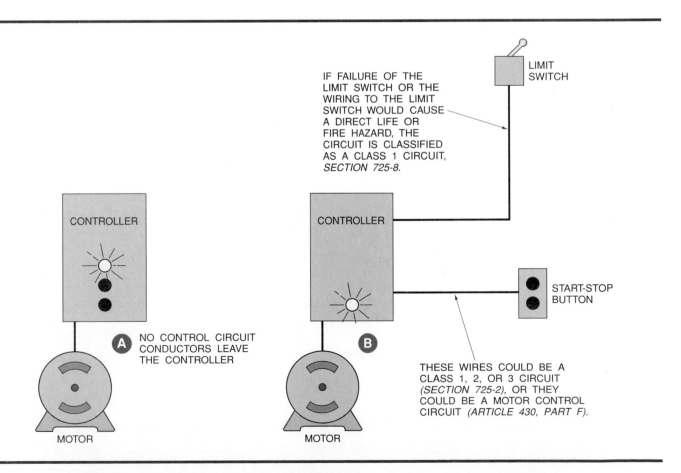

Figure 24-13   Diagram showing how Class 1, 2, and 3 circuits are categorized.

REVIEW

1. The residence in this text is heated with:

   a. Gas

   b. Electricity

   c. Oil

   Circle the correct answer.

2. The *NEC®* in *Sections* _____ , _____ , and _____ requires that a disconnecting means must be _____ _____ of the furnace.

3. The *NEC®* in *Section* _____ requires that a central heating system must be supplied by a _____ branch-circuit.

4. In a home, the low-voltage wiring between the thermostat and furnace is:

   a. Class 1

   b. Class 2

   c. Class 3

   Circle the correct answer.

5. Class 2 circuits are _____-limited and are considered to be _____ from _____ hazard and from _____ _____ .

6. The nameplate on a transformer will indicate whether or not it is a Class 2 transformer.

   (True)   (False)   Circle the correct answer.

7. If Class 2 conductors are insulated for 32 volts, and the power conductors are insulated for 600 volts, does the Code permit pulling these conductors through the same raceway?

   (Yes)   (No)   Circle the correct answer. Give *NEC®* section number. _____ .

8. If you use 600-volt rated conductors for Class 2 wiring, does the Code permit pulling these conductors through the same raceway as the power conductors? (Yes) (No) Circle the correct answer. Give the *NEC®* section number. _____ .

9. The *NEC®* is very strict about securing anything to electrical raceways. Name the Code sections that prohibit this practice. *Sections* _____

   _____

10. Does the Code permit attaching the low-voltage thermostat cable to the EMT that brings power to a furnace? (Yes) (No) Circle the correct answer. Give the *NEC®* section number.

    _____

11. Explain what a thermocouple is and how it operates. _____

    _____

    _____

    _____

    _____

Television, Telephone, and Low-Voltage Signal Systems

OBJECTIVES

After studying this Unit, the student will be able to

- install television outlets, antennas, cables, and lead-in wires.
- list CATV installation requirements.
- describe the basic operation of satellite antennas.
- install telephone conductors, outlet boxes, and outlets.
- define what is meant by a signal circuit.
- describe the operation of a two-tone chime and a four-note chime.
- install a chime circuit with one main chime and one or more extension chimes.

INSTALLING THE WIRING FOR HOME TELEVISION

This Unit discusses the basics of installing outlets, cables, receivers, antennas, amplifiers, multiset couplers, and some of the *NEC®* rules for home television. Television is a highly technical and complex field. A trouble-free installation of a home system should be done by a competent television technician. These individuals receive training and certification through the Electronic Technicians Association in much the same way that electricians receive their training through apprenticeship and journeyman training programs, and learning about the *NEC®* via membership in the International Association of Electrical Inspectors organization. Some cities and states require that, when more than three television outlets are to be installed in a residence, the television technician making the installation be licensed and certified. Just as electricians can experience problems because of poor connections, terminations, and splices, poor reception problems can arise as a result of poor terminations. In most cases, poor crimping is the culprit. Electrical

shock hazards are present when components are improperly grounded and bonded.

According to the plans and for this residence, television outlets are installed in the following rooms:

Front Bedroom	1
Kitchen/Nook	1
Laundry	1
Living Room	3
Master Bedroom	1
Recreation Room	3
Study/Bedroom	1
Workshop	1

Television outlets are usually installed in standard sectional switch (device) boxes, or 4-inch square 1½-inch deep outlet boxes with a single-gang plaster ring, Figure 25-1. The boxes may be metallic or nonmetallic, and are roughed-in in the same manner as regular wiring. A box is installed at each point on the system where a television outlet is to be located. For new construction, shielded

Figure 25-1   Nonmetallic boxes and a nonmetallic raised cover.

coaxial cables are installed concealed in the walls during the normal rough-in stage of construction. For remodel work, television cables can be concealed in the walls by "fishing" the cables through the walls.

Shielded 75-ohm coaxial cable is most often used to hook up television sets to minimize interference and keep the color signal strong. There are different kinds of shielded coaxial cable. Double-shielded cable that has a 100% foil shield covered with a 40% or greater woven braid is recommended. This coaxial cable has a PVC outer jacket. The older style, flat 300-ohm twin lead cable, can still be found on existing installations, but the reception might be poor. To improve reception, 300-ohm cables can be replaced with shielded 75-ohm coaxial cables.

For this residence, 12 television outlets are to be installed. For this many outlets, it is recommended that a shielded 75-ohm coaxial cable be run from each outlet location back to one central point, Figure 25-2. This central point could be where the incoming CATV (Community Antenna Television) cable comes in, where the cable from an antenna comes in, or where the TV output of a satellite receiver is carried. There are so many possibilities that each installation must be analyzed to determine what is best for that particular installation. For rooftop or similar antenna mounting, the cable is run down the outside of the house, then into the basement, garage, or other convenient location where the proper connections are made. CATV companies generally run their incoming coaxial cables underground, again to some convenient point just inside of the house. Here, the

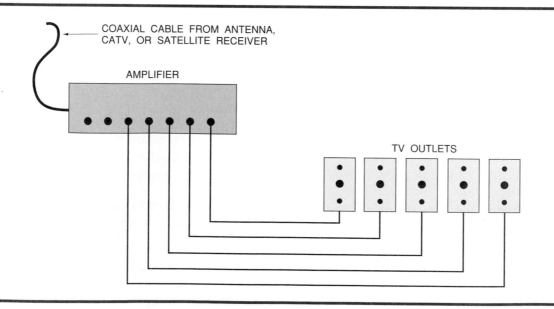

Figure 25-2   A television master amplifier distribution system may be needed where many TV outlets are to be installed. This will minimize the signal loss. In simple installations, a multiset coupler can be used, Figure 25-3.

technician can hook up all of the coax cables using a multiset coupler, Figure 25-3, or an amplifier, Figure 25-2, to boost possible lost signals. An amplifier may not be necessary if *only* those coaxial cables that will be used are hooked up.

Cable television does not require an antenna on the house because the cable company has the proper antenna to receive signals from satellites. The cable company then distributes these signals throughout

the community they have contractually agreed to serve. These contracts usually require the cable company to run their coaxial cable to a point just inside of the house. Inside the house, they will complete the installation by furnishing cable boxes, controls, and the necessary wiring. In geographical areas where cable television is not available, antennas of the types shown in Figure 25-4, or satellite "dishes," as illustrated in Figure 25-6, must be installed.

Faceplates for the TV outlets are available in styles to match most types of electrical faceplates in the home.

Although combination two-gang faceplates can be used so that a 120-volt receptacle and a TV outlet can be mounted on one box, Figure 25-5, most installers prefer to keep the television outlets completely separate from the 120-volt receptacles. Where a combination faceplate is used, the electrical wall box must have a metal barrier between the 120-volt wiring and the television outlet wiring, *Section 810-18(c)*. This barrier prevents line voltage interference with the television signal and keeps the two different voltages separate from each other for safety. More *NEC*® rules are discussed later in this Unit.

Figure 25-3   Multiset coupler.

Figure 25-4   Typical mounting of television antennas.

Figure 25-5 Typical faceplates.

Although shielded coaxial cable minimizes interference, it is good practice to keep the coaxial cables on one side of the stud space, and keep the light and power cables on the other side of the stud space.

Code Rules for Cable Television (CATV)
(Article 820)

Cable television is more correctly called Community Antenna Television, hence the term CATV.

CATV systems are installed both overhead and underground in a community. Then, to supply an individual customer, the CATV company runs coaxial cables through the wall of a residence at some convenient point. Up to this point of entry *(Article 820, Part B)* and inside the building *(Article 820, Part E)*, the cable company must conform to the requirements of *Article 820,* plus local codes if applicable.

Coaxial cable generally used for residential installations is Type CATV. Other types are listed in *Table 820-53* of the *NEC.*®

Here are some of the key rules to follow when making a coaxial cable installation.

1. The outer conductive shield of the coaxial cables must be grounded as close to the point of entry as possible, *Section 820-33.*

2. Coaxial cables shall not be run in the same conduits or box with electric light and power conductors, *Section 820-52(a)(1)(b).*

3. Do not support coaxial cables to raceways that contain electrical light and power conductors, *Section 820-52(e).*

4. Keep the coaxial cable at least 2 inches (50.8 mm) from light and power conductors unless the conductors are in a raceway, non-metallic-sheathed cable, armored cable, or UF cable, *Section 820-52(a)(2).* This clearance requirement really pertains to old knob-and-tube wiring. The 2-inch clearance is not required if the coaxial cable is in a raceway.

5. Where underground coaxial cables are run, they must be separated by at least 12 inches (305 mm) from underground light and power conductors, unless the underground conductors are in a raceway, Type UF cable, or Type USE cable, *Section 820-11(b).* The 12-inch clearance would not be required if the coaxial cable has a metal cable armor.

6. The grounding conductor *(Article 820, Part D)*:

   a. Must be insulated.

   b. Must not be smaller than No. 14 copper or other corrosion-resistant conductive material.

   c. May be solid or stranded.

   d. Must be run in a line as straight as practicable.

   e. Shall be guarded from physical damage.

   f. Shall be connected to the nearest accessible location on one of the following:

      the building or structure grounding electrode, *Section 250-50*

      the grounded interior metal water pipes, *Section 250-104(a)*

      the power service accessible means external to the enclosure, *Section 250-92(b)*

      the metallic power service raceway

      the service equipment enclosure

      the grounding electrode conductor or its metal enclosure

      the grounding electrode conductor

      the grounding electrode of a building's disconnect when these are properly grounded.

   g. If *none* of the options in *f* is available, then ground to any of the electrodes per *Section 250-50,* such as metal underground water pipes, metal frame of building, concrete encased electrode, or ground ring.

h. If *none* of *f* or *g* is available, then ground to an effectively grounded metal structure or to any of the electrodes per *Section 250-52*, such as driven rod or pipe, or a metal plate.

i. *CAUTION:* If any of the preceding results in one grounding electrode for the CATV shield grounding and another grounding electrode for the electrical system, bond the two electrodes together with a bonding jumper no smaller than No. 6 copper or its equal. Grounding and bonding the coaxial cable shield to the same grounding electrode as the main service ground minimizes the possibility that a difference of potential might exist between the shield and other grounded objects, such as a water pipe.

SATELLITE ANTENNAS

A satellite antenna is often referred to as a "dish," Figure 25-6. A satellite has a parabolic shape that concentrates and reflects the signal beamed down from one of the many satellites located 22,247 miles above the equator. The orbit in which satellites travel is called the "Clarke" belt, Figure 25-10.

Satellite reception has made it possible for the homeowner to:

- use satellite reception only, in which case the homeowner would not receive local programs.

- use satellite reception *and* use CATV for local channels. By law, CATV companies must supply this basic service and cannot require the homeowner to sign up for premium programs.

- use satellite reception for all of the programs offered by the satellite company, *and* a roof-top or similar antenna for local channels.

Satellite antennas are available in solid or "see-through" types. See-through types are constructed of perforated metal, and have less wind resistance and less weight than solid dishes. The size of the antenna shown in Figure 25-6(B) can be from 6 to 10 feet (1.83 to 3.05 m) in diameter. Depending upon local codes, satellite antennas can be mounted on the roof, a chimney, a post, a pedestal, or on the side of the

A B

Figure 25-6   A. A digital satellite system 18-inch (457 mm) antenna, a receiver, and a remote control. *Courtesy of Thomson Consumer Electronics.* B. A large six-foot (1.83 m) satellite antenna securely mounted to a post that is anchored in the ground.

house. Some localities do not permit these large size antennas.

The line-of-sight from the antenna to the satellite must not be obstructed by trees, buildings, telephone poles, or other structures.

The latest satellite technology is the *digital satellite system* (DSS). A digital system allows the use of small antennas, approximately 18 inches (457 mm) in diameter, that can easily be mounted on the roof, the chimney, the side of the house, on a pipe, or on a pedestal, using the proper mounting hardware, Figure 25-6(A). A digital system offers clearer pictures and clearer sound, greater choices of channel selection, and neater appearance insofar as the size of the antenna is concerned. The receiver of a digital system requires a connection to the telephone system for program billing, updating the program access, and ensuring continuous program service. Once each month the satellite company will call the receiver and download all of the information that has been stored. Any and all pay-per-view and other premium services used during that month will be recorded during this very brief downloading. The digital satellite receiver has a modular telephone phone jack on the back of the receiver for this connection.

The basic operation of a satellite system is for a transmitter on earth to beam an "uplink" signal to a satellite in space. Electronic devices on the satellite reamplify and convert this signal to a "downlink" signal, then retransmit the downlink signal back to earth, Figure 25-7.

C-band satellites transmit at 3.7 to 4.2 GHz.

KU-band (low and medium power) transmit at 11.7 to 12.2 GHz.

KU-band (high power) transmit at 12.2 to 12.7 GHz. This high frequency makes it possible to use a small 18-inch (457 mm) antenna.

The term *gigahertz* (GHz) means billions of cycles per second.

The term *megahertz* (MHz) means millions of cycles per second.

On earth, the downlink signal is picked up by the "feedhorn" located on the front of a satellite antenna, Figure 25-6.

A "feedhorn" is located at the exact focal point in relationship to the parabolic reflector to gather the maximum amount of signal. The feedhorn funnels the downlink satellite signal into a "low noise amplifier" where the signal is amplified, then fed to a "downconverter" where the signals are converted from high frequency to low frequency (70 MHz)

SATELLITE 22,247 MILES ABOVE THE EARTH

UPLINK SIGNAL

DOWNLINK SIGNAL

EARTH STATION TRANSMITTER

SATELLITE "DISH" ANTENNA

Figure 25-7 **The earth station transmitter beams the signal up to the satellite where the signal is amplified, converted, and beamed back to earth. The power requirements of the satellite are in the range of 5 watts and are provided by the use of solar cells.**

CONNECT TO PROPER TERMINALS ON SATELLITE ANTENNA

5 OR 6 CONDUCTOR CABLE TO DOWN CONVERTER AND LOW-NOISE AMPLIFIER

75-OHM COAXIAL CABLE

STANDARD ANTENNA

5 OR 6 CONDUCTOR CABLE TO MOTOR

A-B SWITCH

CABLE

RECEIVER

75-OHM COAXIAL CABLE

75-OHM COAXIAL CABLE

120-VOLT LINE CORD

VCR

TV

Figure 25-8 Typical satellite antenna/cable/ standard antenna wiring. Connection to a standard antenna cable, CATV cable, or to a satellite cable enables the homeowner to watch one channel while recording another channel. The installation manual for the receiver usually has a number of different hookup diagrams. In this diagram, the A-B switch allows a choice of connecting to a standard antenna or to the CATV cable. This hookup can be desirable should the CATV or satellite reception fall.

SATELLITE RECEIVER

CH 3

CH 4

SATELLITE IN

IN FROM ANT OUT TO TV

PHONE JACK

Figure 25-9 The terminals on the back of a digital satellite receiver. The connections are similar to those in Figure 25-8. Note the difference in that the digital satellite receiver has a modular telephone plug, a feature that uses a toll-free 800 number to update the access card inside the receiver to ensure continuous program service. The telephone can also handle program billing.

bling their uplink signals. The home satellite dish owner needs a special descrambling device to be able to decode and view these scrambled channels.

Small 18-inch (457 mm) antennas are easily focused following the manufacturer's instructions. Large antennas are more difficult to focus, and in some instances are motor driven. One manufacturer utilizes a reversible 36-volt dc motor, controllable from inside the home for fine-tune focusing of the antenna to the satellite.

For the more complicated large dish installations, the cables that are installed between the receiver inside the home and the satellite antenna consist of color-coded conductors, making them easy to identify when doing the hook-up. The number of conductors necessary depends upon circuit designs of the receiver and the satellite. For the newer digital satellite systems, the hook-up is greatly simplified, using 75-ohm coaxial cable between the receiver and the antenna. Always follow the manufacturer's installation instructions.

All satellites in space are in the same orbit, Figure 25-10.

Figure 25-11 shows one method of installing a large satellite antenna to a metal post securely positioned in concrete.

that is proper for the receiver. All of these are mounted in a housing in front of the antenna.

The receiver is located inside the home and is connected to the "downconverter" on the dish with a 75-ohm coaxial cable. The receiver in turn is connected to the television set, Figures 25-8 and 25-9.

More and more satellite programmers are scram-

Figure 25-10   All satellites travel around the earth in the same orbit called the "Clarke" belt. They appear to be stationary in space because they are rotating at the same speed at which the earth rotates. This is called "geosynchronous orbit (geostationary orbit)." The satellite receives the uplink signal from earth, amplifies the signal, and transmits it back to earth. The downlink signal is picked up by the satellite antenna.

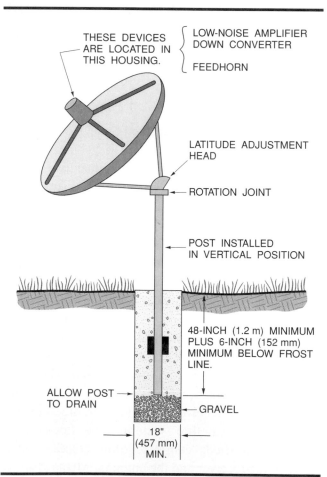

Figure 25-11   Typical satellite antenna of the type installed in remote areas where reception is poor, or in areas served by CATV, but where the homeowner does not wish to subscribe and pay for the service of the CATV company. Always check with the local inspectors to be sure that all installation code rules are met.

CODE RULES FOR THE INSTALLATION OF ANTENNAS AND LEAD-IN WIRES
(ARTICLE 810)

Home television, video cassette recorders (VCRs), compact disc equipment, and AM and FM radios generally come complete with built-in antennas. For those locations in outlying areas on the fringe or out of reach of strong signals, it is quite common to install a separate antenna system.

Either indoor or outdoor antennas may be used with black and white and color televisions. The front of an outdoor antenna is aimed at the television transmitting station. When there is more than one transmitting station and they are located in different directions, a rotor is installed. A rotor turns the antenna on its mast so that it can face in the direction of each transmitter. The rotor is controlled from inside the building. The rotor controller's cord is plugged into a regular 120-volt receptacle to obtain power. A four-wire cable is usually installed between the rotor motor and the control unit. The wiring for a rotor may be installed during the roughing-in stage of construction, running the rotor's four-wire cable into a device wall box, allowing 5 or 6 feet (1.5 to 1.8 m) of extra cable, then installing a regular single-gang switch-plate when finishing.

Article 810 covers this subject. Although instructions are supplied with antennas, the following key points of the Code regarding the installation of antennas and lead-in wires should be followed.

1. Antennas and lead-in wires shall be securely supported, *Section 810-12.*

2. Antennas and lead-in wires shall not be attached to the electric service mast, *Section 810-12.*

3. Antennas and lead-in wires shall be kept away from all light and power conductors to avoid accidental contact with the light and power conductors, *Section 810-13.*

4. Antennas and lead-in wires shall not be attached to any poles that carry light and power wires over 250 volts between conductors, *Section 810-12.*

5. Lead-in wires shall be securely attached to the antenna, *Section 810-12.*

6. Outdoor antennas and lead-in conductors shall not cross over light and power wires, *Section 810-13.*

7. Lead-in conductors shall be kept at least 2 feet (610 mm) away from open light and power conductors, *Section 810-13.*

8. Where practicable, antenna conductors shall not be run under open light and power conductors, *Section 810-13.*

9. On the outside of a building:
   a. Position and fasten lead-in wires so they cannot swing closer than 2 feet (610 mm) to light and power wires having *not* over 250 volts between conductors; 10 feet (3.05 m) if *over* 250 volts between conductors, *Section 810-18(a).*
   b. Keep lead-in conductors at least 6 feet (1.83 m) away from a lightning rod system, or bond together according to *Section 250-60* and *Section 810-18(a).*
   Note: The clearances in (a) and (b) are not required if the light and power conductors or the lead-in conductors are in a raceway or metal cable armor.

10. On the inside of a building:
    a. Keep antenna and lead-in wires at least 2 inches (51 mm) from other open wiring (as in old houses) unless the other wiring is in a raceway or cable, *Section 810-18(b).*
    b. Keep lead-in wires out of electric boxes unless there is an effective, permanently

installed barrier to separate the light and power wires from the lead-in wire, *Section 810-18(c).*

11. Grounding, Figure 25-12:
    a. The grounding wire must be copper, aluminum, copper-clad steel, bronze, or similar corrosion-resistant material, *Section 810-21(a).*
    b. The grounding wire need not be insulated. It must be securely fastened in place, may be attached directly to a surface without the need for insulating supports, shall be protected from physical damage or be large enough to compensate for lack of protection, and shall be run in as straight a line as is practicable, *Section 810-21(b), (c),* and *(d).*
    c. The grounding conductor shall be connected to the nearest accessible location on: the building or structure grounding electrode *(Section 250-50)*; or, the grounded interior water pipe *(Section 250-104(a))*; or, the metallic power service raceway; or, the service equipment enclosure; or, the grounding electrode conductor or its metal enclosure, *Section 810-21(f)(1).*
    d. If none of *c* is available, then ground to any one of the electrodes per *Section 250-50,* such as metal underground water pipe, metal frame of building, concrete-encased electrode, or ground ring, *Section 810-21(f)(2).*

GROUND WIRE FITTING

COAX CABLE
TO TV SET OR
CABLE BOX

COAX CABLE
TO CATV
OR SATELLITE
RECEIVER

GROUNDING
CONDUCTOR
MINIMUM
NO. 10 AWG
COPPER

"LISTED"
GROUND
CLAMP

GROUNDED INTERIOR
METAL WATER PIPE

Figure 25-12   Typical connections for the required grounding per *Section 810-21.*

e. If neither *c* nor *d* are available, then ground to an effectively grounded metal structure or to any one of the electrodes per *Section 250-52*, driven rod or pipe, or a metal plate, *Section 810-21(f)(3)*.

f. The grounding conductor may be run inside or outside of the building, *Section 810-21(g)*.

g. The grounding conductor shall not be smaller than No. 10 copper or No. 8 aluminum, *Section 810-21(h)*.

h. *CAUTION:* If any of the preceding rules results in one grounding electrode for the antenna and another grounding electrode for the electrical system, bond the two electrodes together with a bonding jumper not smaller than No. 6 copper or its equal, *Section 810-21(j)*.

The objective of grounding and bonding to the same grounding electrode as the main service ground is to reduce the possibility of having a difference of voltage potential between the two systems.

TELEPHONE WIRING (SYMBOL ▶|)
(ARTICLE 800)

Since the deregulation of telephone companies, residential "do-it-yourself" telephone wiring has become quite common. Following is an overview of residential telephone wiring.

The telephone company will install the service line to a residence and terminate at a protector device, Figure 25-13. The protector protects the system from hazardous voltages. The protector may be mounted either outside or inside the home. Different telephone companies have different rules. Always check with the phone company before starting your installation.

Figure 25-13   Typical residential telephone installation. The telephone company installs and connects the underground cable to the protector. In this illustration, the protector is mounted to the metal service-entrance raceway. This establishes the ground that offers protection against hazardous voltages such as a lightning surge on the incoming line. A cable is then run to a Standard Network Interface, referred to as SNI or just NI. These are shown in (A) and (B) in the diagram. The interior telephone wiring is then connected to the SNI, either with a modular plug as shown in (A), or "hard-wired" through a neoprene grommet on the bottom of (B). The SNI of the type depicted by (B) are preferred because they are weatherproof and can be mounted on the outside of the house.

The point where the telephone company ends and the homeowner's interior wiring begins is called the *demarcation point*, Figure 25-13. The preferred demarcation point is outside of the home, generally near the electric meter where proper grounding and bonding can be done. A *network interface* might be of the types depicted in Figure 25-13, at (A) and (B). In the past, interior telephone wiring in homes was installed by the telephone company. Today, the responsibility for roughing-in the interior wiring is usually left up to the electrician. Telephone wiring is

not to be run in the same raceway with electrical wiring. In addition, it cannot be installed in the same outlet and junction boxes as electrical wiring. Whatever the case, the rules found in *Article 800* apply. Refer to Figures 25-13, 25-14, and 25-15.

For a typical residence, the electrician will rough-in wall boxes wherever a telephone outlet is wanted. The boxes can be of the "single-gang" types illustrated in Figures 25-1, 2-11, 2-17, 2-19, and 2-32. Concealed telephone wiring in new homes is generally accomplished using multipair cables that are run from each telephone outlet to a common point, Figure 25-14(A). The cable can also take the form of a "loop system," Figure 25-15. A 3-pair, 6-conductor twisted cable is shown in Figure 25-16. There are some localities that require that these cables be run in a metal raceway, such as electrical metallic tubing (EMT).

The outer jacket is usually of a thermoplastic material. This outer jacket most often is a neutral color such as light gray or beige that blends in with decorator colors in the home. This is particularly important in existing homes where the telephone cable may have to be exposed. Cables, mounting boxes, junction boxes, terminal blocks, jacks, adaptors, faceplates, cords, hardware, plugs, etc., are all

STANDARD NETWORK INTERFACE (SNI)

JUNCTION BOX

A

B JUNCTION BOX IT FEATURES A SHORT, PREWIRED CORD THAT PLUGS INTO A STANDARD NETWORK INTERFACE (SNI). IT ALLOWS EASY CONNECTION OF ADDITIONAL TELEPHONE EXTENSIONS.

Figure 25-14 Individual telephone cables run to each telephone outlet from the telephone company's connection point.

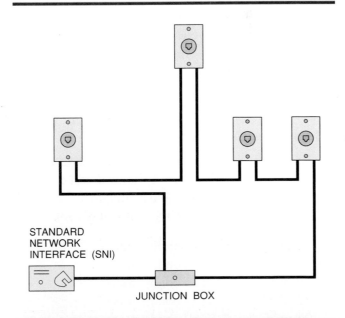

STANDARD NETWORK INTERFACE (SNI)

JUNCTION BOX

Figure 25-15 A complete "loop system" of the telephone cable. If something happens to one section of the cable, the circuit can be fed from the other direction.

❶ *TIP* IS THE CONDUCTOR THAT IS CONNECTED TO THE TELEPHONE COMPANY'S "POSITIVE" TERMINAL. IT IS SIMILAR TO THE NEUTRAL CONDUCTOR OF A RESIDENTIAL WIRING CIRCUIT.

❷ *RING* IS THE CONDUCTOR THAT IS CONNECTED TO THE TELEPHONE COMPANY'S "NEGATIVE" TERMINAL. IT IS SIMILAR TO THE "HOT" UNGROUNDED CONDUCTOR OF A RESIDENTIAL WIRING CIRCUIT.

Figure 25-16   Color coding for a 2-pair and a 3-pair, 6-conductor twisted telephone cable. The color coding for 2-pair telephone cable stands alone. For three or more pair cables, the color coding becomes WHITE/BLUE, WHITE/ORANGE, WHITE/GREEN, WHITE/BROWN, WHITE/SLATE, RED/BLUE, RED/ORANGE, RED/GREEN, RED/BROWN, RED/SLATE, BLACK/BLUE, BLACK/ORANGE, BLACK/GREEN, BLACK/ BROWN, BLACK/SLATE, YELLOW/BLUE, YELLOW/ORANGE, YELLOW/GREEN, YELLOW/BROWN, YELLOW/SLATE, VIOLET/BLUE, VIOLET/ORANGE, VIOLET/GREEN, VIOLET/BROWN, VIOLET/SLATE. In multipair cables, the additional pairs can be used for more telephones, fax machines, security reporting, speaker phones, dialers, background music, etc.

available through electrical distributors, builders' supply outlets, telephone stores, electronic stores, hardware stores, and similar wholesale and retail outlets. The selection of types of the preceding components is endless.

Telephone Conductors

The color coding of telephone cables is shown in Figure 25-16. Figure 25-17 shows some of the many types of telephone cable available.

Cables are available in the following sizes:

2 pair – – – – – (4 conductors)
3 pair – – – – – (6 conductors)
6 pair – – – – – (12 conductors)
12 pair – – – – – (24 conductors)
25 pair – – – – – (50 conductors)
50 pair – – – – – (100 conductors)

Telephone circuits require a separate pair of con-ductors from the telephone all the way back to the phone company's central switching center. To keep interference that could come from other electrical equipment to a minimum, such as electric motors and fluorescent fixtures, each pair of telephone wires (2-wires) is twisted.

Cross-Talk Problems

A common telephone interference problem in homes is "cross-talk." This problem is most often traced to 2-line telephones where flat cords are used. The signals traveling in one pair are picked up by the adjacent pair in the cord. This cross-talk problem is virtually eliminated by using cords that have con-ductors twisted at proper intervals. In Figure 25-16, each pair of conductors is twisted, then all of the pairs are again twisted. Avoid the use of flat 2-line cords in lengths over 7 feet (2.1 m). Flat 1-line cords present no problem.

Figure 25-17 Some of the many types of telephone cable available are illustrated. Note that for ease of installation, the modular plugs and terminals have been attached by the manufacturer of the cables. These cords are available in round and flat configurations, depending upon the number of conductors in the cable.

The recommended circuit lengths are:

No. 24 gauge – – – – not over 200 feet (61 m)

No. 22 gauge – – – – not over 250 feet (76.2 m)

Many telephone companies suggest that not more than five telephones be connected to one line because there may not be enough power to ring more than five ringers. Most newer telephones indicate on the label or on the carton the telephone REN (Ringer Equivalence Number). The REN refers to the older style electromechanical ringer which was considered to be 1-REN. Cordless telephones have a rating of 0.2 to 0.3 RENS. Electronic ringers use very little power. Therefore, many electronic phones can be connected to one line before running into problems. It is a good idea to call the local telephone company and ask if there are limitations concerning the number of phones that may be connected to one line.

EXAMPLE:

Five 1-REN phones = 5 RENS
Three 1-REN and one 2-REN phone = 5 RENS
Ten ½-REN phones = 5 RENS

If you are considering installing more than five telephones in a home, pay attention to the REN numbers. Possible line cut-off can occur when too many phones are connected to one line.

When several telephones are used in a residence, most telephone companies recommend that at least one of these telephones be permanently installed. This would probably be the wall phone in the kitchen. The remaining phones may remain portable. This means that they can be plugged into any of the phone jacks, Figures 25-18 and 25-19. If all of the phones were portable, there might be a time when none of the phones were connected to phone jacks. As a result there would be no audible signal.

Figure 25-18 Telephone modular jack.

Figure 25-19 Three styles of wall plates for modular telephone jacks: rectangular stainless steel, weatherproof for outdoor use, and circular.

The electrical plans show that nine telephone outlets are provided as follows:

Front Bedroom	1
Kitchen/Nook	1
Laundry	1
Living Room	1
Master Bedroom	1
Rear Outdoor Patio	1
Recreation Room	1
Study/Bedroom	1
Workshop	1

Installation of Telephone Conductors
(Sections 800-51 and 800-52)

Telephone conductors:

1. are generally Type CM or Type CMX (for dwellings only), listed for telephone installation as being resistant to the spread of fire. Refer to *Table 800-50* for other types.

2. shall be separated by at least 2 inches (51 mm) from light and power conductors unless the light and power conductors are in a raceway, or in nonmetallic-sheathed cable, Type AC cable, or Type UF cable, *Section 800-52(c)(2)*.

3. shall not be placed in any conduit or boxes with electric light and power conductors unless the conductors are separated by a partition, *Section 800-52(c)(1)*.

4. may be terminated in either metallic or nonmetallic boxes. Check with the local telephone company or local electrical inspector for specific requirements.*

5. should not share the same stud space as the electrical branch-circuit wiring.*

6. should be kept at least 12 inches (305 mm) away from electrical branch-circuit wiring where the telephone cable is run parallel to the branch-circuit wiring. This guards against induction noise.*

7. should not share the same bored holes as electrical wiring, plumbing, or gas pipes.*

8. should be kept away from hot water pipes, hot air ducts, and other heat sources that might harm the insulation.*

9. should be attached to the sides of joists and studs with insulated staples, being careful not to crush the cable.*

10. Do not support telephone conductors from raceways that contain light and power conductors, *Section 800-52(e)*.

Conduit

In some communities, electricians prefer to install regular device boxes at a telephone outlet location, then "stub" a ½-inch conduit to the basement or attic from this box. Then, at a later date, the telephone cable can be "fished" through the conduit.

Grounding *(Article 800, Part D)*

The telephone company will provide the proper grounding of their incoming cable sheath and protector, generally using wire not smaller than No. 14 AWG copper or equal. Or, they might clamp the protector to the grounded metal service raceway conduit to establish "ground." The grounding require-

---

* These are recommendations from residential telephone manufacturers' installation data.

ments are practically the same as for CATV and antennas, which were covered earlier in this Unit.

You can review these requirements in *Section 800-40*. Also see *Section 250-92(b)(3)* and the *Fine Print Notes*.

Safety

Open circuit voltage between conductors of an idle pair of telephone conductors can range from 50 to 60 volts dc. The ringing voltage can reach 90 volts ac. Therefore, always work carefully with insulated tools and stay clear of bare terminals and grounded surfaces. Disconnect the interior telephone wiring if work must be done on the circuit, or take the phone off the hook, in which case the dc voltage level will drop and there should be no ac ringing voltage delivered.

SIGNAL SYSTEM (CHIMES)

A signal circuit is described in the *NEC®* as any electrical circuit that energizes signaling equipment. Signaling equipment includes such devices as chimes, doorbells, buzzers, code-calling systems, and signal lights.

Door Chimes (Symbol CH)

Present-day dwellings often use chimes rather than bells or buzzers to announce that someone is at a door. A musical tone is sounded rather than a harsh ringing or buzzing sound. Chimes are available in single-note, two-note, eight-note (four-tube),

and repeater tone styles. In a repeater tone chime, both notes sound as long as the push button is depressed. In an eight-note chime, contacts on a motor-driven cam are arranged in sequence to sound the notes of a simple melody when the chime button is pushed. This type of chime is usually installed in dwellings having three entrances. The chime can be connected so that the eight-note melody sounds for the front door, two notes sound for the side door, and a single note sounds for the rear door. Chimes are also available with clocks and lights.

Electronic chimes may relay their chime tones through the various speakers of an intercom system. When any chime is installed, the manufacturer's instructions must be followed.

The plans show that two chimes are installed in the residence. Two-note chimes are used. Each chime has two solenoids and two iron plungers. When one solenoid is energized, the iron plunger is drawn into the opening of the solenoid. A plastic peg in the end of the plunger strikes one chime tone bar. When the solenoid is de-energized, spring action returns the plunger where it comes to rest against a soft felt pad so that it does not strike the other chime tone bar. Thus, a single chime tone sounds. As the second solenoid is energized, one chime tone bar is struck. When the second solenoid is de-energized, the plunger returns and strikes the second tone bar. A two-tone signal is produced. The plunger then comes to rest between the two tone bars. Generally, two notes indicate front door signaling, and one note indicates rear or side door signaling.

Figure 25-20 shows various push-button styles used for chimes. Many other styles are available. Figure

Figure 25-20 Push buttons for door chimes. *Courtesy* Nutone, Inc.

Figure 25-21 Typical residential door chimes. *Courtesy* NuTone, Inc.

25-21 shows several types of typical residential-type wall-mounted chimes. The symbols used to indicate push buttons and audible signals on the plans are shown in Figure 25-22.

Figures 25-23 and 25-26 show how to provide proper backing for chimes. This is done during the rough-in stages of the electrical installation.

Figure 25-22 Symbols for chimes and push buttons.

BE SURE TO PROVIDE
ADEQUATE BACKING TO
SUPPORT THE CHIME

RECOMMENDED HEIGHT
"EYE" LEVEL TO CENTER
OF CHIME, BUT NO
HIGHER THAN 6'6" (1.98 m)
TO CENTER OF CHIME.

CHIMES WITH TUBES MIGHT
HAVE A TOTAL LENGTH OF
50" (1.27 m) OR MORE, IN WHICH
CASE "EYE" LEVEL MIGHT BE
TOO LOW. MANUFACTURERS
OF CHIMES OFFER MOUNTING
HEIGHT SUGGESTIONS IN
THEIR CATALOGS.

Figure 25-23   A method of providing support for a surface-mounted chime. This is the rough-in stage, before the walls are closed.

Chimes for the Hearing Impaired

Chimes for the hearing impaired are available with accessory devices that can turn on a dedicated lamp at the same time the chime is sounded. Figure 25-24 shows the devices needed to accomplish this. Figure 25-25 shows the wiring details.

Transformers

Because of the low-voltage ratings and power limitations of chime transformers, the secondary wiring and the transformer itself are classified as Class 2 circuits.

The transformers used to operate door chimes have a greater capacity than the transformers used with bell and buzzer circuits, Figure 25-27. The voltage output of chime transformers ranges between 10 and 24 volts. These transformers are rated from 5 to 30 volt-amperes (watts). Bell transformers have a voltage output range of 6 to 10 volts and a rating of 5 to 20 volt-amperes (watts).

Chime transformers used in dwellings are available with a 16-volt rating. Transformers that give a combination of voltages—such as 4, 8, 12, and 24 volts—also are used.

Underwriters Laboratories (U.L.) lists two types of chime transformers that are normally used for chime wiring. The *inherently limited transformer* is designed to limit the short-circuit current to a maximum of 8 amperes. The *not inherently limited transformer* is rated at 100 volt-amperes or less. This transformer has an overcurrent protective device that limits the voltage and current. See *Section 725-41*. The open circuit voltage limitation of both types of transformers must not exceed 30 volts. Most transformers suitable for use in dwellings have a built-in thermal overload device. Whenever a short circuit occurs in the bell-wire circuit, the overload device opens and closes repeatedly until the short is cleared.

Figure 25-24 Accessory devices can be attached to new or existing chimes to provide visual signals when the chime is sounded. The first photo shows the module that is connected and mounted inside the chime. The second photo shows a transmitter that is plugged into a nearby receptacle. The third photo shows a receiver into which a lamp is plugged.

CHIME

"A" IS A MODULE THAT MOUNTS INSIDE THE CHIME. IT IS CONNECTED TO THE SAME TERMINALS ON THE CHIME AS THE REGULAR CONNECTIONS.

TWO WIRES ARE RUN BETWEEN MODULE "A" AND RECEIVER "B." THESE WIRES ARE LOW VOLTAGE, SAME AS REGULAR CHIME WIRES. RUN CONDUCTORS CONCEALED IN WALL, OR RUN ALONG DOOR FRAMES AND BASEBOARDS.

TRANSMITTER "B" PLUGS INTO A NEARBY 120-VOLT WALL OUTLET.

RECEIVER "C" PLUGS INTO A 120-VOLT WALL OUTLET. LAMP PLUGS INTO RECEIVER. MORE THAN ONE RECEIVER CAN BE PLUGGED IN. THE SIGNAL IS TRANSMITTED THROUGH THE 120-VOLT HOUSE WIRING.

Figure 25-25 Installation diagram for auxiliary devices for use with chimes that can help the hearing impaired. When the chime sounds, the signal is transmitted to the receiver, turning on the lamp that is plugged into the receiver. The "on" time is adjustable at "C". The lamp should be a "dedicated" lamp to light up only when the chime sounds.

BE SURE TO PROVIDE
ADEQUATE BACKING FOR
SUPPORTING THE CHIME.

"ROUGH-IN" BOX FOR
FLUSH-MOUNTED CHIME.

RECOMMENDED HEIGHT
"EYE" LEVEL TO CENTER
OF CHIME, BUT NO HIGHER
THAN 6'6" (1.98 m) TO CENTER
OF CHIME. MANUFACTURERS
OF CHIMES PROVIDE MOUNTING
HEIGHT SUGGESTIONS IN
THEIR CATALOGS.

Figure 25-26   Roughing-in for a flush-mounted chime.

Figure 25-27   Chime transformers. *Courtesy* NuTone.

Additional Chimes

To extend a chime system to cover a larger area, a second or third chime may be added. In the residence, two chimes are used: one chime is mounted in the front hall and a second (extension) chime is mounted in the recreation room. The extension chime is wired in parallel to the first chime. The wires are run from one chime terminal board to the terminal board of the other chime. The terminals are connected as follows: transformer to transformer, front to front, and rear to rear, Figure 25-28.

When chimes are added to a chime circuit, a transformer with a higher wattage rating may be required to energize the greater number of solenoids being used at one time. Also, more wire is used in the circuit and the voltage drop and power loss are greater. This increase is the result of resistance in the wire between the chimes and the transformer. *Section 725-41(b)* does not allow transformers supplying Class 2 systems to be connected in parallel unless listed for interconnection.

If a buzzer (or bell) and a chime are connected to a single transformer and are used at the same time, the transformer will put out a fluctuating voltage. This condition does not allow either the buzzer or the chime to operate properly. The use of a transformer with a larger rating may solve this problem.

The wattage consumption of chimes varies with the manufacturer. Typical ratings are as follows:

TYPE OF CHIME	POWER CONSUMPTION
Standard two-note	10 watts
Repeating chime	10 watts
Internally lighted, two lamps	10 watts
Internally lighted, four lamps	15 watts
Combination chime and clock	15 watts
Motor-driven chime	15 watts
Electronic chime	15 watts

Transformers with ratings of 5, 10, 15, 20, and 30 watts (VA) are available. For a multiple chime installation, wattage ratings for the individual chimes are added. The total value is the minimum transformer rating needed to do the job properly. If there are still technical questions or problems, check the literature supplied by the chime and transformer manufacturers.

Wiring for Chime Installation

Bell Wire and Cable.   The wire used for low-voltage bell and chime circuits, and for connecting low-voltage thermostats, is called *bell wire, annunciator wire,* or *thermostat wire.*

These wires are generally made of copper and have a thermoplastic insulation suitable for use on 30 volts or less.

Because the current required for bell and chime circuits is rather small, No. 18 AWG conductors are quite often used. If the length of the circuit is very long, a larger conductor is suggested.

Multiconductor cables consist of two, three, or more single wires covered with a single protective insulation. This type of cable is often used for electrical installations because there is less danger of damage to individual wires and it gives a neat appearance to the wiring. The conductors within the cables are color coded to make circuit identification easy.

Bell wire and cable may be fastened directly to surfaces with insulated staples or cleats, or may be installed in the raceways. Be careful not to pierce or crush the wires. Do not pull bell wires or other low-voltage cables through the same holes in studs and joists that contain nonmetallic-sheathed cable, BX, conduit, or other piping because they cannot stand the physical abuse that they would be subjected to. The job requirements will determine how the conductors are to be attached. In the residence in the plans, the bell wire is run along the sides of the floor joists in the basement and on the sides of the studs in the walls.

Wiring Circuit.   The circuit shown in Figure 25-28 is recommended for this chime installation because it provides a hot low-voltage circuit at the front hall location. However, it is not the only way in which these chimes may be connected.

Figure 25-28 shows that a two-wire cable runs from the transformer in the utility room to the front hall chime. A two-wire cable then runs from the chime to both the front and rear door push buttons. A three-wire cable also runs between the front hall chime and the recreation room chime. Because of

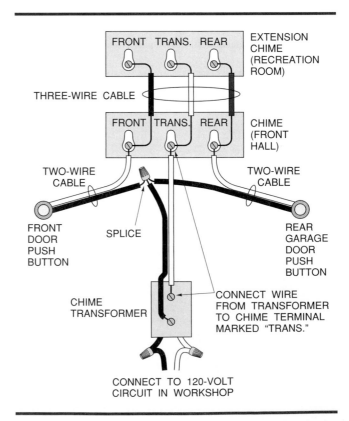

Figure 25-28   Circuit for chime installation.

the hot, low-voltage circuit at the front hall location, a chime with a built-in clock can be used. A four-conductor cable may be run to the extension chime so that the owner may install a clock-chime at this location also.

The plans show that the chime transformer is mounted on one of the ceiling boxes in the utility room. The line voltage connections are easy to make here. Some electricians prefer to mount the chime transformer on the top or side of the distribution panel or load center. The electrician decides where to mount the transformer after considering the factors of convenience, economy, and good wiring practice.

It may seem obvious, but the issue of accessibility bears repeating. Electrical equipment junction boxes and conduit bodies must be accessible. It is permitted to have certain electrical equipment such as a chime transformer, junction boxes and conduit bodies above a dropped lay-in ceiling because these are accessible by dropping out a ceiling panel. Never install electrical equipment, junction boxes, or conduit boxes above a permanently closed-in ceiling or within a wall. There may come a time when this electrical equipment and junction boxes need to be accessed. This includes chime transformers.

NEC® Rules for Signal Systems

Article 725 covers the requirements for low-voltage, Class 2 circuits, discussed in Unit 24.

Repeating some of the most basic Code installation requirements:

- Do *not* install Class 2 wiring in the same raceway or cable as light and power wiring, *Section 725-54(a)(1)*.

- Keep Class 2 bell wires at least 2 inches (50.8 mm) away from light or power wiring. This really pertains to old, open-knob, and tube wiring. Today, where light and power wiring is installed in Type NMC, Type AC, or in conduit, there is no problem, *Section 725-54(a)(3)*.

- Class 2 wiring may not enter the same enclosure, unless there is a barrier in the enclosure that separates the Class 2 wiring from the light and power conductors, *Section 725-54(a)(1)*.

- Chime transformers are listed as Class 2 transformers. Thus, all of the wiring connected to the secondary of the transformer is Class 2 wiring.

- Do *not* run Class 2 cables through the same holes as light and power cables and conduits. There is too much chance of physical damage to the Class 2 cables.

- Do not strap, tape, or attach by any means Class 2 cables to other electrical raceways, *Section 725-54(d)*.

- Class 2 wiring must be installed in a neat and workmanlike manner, and shall be supported by the building structure in such a manner that the cable will not be damaged by normal building use, *Section 725-7*.

- Because the insulation and the jacket of Class 2 cables and conductors are rather thin, be careful during installation to avoid damage. Secure these cables and conductors with care, using the proper type of staples. Staple "guns" are available that use rounded staples that nicely straddle the cable instead of flattening the cable, which might result in shorted out wires.

REVIEW

Note: Refer to the Code or the plans where necessary.

TELEVISION CIRCUIT

1. How many television outlets are installed in this residence? _____

2. What type of television cable is commonly used and recommended? _____

_____

3. What determines the design of the faceplates used? _____

_____

4. What must be provided when installing a television outlet and receptacle outlet in one wall box? _____

_____

_____

_____

_____

_____

5. From a cost standpoint, which system is more economical to install: a master amplifier distribution system or a multiset coupler? Explain the basic differences between these two systems. _____

_____

_____

_____

_____

_____

_____

_____

_____

6. How many wires are in the cable used between a rotor and its controller? _____

7. Digital satellite systems use an antenna that is approximately (18 inches) (36 inches) (72 inches) in diameter. Circle the correct answer.

8. List the requirements for cable television inside the house. _____

_____

_____

_____

_____

_____

9. What article of the Code references the requirements for cable community television installation? _____

10. It is generally understood that grounding and bonding together all metal parts of an electrical system and the metal shield of the cable television cable to the same grounding reference point in a residence will keep both systems at the same voltage level should a surge, such as lightning, occur. Therefore, if the incoming cable to a house that has been installed by the CATV cable company installer has the metal shield grounded to a driven ground rod, does this installation conform to the *NEC®*?

_____

11. The basics of cable television are that a transmitter on earth sends an _____ signal to a _____ where the signal is reamplified and sent back to earth via the _____ signal. This signal is picked up by the _____ on the satellite antenna, which is sometimes referred to as _____ a. The feedhorn funnels the signal into a _____ _____ _____ , then on to a _____ , where the high-frequency signal is converted to a _____-frequency signal suitable for the television receiver.

12. All television satellites rotate above the earth in (the same orbit) (different orbits). Circle correct answer.

13. Television satellites are set in orbit (10,000) (18,000) (22,247) miles above the earth, which results in their rotating around the earth at (precisely the same) (different) rotational speed as the earth rotates. This is done so that the satellite "dish" can be focused on a specific satellite (once) (one time each month) (whenever the television set is used). Circle correct answer.

14. What section of the Code prohibits supporting coaxial cables from raceways that contain light or power conductors? Section _____

15. When hooking up one CATV cable and another cable from an outdoor antenna to a receiver that has only one antenna input terminal, an _____ switch is usually installed.

TELEPHONE SYSTEM

1. How many locations are provided for telephones in the residence? _____

2. At what height are the telephone outlets in this residence mounted? Give measurement to center.

_____

_____

_____

_____

_____

3. Sketch the symbol for a telephone outlet.

4. Is the telephone system regulated by the *NEC*®? _____

5. a. Who is to furnish the outlet boxes required at each telephone outlet? _____

   b. Who is to furnish the faceplates? _____

6. Who is to furnish the telephones? _____

7. Who does the actual installation of the telephone equipment? _____

8. How are the telephone cables concealed in this residence? _____

_____

_____

9. The point where the telephone company's cable ends and the interior telephone wiring meet is called the _____ point. The device installed at this point is called a _____ .

10. What are the colors contained in a four-conductor telephone cable assembly and what are they used for? _____

_____

_____

_____

_____

_____

11. Itemize the Code rules for the installation of telephone cables in a residence. _____

_____

_____

_____

12. If finger contact were made between the red conductor and green conductor at the instant a "ring" occurs, what shock voltage would be felt? _____

13. What section of the Code prohibits supporting telephone wires from raceways that contain light or power conductors? *Section* _____

14. The term *cross-talk* is used to define the hearing of the faint sound of voices in the background when you are using the telephone. Cross-talk can be reduced significantly by running (flat cables) (twisted pair cables) to the telephones. Circle the correct answer.

15. What section of the Code prohibits telephone cables from being installed in the same box or enclosure, or from being pulled into the same raceway as light and power circuits? _____

SIGNAL SYSTEM

1. What is a signal circuit? _____

_____

2. What style of chime is used in this residence? _____

3. a. How many solenoids are contained in a two-tone chime? _____

   b. What closes the circuit to the solenoid of a chime? _____

4. Explain briefly how two notes are sounded by depressing one push button (when two solenoids are provided). _____

_____

_____

_____

_____

_____

5. a. Sketch the symbol for a push button. _____

   b. Sketch the symbol for a chime. _____

6. a. At what voltage do residence chimes generally operate? _____

   b. How is this voltage obtained? _____

7. What is the maximum volt-ampere rating of transformers supplying Class 2 systems?

   _____

8. What two types of chime transformers for Class 2 systems are listed by U.L.?

   _____

   _____

9. Is the extension chime connected with the front hall chime in series or in parallel?

   _____

10. How many bell wires terminate at

    a. the transformer? _____

    b. the front hall chime? _____

    c. the extension chime? _____

    d. each push button? _____

11. a. What change in equipment may be necessary when more than one chime is con-
        nected to sound at the same time on one circuit? _____

    _____

    _____

    b. Why? _____

    _____

    _____

    _____

    _____

12. What type of insulation is usually found on low-voltage wires? _____

13. What size wire is installed for signal systems of the type in this residence? _____

14. a. How many wires are run between the front hall chime and the extension chime in
        the recreation room? _____

    b. How many wires are required to provide a "hot" low-voltage circuit at the exten-
        sion chime? _____

15. Why is it recommended that the low-voltage secondary of the transformer be run to
    the front hall chime location and separate two-wire cables be installed to each push
    button? _____

    _____

16. a. Is it permissible to install low-voltage Class 2 systems in the same raceway or
        enclosure with light and proper wiring? _____

    b. What Code section covers this? _____

17. a. Where is the transformer in the residence mounted? _____

    b. To what circuit is the transformer connected? _____

18. a. How many feet of two-conductor bell wire cable are required? _____

 b. How many feet of three-conductor bell wire cable are required? _____

19. How many insulated staples are needed for the bell wire if it is stapled every 2 feet (610 mm)? _____

20. Should low-voltage bell wire be pulled through the same holes in studs and joists that also contain nonmetallic-sheathed cable? _____

21. What section of the Code prohibits supporting fire-protective signaling circuits from raceways that contain light or power conductors? *Section* _____ .

Heat, Smoke, and Carbon Monoxide Detectors, Fire Alarms, Security Systems

OBJECTIVES

After studying this Unit, the student will be able to

- understand the *National Fire Protection Association Standard No. 72*, which is the National Fire Alarm Code.
- name the two basic types of smoke detectors.
- discuss the location requirements for the installation of heat and smoke detectors for minimum acceptable levels of protection.
- discuss the location requirements for the installation of heat and smoke detectors for increased protection that exceed the minimum acceptable levels of protection.
- list the major components of typical residential smoke, heat, and security systems.
- discuss general *NEC*® requirements for the installation of residential smoke, heat, and security systems.
- understand the basics of carbon monoxide detectors.

THE IMPORTANCE OF HEAT, SMOKE, AND CARBON MONOXIDE DETECTORS

This Unit will touch upon the basic requirements for protection against the hazards of fire, heat, smoke, and carbon monoxide.

Fire is the third leading cause of accidental death. Home fires account for the biggest share of these fatalities, most of which occur at night during sleeping hours.

Heat, smoke, and carbon monoxide detectors are installed in a residence to give the occupants early warning of the presence of fire or toxic fumes. Fires produce smoke and toxic gases that can overcome the occupants while they sleep. Most fatalities result from the inhalation of smoke and toxic gases, rather than from burns. Heavy smoke reduces visibility.

In nearly all home fires, detectable smoke precedes detectable levels of heat. People sleeping are less likely to smell smoke than people who are awake. The smell of smoke probably will not awaken a sleeping person. Therefore, smoke detectors are considered to be the primary devices for protecting lives. Heat detectors should be installed in addition to smoke detectors, but should not replace them. A rather recent entry into the arena are carbon monoxide detectors, which sense dangerous carbon monoxide emitted from a malfunctioning furnace or other source. The *National Fire Protection Association* Standard No. 720 covers carbon monoxide detection and protection. Underwriters Laboratories (U.L.) Standard 2034 covers single and multiple station carbon monoxide detectors.

NATIONAL ELECTRICAL CODE®

Article 760 covers the installation requirements for fire alarm systems that operate at 600 volts or less. Because the topic of fire protective signaling systems is so complex, the *NEC®* refers to a number of National Fire Protection Association standards that contain information and recommendations not found in the *NEC®* The following is a brief overview of some of the key points of *Article 760.*

Fire protective signaling systems are classified into two groups:

1. Nonpower-limited circuits (NPLFA):
   - the conductors require overcurrent protection per the values found in *Section 310-15* (the same as for regular conductors). Maximum overcurrent protection for No. 18 AWG conductors is 7 amperes, and for No. 16 AWG it is 10 amperes, *Section 760-23.*
   - derating the ampacity of the conductors required if the conductors carry continuous loads in excess of 10% of their ampacity, *Section 760-28(b)(1).*
   - may be in the same enclosure, raceway, or cable *only* where connected to the same equipment, *Section 760-26(b).*
   - only copper conductors permitted, *Section 760-31(a).*
   - conductors larger than No. 16 AWG may be types listed in *Section 310-13.* No. 18 and No. 16 AWG conductors must be of the types listed for nonpower-limited signaling circuit use, Types NPLFP, NPLFR, NPLF. See *Section 760-31.*
   - cables must be listed as resistant to the spread of fire. *Table 760-31(g)* shows the marking found on these cables.
   - cables may be installed on the surface or concealed, and shall be adequately supported and protected from physical damage, *Section 760-8.*
   - cables shall not be installed in ducts or plenums unless listed for that purpose, *Section 760-30(b)(1).*

2. Power-limited (PLFA):
   - must have power output capabilities limited to the values specified in *Chapter 9, Tables 12(a)* and *12(b), Section 760-41.*
   - wiring on supply side (line side) to be installed according to nonpower-limited circuit requirements and in accordance with the normal wiring requirements found in *Chapters 1* through *4* of the *NEC.®*
   - wiring on load side to have copper conductors of the types listed in *Table 760-61.*
   - conductors and cables may be run on the surface or be concealed, shall be adequately supported and protected against physical damage, *Section 760-51(b)(1).*
   - power-limited fire alarm conductors shall not be supported by taping, strapping, or attaching by any means to the exterior of any conduit or other raceway, *Section 760-54(c).*
   - conductors and cables shall be listed as suitable for the purpose, Types FPLP, FPLR, FPL, *Section 760-61.*
   - *Table 760-61* provides a table that shows the different types of cable that can be substituted for one another when necessary.
   - conductors and cables shall not be run in the same raceway, or be in the same enclosure as light and power wiring unless separated by a barrier or are in the enclosure to connect that particular piece of equipment.

NATIONAL FIRE PROTECTION ASSOCIATION (NFPA) STANDARD NO. 72

The *National Fire Protection Association (NFPA) Standard No. 72,* "National Fire Alarm Code," presents the minimum requirements for the proper selection, installation, operation, and maintenance of fire warning equipment. The standard also includes many recommendations that exceed the minimum requirements. Chapter 2 specifically covers household fire warning equipment.

NFPA Standard No. 72 defines a household fire alarm system as *"A system of devices that produces an alarm signal in the household for the purpose of notifying the occupants of the presence of a fire so that they will evacuate the premises."*

Fire warning devices commonly used in a residence are heat detectors, Figure 26-1, and smoke detectors, Figure 26-2.

The more elaborate systems connect to a central monitoring customer service center through a telephone line. These systems offer instant contact with police or fire departments, a panic button for

Figure 26-1 Heat detector.

Figure 26-2 Smoke detector.

emergency medical problems, low temperature detection, flood (high-water) detection, perimeter protection, interior motion detection, and other features. All detections are transmitted first to the company's customer service center, where personnel on duty monitor the system 24 hours a day, 7 days a week. They will verify that the signal is valid. After verification, they will contact the police department, the fire department, and individuals whose names appear on a previously agreed upon list that the homeowner prepared and submitted to the company's customer service center.

NFPA standards are *not* a law until adopted by a city, state, or other governmental body. Some NFPA standards are adopted in total, and others are adopted in part. Installers must be aware of all requirements in the locality in which they are doing work. Most communities require a special permit for the installation of fire warning equipment and security systems

in homes, and require registration of the system with the police and/or fire departments. In most instances, fire protection requirements are found in the building codes of a community, and not necessarily spelled out in that community's electrical code. Remember, the *NEC®* sets forth the requirements of how to physically install a fire protective signaling system. It does *not* say where to install the detectors. The "where to install" information is found in *NFPA Standard No. 72*, which is the standard adopted by most building officials.

A sufficient number of detectors, properly located, must be installed to provide adequate protection. These requirements and recommendations are discussed later in this Unit.

TYPES OF SMOKE DETECTORS

Two common types of smoke detectors are the photoelectronic type and the ionization type. They usually contain an indication light to show that the unit is functioning properly. They also may have a test button that simulates smoke, so that when the button is pushed, the detector's smoke detecting ability as well as its circuitry and alarm are tested. Some systems are tested with a magnet.

Smoke detectors generally will not sense heat, flame, or gas. However, some smoke detectors can be set off by acetylene and propane gas.

Photoelectronic Type

The photoelectronic type of smoke detector has a light sensor that measures the amount of light in a chamber. When smoke is present, an alarm is sounded that indicates a reduction in light due to the obstruction of the smoke. This type of sensor detects smoke from burning materials that produce great quantities of smoke, such as furniture, mattresses, and rags. This type of detector is less effective for gasoline and alcohol fires, which do not produce heavy smoke.

Ionization Type

The ionization type of detector contains a low-level radioactive source, which supplies particles that ionize the air in the detector's smoke chamber. Plates in this chamber are oppositely charged. Because the air is ionized, an extremely small amount of current (millionths of an ampere) flows between the plates. Smoke entering the chamber impedes the movement of the ions, reducing the current flow, which causes an alarm to sound.

The ionization type of detector is effective for detecting small amounts of smoke, as in gasoline and alcohol fires, which are fast-flaming.

TYPES OF HEAT DETECTORS

Some of the types of heat detectors are described in the following.

- *Fixed* temperature heat detectors are available that sense a specific *fixed* temperature, such as 135°F or 200°F. They shall have a temperature rating at least 25°F (14°C) above the normal temperature expected, but not to exceed 50°F (28°C) above the expected temperature.

- *Rate-of-Rise* heat detectors sense rapid changes in temperature (12 to 15 degrees per minute), such as those caused by flash fires.

- *Fixed/Rate of Rise* temperature detectors are available as a combination unit.

- *Fixed/Rate of Rise/Smoke* combination detectors are also available in one unit.

INSTALLATION REQUIREMENTS

The following information includes specific recommendations for installing smoke and heat detec-

tors in homes. Some of these requirements are illustrated in Figures 26-3, 26-4, and 26-5. Complete data is found in *NFPA Standard No. 72* and in the instructions furnished by the manufacturer of the equipment.

Individual detectors as well as complete systems can be purchased at electrical supply houses, home centers, hardware stores, and similar retail outlets. These are packaged with the necessary mounting hardware and detailed installation instructions.

Once installed, detectors should be checked periodically to make sure they are operating properly.

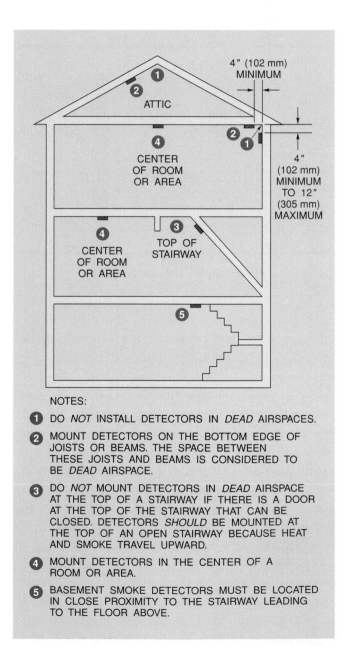

NOTES:

❶ DO *NOT* INSTALL DETECTORS IN *DEAD* AIRSPACES.

❷ MOUNT DETECTORS ON THE BOTTOM EDGE OF JOISTS OR BEAMS. THE SPACE BETWEEN THESE JOISTS AND BEAMS IS CONSIDERED TO BE *DEAD* AIRSPACE.

❸ DO *NOT* MOUNT DETECTORS IN *DEAD* AIRSPACE AT THE TOP OF A STAIRWAY IF THERE IS A DOOR AT THE TOP OF THE STAIRWAY THAT CAN BE CLOSED. DETECTORS *SHOULD* BE MOUNTED AT THE TOP OF AN OPEN STAIRWAY BECAUSE HEAT AND SMOKE TRAVEL UPWARD.

❹ MOUNT DETECTORS IN THE CENTER OF A ROOM OR AREA.

❺ BASEMENT SMOKE DETECTORS MUST BE LOCATED IN CLOSE PROXIMITY TO THE STAIRWAY LEADING TO THE FLOOR ABOVE.

Figure 26-4 **Recommendations for the installation of heat and smoke detectors.**

Figure 26-3 **Recommended location of heat or smoke detector *between* sleeping areas and rest of house. Locate outside of the bedroom, but near the bedrooms. In new construction, a smoke detector must be installed in each bedroom.**

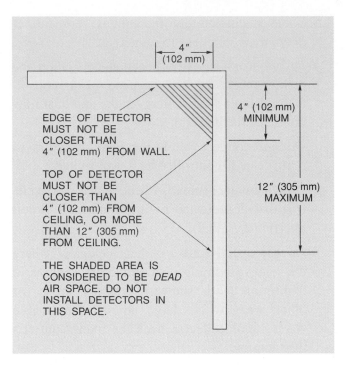

Figure 26-5 Do not mount smoke or heat detectors in the *dead* airspace where the ceiling meets the wall.

Testing every six months is quite common. Refer to the manufacturer's recommendations.

Many companies specialize in installing complete fire and security systems. They also offer system monitoring at a central office and can notify the police or fire department when the system gives the alarm. How elaborate the system should be is up to the homeowner.

Many local building codes do have requirements for the location, number, and type of smoke detectors to be installed in residential occupancies.

The Absolute Minimum Level of Protection

To meet the absolute minimum level of protection, do the following things:

- Install smoke detectors on each level of a residence.

- Install a smoke detector in each bedroom or sleeping area of a residence.

- In new construction, interconnect smoke detectors so that when one operates, all others will also operate. See Figure 26-7.

- Make sure that the detectors are listed by a recognized testing agency.

Exceeding Minimum Levels of Protection

The following are recommendations for attaining levels of protection that exceed the minimum level stated above and include guidelines for installing smoke and heat detectors in a home.

- Install smoke detectors for protection that *exceeds* the minimum acceptable protection levels, in *all* rooms, basements, hallways, heated attached garages, and storage areas. Installing detectors in these locations will increase escape time, particularly if the room or area is separated by a door(s). In some instances, smoke detectors are installed in attics and crawl spaces. Check the manufacturer's recommendations for installing smoke detectors in attics, crawl spaces, and garages.

- Consider special detectors that light up or vibrate when the occupants are hard of hearing.

- Consider low temperature detectors (i.e., 45°F) that can detect low temperatures should the heating system fail. The damage caused by frozen water pipes bursting can be extremely costly.

Installation

Following are general recommendations for installing smoke and heat detectors.

- *Do* install smoke detectors on the ceiling, as close as possible to the center of the room or hallway.

- *Do* install smoke detectors at the high end of a room that has a sloped, gabled, or peaked ceiling where the rise is greater than 1 foot per 8 feet measured horizontally. Do not install in the dead air space, Figures 26-4 and 26-5.

- *Do* install *heat* detectors at the high end of a room that has a sloped, gabled, or peaked ceiling where the rise is greater than 1 foot per 8 feet measured horizontally. Mount them within 3 feet (914 mm) of the peak, but do not install in the dead air space, Figures 26-4 and 26-5.

- *Do* make sure that the path of rising smoke will reach the smoke detector when the detector is installed in a stairwell. This will usually be at the top of the stairway. The detector must not be located in a dead air space created by a closed door at the head of the stairway.

- *Do* install smoke detectors on a basement ceiling close to the stairway to the first floor.

- *Do* install smoke detectors in split-level homes. A smoke detector installed on the ceiling of an upper level can suffice for the protection of an adjacent lower level, provided the two levels are not separated by a door. Better protection is to install detectors for each level.

- *Do* install smoke detectors that are hard-wired directly to a 120 VAC source when installing detectors in *new* construction. In most cities this is mandatory. Dual-powered detectors having both 120 VAC and battery power provide the best protection. In existing homes, battery-powered detectors are the most common, but 120 VAC detectors or dual-powered detectors can be installed. Remember, battery-powered units will not operate with dead batteries. AC powered units will not operate when the power supply is off.

- *Do* install interconnected detectors when required by local codes. When one goes off, all other detectors also are triggered.

- *Do* always consider the fact that doors, beams, joists, walls, partitions, and similar obstructions will interfere with the flow of heat and smoke and, in most cases, create new areas needing additional smoke and heat detectors.

- *Do* consider that the accepted maximum distance between heat detectors mounted on flat ceilings is 50 feet (15 m), and from detector to wall 25 feet (7.6 m). This information is explained in detail by *NFPA 72*. Where beams, joists, etc. will obstruct the flow of heat, the 50-foot distance is reduced to 25 feet (7.6 m), and the 25-foot distance is reduced to 12½ feet (3.81 m).

- *Do not* install smoke or heat detectors in the dead air space at the top of a stairway that can be closed off by a door.

- *Do not* place the edge of a ceiling-mounted smoke or heat detector closer than 4 inches (102 mm) to the wall.

- *Do not* place the top edge of a wall-mounted smoke or heat detector closer than 4 inches (102 mm) to the ceiling and farther than 12 inches (305 mm) down from the ceiling.

Some manufacturers recommend not farther down than 6 inches (152 mm).

- *Do not* install smoke or heat detectors where the ceiling meets the wall because this is considered dead air space where smoke and heat may not reach the detector.

- *Do not* connect smoke or heat detectors to wiring that is controlled by a wall switch.

- *Do not* connect smoke or heat detectors to a circuit that is protected by a ground-fault circuit interrupter (GFCI).

- *Do not* install smoke or heat detectors where the relative humidity exceeds 85%, such as in bathrooms with showers, laundry areas, or other areas where large amounts of visible water vapor collect. Check the manufacturer's instructions.

- *Do not* install smoke or heat detectors in front of air ducts, air conditioners, or any high-draft areas where the moving air will keep the smoke or heat from entering the detector.

- *Do not* install smoke detectors in kitchens where the accumulation of household smoke can result in setting off an alarm, even though there is no real hazard. The person in the kitchen will know why the alarm sounded, but other people in the house may panic. This problem exists in multifamily dwelling units, where unwanted triggering of the alarm in one dwelling unit might cause panic for people in the other units. The photoelectronic type *may* be installed in kitchens, but must *not* be installed directly over the range or cooking appliance. A better choice in a kitchen would be to install a heat detector. Consult the manufacturer's recommendations.

- *Do not* install smoke detectors where the temperature can fall below 32°F or rise above 120°F unless the detector is specifically identified for this application.

- *Do not* install smoke detectors in garages where vehicle exhaust might set off the detector. Instead of a smoke detector, install a heat detector.

- *Do not* install smoke detectors in airstreams that will pass air originating at the kitchen cooking appliances across the detector. False alarms will result.

- *Do not* install smoke detectors near fluorescent fixtures. Install at least 6 feet (1.8 m) away.

- *Do not* install smoke or heat detectors on a ceiling where the ceiling will be excessively cold or hot. Smoke and heat will have difficulty reaching the detectors. This could be the case in older homes that are not insulated—or are poorly insulated. Instead, mount the detectors on the side wall, Figure 26-5.

- *Do not* install smoke or heat detectors on an outside wall that is not insulated, or poorly insulated. Instead, mount the detectors on an inside wall.

- *Do not* install smoke or heat detectors on a ceiling that employs radiant heating.

Manufacturers' Requirements

Here are a few of the responsibilities of the manufacturer of smoke and heat detectors. Complete data is found in *NFPA Standard No. 72*.

- The power supply must be capable of operating the signal for at least 4 minutes continuously.

- Battery-powered units must be capable of a low battery warning "beep" of at least one beep per minute for seven consecutive days.

- Direct-connected 120V ac detectors must have a visible indicator that shows "power-on."

- Detectors must not signal when a power loss occurs or when power is restored.

Wiring Requirements

Direct-Connected Units. These units are connected to a 120-volt circuit. The black and white wires on the unit are spliced to the black and white wires of the 120-volt circuit, Figure 26-6.

Feed-through (Tandem) Units. These units may be connected in tandem up to 10 units. Any one of these detectors can sense smoke, then send a signal to the remaining detectors, setting them off so that all detectors sound an alarm. These units contain three wires: one black, one white, and one yellow, Figure 26-7. The yellow wire is the interconnecting wire.

Cord-Connected Units. These units operate in the same way as direct-connected units, except that they are plugged into a wall outlet that is not controlled by a switch. These units are mounted to the wall on a bracket or screws. A clip on the cord is attached to the wall outlet with the same screw that fastens the faceplate to the outlet, Figure 26-8. This prevents the cord from being pulled out, rendering the unit inoperative.

Battery-Operated Units. These units require no wiring; they are simply mounted in the desired locations. Batteries last for about a year. Battery-operated units should be tested periodically, as recommended by the manufacturer.

Figure 26-7 Wiring of feed-through (tandem) detector units. These detectors are "interconnected."

Figure 26-6 Wiring of direct-connected detector units.

Figure 26-8 Cord-connected detector units are plugged into wall outlet.

COMBINATION DIRECT/BATTERY/FEED-THROUGH DETECTORS

Combination direct/battery/feed-through units are connected directly to a 120 V ac circuit (not GFCI protected) and are equipped with a 9-volt dc battery, Figure 26-9. Thus, in the event of a power outage, these detectors will continue to operate on the 9-volt dc battery. The ability to interconnect a number of detectors offers "whole house" protection, because when one unit senses smoke, all the other interconnected units will sound an alarm.

Read carefully the application and installation manuals that are prepared by the manufacturers of the smoke and heat detectors.

In this residence we find four smoke detectors connected to Circuit A17. These units are hard-wired directly to the 120-volt ac branch-circuit. They are interconnected feed-through units that will set off signaling in all units should one unit trigger. These interconnected smoke detectors are also powered by a 9-volt dc battery that allows the unit to operate in the event of a power failure in the 120-volt ac supply.

POWER SUPPLIES

For new homes, *NFPA Standard No. 72, Section 2-3.1.2* requires that smoke detectors be powered by 120-volt ac supply as the main power source, and by a battery as the secondary (standby) power source. In existing homes, *Exception No. 5* to *Section 2-3.1.2* permits the detectors to be either ac only or battery

Figure 26-9 An AC/DC smoke alarm that operates on both 120 volts AC power and a 9-volt battery. In the event of a power outage, the detector continues to operate on the 9-volt battery. A constant green LED indicates that the unit is operating on 120-volt AC power. An intermittent red LED indicates that the unit has switched to the 9-volt battery back-up mode. *Courtesy of* BRK Electronics.

powered only. Check with your local building officials to find out what they require in your community.

CARBON MONOXIDE DETECTORS

When a heat or smoke detector goes off, it is generally not difficult to determine the source of the problem. But when a carbon monoxide detector goes off, it is hard to determine the source.

Carbon monoxide is odorless, invisible, and tasteless. It is undetectable to any of a person's five senses —taste, smell, sight, touch, hear. Carbon monoxide replaces oxygen in the bloodstream, resulting in brain damage or total suffocation.

Carbon monoxide results when fuel burns incompletely. Common sources of carbon monoxide in a home can come from a malfunctioning furnace, gas appliances, kerosene heaters, automobile or other gas engines, charcoal grills, fireplaces, or a clogged chimney.

Symptoms of carbon monoxide poisoning are similar to flulike illnesses such as dizziness, fatigue, headaches, nausea, and irregular breathing.

Carbon monoxide (CO) detectors detect carbon monoxide from any source of combustion. They are not designed to detect smoke, heat, or gas. The Consumer Product Safety Commission (CPSC) recommends that at least one carbon monoxide detector be installed in each home. If only one carbon monoxide detector is installed, it should be in the area just outside of individual bedrooms. The CPSC recommends that a carbon monoxide detector be installed on each floor of a multilevel house.

Carbon monoxide detectors monitor the air in the house, and sound a loud alarm when carbon monoxide above a predetermined level is detected. They provide early warning before deadly gases build up to dangerous levels.

For residential applications, carbon monoxide detectors are available for hard wiring (direct connection to the branch-circuit wiring), plug-in (the male attachment plug is built into the back of the detector for plugging directly into a 120-volt wall receptacle outlet), cord connected (has a cord and plug), battery operated, and combination 120-volt and battery-operated units.

NFPA Standard No. 720 is the *Household Carbon Monoxide (CO) Warning Equipment Standard.* This standard contains the installation recommendations for carbon monoxide detectors. Here is a summary of these recommendations.

Installation

- *Do* buy carbon monoxide detectors that are tested and listed in conformance to U.L. Standard 2034.

- *Do* install carbon monoxide detectors according to the manufacturer's instructions. This is a requirement of *Section 110-3(b)* of the *NEC.®*

- *Do* install carbon monoxide detectors in or near bedrooms and living areas.

- *Do* install carbon monoxide detectors in locations where smoke detectors are installed.

- *Do* test carbon monoxide detectors once each week, or as recommended by the manufacturer.

- *Do* remove carbon monoxide detectors before painting, stripping, wall-papering, or using aerosol sprays. Store in a plastic bag until project is completed.

- *Do not* connect carbon monoxide detectors to a switched circuit, or a circuit controlled by a dimmer.

- *Do not* install carbon monoxide detectors closer than 6 inches from a ceiling. This is considered dead air space.

- *Do not* install carbon monoxide detectors in garages, kitchens, or furnace rooms. This could lead to nuisance alarms, may subject the detector to substances that can damage or contaminate the detector, or the alarm might not be heard.

- *Do not* install carbon monoxide detectors within 15 feet of a cooking or heating gas appliance.

- *Do not* install carbon monoxide detectors in dusty, dirty, or greasy areas. The sensor inside of the detector can become coated or contaminated.

- *Do not* install carbon monoxide detectors where the air will be blocked from reaching the detector.

- *Do not* install carbon monoxide detectors in dead air spaces such as in the park of a vaulted ceiling, or where a ceiling and wall meet.

- *Do not* install carbon monoxide detectors where the detector will be in the direct air stream of a fan.

- *Do not* install carbon monoxide detectors where temperatures are expected to drop below 40°F

or get hotter than 100°F. Its sensitivity will be affected.

- *Do not* install carbon monoxide detectors in damp or wet locations such as showers and steamy bathrooms. Its sensitivity will be affected.

- *Do not* mount carbon monoxide detectors directly above or near a diaper pail. The high methane gas will cause the detector to register, and this is not carbon monoxide.

- *Do not* clean carbon monoxide detectors with detergents or solvents.

- *Do not* spray carbon monoxide detectors with hair spray, paint, air fresheners, or other aerosol sprays.

- *Do* carefully vacuum carbon monoxide detectors once each month or as recommended by the manufacturer.

U.L. Standard 2043 *Single and Multiple Station Carbon Monoxide Alarms* contains the requirements that manufacturers must conform to for household carbon monoxide alarms.

SECURITY SYSTEMS

It is beyond the scope of this text to cover all types of residential security systems. Instead, we will focus on some of the features available today from various manufacturers of these systems.

Professionally installed and homeowner-installed security systems can range from a simple system to a very complex system. Features and options include master control panels, remote controls, perimeter sensors for doors and windows, motion sensors, passive infrared sensors, wireless devices, interior and exterior sirens, bells, electronic buzzers, strobe lights to provide audio (sound) as well as visual detection, heat sensors, carbon monoxide sensors, smoke detectors, glass break sensors, flood sensors, low temperature sensors, and lighting modules. Figure 26-10 shows some of these devices. Figure 26-11 shows a schematic one-line diagram for a typical security system.

Monitored security systems are connected by telephone lines to a central station, a 24-hour monitoring facility. If the alarm sounds, security professionals will contact the end user to verify the alarm before sending the appropriate authorities (fire, police, ambulance) to the home.

The wireless key allows the end user to disarm and arm the system with the press of a button. There is no need to remember a security code and it can also be used to operate lights and appliances.

Monitored smoke detectors are on 24 hours a day, even if the burglary protection is turned off.

Magnetic contacts are often located at vulnerable entry points such as doors.

Sirens, such as these placed inside the home, alert families to emergencies.

Motion detectors are interior security devices that sense motion inside the home.

A wireless panic button allows the user to press a button to signal for help.

The wireless keypad is portable so you can arm or disarm the system from the yard or driveway.

Glassbreak detectors, available in a variety of sizes, will sound the alarm when they detect the sound of breaking glass.

Figure 26-10 *Photos courtesy of* First Alert Professional Security Systems.

The decision about how detailed an individual residential security system should be generally begins with a meeting between the homeowner and the security consultant or licensed electrical contractor before the actual installation. Most security system installers, consultants, and licensed electrical contractors are familiar with a particular manufacturer's system.

The systems are usually explained and even demonstrated during an in-home security presentation. They are then professionally installed by the security system company or licensed electrical contractor.

Do-It-Yourself kits that contain many of the

Figure 26-11   Diagram of typical residential security system showing some of the devices available. Complete wiring and installation instructions are included with these systems. Check local code requirements in addition to following the detailed instructions furnished with the system. Most of the interconnecting conductors are No. 18 AWG.

features provided by the professionally installed systems can be found at home centers, hardware stores, electrical distributors, and electronic stores. Depending on the system, it may or may not be possible to connect to a central monitoring facility.

The wiring of a security system consists of small, easy to install, low-voltage, multiconductor cables made up of No. 18 AWG conductors. These conductors should be installed after the regular house wiring is completed to prevent damage to these smaller cables. Usually the wiring can be done at the same time as the chime wiring is being

installed. Security system wiring comes under the scope of *Article 725*.

When wiring detectors such as door entry, glass break, floor mat, and window foil detectors, circuits are electrically connected in series so that if any part of the circuit is opened, the security system will detect the open circuit. These circuits are generally referred to as "closed" or "closed loop." Alarms, horns, and other signaling devices are connected in parallel, because they will all signal at the same time when the security system is set off. Heat detectors and smoke detectors are generally connected in parallel because all of these devices will "close" the circuit to the security master control unit, setting off the alarms.

The instructions furnished with all security systems cover the installation requirements in detail, alerting the installer to Code regulations, clearances, suggested locations, and mounting heights of the systems components.

Always check with the local electrical inspector to determine if there are any special requirements in your locality relative to the installation of security systems.

Do not support power-limited fire-protective signaling circuit conductors from raceways that contain light and power conductors, *Section 760-54(c)*.

REVIEW

Note: Refer to the Code or the plans where necessary.

1. What article in the *NEC*® covers fire protective signaling systems? _____

   _____

2. Is it permissible to connect smoke and fire detectors with the same kind of low-voltage wire that is used to hook up door chimes, or must cables that are listed for the purpose be used? _____

   _____

3. Find in *Article 760* the section that prohibits electrical raceways from being used to support the cables of a fire protective signaling system installation. *Section* _____

4. Find in *Article 760* the sections that prohibit fire protective system wires and cables from being installed in the same raceway as light and power conductors. *Section*

   _____

5. If it is necessary to install fire protective system cables through a duct, the cables (circle the correct answer):

   a. must be listed for the purpose.

   b. must be at least No. 14 AWG.

   c. are not permitted to be run through ducts.

6. What is the name and number of the NFPA standard that covers smoke and heat detectors for homes? _____

7. Has the standard that covers smoke and heat detectors for homes been adopted by the community in which you live? _____

8. a. Name the two basic types of heat detectors. _____

   _____

   b. Name the two basic types of smoke detectors. _____

   _____

9. Why is it important to mount a smoke or heat detector on the ceiling not closer than 4 inches (102 mm) from a wall? _____

   _____

   _____

10. In new residential construction, a smoke detector must be installed (circle the correct answer):
    a. in each bedroom or sleeping area.
    b. at the top of a basement stairway that has a closed door at the top.
    c. in a garage that is subject to subzero temperatures.

11. Always follow the installation instructions of (circle the correct answer):
    a. the manufacturer of the system.
    b. your neighbor.
    c. the man at the hardware store.

12. Smoke detectors, heat detectors, and security system wires and cables are much smaller than regular house wiring. Would it be better to install these wires (before) or (after) installing the regular house wiring to avoid possible damage to the smaller cables? Circle the correct answer.

13. Fire protection system conductors are generally No. _____ AWG.

14. When more than one smoke detector is installed in a new home, these units must be
    a. connected to separate 120-volt circuits.
    b. connected to one 120-volt circuit.
    c. interconnected so that if one detector is set off, the others would also sound a warning.
    Circle the correct answer(s).

15. It is advisable to connect smoke and heat detectors to circuits that are protected by GFCI devices. (True) (False)   Circle the correct answer.

16. Because a smoke detector might need servicing or cleaning, be sure to connect it in such a manner that it is controlled by a wall switch. (True) (False)   Circle the correct answer.

17. Smoke detectors of the type installed in homes must be able to sound a continuous alarm when set off for at least
    a. 4 minutes
    b. 30 minutes
    c. 60 minutes

18. In addition to detecting carbon monoxide, these detectors also detect heat and smoke. (True) (False)   Circle the correct answer.

19. Carbon monoxide is the result of fuel burning _____ .

20. In a small home, the best place to install a carbon monoxide detector is:
    a. in the basement.
    b. near the bedrooms.
    c. in the kitchen.

21. It is a good idea to connect a carbon monoxide detector to a switched receptacle so if it becomes necessary to work on it, it will be easy to shut off the power. (True) (False) Circle the correct answer.

22. Where would you look to learn all of the installation requirements for carbon monoxide detectors? Circle the correct answer(s).
    a. NFPA 70
    b. NFPA 720
    c. U.L. Standard 2034

Remote-Control Systems– Low Voltage

OBJECTIVES

After studying this Unit, the student will be able to

- explain the operation of a low-voltage, remote-control system for lighting circuits.
- interpret the wiring diagrams of various types of low-voltage, remote-control systems.
- install a low-voltage, remote-control system to comply with the requirements of the *NEC*.®

The general lighting circuits described in this text are wired using standard methods. Another way of installing lighting circuits is to use a remote-control or low-voltage switching system. This system provides very flexible switching.

The remote-control wiring discussed in this Unit is one manufacturer's system. In general, it is difficult to mix components of one manufacturer with those of another.

The following text is an overview of how typical remote-control, low-voltage systems operate. There are differences in the form, fit, and function of systems among the many manufacturers of these systems. Electrical wholesalers and distributors usually carry one or two brands of these systems. Once a decision has been reached to install a remote-control, low-voltage system, it is best to stay with one manufacturer's components.

There are many other types of advanced home automation systems available today. An overview of these systems are discussed in Unit 31.

A remote-control wiring system uses controlling devices (such as relays) at the equipment to be controlled. An auxiliary means is provided to operate the controlling device. A room thermostat in a residence is an example of a low-voltage remote-control system. For example, when calling for cooling, the low-voltage contacts in the thermostat "pull-in" a relay. This closes the line-voltage contacts in the relay, turning on the air conditioner. When the thermostat is satisfied, the low-voltage contacts in the thermostat open, dropping out the relay. The line-voltage contacts in the relay open, turning off the air conditioner.

A low-voltage, remote-control system can be installed in a home for the control of lighting and receptacle outlets. The system is generally more expensive than conventional wiring, but offers tremendous flexibility for the control of lighting and receptacle outlets. This type of installation lends itself to large upscale residences as well as commercial and industrial buildings.

A remote-control system for lighting circuits consists of low-voltage cable, low-voltage relays, a transformer(s), low-voltage switches, and master controls for the more elaborate systems. Because the system operates on 24 volts, the shock hazard is less than for conventional 120-volt circuitry.

REMOTE-CONTROL SWITCHES

The low-voltage, remote-control system uses special low-voltage switches that are normally open, single-pole, double-throw momentary contact switches. These switches are smaller than conventional switches and mount on yokes provided with the switches. They are also available in the same configuration as the interchangeable line of switches. They may have color-coded lead wires, screw terminals, or quick-connect terminals. Figure 27-1 shows a wiring diagram of a low-voltage switch. The white lead of the switch is connected to the transformer. The red lead of the switch is connected to the red lead of the relay (on) and the black lead is connected to the black lead of the relay (off).

Pilot light switches have a fourth wire, which is yellow. The pilot light is energized through the yellow wire returning from a relay that has the necessary auxiliary contacts.

Figure 27-2 shows various types of low-voltage, remote-control switches.

Figure 27-3 shows a master selector switch module that has a combined total of eight switches, pilot lights, or blanks for future use. In the residence, for example, master selector switch modules could be located in one of the bedrooms, the kitchen, the living room, or any other location. If one or more of these switches is used to control a master group control or master sequencer, Figure 27-4, the number of lighting and receptacle outlets to be controlled is endless.

For example, one switch in a cluster of eight switches could be hooked up to control a group controller that turns on all of the outside lights should the homeowner hear prowlers. Using a master is faster and easier than operating a number of individual switches. For more than eight switches in one location, additional master modules can be installed in a group.

Low-voltage switches may have pilot lights that indicate when a circuit is on, or locator lights for ease in finding the switch in the dark. Key-operated switches are shown in Figure 27-2.

Figure 27-1 Wiring diagram of a low-voltage switch.

Figure 27-2 Types of low-voltage switches. From left to right: (A) switch with location light, (B) two unlighted switches, (C) unlighted key operated switch, pilot light key operated switch, blank filler, pilot light, (D) a key, and (E) switch sized the same as the interchangeable line of switches shown in Figure 5-27. *Courtesy* **GE Total Lighting Control.**

These low-voltage switches are mounted in switch (device) boxes, on standard raised plaster rings selected for the thickness of the finished wall, or on brackets supplied by the manufacturer of the system. Single-gang mounting can accommodate one or two devices. Two-gang mounting can accommodate three or four devices. Master switch modules that can accommodate one through eight devices, as shown in Figure 27-3, require a 4¹¹/₁₆-inch square box.

Installing switch boxes (single-gang, two-gang, or 4-inch square with suitable plaster ring) at switch locations provides the room needed to make splices and connections, rather than merely pushing the wires back into the insulation or vapor barrier if plaster rings alone were used. It is a matter of choice —and it might be a code issue in your community. Check this out before beginning an installation.

LOW-VOLTAGE RELAYS

Figure 27-5 shows a low-voltage relay. Figure 27-6 shows the internal wiring of the relay. The relay operates on 24 volts, supplied by a transformer, Figure 27-7. The relay opens and closes the 120-volt power contacts. Relays with four low-voltage leads have a separate set of low-voltage "pilot" contacts connected to the relay's 24-volt supply. Relays with five low-voltage leads have the separate set of low-voltage "pilot" contacts connected by a 24-volt circuit independent of the relay's 24-volt supply. The four and five conductor relays are used to control other circuits, relays, master controls, pilot lights, etc.

The relay illustrated in Figure 27-5 is rated for 1 hp, 20 amperes, 125 and 277 volts. This type of relay can be used to control ac loads, including

Figure 27-3 A master selector switch that has up to eight switches, pilot lights, or blank spaces. Note the key slot. This is a key-operated lock-out switch that can be connected to deactivate one or all of the switches in the plate. *Courtesy* GE Total Lighting Control.

Figure 27-4 Where simultaneous control of more than one relay is needed, a master group control can be installed. This provides group control, but still permits control by other individual switches. This master group control has four master switch inputs. Each master switch input is connected to a bank of three relay outputs. This type of master group control is more than adequate for most residential applications. For very involved installations, a master sequencer that contains a microprocessor can be installed. A master sequencer provides a virtually unlimited amount of control. *Courtesy* GE Total Lighting Control.

Figure 27-5 A low-voltage relay. It has three leads: Red (on), black (off), and blue (common). The 3-wire relay is the basic type. Relays for pilot control of other relays, master controls, and lights contain an extra set of internal contacts. These relays will have four leads, if the pilot contacts are fed by the same circuit as the relay. Relays will have five leads if the pilot contacts are totally isolated from the circuit that supplies the relay, which allows the pilot contacts to control some other low-voltage circuit. The relay shown requires splicing of the leads. Relays are available that have quick-connect terminals, making it easier to make up the connections where many relays are mounted in a control cabinet. *Courtesy* GE Total Lighting Control.

tungsten filament and fluorescent loads, up to the full ampere rating of the relay.

Relays can be mounted through a ½-inch knockout opening of a standard outlet box. For quiet operation, they may be mounted through a ¾-inch knockout using a rubber grommet. When the relay is mounted from the inside of the outlet box through

the knockout, the two 120-volt leads of the relay remain inside the box. The 24-volt leads are kept outside the box, Figures 27-11 and 27-12.

For complex systems, component cabinets are used, Figure 27-10. These cabinets will contain conductor termination strips, relays, transformer, master switch controls, and master sequencers. These cabinets can be flush or surface mounted, and are generally installed near the fuse/circuit breaker distribution panel. A barrier in the box separates the 120-volt and 24-volt wiring.

Basically, the relay is a double solenoid. Touching the ON position of a switch momentarily energizes the coil connected to the red and blue leads. The iron core tries to center itself within the coil. In Figure 27-6, the core moves to the left. This mechanically latches the 120-volt contacts in the ON position.

Touching the OFF position of the switch momentarily energizes the coil connected to the black and blue leads. The iron core again tries to center itself within that particular coil. In Figure 27-6, the core moves to the right. This mechanically latches the 120-volt contacts in the OFF position.

For complex installations, control of relays is also possible using photo cells and touch-tone telephones. These are used in combination with master controls and sequencers, and can be connected to override the manual switches for selected circuits. In other words, you can use a touch-tone phone to turn lights on and off.

Figure 27-6 A cut-a-way view of a low-voltage relay showing the internal mechanism as well as the external line voltage and low voltage leads. Some relays have screw terminals instead of conductor leads.

Figure 27-7 A Class 2 transformer for remote-control systems. *Courtesy* GE Total Lighting Control.

Figure 27-8 A silicon rectifier that converts an ac full sine wave to a half-wave. Although not absolutely required, they are recommended for use in low-voltage systems to extend the life of relays, pilot lamps, and locator lamps. The diagram in Figure 27-12 shows a rectifier connected to the secondary of the transformer. *Courtesy* GE Total Lighting Control.

TRANSFORMERS

A low-voltage, remote-control system typically requires just one transformer, Figure 27-7. The transformer steps down the lighting circuit voltage from 120 volts to 24 volts for the low-voltage control circuit. The transformer rating is no larger than 100 volt-amperes. These transformers are listed as Class 2 transformers. When the transformer is overloaded, its output voltage decreases and there is less current output. In other words, the transformer has energy-limiting characteristics that counteract any overload. A specially designed core limits the amount of electrical energy that can be delivered at the secondary or output terminals. Some transformers have an internal thermal breaker that opens the primary circuit to protect the transformer from overheating. As soon as it cools down, the thermal breaker resets itself and the transformer is automatically reconnected to the line.

Rarely is more than one switch pressed at the same time in a low-voltage, remote-control system. Although many relays may be connected to the transformer, it is the same as if there were only one. The transformer seems to be underrated for the connected load. Actually, it can control many relays. The manufacturer will specify the maximum number of relays that a transformer can supply. One manufacturer limits the number of simultaneously operated relays to five. When the number of relays to be operated at the same time exceeds the transformer's recommended limit, master controls are used.

Some manufacturers of remote-control systems suggest that a rectifier be added to the secondary circuit, Figure 27-8. This device changes the ac supply to direct or pulsating direct current, depending upon whether a half-wave or a full-wave rectifier is used. If a relay is energized for a long time, the ac supply will produce eddy currents inside the laminations. As a result, there is a temperature buildup that may damage the relay. Recall that this type of relay is meant to provide momentary contact operation only. Direct current or pulsating direct current does not produce as much heat in this type of relay as does alternating current.

CONDUCTORS

No. 20 AWG stranded copper conductors are generally used for low-voltage, remote-control systems, Figure 27-9. To minimize voltage drop, larger conductors may be required for long runs, or where a number of relays are operated at the same time. The equipment manufacturers' technical manuals and installation manuals must be consulted to be sure the completed installation will perform as intended.

Figure 27-9 Typical cable used for remote-control wiring. *Courtesy* GE Total Lighting Control.

The thermoplastic insulation on the individual conductors is color coded for ease in identification. The conductors are protected by an outer jacket. Cables come in:

- 3-conductor cable (red, black, white) for connecting switches.

- 4-conductor cable (red, black, yellow, white) for connecting switches with pilot and locator lights.

- 4-conductor cable (red, black, blue, white) for connecting sensors.

- 5-conductor cable (red, black, yellow, yellow, blue) for connecting relays that have auxiliary contacts.

- 22-conductor cable, for connecting remote groups of six relays, has six sets of 3-conductor (red, black, yellow), two whites, a blue, and a yellow.

- 25-conductor cable, for connecting eight-gang master switch modules, has eight sets of 3-conductor (red, black, yellow) plus a No. 18 AWG white common conductor.

Individual color-coded No. 18 AWG conductors for miscellaneous connections should be black, red, blue, blue/white, and white (common) to correspond to the color coding of the multiconductor cable mentioned earlier.

The cables discussed here are suitable for use in Class 2 circuits. See *Table 725-61* of the *NEC®* for cable markings and their permitted uses. Cables marked CL2 are for general-purpose use. Cables marked CL2P are Class 2 plenum cables that can be used in air-handling plenums. Cables marked CL2X are limited to residential installations. These different markings are related to the cables' fire resistance capabilities.

Later in this Unit you will learn that Class 2 conductors are not permitted to be run together or mixed with 120-volt wiring.

Because the jacket on low-voltage cables is rather thin, the electrician must be careful not to damage the jacket during the rough-in stages of the installation. *Do not* install low-voltage wiring through the same holes that the power and lighting 120-volt cables or conduits are installed. Physical damage to the jacket and individual conductors of the low-voltage cables is possible. The possibility of short-circuits can result in relay malfunction or failure of other system components.

INSTALLATION PROCEDURE

The installation of a remote-control system in new construction begins with the roughing-in of the 120-volt system. All of the switch legs or switch loops are omitted for the system. In other words, the 120-volt supply conductors are brought directly to each outlet or switch box that will have a switched lighting fixture or switched outlet attached to it. Although the various devices used may be numbered or color coded in different ways, the electrician must remember that only three low-voltage conductors are required between the relay and the switch. In addition, two low-voltage conductors must be run from the transformer to either the relay or the switch to carry the low-voltage supply. These wires are spliced like color to like color. The wiring is then completed by slipping the relay through the ½-inch knockout in the outlet or switch box. The 120-volt connections are made in the standard way.

The low-voltage cables are stapled in place in new construction. When existing buildings are rewired, the low-voltage cables may be fished through the walls and ceiling or run behind moldings and baseboards.

Figure 27-10 **For complex low-voltage, remote-control installations, a component cabinet is installed near the electrical distribution panel. This cabinet provides adequate space and hardware for the proper mounting of relays, transformer, master switch(es), rectifier. Terminal strips are provided for ease in making up the conductor connections and terminations. The wiring trough ties the component cabinet and the electrical distribution panel together for routing of the 120-volt conductors. An equally satisfactory installation would have been to install nipples between the cabinet and the panel.**

For some installations, many or all of the relays should be placed at one location using a component cabinet, Figure 27-10. The 120-volt connections are done on one side of the barrier; the 24-volt connections are made on the other side of the barrier. Thus, the 120-volt wiring is separated from the 24-volt wiring in conformance to *Section 725-54(a)(1).*

One method of installing remote-control, low-voltage switches is shown in Figure 27-11. Multipoint control of the light results when the low-voltage, remote-control switches are connected in parallel.

Figure 27-12 shows one method of connecting a master selector module with individual low-voltage switches. The lighting outlets can be placed in different parts of the building and can be connected to operate 120-volt circuits. These outlets are then controlled by their own individual low-voltage switches or the master selector control. For simplicity, only three switches are shown connected. This particular master module can have as many as eight switches, pilot lights, or blank spaces.

In this residence, an eight-switch module could be installed in the master bedroom to control the weatherproof receptacle(s) on the front porch and on the outside wall of the front bedroom, the front entrance and garage lights (outside front), post light (front), rear yard floods, side garage door light, and

Figure 27-11 Typical cable layout and wiring diagram for low-voltage switches.

Figure 27-12 Cable layout and wiring diagram for a low-voltage remote-control system that has individual switches as well as a master selector module. The master module has eight switches. For simplicity, the diagram shows only three switches connected. Note that the individual switches and the corresponding switches in the master module are connected in parallel. The blue conductor is carried to the master module in the event that pilot light or locator switches are installed. For a more elaborate system, one or more of the switches in the master module could control a master group control or a master sequencer, as illustrated in Figure 27-4.

other lighting in the home that would be desirable to turn on from a single location. Eight-switch modules could be installed in other areas in the home. If eight switches in one location is not enough, then a master sequencer can be installed, Figures 27-4 and 27-10. The possibilities of remote control are endless, all dependent upon the complexity needed to fulfill the homeowner's needs. Cost can become the deciding factor.

CLASS 2 SYSTEMS REVIEW

We have already discussed Class 2 wiring in Units 24 and 25. Here is a review of some of the key installation issues.

Wiring on the line side (supply side) of a Class 2 transformer must conform to the wiring methods given in *Chapters 1* through *4* of the *NEC.* All wiring on the load side (secondary) of a Class 2 transformer must conform to *Section 725-52*.

Class 2 circuits are not as dangerous as standard 120-volt wiring. U.L. sets 30 volts RMS as the maximum secondary voltage for Class 2 systems. Because Class 2 circuits are power-limited, there is a lesser safety issue from both fire and electric shock standpoint. These power limitations are spelled out in *Table 11(a)*, *Chapter 9*, *NEC.*

Conductors for Class 2 circuits shall not be run in the same raceway as regular light and power conductors, *Section 725-54(a)(1)*. Class 2 wiring may enter the same compartment or enclosure as light and power conductors if separated by a barrier or if introduced solely to connect the equipment, *Section 725-54(a)(1)*, *Exceptions 1* and *2*. See Figure 27-13.

Class 2 wiring must be kept at least 2 inches (50.8 mm) away from light and power conductors, unless the light and power conductors are in a raceway, nonmetallic-sheathed cable, or UF cable,

Figure 27-13 Low-voltage Class 2 wiring *shall not* be run in the same raceway or be in the same outlet box or enclosure with light and power circuits unless separated by a barrier, *Section 725-54(a)(1)*. In Figure 27-10, the barrier serves as a mounting means for the relays, keeping the 120-volt wiring to the right of the barrier, and the low-voltage Class 2 wiring to the left of the barrier.

Section 725-54(a)(3). This clearance requirement would only be an issue if the wiring in the home was open type knob-and-tube wiring.

Class 2 wiring must not be strapped, taped, or attached by any means to conduits or other raceways, *Section 725-54(d)*. Class 2 wiring must be independently supported. *Section 725-7* tells us that Class 2 circuits shall be installed in a neat and workmanlike manner, and that the cables shall be supported by the building structure in such a manner that the cables will not be damaged by normal building use.

Because Class 2 cables are somewhat fragile, they should not be run through the same holes in the joists, studs, and other framing members with nonmetallic-sheathed cable or armored cable. They could easily become damaged, rendering the system inoperative.

REVIEW

Note: Refer to the Code or the plans where necessary.

1. What is the approximate voltage used on low-voltage remote-control systems? _____

_____

2. What are the advantages of a low-voltage system? _____

_____

_____

_____

_____

_____

3. What type of switch is used to control the relays of a low-voltage system? _____

_____

4. Are switch boxes recommended at all low-voltage switch locations? Explain. _____

_____

5. Are three-way and four-way switches used in remote-control circuits? Explain your
   answer. _____

_____

_____

_____

_____

_____

_____

6. What is a master selector switch module? _____

_____

_____

_____

_____

_____

7. What is a master group control? _____

_____

_____

_____

_____

_____

_____

8. Regarding the transformer for a low-voltage, remote-control system, what must be
   taken into consideration? _____

_____

_____

_____

_____

9. Explain briefly the operating principle of a low-voltage relay. _____

   _____

   _____

   _____

   _____

   _____

   _____

   _____

10. What is the electrical rating of relays discussed in this Unit? _____

    _____

11. What will happen if the common lead from the transformer is connected to the red lead of the relay instead of the blue lead, and the blue and black leads are connected to the switch leads? _____

    _____

    _____

    _____

    _____

12. a. Is it permitted to run low-voltage conductors and line-voltage (120-volt) conductors in the same raceway or into the same box? Explain your answer. _____

    _____

    _____

    _____

    b. Low-voltage, remote-control wires and cables (should) (should not) be installed through the same bored holes that contain nonmetallic-sheathed cables, armored cables, electrical conduits, or water pipes. Circle the correct answer.

13. The maximum nameplate rating of Class 2 transformers is _____ volt-amperes.

14. Because the insulation on low-voltage wire is much thinner than the insulation on regular building wire, the electrician should _____ when installing the conductors to avoid _____ the insulation.

15. a. What is the purpose of a rectifier? _____

    _____

    b. Must rectifiers be used on low-voltage systems? _____

    _____

16. What type of conductors are generally used for a low-voltage remote-control system?

    _____

    _____

    _____

    _____

    _____

17. a. Why are color-coded wires and cable used in low-voltage control systems? _____

_____

_____

b. What is the color coding for three-conductor low-voltage cable that is used for connecting switches? _____

_____

18. The type of low-voltage wiring discussed in this Unit is classified as (Class 1) (Class 2). Circle one.

19. When overcurrent protection is not provided in the secondary circuits of the transformers described in this Unit, the transformers must _____ their power output.

20. When installing a Class 2 transformer, the wiring on the primary side of the transformer must conform to *Chapters* _____ through _____ of the *NEC.* The maximum size overcurrent protection on the supply side of the Class 2 transformer is _____ amperes. These requirements are found in *Section* _____ of *Article 725.*

21. Complete all line voltage and low-voltage connections in the following diagram. Use colored pencils to indicate conductors. Show low-voltage connections with a dot. Switch No. 1 controls lamps A and C, switch No. 2 controls lamp B, and switch No. 3 controls the bottom of each convenience receptacle. The top of each receptacle is to be hot at all times. The line-voltage wiring is in armored cable.

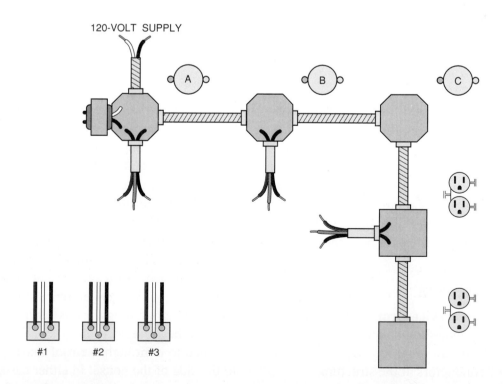

UNIT 28

Service-Entrance Equipment

OBJECTIVES

After studying this Unit, the student will be able to

- define electrical service, overhead service, service drop, and underground service.
- list the various Code sections covering the installation of a mast-type overhead service and an underground service.
- discuss the Code requirements for disconnecting the electrical service using a main panel and load centers.
- discuss the grounding of interior ac systems and the bonding of all service-entrance equipment.
- discuss the "UFER" ground method.
- describe the various types of fuses.
- select the proper fuse for a particular installation.
- explain the operation of fuses and circuit breakers.
- explain the term *interrupting rating*.
- determine available short-circuit current using a simple formula.

An electric service is required for all buildings containing an electrical system and receiving electrical energy from a utility company. ►The *NEC*® defines the term *service* as "the conductors and equipment for delivering electric energy from the serving utility to the wiring system of the premises served."◄ The utility company generally must be contacted to determine where they want the meter to be located.

OVERHEAD SERVICE

The Code defines a **service drop** as the overhead service conductors, including splices, if any, which are connected from an outdoor support to the service-entrance conductors at the structure.

The overhead service includes all of the service equipment and installation means from the attach-

ment of the service-drop wires on the outside of the building to the point where the circuits or feeders are tapped to supply specific loads or load centers. Usually, residential type watt-hour meters are located on the exterior of a building. In some cases, the entire service-entrance equipment may be mounted outside the building. This includes the watt-hour meter and the disconnecting means. Figure 28-1 illustrates the Code terms for the various components of a service entrance. Figure 28-2 illustrates the service entrance, main panel, sub-panel, and grounding for this service.

Overhead service drop conductors might be attached to a through-the-roof mast-type service, or to the side of the house. In either case, proper clearance is required aboveground, roof, fences, windows, sidewalks, decks, etc.

Figure 28-1   Code terms for services. See Figures 28-2 and 28-7 for grounding requirements.

MAST-TYPE SERVICE ENTRANCE

The mast service, Figure 28-1, is a commonly used method of installing an overhead service entrance. The mast service is often used on buildings with low roofs, such as ranch-style dwellings, to ensure adequate clearance between the ground and the lowest service conductor.

The service raceway is run through the roof, as shown in Figure 28-1, using a roof flashing and neoprene seal fitting, as illustrated in Figure 28-3(A). This fitting keeps water from seeping in where the raceway penetrates the roof. The conduit is securely fastened to comply with code requirements. Many types of fittings are readily available at electrical supply houses for securely fastening raceways and other electrical equipment to wood, brick, masonry, and other surfaces. Figure 28-3(B) shows a typical through-the-wall support fitting.

Clearance Requirements for Mast Installations

Several factors determine the maximum length of conduit that can be installed between the roof support and the point where the service-drop conductors are attached. These factors are the service-drop length, the system voltage, the roof pitch, and

METER PEDESTAL AND COPPER
S.E. CONDUCTORS TWO 2/0 THW,
THWN, XHHW ONE NO. 1 BARE
NEUTRAL FURNISHED AND INSTALLED
BY ELECTRICAL CONTRACTOR

METER

THESE
CONDUCTORS
FURNISHED
AND INSTALLED
BY UTILITY

PAD-MOUNT
TRANSFORMER

COPPER CONDUCTORS–
THREE NO. 3 THHN OR THWN

1" CONDUIT

PANEL B

PANEL A

125 A
M.L.O.

200
A
MAIN

EQUIPMENT
GROUNDING
CONDUCTOR
BUS

MAIN
BONDING
JUMPER

NEUTRAL BUS

NEUTRAL
BUS

GROUNDING ELECTRODE CONDUCTOR–
NO. 4 AWG COPPER, *TABLE 250-66.*

GROUNDING ELECTRODE CONDUCTOR
TO A GROUND ROD NEED NOT BE
LARGER THAN NO. 6 AWG COPPER,
SECTION 250-50(a)(2).

CONNECT GROUNDING ELECTRODE CONDUCTOR
TO FIRST 5 FEET (1.5 m) OF UNDERGROUND
METAL WATER PIPE. THE FIRST 5 FEET MAY
INCLUDE THE WATER METER. BOND AROUND
THE WATER METER. *SECTION 250-50.*

SUPPLEMENTAL GROUNDING ELECTRODE –
MINIMUM 20 FT. (6.1 m) OF NO. 4 AWG
BARE COPPER ENCASED IN BOTTOM OF
CONCRETE FOOTING, *SECTION 250-50(c).*

5' (1.5 m)

DRIVEN GROUND ROD SERVING
AS ALTERNATIVE SUPPLEMENTAL
GROUNDING ELECTRODE,
SECTIONS 250-50(a)(2) AND 250-52(c).

DO NOT USE THE INTERIOR METAL
WATER PIPING BEYOND THE FIRST
5 FEET (1.5 m) (FROM WHERE IT
ENTERS THE BUILDING) AS A
"CONDUCTOR" TO INTERCONNECT
THE SUPPLEMENTAL ELECTRODES
(ROD OR CONCRETE ENCASED)
AND THE UNDERGROUND METAL
WATER PIPE ELECTRODE,
SECTIONS 250-50 AND 250-50(a).

A SUPPLEMENTAL ELECTRODE MAY BE
BONDED TO THE GROUNDING ELECTRODE
CONDUCTOR, THE GROUNDED SERVICE-
ENTRANCE CONDUCTOR, THE GROUNDED
SERVICE RACEWAY, OR ANY GROUNDED
SERVICE ENCLOSURE, *SECTION 250-50(a)(2).*

Figure 28-2 **The service-entrance, main panel, subpanel, and grounding for the service in this residence. Note that because the integrity of underground water pipes is questionable, the Code in *Section 250-50(a)* requires a supplemental grounding electrode. Shown are two common types of supplemental grounding electrodes: a ground rod or a concrete-encased grounding electrode. In many instances, the concrete-encased grounding electrode is better than the water pipe ground, particularly since the advent of nonmetallic (PVC) water piping systems.**

the size and type of conduit (aluminum or steel), Figure 28-4.

Service Mast as Support *(Section 230-28)*

The bending force on the conduit increases with an increase in the distance between the roof support and the point where the service-drop conductors are attached. The pulling force of a service

drop on a mast service conduit increases as the length of the service drop increases. As the length of the service drop decreases, the pulling force on the mast service conduit decreases.

If extra support is not to be provided, the mast service conduit must be at least 2 inches in diameter. This size prevents the conduit from bending due to the strain of the service-drop conductors.

If extra support is needed, it is usually in the

Figure 28-3 (A) a neoprene seal with metal flashing to use where the service raceway comes through the roof. (B) a through-the-wall support for securing the service raceway. (C) a clamp that fastens to the service raceway to which a guy wire can be attached for support of the service riser. (D) an insulator that can be screwed through the roof into a roof truss or rafter. A guy wire is attached to it and to the clamp (C). Commonly referred to as a "sill plate," (E) is used to protect service-entrance cable where it enters the building.

Figure 28-4 Clearance requirements for service-drop conductors passing over residential roofs, *Section 230-24(a)*, where voltage between conductors does not exceed 300 volts.

form of a guy wire attached to a mast fitting, Figure 28-3(C), and to a roof fitting, Figure 28-3(D). Again, many support fittings and devices are available at electrical supply houses.

When service-entrance cable is used, a fitting of the type shown in Figure 28-3(E) can be used to protect the cable from physical damage where it penetrates the outside wall of the house. When used as a through-the-roof service mast, rigid metal conduit or intermediate metal conduit is not required to be supported within 3 feet (914 mm) of the service head, *Section 346-12(a)* and *Section 345-12(a)*. The roof mast kit provides adequate support when properly installed. A guy wire might be necessary if the service-drop conductors are long or heavy.

Intermediate metal conduit is also permitted to be used as a service mast. It *does not* require secure fastening within 3 feet (914 mm) of the service head. To use intermediate metal conduit as a service mast requires special permission from the electrical inspector, *Section 345-12(a)*.

Only power service-drop conductors are permitted to be attached to and supported by the service mast, *Section 230-28*. Refer to Figure 28-5.

All fittings used with raceway-type service masts must be identified for use with service masts, *Section 230-28*.

Consult the utility company and electrical inspection authority for information relating to their specific requirements for clearances, support, and so on, for service masts.

The *NEC®* rules for insulation and clearances apply to the service-drop and service-entrance conductors. For example, the service conductors must be insulated, except where the voltage to ground does not exceed 300 volts. In this case, the grounded neutral conductors are not required to have insulation.

Section 230-24(a) gives clearance allowances for the service drop passing over the roof of a dwelling, Figure 28-4.

The installation requirements for a typical service entrance are shown in Figures 28-6 and 28-7. Figure 28-6 shows the required clearances above the ground. The wiring connections, grounding requirements and Code references are given in Figure 28-7.

UNDERGROUND SERVICE

Underground service means the cable installed underground from the point of connection to the system provided by the utility company.

New residential developments often include underground installations of the high-voltage electrical systems. The conductors in these distribution

Figure 28-5   Only power service-drop conductors are permitted to be attached to and supported by the service mast. Do *not* attach or support television cables, telephone cables, or anything else to the mast, *Section 230-28*.

NOTE:
CLEARANCES
ARE DESIGNATED
BY LETTERS
A THROUGH G

CLEARANCES FOR SERVICE
DROPS – RESIDENTIAL –
120/240 VOLT SINGLE PHASE:
A = 10 FT (3.05 m) MINIMUM
B = 12 FT (3.66 m) MINIMUM
C = 10 FT (3.05 m) MINIMUM

D = 18 FT (5.49 m) MINIMUM
E = 3 FT (914 mm) MINIMUM
F = CONDUCTORS RUN ABOVE TOP LEVEL OF WINDOW
ARE CONSIDERED OUT OF REACH FROM THAT WINDOW.
3 FT (914 mm) CLEARANCE NOT REQUIRED.
G = 10 FT (3.05 m) MIN. FROM DRIP LOOP TO FINISH GRADE

Figure 28-6 Clearances for a typical residential service in accordance with *Sections 230-24* and *230-26*. Clearances (A) and (C) of 10 feet (3.05 m) permitted only if the service-entrance drop conductors are supported on and cabled together with a grounded bare messenger wire where the voltage to ground *does not* exceed 150 volts. This 10-foot (3.05 m) clearance must also be maintained from the lowest point of the drip-loop. Clearance (B) of 12 feet (3.66 m) permitted over residential property and driveways where the voltage to ground *does not* exceed 300 volts. The circled insert shows clearances from building openings, *Section 230-9*.

systems end in the bases of pad-mounted transformers, Figure 28-8. These transformers are placed at the rear lot line or in other inconspicuous locations in the development. The transformer and primary high-voltage conductors are installed by, and are the responsibility of, the utility company.

The conductors installed between the pad-mounted transformers and the meter are called **service lateral conductors**. Normally, the electrical utility supplies and installs the service laterals. Figure 28-8 shows a typical underground installation.

The wiring from the external meter to the main service equipment is the same as the wiring for a service connected from overhead lines, Figure 28-7. Some local codes may require conduit to be installed underground from the pole to the service-entrance equipment. *NEC*® requirements for underground

services are given in *Sections 230-30, 230-31* and *230-32*. The underground conductors must be suitable for direct burial in the earth, Type USE single-conductor or Type USE cable containing more than one conductor, *Article 338*.

If the electric utility installs the underground service conductors, the work must comply with the rules established by the utility. Utilities follow the National Electrical Safety Code, NESC. These rules may not be the same as those given in *NEC*® *Section 90-2*.

When the underground conductors are installed by the electrician, *Section 230-49* applies. This section deals with the protection of conductors against physical damage. *Section 230-49* also refers to *Section 300-5*, which covers all situations involving underground wiring, such as the sealing of raceways where they enter a building, Figure 28-17.

POINT OF ATTACHMENT TO BUILDINGS,
SECTION 230-26, SHALL BE NOT LESS THAN
10 FEET (3.05 m) ABOVE FINISHED GRADE

POINT OF ATTACHMENT TO
BUILDINGS, *SECTION 230-26*

OVERHEAD SERVICE
LOCATIONS, *SECTION 230-54*

CLEARANCES OF SERVICE
DROPS, *SECTION 230-24*

THREADED HUB

ONLY POWER SERVICE-DROP CONDUCTORS ARE PERMITTED
TO BE ATTACHED TO A SERVICE MAST, *SECTION 230-28.*

SERVICE-ENTRANCE CONDUCTORS:
• 100 AMPERE MINIMUM, *SECTION 230-42.*
• GROUNDED NEUTRAL MAY BE INSULATED OR BARE,
  SECTIONS 230-30 AND *230-41.*
• SEE *TABLE 310-15(b)(6)* FOR SPECIAL AMPACITY RATINGS.

CONDUIT SIZE: *CHAPTER 9. TABLES 1, 4, 5,* AND *8.*
ALSO *TABLES C2* THROUGH *C12A. APPENDIX C TO NEC.*

C_L

MAX. 60" (1.5 m)
MIN. 30" (914 mm)

RECOMMENDED
HEIGHT OF
METER SOCKET.
VERIFY WITH
UTILITY COMPANY.

MOUNT METER AT EYE LEVEL
NEUTRAL BONDED TO METER ENCLOSURE

BONDING-BUSHING
NOT REQUIRED
HERE IF
THREADED
HUB

MAIN SWITCH AND OVERCURRENT PROTECTION,
SECTIONS 230-70 THROUGH *230-95.* DO NOT INSTALL
IN BATHROOMS, *SECTIONS 240-24(e)* AND *230-70(a).*

PRESSURE CONNECTORS, NO SOLDER.

DISCONNECT MEANS LOCATED AT READILY ACCESSIBLE
LOCATION NEAREST POINT OF ENTRANCE OF SERVICE
CONDUCTORS, *SECTION 230-70(a).*

BONDING, *SECTIONS
250-92* THROUGH
250-104, ALSO
INSULATING BUSHINGS,
SECTION 373-6(c)

EQUIPMENT
GROUNDING
CONDUCTOR BUS,
SECTION 384-20

FUSES OR CIRCUIT BREAKERS MUST HAVE
INTERRUPTING RATING SUFFICIENT FOR VOLTAGE
EMPLOYED AND CURRENT WHICH MUST BE
INTERRUPTED, *SECTIONS 110-9* AND *110-10.*

MAIN BONDING JUMPER, *SECTIONS 250-28,
250-102,* AND *384-3(c).*

NEUTRAL BAR GROUNDED TO ENCLOSURE,
SECTIONS 250-142 AND *384-20.*

SERVICE GROUNDING ELECTRODE CONDUCTOR,
TABLE 250-66, SECTIONS 250-62, AND *250-64.*

GROUND
LEVEL

► GROUNDING ELECTRODE CONDUCTER MUST
BE ONE CONTINUOUS LENGTH WITHOUT SPLICE.
A SPLICE WITH AN IRREVERSIBLE COMPRESSION
TYPE CONNECTOR OR EXOTHERMIC WELDING
IS PERMITTED, *SECTION 250-64(b).* ◄

IF GROUNDING ELECTRODE CONDUCTOR
IS IN METAL RACEWAY, BOND BOTH ENDS.
SECTION 250-64(c)

BONDING, *ARTICLE 250, PART E.* BONDING JUMPERS
MUST BE LONG ENOUGH TO ALLOW REMOVAL OF METER
WITHOUT DISCONNECTING BOND, *SECTION 250-68.*

GROUND CLAMPS

GROUND
CLAMPS

COLD
WATER
PIPE

HOT WATER PIPE

WATER
HEATER

BONDING JUMPER,
SECTION 250-104.

CONNECTION TO GROUNDING
ELECTRODE, *SECTION 250-68*
AND *SECTION 250-70.*

GROUND ROD

GROUND
ROD

THE GROUNDING ELECTRODE CONDUCTOR
MUST BE CONNECTED SOMEWHERE ON THE
FIRST 5 FEET OF METAL WATER PIPE
SERVICE WHERE IT ENTERS THE BUILDING.

SUPPLEMENTAL GROUNDING ELECTRODE,
SECTIONS 250-50 AND *250-52.* MINIMUM
LENGTH 8 FT (2.44 m). THIS GROUNDING
ELECTRODE CONDUCTOR NEED
NOT BE LARGER THAN NO. 6 COPPER,
SECTION 250-50(a).

CONCRETE ENCASED GROUNDING
CONDUCTOR ALSO PERMITTED.

**Figure 28-7   The wiring of a typical service-entrance installation and Code rules for the system grounding. Refer
also to Figures 28-2 and 28-21. Today, most services are installed underground as shown in Figure 28-8.**

► SERVICE LATERALS NOT ENCASED IN CONCRETE THAT ARE BURIED 18 IN. (457mm) OR MORE BELOW GRADE SHALL HAVE THEIR LOCATION IDENTIFIED BY A WARNING RIBBON THAT IS PLACED IN THE TRENCH AT LEAST 12 IN. (305 mm) ABOVE THE UNDERGROUND INSTALLATION, *SECTION 300-5(d)*. ◄

METER

METER ENCLOSURE

METER ENCLOSURE MOUNTED ON SIDE OF RESIDENCE

MAIN SERVICE-ENTRANCE PANEL

MAIN SERVICE-ENTRANCE PANEL

METER ENCLOSURE TO BELOW GROUND LINE

CAUTION: BURIED ELECTRIC LINE

PAD-MOUNTED TRANSFORMER

CONDUIT RUN FROM METER ENCLOSURE TO MAIN SERVICE-ENTRANCE PANEL

TYPE "USE" SINGLE-CONDUCTOR OR TYPE "USE" CABLE

UNDERGROUND SERVICE-ENTRANCE CONDUCTORS (SERVICE LATERAL) RUN FROM THE METER TO THE PAD-MOUNTED TRANSFORMER PLACED ON LOT LINE AT REAR OF RESIDENCE: THESE CONDUCTORS USUALLY ARE INSTALLED AND FURNISHED BY ELECTRIC UTILITY COMPANY.

Figure 28-8   Underground service.

MAIN SERVICE DISCONNECT LOCATION

The main service disconnecting means shall be installed at a readily accessible location outside or inside of the building so that the service conductors inside the building are as short as possible, *Section 230-70(a)*, Figure 28-9.

If for some reason it is necessary to locate service-entrance equipment some distance inside the building, *Section 230-6* considers service-entrance conductors to be outside of a building if the service-entrance conductors are:

- installed under not less than 2 inches (50.8 mm) of concrete beneath the building

  or

- installed within the building in a raceway encased in at least 2 inches (50.8 mm) of concrete or brick.

THIS SWITCH *IS* AS CLOSE AS POSSIBLE TO THE POINT WHERE THE SERVICE-ENTRANCE CONDUCTORS ENTER THE BUILDING

THIS SWITCH *IS NOT* AS CLOSE AS POSSIBLE TO THE POINT WHERE THE SERVICE-ENTRANCE CONDUCTORS ENTER THE BUILDING

Figure 28-9   Main service disconnect location. Do *not* install in a bathroom. Refer to *Sections 230-70(a)* and *240-24(e)*.

Panelboard Directory

Section 384-13 states that all panels must have a "legibly identified" circuit directory indicating what the circuits are for. *Section 110-22* also requires that disconnect switches be "legibly marked" as to what the disconnect is for. The directory is furnished and attached to the inside of the cover by the panelboard manufacturer. Figure 28-10 shows a typical 240/120-volt, single-phase, 3-wire load center. Figure 28-11 shows the circuit identification for Main Service Panel A. Figure 28-14 shows the circuit identification for Panel B. Many times, the electrician will find the directory furnished by the manufacturer too small, and will type a directory on a larger size card, and will attach this to the inside of the panel cover.

Disconnect Means (Panel A)

The requirements for disconnecting electrical services are covered in *Sections 230-70* through *230-82*. *Section 230-71(a)* requires that the service disconnecting means consist of not more than six switches or six circuit breakers mounted in a single enclosure, in a group of separate enclosures in the same location, or in or on a switchboard. This permits the disconnection of all electrical equipment in the house with not

Figure 28-10 A typical 120/240-volt, single-phase, 3-wire load center. This load center has a main circuit breaker. Note the neutral terminal strips on which to terminate the white branch-circuit grounded conductors, and the equipment grounding terminal strips on which to terminate the equipment grounding conductors of nonmetallic-sheathed cable. When used as service equipment, the load center must be "listed" as suitable for use as service equipment, *Section 230-70*. The panel must also be marked to identify it as being suitable for use as service equipment, *Section 230-66*. *Courtesy* Square D Company.

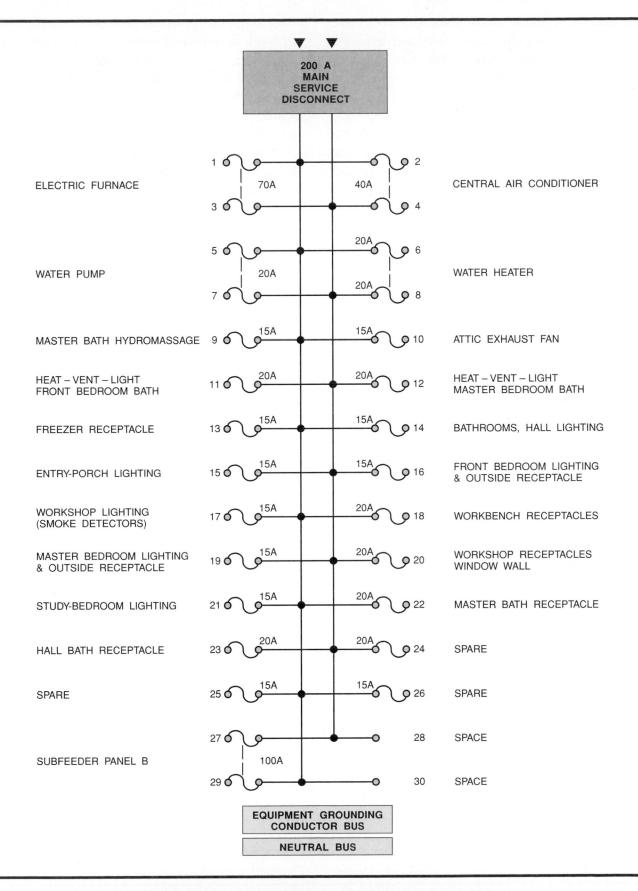

Figure 28-11 Circuit schedule of Main Service Panel "A."

more than six hand operations. Some local codes take exception to the six disconnect rule, and require a single main disconnect. Check this with the local authority that has jurisdiction.

In damp and wet locations, to prevent the accumulation of moisture that could lead to rusting of the enclosure and damage to the equipment inside the enclosure, *Section 373-2(a)* specifies that there be at least a ¼-inch (6.35 mm) air space between the wall and a surface-mounted enclosure. Most disconnect switches, panelboards, meter sockets, and similar equipment have raised mounting holes that provide the necessary clearance. This is clearly visible in the open-meter socket in Figure 28-16. The *NEC®* (in Definitions) considers masonry in direct contact with the earth to be a wet location. The *Exception* to *Section 373-2(a)* allows nonmetallic enclosures to be mounted without an air space on concrete, masonry, tile, and similar surfaces.

Section 384-14 defines a lighting and appliance branch-circuit panelboard as one that has not more than 10% of the overcurrent devices rated less than 30 amperes with provisions for the connection of neutral conductors. *Section 384-16(a)* states that a lighting and appliance branch-circuit panelboard *shall not* have more than two main circuit breakers or two sets of fuses for the panelboards overcurrent protection and disconnecting means.

A lighting and appliance branch-circuit panelboard *shall not* have more than 42 overcurrent devices in it, *Section 384-15*. A two-pole breaker is counted as two overcurrent devices.

▶ *Section 230-79* tells us that the service disconnect shall have a rating not less than the load to be carried, determined in accordance with *Article 220*.◀ We show how these calculations are made in Unit 29.

For a one-family residence, the minimum size main service disconnecting means is 100 amperes, three-wire, *Section 230-79(c)*.

Service equipment must be marked that it is suitable for use as service equipment, *Section 230-66*.

▶ *Section 230-70(b)* requires that the main service disconnect must be permanently marked to identify it as service disconnect.◀

When nonmetallic-sheathed cable is used, a panelboard must have a terminal bar for attaching the equipment grounding conductors, *Section 384-20*. See Figure 28-12.

TERMINAL FOR CONNECTION OF EQUIPMENT GROUNDING CONDUCTOR (BARE) OF NON-METALLIC SHEATHED CABLE (TYPE NM, NMC, UF). NOT MORE THAN THREE EQUIPMENT GROUNDING CONDUCTORS PER SET-SCREW, PER U.L. PANELBOARD STANDARD NO. 67. TERMINAL BAR MUST BE GROUNDED TO PANEL ENCLOSURE. THE EQUIPMENT GROUND BAR SHALL NOT BE CONNECTED TO THE NEUTRAL BAR EXCEPT AT SERVICE EQUIPMENT, OR IF PANEL IS LOCATED IN A SEPARATE BUILDING.

AT SERVICE-ENTRANCE EQUIPMENT, GROUNDING CONDUCTOR BUS AND NEUTRAL CONDUCTOR BUS MUST BE BONDED TOGETHER; THIS IS FURNISHED BY THE MANUFACTURER OF THE PANEL.

TERMINAL FOR CONNECTION OF NEUTRAL CONDUCTORS. MUST BE WHITE OR GRAY INSULATED EXCEPT NEUTRAL SERVICE-ENTRANCE CONDUCTOR MAY BE BARE. ONLY ONE NEUTRAL CONDUCTOR PER SET-SCREW. MUST BE INSULATED FROM ENCLOSURE EXCEPT AT SERVICE-ENTRANCE EQUIPMENT.

SERVICE-ENTRANCE NEUTRAL CONDUCTOR

DO NOT CONNECT EQUIPMENT GROUNDING CONDUCTORS TO NEUTRAL BAR UNLESS BAR IS IDENTIFIED FOR THE PURPOSE AND IS LOCATED IN MAIN PANEL WHERE NEUTRAL AND GROUNDING ELECTRODE CONDUCTOR ARE PERMITTED OR REQUIRED TO BE INTERCONNECTED.

BRANCH-CIRCUIT NEUTRAL CONDUCTORS

MAIN BONDING JUMPER TO ENCLOSURE. IF SCREW, MUST BE GREEN, *SECTION 250-28(b)*.

GROUNDING ELECTRODE CONDUCTOR

Figure 28-12 Connections of service neutral, branch-circuit neutral, and equipment grounding conductors at panelboards, *Section 384-20*.

For this residence, Panel A serves as the main disconnecting means for the service. This load center consists of a main circuit breaker and many branch-circuit breakers. Figure 28-10 is a photo of this type of load center. Figure 28-11 is the schedule of the circuits connected to Panel A. Panel A is located in the Workshop.

Working Space. To provide for safe working conditions around electrical equipment, *Section 110-26* contains a number of rules that must be followed. For residential installations, there are some guidelines for working space.

- A working space of not less than 30 inches (762 mm) wide and 3 feet (914 mm) deep must be provided in front of electrical equipment, such as Panels A and B.

- This working space must be kept clear and not used for storage.

- Do not install electrical panels inside of cabinets, nor above shelving, washers, dryers, freezers, work benches, etc.

- Do not install electrical panels in bathrooms.

- Do not install electrical panels above or close to sump pump holes.

- The hinged cover (door) on the panel shall be able to be opened to a full 90 degrees.

- ▶ A "zone" equal to the width and depth of the electrical equipment, from the floor to a height of 6 feet (1.83 m) above the equipment or to the structural ceiling, whichever is lower, shall be dedicated to the electrical installation. No piping, ducts, or equipment foreign to the electrical installation shall be located in this zone. This zone is dedicated space intended for electrical equipment only! See *Section 110-26(f)*. ◀

Mounting Height. Disconnects and panelboards must be mounted so the center of the disconnect handle in its highest position is not higher than 6'7" (2.0 m) above the floor, *Section 380-8*.

Headroom. To provide for safe adequate working space, and easy access to main service equipment and electrical panels, the *NEC®* requires that there be at least 6½ feet (2 m) of headroom around this equipment. In the past, many service panels were installed in crawl spaces where the home had no basement. This is no longer permitted, *Section 110-26(e)*.

Lighting Required for Service Equipment and Panels. Working on electrical equipment with inadequate lighting can result in injury or death. In *Section 110-26(d)*, there is a requirement that illumination be provided for service equipment and panelboards. But, the Code does not spell out how much illumination is required. It becomes a judgment call on the part of the electrician and/or electrical inspector. Note that the electrical plans for the workshop show a porcelain pull-chain lampholder to the front of and off to one side of the main service Panel A. In many cases, adjacent lighting, such as the fluorescent ceiling fixtures in the recreation room, is considered to be the required lighting for equipment such as Panel B.

Load Center (Panel B)

When a second load center is installed, it is oftentimes called a subpanel.

For installation where the main service panel is a long distance from areas that have many circuits and/or heavy load concentration, as in the case of the kitchen and laundry of this residence, it is recommended that at least one additional panel be installed near these loads. This results in the branch-circuit wiring being short. Line losses (voltage and wattage) are considerably less than if the branch-circuits had been run all the way back to the main panel.

The cost of material and labor to install an extra panel should be compared to the cost of running all of the branch-circuit "home runs" back to the main panel.

Panel B is located in the recreation room. It is fed by three No. 3 THHN or THWN conductors run in a 1-inch conduit originating at Panel A. This feeder is protected by a 100-ampere, 2-pole circuit breaker located in Panel A.

Do not install panelboards or load centers in bathrooms, *Section 240-24(e)*. Damaging moisture problems may result from the presence of water, and shock hazard may result from both the presence of water and the close proximity of metal faucets, etc.

Do not install panelboards or load centers in closets, *Section 240-24(d)*. There is a fire hazard because of the presence of ignitable materials.

Panel B is shown in Figure 28-13. The circuit schedule for Panel B is found in Figure 28-14.

Figure 28-13 Typical 240/120-volt, single-phase, 3-wire panelboard, sometimes called a load center. This panel does not have a main disconnect, and is referred to as a *main lug only (MLO)* panel. The branch circuit-breakers are not shown. This is the type of panel that could be used as Panel B in the residence discussed in this text. Most panels have a separate terminal bus on which to terminate the equipment grounding conductors of nonmetallic-sheathed cable, Figure 28-12. *Courtesy* **Square D Company,** *Group Schneider.*

SERVICE-ENTRANCE CONDUCTOR SIZING

Section 230-42(a) states that the minimum size service-entrance conductors must be of sufficient size to carry the load as computed according to *Article 220.*

▶ *Section 230-42(b)* states that the ungrounded service-entrance conductors must have an ampacity of not less than the minimum rating of the service disconnect. ◀

The calculations for sizing service-entrance conductors are discussed in detail in Unit 29.

SERVICE-ENTRANCE OVERCURRENT PROTECTION

After the ampacity requirements for the service-conductors have been calculated, we need to determine the overload protection required for the service-entrance conductors. We find the basic overcurrent protection requirements in *Part G* of *Article 230.*

In *Section 230-90(a)*, we find that:

* each ungrounded service conductor must have overload protection. You may have already taken note that the overcurrent protection for service-entrance conductors is on the load end of the conductors. This is acceptable, since the *NEC®* accepts the fact that short-circuit overcurrent protection of service-entrance conductors is provided by the utility's primary transformer fuse. For most conductor overcurrent protection, the overcurrent protection is placed where the conductor begins.

* the ampere rating of the overcurrent protective device (fuses or circuit breakers) must not be higher than the ampacity of the service conductors.

* where the standard ampere rating of the fuses or circuit breaker does not match the ampacity of the conductor, then the next higher ampere rated fuse of circuit breaker is permitted. This is based on *Sections 240-3(b)* and *240-6.*

* where the special ampacity values permitted in *Table 310-15(b)(6)* are used for single-phase, 120/240-volt services on houses, the overcurrent protection for the service-entrance conductors is based on the ampere ratings shown in Column 3 of this table. Some local codes do not allow the use of this table.

In *Section 230-91* we find that:

* the overcurrent devices for services are usually an integral part of the service disconnect.

* if the building is a multi-occupancy building, such as a duplex or fourplex apartment or condo, each tenant must have access to the service disconnect and overcurrent devices for his or her particular unit.

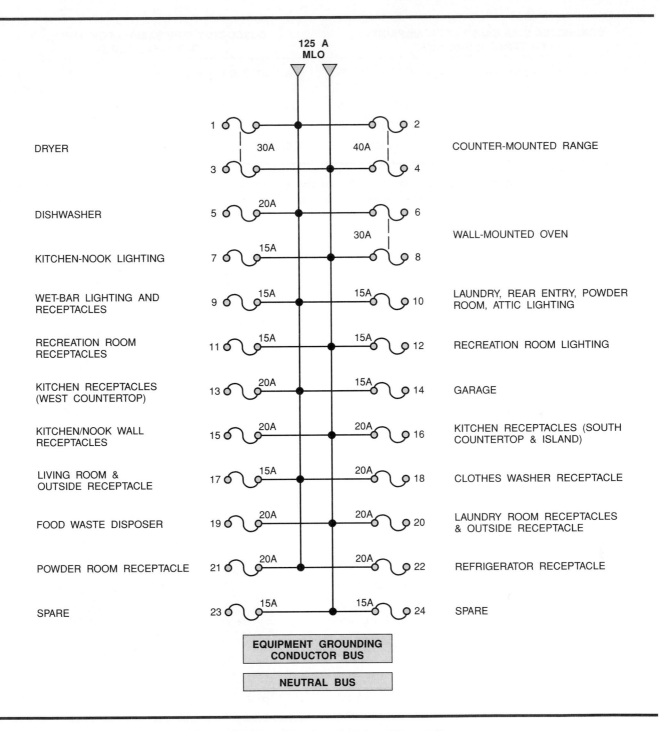

Figure 28-14   Circuit schedule of Panel B.

SERVICE-ENTRANCE RACEWAY SIZING

This residence is served with an underground service. The conduit running from the meter to Main Panel A must be sized correctly for the conductors it will contain. In the case of a through-the-roof mast service, the conduit might be sized for mechanical strength reasons rather than conductor fill. All utilities have requirements for size and guying of mast services. Figure 28-15 shows the calculation for sizing the service-entrance conduit. This calculation is similar to other examples in this text. The conductor dimensions are found in *Chapter 9* of the *NEC.*®

CONDUCTOR SIZE BASED UPON AMPACITY PER *TABLE 310-15(b)(6)*		CONDUCTOR SIZE BASED UPON AMPACITY PER *TABLE 310-16*	
TWO NO. 2/0 THWN COPPER	0.2223 SQ. IN. 0.2223 SQ. IN.	TWO NO. 3/0 THWN COPPER	0.2679 SQ. IN. 0.2679 SQ. IN.
ONE NO. 1 BARE COPPER	0.0870 SQ. IN.	ONE NO. 1 BARE COPPER	0.0870 SQ. IN.
TOTAL	0.5316 SQ. IN.	TOTAL	0.6228 SQ. IN.
CONDUIT SIZE	1¼ INCH	CONDUIT SIZE	1½ INCH

Figure 28-15 Calculations for determining the size of the electrical metallic tubing service-entrance raceway between Main Panel A and the meter base. Dimensional data found in *Chapter 9, Tables 1, 4, 5,* and *8* of the *NEC.*®

METER/METER BASE

The electric utility must be contacted before the installation of the service equipment begins. They will determine when, where, and how they intend to connect to the homeowner's service. Because of the ever-increasing number of wood decks and concrete patios being added to homes, many utilities prefer that the meter be mounted on the side of the house rather than in the rear. If there is a raised deck, the meter could be only a short distance above the deck, making it both subject to physical damage and hard to read. If concrete patios are poured over underground utility cables, it is very difficult to repair or replace these cables should there be problems. Utilities are happy to furnish manuals and brochures detailing their requirements for services that are not found in the *NEC.*®

The electric utility furnishes the watt-hour meter. The meter base, also referred to as a meter socket, is usually furnished and installed by the electrical contractor, although some utilities furnish the meter base to the electrical contractor for installation.

Figure 28-8 shows a service supplied from an underground service. Figure 28-9 discusses the requirement that the main disconnect must be as close as possible to the point of entrance of the service-entrance conductors. Figure 28-16 shows a typical meter socket for an overhead service, usually mounted eye level on the outside wall of the house. The service raceway or service-entrance cable is connected to the threaded hub (boss) on top of the meter socket. Figure 28-17 shows how to seal the raceway where it enters the building to keep out moisture.

Figure 28-16 Typical meter socket. *Courtesy of Milbank Mfg. Co.*

Probably more common for residential services is the "pedestal," as shown in Figure 28-18. Usually mounted on the outside wall of the house, they could also be installed on the lot line between residential properties. Contact the electric utility on this issue.

SECTION 300-5(g) AND SECTION 300-7 STATE THAT WHEN RACEWAYS PASS THROUGH AREAS HAVING GREAT TEMPERATURE DIFFERENCES, SOME MEANS MUST BE PROVIDED TO PREVENT PASSAGE OF AIR BACK AND FORTH THROUGH THE RACEWAY. NOTE THAT OUTSIDE AIR IS DRAWN IN THROUGH THE CONDUIT WHENEVER A DOOR OPENS. COLD OUTSIDE AIR MEETING WARM INSIDE AIR CAUSES THE CONDENSATION OF MOISTURE. THIS CAN RESULT IN RUSTING AND CORROSION OF VITAL ELECTRICAL COMPONENTS. EQUIPMENT HAVING MOVING PARTS, SUCH AS CIRCUIT BREAKERS, SWITCHES, AND CONTROLLERS, IS ESPECIALLY AFFECTED BY MOISTURE. THE SLUGGISH ACTION OF THE MOVING PARTS IN THIS EQUIPMENT IS UNDESIRABLE.

INSULATION OR OTHER TYPE OF SEALING COMPOUND CAN BE INSERTED AS SHOWN TO PREVENT THE PASSAGE OF AIR.

SECTION 230-8 REQUIRES SEALS WHERE SERVICE RACEWAYS ENTER FROM AN UNDERGROUND DISTRIBUTION SYSTEM.

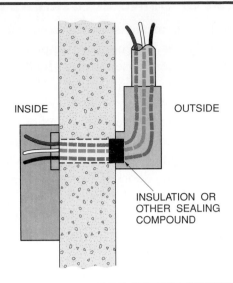

INSIDE   OUTSIDE

INSULATION OR OTHER SEALING COMPOUND

Figure 28-17   Installation of conduit through a basement wall.

THE ELECTRIC UTILITY PROVIDES THE UNDERGROUND SERVICE LATERAL

Figure 28-18   Installing a metal "pedestal" allows for ease of installation of the underground service-entrance conductors by the utility. *Photo Courtesy of* **Milbank Manufacturing Co.**

For underground services, the utility generally runs the underground lateral service-entrance conductors in a trench from a pad-mount transformer to the line-side terminals of the meter base. The electrician then takes over to complete the service installation by running service-entrance conductors from the load side of the meter to the line side of the main service disconnect.

Quite often, the electric utility and the telephone company have agreements whereby the electric utility will "lay" both power cables and telephone cables in the same trench. The electric utility carries both types of cables on their underground installation vehicles. This is a much more cost effective way to do this, rather than having each utility dig and then backfill their own trenches.

Often overlooked is a requirement in *NFPA Standard No. 54* (Fuel Gas Code), *Section 2.7.2(c)* requiring that gas meters be located at least 3 feet (914 mm) from sources of ignition. Examples of sources of ignition might be an air-conditioner disconnect where an arc can be produced when the disconnect is opened under load. Furthermore, some electric utilities require a minimum of 3 feet (914 mm) clearance between electric metering equipment and gas meters and gas regulating equipment. It is better to be safe than sorry. Check this issue with the local electrical inspector if there might be a problem with the installation you are working on.

COST OF USING ELECTRICAL ENERGY

A watt-hour meter is always connected into some part of the service-entrance equipment to record the amount of energy used. Billing by the utility is generally done on a monthly basis. The meter might be mounted on the side of the house, or on a pedestal somewhere on the premise on the lot line. The utility makes this decision. The residence discussed in this text has one meter, mounted on the back of the house near the sliding doors of the master bedroom.

The kilowatt (kW) is a convenient unit of electrical power. One thousand watts (w) is equal to one kilowatt. The watt-hour meter measures and records both wattage and time.

For residential metering, most utilities have rate schedules based on "cents per kilowatt-hour." Stated another way: how much wattage is being used and for how long?

Burning a 100-watt light bulb for 10 hours is the same as using a 1,000-watt electric heater for 1 hour. Both equal one kilowatt-hour.

$$kWh = \frac{watts \times hours}{1,000} = \frac{100 \times 10}{1,000} = 1 \text{ kWh}$$

$$kWh = \frac{watts \times hours}{1,000} = \frac{1,000 \times 1}{1,000} = 1 \text{ kWh}$$

In these examples, if the electric rate is 8 cents per kilowatt-hour, the cost to operate the 100-watt light bulb for ten hours and the cost to operate the electric heater for one hour are the same—8 cents. Both loads use one kilowatt-hour of electricity.

Simple Formula for Calculating the Cost of Using Electricity

The cost of electrical energy used can be calculated as follows:

$$Cost = \frac{watts \times hours \text{ used} \times cost \text{ per kWh}}{1,000}$$

EXAMPLE: Find the cost of operating a color television set for 8 hours. The label on the back of the television set indicates 175 watts. The electric rate is $0.10494 per kilowatt-hour.

$$Cost = \frac{175 \times 8 \times \$0.10494}{1,000} = \frac{\$0.1469}{(approx. 15¢)}$$

EXAMPLE: Find the approximate cost of operating a central air conditioner per day that on average runs 50% of the time during a 24-hour period on a typical hot summer day. The unit's nameplate is marked 240 volts, single-phase, 23 amperes. The electric rate is $.09 cents per kilowatt-hour. The steps are:

1. The time the air conditioner operates each day is: $24 \times 0.50 = 12$ hours.

2. Convert the nameplate data to use in the calculations:

$$Watts = volts \times amperes = 240 \times 23$$
$$= 5,520 \text{ watts}$$

3. $Cost = \dfrac{watts \times hours \text{ used} \times cost \text{ per kWh}}{1,000}$

$$= \frac{5,520 \times 12 \times \$0.09}{1,000} = \$5.96$$

Note: The answers to these examples are approximate because they were based on watts. Power factor and efficiency factors were not included as they generally are unknown when coming up with rough estimates of an electric bill. For all practical purposes, the answers are acceptable.

Here is an example of what a typical monthly electric bill might look like.

GENERIC ELECTRIC COMPANY			
Anyplace, USA			
Days of Service	from 02-01-99	to 02-28-99	Due Date 03-25-99
Present reading			84,980
Previous reading			83,655
Kilowatt-hours used			1,325
Rate/kWh 1st 400 kWh @ $0.10494			$41.98
remaining 925 kWh @ $0.06168 . .			57.05
Energy charge			99.03
Basic service charge (single-dwelling)			8.91
State tax			4.24
Total current charges			$112.18
Total amount due by 03-25-99			$112.18
Total amount due after 03-25-99			$113.30

Some utilities increase their rates during the hot summer months when the air-conditioning load is high. Other utilities provide a second meter for specific loads such as electric water heaters, air conditioners, heat pumps, or total electric heat. Other utilities use electronic watt-hour meters that have the capability of registering kilowatt-hour consumption during specific "time-of-use." These electronic watt-hour meters might have up to four different "time-of-use" periods, each period having a different "cents per kilowatt-hour" rate.

Other charges that might appear on a "light bill" might be a fuel adjustment charge based on a "per kWh" basis. Such charges enable a utility to recover from the consumer extra expenses it might incur for fuel costs used in generating electricity. Fuel charges can vary with each monthly bill without the utility having to apply to the regulatory agency for a rate change.

GROUNDING–WHY GROUND

Article 250 covers grounding. We find in *Section 250-2* the fundamental reasons for grounding electrical systems and electrical equipment.

- *Section 250-2(a):* ▶ ***Grounding of Electrical Systems.*** *Electrical systems required to be grounded shall be connected to earth in a manner that will limit the voltage imposed by lightning, line surges, or unintentional contact with higher voltage lines and that will stabilize the voltage to earth during normal operation.* ◀

- *Section 250-2(b):* ▶ ***Grounding of Electrical Equipment.*** *Conductive materials enclosing electrical conductors or equipment, or forming part of such equipment, shall be connected to earth so as to limit the voltage to ground on these materials. Where the electric system is required to be grounded, these materials shall be connected together and to the supply system grounded conductor as specified in this article. Where the electric system is not solidly grounded, these materials shall be connected together in a manner that establishes an effective path for fault current.* ◀

- *Section 250-2(c):* ▶ ***Bonding of Electrically Conductive Materials.*** *Electrically conductive materials, such as metal water piping, metal gas piping, and structural steel members, that are likely to become energized shall be bonded as specified in this article to the supply system grounded conductor, or in the case of an ungrounded electrical system to the electrical system grounded equipment, in a manner that establishes an effective path for fault current.* ◀

- *Section 250-2(d):* ▶ ***Performance of Fault Current Path.*** *The fault current path shall be permanent and electrically continuous, shall be capable of safely carrying the maximum fault likely to be imposed on it, and shall have sufficiently low impedance to facilitate the operation of overcurrent devices under fault conditions.* ◀

  ▶ *The earth shall not be used as the sole equipment grounding conductor or fault-return path.* ◀

Proper grounding means that overcurrent protective devices will operate fast when responding to ground faults. Effective grounding occurs when a low-impedance (ac resistance) ground path is provided. Ohm's law verifies that, in a given circuit, "the lower the impedance, the higher the value of current." As the value of ground-fault current increases, there is an increase in the speed with which a fuse will open or a circuit breaker will trip. This is called an *inverse time* relationship.

Clearing a ground fault or short-circuit is important because arcing damage to electrical equipment as well as conductor insulation damage is closely related to a value called *ampere-squared-seconds* (I^2t), where

$I =$ the current in amperes flowing phase to ground, or phase to phase.

$t =$ the time in seconds that the current is flowing.

Thus, there will be less equipment and/or conductor damage when the fault current is kept to a low value, and when the time that the fault current is allowed to flow is kept to a minimum. The impedance of the circuit determines the *amount* of fault current that will flow. The speed of operation of a fuse or circuit breaker determines the amount of *time* the fault current will flow.

Grounding electrode conductors and equipment grounding conductors carry an insignificant amount of current under normal conditions. However, when a ground fault occurs, these grounding conductors as well as the actual circuit conductors must be capable of carrying whatever value of fault current might flow (how much?) for the time (how long?) it takes for the overcurrent protective device to clear the fault, *Section 250-2(d).*

▶ This potential hazard is recognized in the *Note* below *Table 250-122*, which states that *"Where necessary to comply with Section 250-2(d), the equipment grounding conductor shall be sized larger than this table."* ◀ This note calls attention to the fact that equipment grounding conductors may need to be sized larger than indicated in the table when high-level ground-fault currents are possible.

What is a high-level ground-fault? To determine the available short-circuit and ground-fault current, a short-circuit calculation is necessary. It is also necessary to know the time/current characteristic of the overcurrent protective device. This is discussed in *Electrical Wiring— Commercial* (Delmar Publishers).

This is referred to as the conductor's "withstand rating."

Tables 250-66 and *250-122* are based upon the fact that copper conductors and their bolted connections can withstand:

- one ampere
- for 5 seconds
- for every 30 circular mils

Thermoplastic insulation on a copper conductor rated 75°C can safely withstand:

- one ampere
- for 5 seconds
- for every 42.25 circular mils

To exceed the previous values will result in damage to the conductors, with possible burn-off and loss of the grounding or bonding path.

When bare equipment grounding conductors are in the same raceway or cable as insulated conductors, always apply the insulated conductor withstand rating limitation.

Properly selected fuses and circuit breakers for normal residential installations generally will protect conductors and other electrical equipment against the types of ground faults and short-circuits to be expected in homes. Fault currents can be quite high in multifamily dwellings (apartments, condos, town houses) making it necessary to take equipment and conductor withstand rating into consideration.

The subject of conductor and equipment withstand rating is covered in greater depth in the *Electrical Wiring— Commercial* (Delmar Publishers).

Grounding Electrode System

Article 250, Part C covers the requirements that establish a *grounding electrode system*. Simply stated, this means that everything is "tied" together. The service-entrance neutral conductor is grounded, this neutral is bonded to the main panel, a grounding electrode conductor is connected to a grounding electrode, and all metal water pipes and gas pipes are bonded together. Should any one of the previous paths become disconnected, the integrity of the grounding system is maintained through the other paths.

Figure 28-7 shows proper grounding and bonding in accordance with *Sections 250-104* and *250-50*.

By having all metal parts bonded together for a grounding electrode system, potential voltage differences between noncurrent-carrying metal parts is virtually eliminated. In addition, the grounding electrode system serves as a means to bleed off lightning, stabilize voltage of the system, and ensures that the overcurrent protective devices will operate.

Figure 28-19 and the following steps illustrate what can happen if the electrical system is not properly grounded and bonded:

1. A "live" wire contacts the gas pipe. The bonding jumper "A" has not been installed.

2. The gas pipe now has 120 volts on it. The pipe is energized. It is "hot."

3. The insulating joint in the gas pipe results in a poor path to ground. Assume a resistance (impedance) of 8 ohms.

4. The 20-ampere overcurrent device does not open:

$$I = \frac{E}{R} = \frac{120}{8} = 15 \text{ amperes}$$

5. If a person touches the "live" gas pipe and the water pipe at the same time, current flows through the person's body. If the body resistance is 12,000 ohms, the current is:

$$I = \frac{E}{R} = \frac{120}{12,000} = 0.01 \text{ ampere}$$

This amount of current passing through a human body can cause death. See Unit 6 for a discussion regarding electric shock.

Figure 28-19   System grounding.

6. The overcurrent device is now "seeing" $15 + 0.01 = 15.01$ amperes and does not open.

7. If the grounding electrode system concept had been used, bonding jumper "A" would have kept the voltage difference between the water pipe and the gas pipe at zero. The overcurrent device would have opened. Checking *Table 8, Chapter 9, NEC,*® we find the dc resistance of 1,000 feet (305 m) of uncoated No. 4 copper wire is 0.308 ohms. The resistance of 100 feet (30.5 m) is 0.0308 ohms. The resistance of 10 feet (3.05 m) is 0.00308 ohms. The current would be:

$$I = \frac{E}{R} = \frac{120}{0.00308} = 38{,}961 \text{ amperes}$$

In an actual installation, the total impedance of *all* parts of the circuit would be much higher than these simple calculations. A much lower current would result. The value of current would be enough to cause the overcurrent device to open.

▶ *Section 250-2(d)* states that the fault-return path:

• shall be permanent and electrically continuous,

• shall be capable of safely carrying the maximum fault current likely to be imposed on it, and

• shall have sufficiently low impedance to facilitate the operation of overcurrent devices under fault conditions. ◀

Section 250-90 states that "bonding shall be provided where necessary to assure electrical continuity and the capacity to conduct safely *any* fault current likely to be imposed."

Section 250-96(a) states in part that the bonding of metal raceways, enclosures, fittings, etc. that serve as the grounding path "shall be effectively bonded where necessary to assure electrical continuity and the capacity to conduct safely *any* fault current likely to be imposed on them."

The Grounding Electrode. *Section 250-50* covers the rules for establishing a good grounding electrode system.

Section 250-50(a) states that a metal underground water piping system 10 feet (3.05 m) or longer is an acceptable grounding electrode. The water pipe grounding electrode *must* be supplemented by at least one additional grounding electrode consisting of any one of those listed in *Sections 250-50* and *250-52*. In this residence, a driven ground rod is the supplemental grounding electrode, as permitted by *Section 250-52(c)*. ▶ If the resistance of this required supplemental ground rod exceeds 25 ohms, then one additional ground rod must be driven. See *Sections 250-50(a)(2)* and *250-56.*◄

This tough requirement for the maximum resistance of the supplemental ground rod(s) was put in place because of the prolific use of insulating fittings, and nonmetallic water piping services – either as an initial installation or replacement of the underground metal water pipe service.

Section 250-50(a) requires that a grounding electrode conductor that is the sole (only) connection to a driven ground rod need not be larger than No. 6 AWG copper or No. 4 AWG aluminum.

If the driven ground rod is not selected as the required supplemental grounding electrode, then any one of the following is acceptable:

- the metal frame of a building (this residence is constructed of wood), *Section 250-50(b)*.

- at least 20 feet (6.1 m) of steel reinforcing bars, ½ inch (12.7 mm) minimum diameter, or at least 20 feet (6.1 m) of bare copper conductor not smaller than No. 4 AWG. These must be encased in concrete at least 2 inches (50.8 mm) thick that is in direct contact with the earth, such as near the bottom of a foundation or footing. This is a concrete-encased ground and is commonly referred to as a UFER ground, named after the gentleman who developed the concept, *Section 250-50(c)*. ► Reinforcing bars are permitted to be bonded together by the usual steel tie wires or other effective means.◄

- at least 20 feet (6.1 m) of bare copper wire, minimum size of No. 2 AWG, buried directly in the earth at least 2½ feet (762 mm) deep. This is called a *ground ring*. See *Section 250-50(d)*.

- ground plates must be at least 2 square feet (0.186 m²), *Section 250-50(d)*. At least one additional plate must be laid if the resistance of one plate is found to exceed 25 ohms, *Section 250-56*.

Some Issues Regarding Bonding of Metal Piping Inside the Home

Metal underground gas piping is not permitted to be used as a grounding electrode, *Section 250-52(a)*.

Many homes are served with copper underground gas pipes. Over the years, many of these underground metal pipes have corroded and leaked gas, thereby setting up a hazardous situation. Gas

utilities now install insulated fittings where the piping enters the building. This breaks the metallic connection, and prevents the galvanic corrosion of the pipe. Many homes are served by PVC underground gas piping from the street to the home. ►However, since the metal gas piping inside of the home could inadvertently become energized, each aboveground portion of a gas-piping system upstream from the equipment shutoff valve shall be electrically continuous and bonded to the grounding electrode system, *Section 250-104(b)*.◄

Section 250-104(c) states that the bonding conductor for the metal piping is to be sized according to *Table 250-122* for the ampere rating of the circuit that might energize the metal piping. It is difficult to determine what circuit might energize the gas piping. The *Fine Print Note* to *Section 250-104(c)* states that "Bonding all piping and metal air ducts within the premises will provide additional safety."

Section 250-104(a) requires that interior metal water piping be bonded, and that the bonding conductor is to be sized according to *Table 250-66*.

Some electrical inspectors will accept metallic mixer faucets as the bonding jumper between the metal hot and cold water pipes. Other do not.

Figure 28-21 illustrates the bonding together of the cold and hot water pipes, and the metal gas piping. The bonding jumper is sized per *Table 250-66*. This installation eliminates the questionable bonding/grounding of the many interconnections of the metal water and gas piping through various appliances.

Ground Rods

Copper ground rods are commonly used as a grounding electrode as well as stainless steel rods. Galvanized iron or steel pipe and conduit, and nonferrous metal plates are also permitted. Aluminum rods are *not* permitted. *Section 250-52* covers ground rods. The following text discusses the Code requirements for copper ground rods. Copper ground rods shall:

- be at least ½ inch (12.7 mm) in diameter.

- be installed below the permanent moisture level, if practical.

- not be less than 8 feet (2.44 m) in length.

- be driven to a depth so that at least 8 feet (2.44 m) is in contact with the soil.

- if solid rock is encountered, be driven at not greater than a 45-degree angle from the vertical or laid in a trench that is at least 2½ feet (762 mm) deep.

- be driven so that the upper end of the rod is flush with or just below ground level. If the upper end is exposed, the ground rod, the ground clamp, and the grounding electrode conductor must be protected against physical damage per *Section 250-10*.

- have a resistance to ground of not less than 25 ohms. If the one rod has more than 25 ohms to ground, drive a second rod. See *Sections 250-50(a)(2)* and *250-56*.

- be at least 6 feet (1.83 m) apart if more than one rod is driven. Actually, it is better to space multiple rods not less than the length of the rod. For example, when driving two 8-foot (2.44 m) rods, space them 8 feet (2.44 m) apart.

Grounding Conductors

A *grounding conductor* is defined as "a conductor used to connect equipment or the grounded circuit of a wiring system to a grounding electrode or electrodes." There are two types of grounding conductors.

1. An *equipment grounding conductor* is the conductor that is used to connect noncurrent-carrying metal parts of equipment, raceways, and other enclosures to the system grounded conductor, to the grounding electrode conductor, or to both—at the service equipment. *Section 250-118* recognizes many types of equipment grounding conductors. Copper or aluminum conductors, rigid metal conduit, electrical metallic tubing, the metal armor of armored cable, and flexible metal conduit all qualify as an equipment grounding conductor. Figures 5-34, 5-35, 5-36, 28-2, 28-12, and 30-3 illustrate equipment grounding conductors.

   Equipment grounding is discussed throughout this text in the appropriate Unit where a specific piece of electrical equipment or appliance requires grounding. Some examples of this are the range, cooktop, dryer, food waste disposer, dishwasher, and water pump.

Table 250-66. Grounding Electrode Conductor for Alternating-Current Systems

Size of Largest Service-Entrance Conductor or Equivalent Area for Parallel Conductors[1]		Size of Grounding Electrode Conductor	
Copper	Aluminum or Copper-Clad Aluminum	Copper	Aluminum or Copper-Clad Aluminum[2]
2 or smaller	½ or smaller	8	6
1 or 1/0	2/0 or 3/0	6	4
2/0 or 3/0	4/0 or 250 kcmil	4	2
Over 3/0 through 350 kcmil	Over 250 kcmil through 500 kcmil	2	1/0
Over 350 kcmil through 600 kcmil	Over 500 kcmil through 900 kcmil	1/0	3/0
Over 600 kcmil through 1100 kcmil	Over 900 kcmil through 1750 kcmil	2/0	4/0
Over 1100 kcmil	Over 1750 kcmil	3/0	250 kcmil

Notes:

1. Where multiple sets of service-entrance conductors are used as permitted in Section 230-40, Exception No. 2, the equivalent size of the largest service-entrance conductor shall be determined by the largest sum of the areas of the corresponding conductors of each set.

2. Where there are no service-entrance conductors, the grounding electrode conductor size shall be determined by the equivalent size of the largest service-entrance conductor required for the load to be served.

[1]This table also applies to the derived conductors of separately derived ac systems.

[2]See installation restrictions in Section 250-64(a).

FPN: See Section 250-24(b) for size of ac system conductor brought to service equipment.

(Reprinted with permission from NFPA 70-1999)

Equipment grounding conductors are sized according to *Table 250-122*.

2. A *grounding electrode conductor* is the conductor that connects the grounding electrode (i.e., metal underground water pipe supply) to the equipment grounding conductor and to the grounded conductor of the circuit, or to both—at the service equipment. Reference to grounding electrode conductors is made in Figures 28-1, 28-2, 28-7, 28-12, 28-20, 28-21, and 30-3. Grounding electrode conductors are sized according to *Table 250-66*.

In this residence, a No. 4 AWG armored copper grounding electrode conductor runs from Main Panel A, across the workshop ceiling to the water pipe, where it is terminated with a "listed" ground clamp.

Figure 28-20 *Section 250-50* states that if available, an underground metallic water pipe 10 feet (3.05 m) or longer in direct contact with the earth shall be used as the grounding electrode. This water pipe ground *must be supplemented* by at least one additional electrode. Here we show a concrete-encased No. 4 AWG bare copper conductor laid near the bottom of the footing, encased by at least 2 inches (51 mm) of concrete. The *minimum* length of the conductor is 20 feet (6.1 m), *Section 250-50(c)*.

Ground Clamps

Ground clamps used for bonding and grounding must be "listed" for the purpose. The use of solder to make up bonding and grounding connections is not acceptable because under high levels of fault current, the solder would probably melt, resulting in loss of the integrity of the bonding and/or grounding path.

Various types of ground clamps are shown in Figures 28-22, 28-23, 28-24, and 28-25. These clamps and their attachment to the grounding electrode must conform to *Section 250-70*.

STREET SIDE OF
WATER METER

SECTION 250-68 REQUIRES THAT
THESE BONDING JUMPERS BE OF
SUFFICIENT LENGTH TO PERMIT
REMOVAL OF THE METER OR WATER
HEATER WITHOUT LOSING THE
INTEGRITY OF THE BONDING PATH.

GROUNDING ELECTRODE
CONDUCTOR FROM MAIN
SERVICE PANEL NEUTRAL BAR
TO CONNECTION ANYWHERE
ON THE FIRST FIVE FEET (1.5 m)
OF METAL WATER PIPE AFTER
IT ENTERS THE BUILDING. THE
FIRST FIVE FEET MAY INCLUDE
THE WATER METER.

COLD WATER
SUPPLY

HOT
WATER
SUPPLY

GAS
SUPPLY

ANODE
ROD

BOND TOGETHER
THE COLD WATER
SUPPLY, THE HOT
WATER SUPPLY,
AND THE GAS PIPE.

DIELECTRIC
FITTING

DIP TUBE

MAIN
SERVICE
LOCATED
INSIDE
RESIDENCE

METER LOCATED
ON OUTSIDE OF
RESIDENCE

MAIN
BONDING
JUMPER

WATER
HEATER

THE SUPPLEMENTAL GROUNDING ELECTRODE
CONDUCTOR MAY BE CONNECTED TO THE
NEUTRAL BAR IN THE MAIN SERVICE.
IF ACCEPTABLE TO THE POWER COMPANY,
IT MIGHT BE ADVANTAGEOUS TO RUN THE
GROUNDING ELECTRODE CONDUCTOR
FROM THE NEUTRAL BAR OF THE METER
BASE TO A DRIVEN GROUND ROD OR
OTHER ACCEPTABLE GROUNDING ELECTRODE,
SECTION 250-50.

NOTES: * GROUND CLAMPS SIZING OF
GROUNDING CONDUCTORS
AS PER NEC® ARTICLE 250

Figure 28-21   One method that may be used to provide proper grounding and bonding of service-entrance equipment for a typical one-family residence.

Connection of Grounding and Grounded Conductors in Main Panel and Subpanel

According to the second paragraph of *Section 384-20*, equipment *grounding* conductors (the bare copper equipment grounding conductors found in nonmetallic-sheathed cable) shall not be connected to the *grounded* conductor (the neutral conductor) terminal bar (neutral bar) unless the bar is identified for the purpose, and is located where the *grounded* conductor is connected to the *grounding* electrode conductor. In this residence, this occurs in Main Panel A. The green bonding screw furnished with Panel A is installed in Panel A. This bonds the neutral bar, the ground bus, the grounded neutral conductor, the grounding electrode conductor, and

the panel enclosure together, Figure 28-12. *Grounding* conductors and *grounded* conductors are not to be connected together anywhere on the load side of the main service disconnect, *Section 250-142(b)*. The green bonding screw furnished with Panel B will *not* be installed in Panel B. More on this a little later in this Unit.

Sheet Metal Screws Not Permitted for Connecting Grounding Conductors

Sheet metal screws are *not* permitted to be used as a means of connecting grounding conductors to enclosures, *Section 250-8*. Sheet metal screws do not have the same fine thread that 10/32 machine screws have, which match the tapped 10/32 threaded holes in outlet boxes, device boxes, and other enclosures.

Figure 28-22 Typical ground clamps used in residential systems.

Figure 28-23 Armored grounding conductor connected with ground clamp to water pipe. *Courtesy* of Thomas & Betts.

Figure 28-24 Ground clamp of the type used to bond (jumper) around water meter. *Courtesy* of Thomas & Betts.

Sheet metal screws "force" themselves into a hole instead of nicely threading themselves into a pretapped matching hole. Sheet metal screws have *not* been tested for their ability to safely carry ground-fault currents as required by *Section 250-2(d)*.

Figure 28-25 Ground clamp of the type used to attach ground wire to well casings. *Courtesy* of Thomas & Betts.

BONDING

▶ The *NEC®* defines **bonding** as "the permanent joining of metallic parts to form an electrically conductive path that will ensure electrical continuity and the capacity to conduct safely any current likely to be imposed." ◀

Bonding jumpers accomplish this in a number of ways:

1. Equipment bonding "ties" equipment together to keep voltage buildup on the equipment the same, so there will be no difference in potential between the various pieces of equipment. For example, to bond two metal raceways together, an **equipment bonding jumper** is installed where there is a break in the metal conduit run, Figure 28-26. Bonding equipment together does *not* necessarily mean that the equipment is properly grounded.

2. A **circuit bonding jumper** connects conductors of a circuit together. This jumper must have the same or greater current-carrying ability than the circuit conductors, Figure 28-27.

3. A **main bonding jumper** is defined as "the connection between the grounded circuit conductor and the equipment grounding conductor at the service." Main bonding jumpers are clearly illustrated in Figures 28-2, 28-7, 28-12, 28-21, and 30-3.

4. Bonding for swimming pools is shown in Figure 30-4.

Figure 28-26 Bonding jumper (equipment).

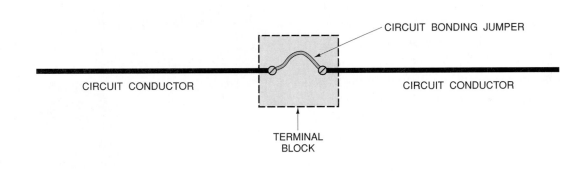

Figure 28-27 Bonding jumper (circuit).

At the main service-entrance equipment, the grounded neutral conductor must be bonded to the metal enclosure. For most residential panels, this main bonding jumper is a bonding screw that is furnished with the panel. This bonding screw is inserted through the neutral bar into a threaded hole in the back of the panel. This bonding screw must be green and must be clearly visible after it is in place, *Section 250-28(b)*.

The required main bonding jumpers for services must not be smaller than the grounding electrode conductor, *Section 250-28(d)*. The lugs on bonding bushings are based on the ampacity of the conductors that would normally be installed in that particular size of raceway. The lugs become larger as the trade size of the bushing increases.

Bonding at service-entrance equipment is very important because service-entrance conductors do not have overcurrent protection at their line side,

other than the electric utility company's primary transformer fuses. Overload protection for service-entrance conductors is at their load end. High available fault currents can result in severe arcing, which is a fire hazard. For all practical purposes, available short-circuit current is limited only by (1) the kVA rating and impedance of the transformer that supplies the service equipment, and (2) the size, type, and length of the service conductors between the transformer and service equipment. Fault currents can easily reach 20,000 amperes or more at the main service in residential installations. Much higher fault current is available in multifamily dwellings such as apartments and condominiums.

Figures 28-7 and 28-21 show the hot and cold metal water pipes bonded together. Some electrical inspectors will consider the hot/cold mixing faucets and the water heater as an acceptable means of bonding the hot and cold water metal pipes together.

Most, however, will require a bonding jumper as illustrated in the figures, consistent with the thinking that electrical grounding and bonding shall not be dependent upon the plumbing trade.

Fault current calculations are presented later in this Unit. The text *Electrical Wiring—Commercial* (Delmar Publishers) covers fault current calculations in greater depth.

Section 250-92(a) lists the parts of the service-entrance equipment that must be bonded. These include the meter base, service raceways, service cable armor, and main disconnect. *Section 250-94* lists the methods acceptable for bonding all of the previous equipment. Bonding bushings, Figure 28-28, and bonding jumpers are installed to ensure a low-impedance path to ground if a fault occurs. Bonding jumpers are required where there are concentric or eccentric knockouts, or where reducing washers are installed. *Sections 300-4(f)* and *373-6(c)* state that if No. 4 AWG or larger conductors are installed in a raceway, an insulating bushing or equivalent must be used, Figure 28-29. Insulating bushings protect the conductors from abrasion where they pass through the bushing. Combination metal/insulating bushings can be used. Some electrical metallic tubing connectors have an insulated throat.

If the bushing is made of insulating material only, as in Figure 28-29, then two locknuts must be used, Figure 28-30.

Figure 28-28 Insulating bonding bushing with grounding lug.

Figure 28-29   Insulating bushings.

LOCKNUTS, ONE INSIDE AND ONE OUTSIDE

JUNCTION BOX

INSULATING BUSHING

(USE DOUBLE LOCKNUTS WHEN INSULATING BUSHING IS NONMETALLIC)

Figure 28-30   The use of double locknuts plus an insulating bushing. See *Sections 300-4(f)* and *373-6(c).*

Installing a Grounding Electrode Conductor from Main Service Disconnect to Metal Underground Water Pipe that Will Serve as the Grounding Electrode

The Code specifies in *Section 250-64(a)* how a grounding electrode conductor is to be installed.

NO. 4 OR LARGER	NO. 6	SMALLER THAN NO. 6
may be attached to the surface on which it is carried. does not need additional mechanical protection unless exposed to severe physical damage, which is highly unlikely in the case of residential installations.	if not subject to physical damage, may be run along the surface of the building structure without additional protection. if subject to physical damage, install in rigid metal conduit, electrical metallic tubing, intermediate metal conduit, rigid nonmetallic conduit, or cable armor.	must be installed in rigid metal conduit, electrical metallic tubing, intermediate metal conduit, rigid nonmetallic conduit, or cable armor.

A TYPICAL ARMORED GROUNDING ELECTRODE CONDUCTOR.

In order to carry high values of ground-fault current, the ground path must have as low an impedance as possible. *Section 250-2* requires that the ground path be *effective*. This means that the path must be permanent and continuous, must have the capability of carrying any value of fault current that it might be called upon to carry, and shall have an impedance low enough to limit the voltage to ground and to ensure that the protective overcurrent device will operate. *Sections 250-90* and *250-96* have similar requirements. For example, the metal raceway enclosing a grounding electrode conductor must be continuous from the main disconnect to the ground clamp, or be made continuous by proper bonding.

Section 250-92(a)(3) requires that the metal raceway be bonded at both ends. At first thought, it might appear that simply installing a grounding electrode conductor in a metal raceway makes for a neat and workmanshiplike installation. However, unless the enclosing metal raceway is properly bonded on both ends, the ground path will be a high impedance path. Bonding the enclosing metal raceway at one end only will result in a ground path impedance approximately twice that of bonding the metal raceway on both ends. Remember—the higher the impedance, the lower the current flow and the longer it takes the overcurrent device to clear the fault.

Past testing (Soares, IAEI, and IEEE data) has shown that for a 300-ampere ground fault, approximately 295 amperes flowed in the metal raceway, and 5 amperes flowed in the enclosed grounding electrode conductor. Another test revealed that for a 590-ampere ground fault, 585 amperes flowed in the metal raceway, and 5 amperes flowed in the enclosed grounding electrode conductor. The tendency for current to flow along the surface of a conductor is called "skin effect." Proper bonding at both ends of a metal raceway is necessary to ensure a low impedance path that will safely transfer the high magnitudes of ground-fault current flowing in the metal conduit to the ground bus in the main panel, and/or to the ground clamp.

Figures 28-31, 28-32, 28-33, 28-34, and 28-35 illustrate accepted methods of installing a grounding electrode conductor between the main panel and the ground clamp. Additional information concerning grounding electrode conductors can be found in Unit 4.

Grounding the Electrical Service When Nonmetallic Water Pipe Is Used

Today it is quite common to find the main water supply to a residence and the interior water piping installed with plastic (PVC) piping.

As previously mentioned, one of the methods used to provide for adequate grounding of services where nonmetallic water supplies are encountered,

Figure 28-31 This installation "meets Code."

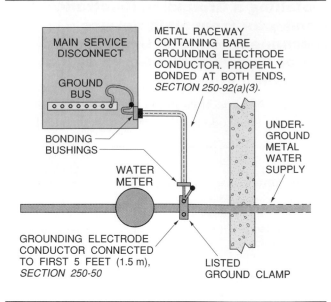

Figure 28-33 This installation "meets Code."

Figure 28-32 This installation "meets Code."

Figure 28-34 This installation *does not* "meet Code."

or where the soil is so dry (sand) that driven ground rods cannot attain a 25-ohm or less resistance ground, *Section 250-56*, or where the metallic underground water piping in direct contact with the earth is less than 10 feet (3.05 m), is to install a concrete-encased electrode, commonly referred to as a UFER ground, Figure 28-20.

A UFER ground consists of a bare copper conductor:

• not smaller than No. 4 AWG

• not less than 20 feet (6.1 m) long encased by at least 2 inches (50.8 mm) of concrete in the footing or foundation

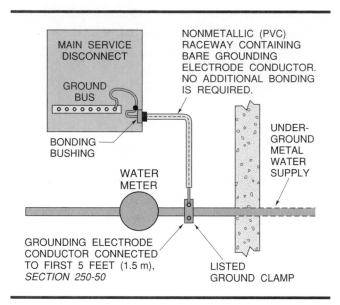

MAIN SERVICE DISCONNECT

GROUND BUS

BONDING BUSHING

NONMETALLIC (PVC) RACEWAY CONTAINING BARE GROUNDING ELECTRODE CONDUCTOR. NO ADDITIONAL BONDING IS REQUIRED.

WATER METER

UNDER-GROUND METAL WATER SUPPLY

GROUNDING ELECTRODE CONDUCTOR CONNECTED TO FIRST 5 FEET (1.5 m), *SECTION 250-50*

LISTED GROUND CLAMP

Figure 28-35   This installation "meets Code."

- in concrete that is in direct contact with the earth
- lying within and near the bottom of the concrete
- solid or stranded

Some soil conditions are so poor for grounding purposes that the UFER ground becomes the primary grounding electrode, as opposed to merely being a supplemental ground.

SUMMARY—SERVICE-ENTRANCE EQUIPMENT GROUNDING

Following are some of the most important requirements for service-entrance equipment grounding. Refer to Figures 28-20 and 28-21.

- The system must be grounded so the maximum voltage to ground on the ungrounded conductors does not exceed 150 volts, *Section 250-20(b)(1)*.

- All grounding schemes shall be installed so that no objectionable currents will flow over the grounding conductors and other grounding paths, *Section 250-6(a)*.

- The grounding electrode conductor must be connected to the supply side of the service disconnecting means. It must not be connected to any grounded circuit conductor on the load side of the service disconnect, *Section 250-24(a)(5)*.

- The neutral conductor must be grounded, *Section 250-26(2)*.

- Tie (bond) everything together, *Sections 250-50, 250-52, 250-92*, and *250-104(a)(b)(c)(d)*.

- The grounding electrode conductor used to connect the grounded neutral conductor to the grounding electrode must not be spliced, *Sections 250-28* and *250-50*.

- Size the grounding electrode conductor according to *Table 250-66*.

- Bond all metal piping together, *Section 250-104*.

- The grounding electrode conductor must be connected to the metal underground water pipe when 10 feet (3.05 m) long or more, including the well casing, *Section 250-50(a)*.

- Connect the grounding electrode conductor to the first 5 feet (1.5 m) of metal underground water piping after it enters the building. This will generally be ahead of the water meter, unless the water meter is located in the street outside of the dwelling. The *NEC®* does not permit interior metal water piping to be used to interconnect electrodes and the grounding electrode conductor, *Section 250-50*.

- A metal water pipe ground must be supplemented by another electrode, such as a bare conductor in the footing, *Section 250-50(c);* a grounding ring, *Section 250-50(d);* the metal frame of the building *if* the metal frame is effectively grounded, *Section 250-50(c);* a rod or pipe electrodes, *Section 250-52(c);* or plate electrodes, *Section 250-52(d)*.

- The grounding electrode conductor shall be copper, aluminum, or copper-clad aluminum. It may be solid or stranded, insulated, covered or bare, *Section 250-62*.

▶• The grounding electrode conductor shall be one continuous length, and not be spliced, unless done so with an irreversible compression type connector "listed" for the purpose, or by exothermic welding, *Section 250-64(b).* ◀

- Bonding shall be provided around all insulating joints or sections of the metal piping system that may be disconnected, *Section 250-68*.

►• The connection of the grounding electrode conductor to the grounding electrode shall be accessible. The connection to a concrete-encased, driven, or buried grounding electrode does not have to be accessible. See *Section 250-68.*◄

• It is acceptable to bury the connection in the ground using clamps listed for direct soil burial, *Section 250-70.*

• The grounding electrode conductor and bonding jumpers shall be connected tightly using the proper lugs, connectors, clamps, exothermic welding, or other "listed" means. Solder connections are not permitted, *Section 250-70.*

MAIN, FEEDER, AND BRANCH-CIRCUIT OVERCURRENT PROTECTION

The *NEC®* covers overcurrent protection in *Article 240.* Overcurrent protection for residential main services, feeders, and branch-circuits comes in the form of fuses and circuit breakers, often called the "safety valves" of an electrical circuit.

Fuses and circuit breakers are sized by matching their ampere ratings to conductor ampacities and connected load currents. They sense overload, short-circuit, and ground-fault conditions, and protect the wiring and equipment from reaching dangerous temperatures.

A fuse will function only one time. It does its job of protecting the circuit.

When a fuse opens, find out what caused it to blow, fix the problem, and replace it with the size and type that will provide proper overcurrent protection. Do not keep replacing a fuse without finding out why the fuse blew in the first place.

When a circuit breaker trips, find out what caused it to trip, fix the problem, then reset it. Do not repeatedly reset the breaker again and again into a fault, as further damage can result.

Overcurrent protective devices are connected in a circuit where the conductors receive their supply, *Section 240-21.* There are some exceptions to this basic rule, also found in *Section 240-21.* A typical example for house wiring is for "taps" that supply built-in ovens and cook-tops. This is covered in Unit 20.

Overcurrent devices are connected "in series" with the ungrounded (hot) conductors of a circuit, *Section 240-20.*

The *NEC®* does not allow overcurrent devices to be connected in the grounded conductor, *Section 240-22.* The two exceptions are: (1) where both grounded and ungrounded conductors open simultaneously and (2) where required in *Sections 430-36* and *430-37* for motor overload protection.

Normal loading of a circuit is when the current flowing is within the capability of the circuit and/or the connected equipment. In Figure 28-36 we have a 15-ampere circuit carrying 10 amperes.

An overload is a condition where the current flowing is more than the circuit and/or connected equipment is designed to safely carry. Figure 28-37 shows a 15-ampere circuit carrying 20 amperes. A momentary overload is harmless, but a continuous overload condition will cause the conductors and/or equipment to overheat—a potential cause for fire. The current flows through the "intended path"—the conductors and/or equipment.

A short-circuit is a condition when two or more conductors come in contact with one another, resulting in a current flow that bypasses the connected load. A short-circuit might be two "hot" conductors coming together, or it might be a "hot" conductor and a "grounded" conductor coming together. In either case, the current flows outside of the "intended path." The only resistance (impedance) is that of the conductors, the source, and the arc. This low resistance results in high levels of short-circuit current. The heat generated at the point of the arc can result in a fire, Figure 28-38.

A ground fault is a condition when a "hot" conductor comes in contact with a grounded surface, such as a grounded metal raceway, metal water pipe, sheet metal, etc., Figure 28-39. The current flows outside of the "intended path." A ground fault can result in a flow of current greater than normal, in which case the overcurrent device will open. A ground fault can also result in a current flow less than normal, in which case the overcurrent device may not open.

Ground-fault circuit interrupters (GFCIs) are an example of a device that protects against very low levels of ground fault passing through the human body when a person touches a "live" wire and a grounded surface. See Unit 6.

Large services and feeders require ground-fault protection for certain ampere ratings, voltage ratings, and type of connection. This is discussed in *Electrical Wiring—Commercial* (Delmar Publishers).

Ⓐ NORMAL CIRCUIT

THE CONDUCTORS CAN SAFELY CARRY THE CURRENT, THEY DO NOT GET HOT. THE 15 AMPERE FUSES DO NOT OPEN.

CONDUCTORS RATED 15 AMPERES
CONDUCTOR RESISTANCE = 0.001 OHMS

15 A FUSE   9.99875 AMPERES

240 VOLT SOURCE

LOAD RESISTANCE
= 24 OHMS

24 Ω

SOURCE RESISTANCE
= 0.001 OHMS

15 A FUSE

CONDUCTORS RATED 15 AMPERES
CONDUCTOR RESISTANCE = 0.001 OHMS

$$I \text{ (THROUGH CIRCUIT)} = \frac{E}{R} = \frac{240}{24 + 0.001 + 0.001 + 0.001} = \frac{240}{24.003} = 9.99875 \text{ AMPERES}$$

Figure 28-36   A normally loaded circuit.

Ⓑ OVERLOADED CIRCUIT

THE CONDUCTORS BEGIN TO GET HOT, BUT THE 15 AMPERE FUSES WILL OPEN IN SOME PERIOD OF TIME BEFORE THE CONDUCTORS ARE DAMAGED.

CONDUCTORS RATED 15 AMPERES
CONDUCTOR RESISTANCE = 0.001 OHMS

15 A FUSE   19.995 AMPERES

240 VOLT SOURCE

LOAD RESISTANCE =

24 Ω   24 Ω

$$\frac{R1}{N} = \frac{24}{2} = 12 \text{ OHMS}$$

SOURCE RESISTANCE
= 0.001 OHMS

15 A FUSE

CONDUCTORS RATED 15 AMPERES
CONDUCTOR RESISTANCE = 0.001 OHMS

$$I \text{ (THROUGH CIRCUIT)} = \frac{E}{R} = \frac{240}{12 + 0.001 + 0.001 + 0.001} = \frac{240}{12.003} = 19.995 \text{ AMPERES}$$

Figure 28-37   An overloaded circuit.

Ⓒ SHORT CIRCUIT

THE CONDUCTORS GET EXTREMELY HOT. THE INSULATION WILL MELT OFF AND THE CONDUCTORS WILL MELT UNLESS THE FUSES OPEN IN A VERY SHORT (FAST) PERIOD OF TIME. CURRENT-LIMITING OVERCURRENT DEVICES WILL LIMIT THE AMOUNT OF "LET-THROUGH" CURRENT, AND WILL LIMIT THE TIME THE FAULT CURRENT IS PERMITTED TO FLOW.

CONDUCTORS RATED 15 AMPERES
CONDUCTOR RESISTANCE = 0.001 OHMS

240 VOLT SOURCE

SOURCE RESISTANCE = 0.001 OHMS

15 A FUSE   80,000 AMPERES

24 Ω

RESISTANCE OF SHORT CIRCUIT = ZERO OHMS

$$\frac{R_1 \times R_2}{R_1 + R_2} = \frac{24 \times 0}{24 + 0} = \frac{0}{24} = 0$$

15 A FUSE

CONDUCTORS RATED 15 AMPERES
CONDUCTOR RESISTANCE = 0.001 OHMS

$$I\,(\text{THROUGH CIRCUIT}) = \frac{E}{R} = \frac{240}{0.001 + 0.001 + 0.001} = \frac{240}{0.003} = 80,000 \text{ AMPERES}$$

Figure 28-38   Note that the connected load is short-circuited.

Ⓓ GROUND FAULT

"HOT" CONDUCTOR

SOURCE

GROUNDED CIRCUIT CONDUCTOR

"HOT" CONDUCTOR COMES IN CONTACT WITH METAL RACEWAY OR OTHER METAL OBJECT. IF THE RETURN GROUND PATH HAS LOW RESISTANCE (IMPEDANCE), THE OVERCURRENT DEVICE PROTECTING THE CIRCUIT WILL CLEAR THE FAULT. IF THE RETURN GROUND PATH HAS HIGH RESISTANCE (IMPEDANCE), THE OVERCURRENT DEVICE WILL NOT CLEAR THE FAULT. THE METAL OBJECT WILL THEN HAVE A VOLTAGE TO GROUND THE SAME AS THE "HOT" CONDUCTOR HAS TO GROUND. IN HOUSE WIRING, THIS VOLTAGE TO GROUND IS 120 VOLTS. PROPER GROUNDING AND GROUND-FAULT CIRCUIT INTERRUPTER PROTECTION IS DISCUSSED ELSEWHERE IN THIS TEXT. THE CALCULATION PROCEDURE FOR A GROUND FAULT IS THE SAME AS FOR A SHORT CIRCUIT, HOWEVER, THE VALUES OF "R" CAN VARY GREATLY BECAUSE OF THE UNKNOWN IMPEDANCE OF THE GROUND RETURN PATH. LOOSE LOCKNUTS, BUSHINGS, SET SCREWS ON CONNECTORS AND COUPLINGS, POOR TERMINATIONS, RUST, ETC. ALL CONTRIBUTE TO THE RESISTANCE OF THE RETURN GROUND PATH, MAKING IT EXTREMELY DIFFICULT TO DETERMINE THE ACTUAL GROUND-FAULT CURRENT VALUES.

Figure 28-39   The insulation on the "hot" conductor has come in contact with the metal conduit. This is termed a "ground fault."

Plug Fuses, Fuseholders, and Adapters *(Article 240, Part E)*

Fuses are a reliable and economical form of overcurrent protection. *Sections 240-50* through *240-54* give the requirements for plug fuses, fuseholders, and adapters. These requirements include the following:

- ▶ plug fuses may be used (1) in circuits that do not exceed 125 volts between conductors, and (2) in circuits having a grounded neutral where no conductor operates at over 150 volts to ground. ◀ In a residential 120/240-volt electrical system, the voltage between the two "hot" conductors is 240 volts, and the voltage from either "hot" conductor to ground is 120 volts.

- the fuses shall have ampere ratings of 0 to 30 amperes.

- plug fuses shall have a hexagonal configuration somewhere on the fuse when rated at 15 amperes or less.

- the screw shell of the fuseholder must be connected to the load side of the circuit.

- Edison-base plug fuses may be used only to replace fuses in existing installations where there is no sign of overfusing or tampering.

- all new installations shall be in Type S fuses.

- Type S fuses are classified at 0 to 15 amperes, 16 to 20 amperes, and 21 to 30 amperes. The reason for this classification is given in the following paragraph.

When the electrician installs fusible equipment, the ampere rating of the various circuits must be determined. Based on this rating, an adapter of the proper size is inserted into the Edison-base fuseholder. The proper Type S fuse is then screwed into the adapter. Because of the adapter, the fuseholder is nontamperable and noninterchangeable. For example, assume that a 15-ampere adapter is inserted for the 15-ampere branch-circuits in the residence. It is impossible to substitute a Type S fuse with a larger rating without removing the 15-ampere adapter.

Type S fuses and adapters, Figure 28-40, are available with ratings from 3/10 of an ampere to 30 amperes. The Type S fuse may have a time-delay feature. If it does, it is known as a dual-element fuse. Momentary overloads do not cause a dual-element

Figure 28-40   Type S fuse and adapter. *Courtesy* of Cooper Bussmann, Inc.

Figure 28-41   A 15-ampere Type W Edison base plug fuse and a 20-ampere Type T time-delay Edison base plug fuse. *Courtesy* of Cooper Bussmann, Inc.

fuse to blow. An example of such an overload is the current surge as an electric motor is started. However, a dual-element fuse opens rapidly if there is a heavy overload or a short-circuit. Figure 28-41 shows a 15-ampere Edison-base plug fuse and a 20-ampere time-delay Edison-base plug fuse.

Dual-Element Fuse (Cartridge and Plug Type)

The most common type of dual-element cartridge fuse is shown in Figure 28-42. These fuses are available in 250-volt and 600-volt sizes with ratings from 0 to 600 amperes, and have an interrupting rating of 200,000 amperes.

A dual-element fuse has two fusible elements connected in series. These elements are known as the *overload element* and the *short-circuit element*. When an excessive current flows, one of the elements opens. The amount of excess current determines which element opens.

Figure 28-42   Cartridge-type dual-element fuse (A) is a 250-volt, 100-ampere fuse. The cutaway view in (B) shows the internal parts of the fuse. When this type of fuse has a rejection slot in the blades of 70–600 ampere sizes, or a rejection ring in the end ferrules of the 0–60 ampere sizes, they are U.L. Class RK1 or RK5 fuses. *Courtesy* **of Cooper Bussmann, Inc.**

The electrical characteristics of these two elements are very different. Thus, a dual-element fuse has a greater range of protection than the single-element fuse.

Overload Element.   The overload element opens when excessive heat is developed in the element. This heat may be the result of a loose connection or poor contact in the fuseholder or excessive current. When the temperature reaches 280°F, a fusible alloy melts and the element opens. However, the mass of the element absorbs a great deal of heat before the cutout opens. A small excess of current will cause this element to open if it continues for a long period of time. This characteristic gives the thermal cutout element a large time-delay on low overloads of about 10 seconds at a current of 500% of the fuse rating. In addition, it provides very accurate protection for prolonged overloads.

Short-Circuit Element.   The capacity of the short-circuit element is high enough to prevent it from opening on low overloads. This element is designed to clear the circuit quickly of short-circuits or heavy overloads above 500% of the fuse rating.

Dual-element fuses are used on motor and appliance circuits where the time-lag characteristic of the fuse is required. Single-element fuses do not have this time lag. Such fuses blow as soon as an overcurrent condition occurs. The homeowner may be tempted to install a fuse with a higher ampere rating. Type S fuseholder adapters prevent the use of a higher ampere rating Type S fuse. Thus, the dual-element fuse is recommended for this type of situation.

Sometimes poor connections at the terminals or loose fuse clips can be found because the fuse tubing will appear to be charred. This problem can be dangerous and should be corrected as soon as possible. Quite often, the heat resulting from poor connections in the fuseclips or fuseholder will cause the brass or copper fuseclips or fuseholders to turn from a bright, shiny appearance to that of antique brass or copper.

Class G Fuses

Class G fuses are rated 480 volts or less. They are available in many ampere ratings from $\frac{1}{2}$ through 60 amperes. The physical size of these fuses is smaller than that of standard cartridge fuses, Figure 28-43.

At 200% load, they have a minimum opening time of 25 seconds. This time-delay characteristic means that they can handle harmless current surges of momentary overloads without nuisance blowing. However, these fuses open very quickly under short-circuit or ground-fault conditions.

These fuses prevent overfusing because they are size limiting in four different case sizes – 1/2 through 15, 20, 25 and 30, and 35 through 60 amperes. For example, a fuseholder designed to accept a 15-ampere, Class G fuse will reject a 20-ampere, Class G fuse. In the same manner, a fuseholder designed to accept a 20-ampere, Class G fuse will reject a 30-ampere, Class G fuse.

Class T Fuses

Another type of physically small fuse is the Class T fuse, Figure 28-44. This type of fuse has a high-interrupting rating and is current limiting. They cannot be interchanged with standard cartridge fuses because of their unique physical size. Available in ampere ratings through 1,200 amperes in 300-volt ratings, Class T fuses open extremely fast under short-circuit or ground-fault conditions. In "series-rated" load centers, Class T fuses are used as main fuses in panels that have circuit breakers for the branch-circuits.

Figure 28-43 Class G fuses. *Courtesy* **of Cooper Bussmann, Inc.**

These fuses protect the low-interrupting rating branch-circuit breakers against high-level short-circuit or ground-faults, Figure 28-46.

Most manufacturers list service equipment, metering equipment, and disconnect switches that make use of Class T fuses. This allows the safe installation of this equipment where fault-currents exceed 10,000 amperes, such as on large services, and any service equipment located close to pad-mount transformers, such as are found in apartments, condominiums, shopping centers, and similar locations.

Fuse Ampere Ratings

The standard ampere ratings for fuses are given in *Section 240-6(a)*. In addition to the standard ratings listed, there are many other "in between" ampere ratings available to match specific load requirements. Check a fuse manufacturer's catalog for other available ampere ratings.

Circuit Breakers

Installations in dwellings normally use thermal-magnetic circuit breakers. On a continuous overload, a bimetallic element in such a breaker moves until it unlatches the inner tripping mechanism of the breaker. Momentary small overloads do not cause the element to trip the breaker. If the overload is heavy or if there is a short circuit, a magnetic coil in the breaker causes it to interrupt the branch-circuit instantly. Unit 13 covers the effect of tungsten lamp loads on circuit breakers.

Sections 240-80 through *240-83* give the requirements for circuit breakers. The following points are taken from these sections.

- Circuit breakers shall be trip free so that even if the handle is held in the ON position, the internal mechanism will trip to the OFF position.

- Breakers shall indicate clearly whether they are on or off.

- A breaker shall be nontamperable so that it cannot be readjusted (trip point changed) without dismantling the breaker or breaking the seal.

Figure 28-44 Class T fuses. Illustrated are the 30-, 60-, 100-, 200-, 400-, and 600-ampere, 300-volt ratings. *Courtesy* **of Cooper Bussmann, Inc.**

- The rating shall be durably marked on the breaker. For small breakers rated at 100 amperes or less and 600 volts or less, the rating must be molded, stamped, or etched on the handle or on another part of the breaker that will be visible after the cover of the panel is installed.

- Every breaker with an interrupting rating other than 5,000 amperes shall have this rating marked on the breaker.

- Circuit breakers rated at 120 volts and 277 volts and used for fluorescent loads, shall not be used as switches unless marked "SWD."

Most circuit breakers are ambient temperature compensated. This means that the tripping point of the breaker is not affected by an increase in the surrounding temperature. An ambient-compensated breaker has two elements. One element heats up due to the current passing through it and the heat in the surrounding area. The other element heats up because of the surrounding air only. The actions of these elements oppose each other. Thus, as the tripping element tends to lower its tripping point because of external heat, the second element opposes the tripping element and stabilizes the tripping point. As a result, the current through the tripping element is the only factor that causes the element to open the circuit. It is a good practice to turn the breaker on and off periodically to "exercise" its moving parts.

Section 240-6(a) gives the standard ampere ratings of circuit breakers.

One Two-Pole or Two Single-Pole Circuit Breakers

Use two-pole circuit breakers for 120/240-volt branch-circuits such as electric ranges, electric ovens, and electric clothes dryers. Use two-pole circuit breakers for straight 240-volt branch-circuits such as electric water heaters, electric furnaces, electric baseboard heaters, air conditioners, and heat pumps. For these situations, *Section 240-20(b)* permits the use of "handle ties" that connect the handles of two single-pole circuit breakers together so both will trip together should a fault occur. But the preference is to use two-pole circuit breakers instead of handle ties.

INTERRUPTING RATINGS FOR FUSES AND CIRCUIT BREAKERS

Here are some very important *NEC®* code sections regarding interrupting ratings.

▶• *Section 110-9:* **Interrupting Rating.** *"Equipment intended to interrupt current at fault levels shall have an interrupting rating sufficient for the nominal circuit voltage and the current that is available at the line terminals of the equipment. Equipment intended to interrupt current at other than fault levels shall have an interrupting rating at nominal circuit voltage sufficient for the current that must be interrupted."*◀

▶• *Section 110-10:* **Circuit Impedance and Other Characteristics.** *"The overcurrent protective devices, the total impedance, the component short-circuit current ratings, and other characteristics of the circuit to be protected shall be selected and coordinated to permit the circuit protective devices used to clear a fault to do so without extensive damage to the electrical components of the circuit. This fault shall be assumed to be either between two or more of the circuit conductors, or between any circuit conductor and the grounding conductor or enclosing metal raceway. Listed products applied in accordance with their listing shall be considered to meet the requirements of this section."*◀

- U.L. Standard No. 67 requires that panelboards be marked with their short-circuit current rating.

Panelboards are available in two types: "fully-rated" and "series-rated." The figures that follow explain the meaning of these terms.

The overcurrent protective device must be able to interrupt the current that may flow under any condition (overload or short circuit). Such interruption must be made with complete safety to personnel and without damage to the panel or switch in which the overcurrent device is installed.

CAUTION: Overcurrent devices with inadequate interrupting ratings are, in effect, bombs waiting for a short-circuit to trigger them into an explosion. Personal injury may result and serious damage will be done to the electrical equipment.

Class G fuses have an interrupting rating of 100,000 amperes root mean square (rms) symmetrical. Plug fuses can interrupt no more than 10,000 amperes rms symmetrical. Cartridge dual-element fuses, Figure 28-42, and Class T fuses, Figure 28-44, have interrupting ratings of 200,000 amperes rms symmetrical. These interrupting ratings are listed by Underwriters Laboratories (U.L.). The interrupting rating of a fuse is marked on its label when the rating is other than 10,000 amperes.

Short-Circuit Currents

This text does not cover in detail the methods of calculating short-circuit currents. For a detailed discussion, see *Electrical Wiring—Commercial* (Delmar Publishers). The ratings required to determine the maximum available short-circuit current delivered by a transformer are the kVA and im-

pedance values of the transformer. The size and length of wire installed between the transformer and the overcurrent device must be considered as well.

The transformers used in modern electrical installations are efficient and have very low impedance values. A low-impedance transformer having a given kVA rating delivers more short-circuit current than a transformer with the same kVA rating and a higher impedance. When an electrical service is connected to a low-impedance transformer, the problem of available short-circuit current is very serious.

The examples given in Figures 28-45, 28-46, 28-47, and 28-48 show that the available short-circuit

Figure 28-45 Fused main/fused branch-circuits, fully-rated.

Figure 28-46 Fused main/breaker branch-circuits, series-rated.

AVAILABLE FAULT CURRENT
20,800 AMPERES

SERVICE EQUIPMENT MUST
HAVE A SHORT-CIRCUIT RATING
OF AT LEAST 20,800 AMPERES

STANDARD MAIN BREAKER
MUST HAVE AT LEAST
20,800-AMPERE INTERRUPTING
RATING

STANDARD BRANCH-
CIRCUIT BREAKERS
MUST HAVE AT LEAST
20,800-AMPERE
INTERRUPTING RATING

MEETING REQUIREMENTS OF *SECTIONS 110-9* AND *110-10*
FOR BREAKER MAIN/BREAKER BRANCH-CIRCUITS.
THIS PANEL IS "FULLY-RATED," NOT "SERIES-RATED."

Figure 28-47   Breaker main/breaker branch-circuits, fully-rated.

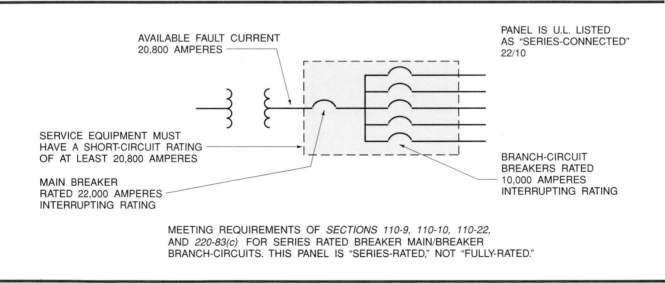

PANEL IS U.L. LISTED
AS "SERIES-CONNECTED"
22/10

AVAILABLE FAULT CURRENT
20,800 AMPERES

SERVICE EQUIPMENT MUST
HAVE A SHORT-CIRCUIT RATING
OF AT LEAST 20,800 AMPERES

MAIN BREAKER
RATED 22,000 AMPERES
INTERRUPTING RATING

BRANCH-CIRCUIT
BREAKERS RATED
10,000 AMPERES
INTERRUPTING RATING

MEETING REQUIREMENTS OF *SECTIONS 110-9, 110-10, 110-22,*
AND *220-83(c)* FOR SERIES RATED BREAKER MAIN/BREAKER
BRANCH-CIRCUITS. THIS PANEL IS "SERIES-RATED," NOT "FULLY-RATED."

Figure 28-48   Breaker main/breaker branch-circuits, series-rated.

current at the transformer secondary terminals is 20,800 amperes.

Determining Short-Circuit Current

The local electric utility and the electrical inspector are good sources of information when short-circuit current is to be determined.

A simplified method is given as follows to determine the approximate available short-circuit current at the terminals of a transformer.

1. Determine the normal full-load secondary current delivered by the transformer.

For single-phase transformers:

$$I = \frac{kVA \times 1000}{E}$$

For three-phase transformers:

$$I = \frac{kVA \times 1000}{E \times 1.73}$$

Where I = current, in amperes
kVA = kilovolt-amperes (transformer nameplate rating)
E = secondary line-to-line voltage (transformer nameplate rating)

2. Using the impedance value given on the transformer nameplate, find the multiplier to determine the short-circuit current.

$$\text{multiplier} = \frac{100}{\text{percent impedance}}$$

3. The short-circuit current = normal full-load secondary current × multiplier.

EXAMPLE: A transformer is rated at 100 kVA and 120/240 volts. It is a single-phase transformer with an impedance of 1% (from the transformer nameplate). Find the short-circuit current.

For a single-phase transformer:

$$I = \frac{\text{kVA} \times 1000}{E}$$

$$I = \frac{100 \times 1000}{240}$$

= 416 amperes, full-load current

The multiplier for a transformer impedance of one percent is:

$$\text{multiplier} = \frac{100}{\text{percent impedance}}$$

$$= \frac{100}{1}$$

multiplier = 100

The short-circuit current = I × multiplier

= 416 amperes × 100

= 41,600 amperes

Thus, the available short-circuit current at the terminals of the transformer is 41,600 amperes.

This value decreases as the distance from the transformer increases.

If the transformer impedance is 1.5%, the multiplier is:

$$\text{multiplier} = \frac{100}{1.5} = 66.6$$

The short-circuit current = 416 × 66.6

= 27,706 amperes

If the transformer impedance is 2%:

$$\text{multiplier} = \frac{100}{2} = 50$$

The short-circuit current = 416 × 50

= 20,800 amperes

(*Note:* A short-circuit current of 20,800 amperes is used in Figures 28-45, 28-46, 28-47, and 28-48.)

CAUTION: The line-to-neutral short-circuit current at the secondary of a single-phase transformer is approximately 1½ times greater than the line-to-line short-circuit current. For the previous example,

Line-to-line short-circuit current = 20,800 amperes

Line-to-neutral short-circuit current
= 20,800 × 1.5 = 31,200 amperes

It is significant to note that U.L. Standard 1561 allows the marked impedance on the nameplate of a transformer to vary plus or minus (±) 10% from the actual impedance value of the transformer. Thus, for a transformer marked "2%Z," the actual impedance could be as low as 1.8%Z to as much as 2.2%Z. Therefore, the available short-circuit current at the secondary of a transformer as calculated previously should be increased by 10% if one wishes to be on the safe side when considering the interrupting rating of the fuses and breakers to be installed.

An excellent point-to-point method for calculating fault currents at various distances from the secondary of a transformer that uses different size conductors is covered in detail in *Electrical Wiring—Commercial* (Delmar Publishers). It is simple and accurate. The point-to-point method can be used to compute both single-phase and three-phase faults.

Calculating fault-currents is just as important as calculating load-currents. Overloads will cause conductors and equipment to run hot, will shorten their life, and will eventually destroy them. Fault currents that exceed the interrupting rating of the overcurrent protective devices can cause violent electrical explosions of equipment with the potential of serious injury to people standing near it. Fire hazard is also present.

PANELS AND LOAD CENTERS

Fused Main/Fused Branches
(Fully-Rated)

The following requirements apply to fused main/fused branches, Figure 28-45.

- must be listed and marked as suitable for use as service equipment when used as service equipment. Does not need this service equipment marking when used as a subpanel that has overcurrent protection ahead of the panel.

- must be sized to satisfy the required ampacity as determined by service-entrance calculations and/or local codes.

- main and branch fuses must have interrupting rating adequate for the available fault current.

- panel must have short-circuit current rating adequate for the available fault current.

- for overload conditions on a branch-circuit, only the branch-circuit fuse(s) will open. All other circuits remain energized. System is selective.[1]

- for short-circuits (line-to-line or line-to-neutral) and for ground faults (line-to-ground) only the fuse protecting the faulted circuit opens. All other circuits remain energized. System is selective.[1]

Fused Main/Breaker Branches
(Series-Rated)

The following requirements apply to fused main/breaker branches, Figure 28-46. Fuse main/breaker branches:

- must be listed and marked as suitable for use as service equipment when used as service equipment. Does not need this service equipment marking when used as a subpanel that has overcurrent protection ahead of the panel.

- must be marked and listed "Series-Rated."

- must be sized to satisfy the required ampacity as determined by service-entrance calculations and/or local codes.

- main fuses must have interrupting rating adequate for the available fault circuit.

- panel must have short-circuit current rating adequate for the available fault current.

- main fuses must be current-limiting, such as Class T fuses, so that if a fault exceeding the branch breaker's interrupting rating occurs, the fuse will limit the let-through fault current to a value less than the maximum interrupting rating of the branch breaker. Branch-circuit breakers generally have 10,000 amperes interrupting rating.

- for branch-circuit faults (L-N or L-G) exceeding the branch breaker's interrupting rating, one fuse opens, deenergizing one-half of the panel. System is nonselective.[1]

- for branch-circuit faults (L-L) exceeding the branch breaker's interrupting rating, both main fuses open, deenergizing the entire panel. System is nonselective.[1]

- check time-current curves of fuses and breakers and check the unlatching time of breakers to determine at what point in "time" and "current" the breakers are slower than the current-limiting fuses.

- for normal overloads and low-level faults, this system is selective.[1]

Breaker Main/Breaker Branches
(Fully-Rated)

The following requirements apply to breaker main/breaker branches, Figure 28-47. Breaker main/breaker branches:

- must be listed and marked as suitable for use as service equipment when used as service equipment. Does not need this service equipment marking when used as a subpanel that has overcurrent protection ahead of the panel.

- must be sized to satisfy the required ampacity as determined by service-entrance calculations and/or local codes.

---

[1] The terms and theory behind selective and nonselective systems are covered in detail in *Electrical Wiring—Commercial* (Delmar Publishers). In all instances, it is advisable to check the time-current curves of fuses and circuit breakers to be installed. Also check the circuit breaker manufacturer's "unlatching time" data, as this will help predetermine at what values of current the system will be selective or nonselective.

- main and branch breakers must have interrupting rating adequate for the available fault current.

- panel must have short-circuit current rating adequate for the available fault current.

- for overload conditions and low-level faults on a branch-circuit, only the branch breaker will trip off. All other circuits remain energized. System is selective.[1]

- for branch-circuit faults (short-circuits or ground faults) that exceed the instant trip setting of the main breaker, both the branch breaker *and* the main breaker trip off. Entire panel is deenergized. System is nonselective.[1]

- check manufacturer's data for time-current characteristic curves and unlatching times of the breakers.

Breaker Main/Breaker Branches
(Series-Rated)

The following requirements apply to breaker main/breaker branches, Figure 28-38. Breaker main/breaker branches:

- must be listed and marked as suitable for use as service equipment when used as service equipment. Does not need this service equipment marking when used as a subpanel that has overcurrent protection ahead of the panel.

- must be sized to satisfy the required ampacity as determined by service-entrance calculations and/or local codes.

- main breaker will have interrupting rating higher than the branch breaker. Example: main C/B has 22,000-ampere interrupting rating, whereas branch C/B has 10,000-ampere interrupting rating.

- must be marked and listed "Series-Rated."

- panel must be listed and marked with its maximum short-circuit current rating.

- for branch-circuit faults (short-circuits or ground faults) that exceed the instant trip setting of the main breaker, both the branch breaker *and* main breaker trip off. Entire panel is deenergized. System is nonselective.[1]

- for overload conditions and low-level faults on a branch, only the branch breaker will trip. System is selective.[1]

- obtain time-current curves and unlatching time data from the manufacturer.

- the panel's series rating is attained because when a heavy fault occurs, both main and branch breakers open. The two arcs in series add impedance; thus the fault current is reduced to a level less than the 10,000 A.I.C. branch breaker.

- when panels containing low interrupting rating circuit breakers are separated from the equipment in which the higher interrupting rating circuit breakers protecting the panel are installed, the system becomes a field-installed "Series-rated System." This panel must be field marked by the installer electrician, and this marking must be readily visible, *Section 110-22.* This marking calls attention to the fact that the actual available fault current exceeds the interrupting rating of the breakers in the panel, and that the panel is protected by properly selected circuit breakers back at the main equipment. Also see *Section 240-83(c).*

**Caution—Series Combination System
Rated _____ Amperes.
Identified Replacement Components Required.**

- CAUTION: The above field labeling is required for any series rated systems—fuse/breaker, or breaker/breaker. Closely check the manufacturer's catalog numbers to be sure that the circuit breaker combination you are intending to install "in series" is listed by U.L. for that purpose. This is extremely important because there are many circuit breakers that have the same physical size and dimensions, yet have different interrupting and voltage ratings, and have not been tested as "series-rated" devices. To install these as a series combination constitutes a violation of *Section 110-3(b)* and could be dangerous.

REVIEW

Note: Refer to the Code or the plans where necessary.

1. Where does an overhead service start and end? _____

    _____

    _____

    _____

2. What are service-drop conductors? _____

    _____

    _____

    _____

3. Who is responsible for determining the service location? _____

4. a. The service head must be located (above) (below) the point where the service-drop conductors are spliced to the service-entrance conductors. Circle one.

    b. What Code section provides the answer to part (a)? _____

5. What is a mast-type service entrance? _____

    _____

    _____

    _____

    _____

    _____

    _____

6. a. What size and type of conductors are installed for this service? _____

    _____

    b. What size conduit is installed? _____

    _____

    c. What size grounding electrode conductor is installed? (not neutral) _____

    d. Is the grounding electrode conductor insulated, armored, or bare? _____

7. How and where is the grounding electrode conductor attached to the water pipe?

    _____

    _____

8. When a conduit is extended through a roof, must it be guyed? _____

    _____

    _____

    _____

9. What are the minimum distances or clearances for the following?

   a. Service-drop clearance over private driveway _____

   b. Service-drop clearance over private sidewalks _____

   c. Service-drop clearance over alleys _____

   d. Service-drop clearance over a roof having a roof pitch of not less than 4/12. (Voltage between conductors does not exceed 300 volts.) _____

   _____

   e. Service-drop horizontal clearance from a porch _____

   f. Service-drop clearance from a fence that can be climbed _____

10. What are the minimum size ungrounded conductors using Type THW copper for the following residential electrical services? The panels are marked 75°C. For the 400-ampere service, install two conductors in parallel per phase. Use two raceways. Install three conductors in each raceway.

AMPERE RATING OF SERVICE	SERVICE-ENTRANCE CONDUCTORS SIZED PER *TABLE 310-16.*	SERVICE-ENTRANCE CONDUCTORS SIZED PER *TABLE 310-15(b)(6).*
100		
200		
400		

Table 310-15(b)(6) is only for 120/240-volt, single-phase residential services and feeders. This table does not apply to services and feeders other than residential.

11. What is the minimum size copper grounding electrode conductor for each of the following residential electrical services? Refer to *Table 250-66.* The ungrounded conductors are Type THW copper.

AMPERE RATING OF SERVICE	GROUNDING ELECTRODE CONDUCTOR WHEN THE SERVICE-ENTRANCE CONDUCTORS ARE SIZED PER *TABLE 310-16.*	GROUNDING ELECTRODE CONDUCTOR WHEN THE SERVICE-ENTRANCE CONDUCTORS ARE SIZED PER *TABLE 310-15(b)(6).*
100		
200		
400		

Table 310-15(b)(6) is only for 120/240-volt, single-phase residential services and feeders. This table does not apply to services and feeders other than residential.

12. What is the recommended height of a meter socket from the ground? _____

   _____

   _____

13. a. May the bare grounded neutral conductor of a service be buried directly in the ground? _____

    b. What section of the Code covers this? _____

14. What exceptions are made regarding the use of bare neutral conductors installed underground? _____

_____

_____

_____

_____

15. How far must mechanical protection be provided when underground service conductors are carried up a pole? _____

16. a. A method of service disconnect may consist of how many switches or circuit breakers? _____

    b. Must these devices be in one enclosure? _____

    c. What type of main disconnect is provided in this residence? _____

_____

17. Complete the following table by filling in the columns with the appropriate information.

	CIRCUIT NUMBER	AMPERE RATING	POLES	VOLTS	WIRE SIZE
A. Living room receptacle outlets					
B. Workbench receptacle outlets					
C. Water pump					
D. Attic exhaust fan					
E. Kitchen lighting					
F. Hydromassage tub					
G. Attic lighting					
H. Counter-mounted cooking unit					
I. Electric furnace					

18. a. What size conductors supply Panel B? _____

    b. What size conduit? _____

    c. Is this conduit run in the form of electrical metallic tubing or rigid conduit?

_____

    d. What size overcurrent device protects the feeders to Panel B? _____

19. How many electric meters are provided for this residence? _____

20. a. According to the Code, is it permissible to ground rural service-entrance systems and equipment to driven ground rods only when a metallic water system is available? _____

  b. What section of the Code applies? _____

21. What table of the Code lists the sizes of grounding electrode conductors to be used for service entrances of various sizes similar to the type found in this residence?

  _____

22. *Section 250-50* requires that a supplemental ground be provided if the available grounding electrode is: (Circle correct answer.)

  a. metal underground water pipe

  b. building steel

  c. concrete encases ground

23. Do the following conductors require mechanical protection?

  a. No. 8 grounding electrode conductor _____

  b. No. 6 grounding electrode conductor _____

  c. No. 4 grounding electrode conductor _____

24. Why is bonding of service-entrance equipment necessary? _____

  _____

  _____

  _____

  _____

  _____

  _____

  _____

25. What special types of bushings are required on service entrances? _____

  _____

26. When No. 4 AWG conductors or larger are installed in conduit, what additional provision is required on the conduit ends? _____

  _____

  _____

  _____

27. What minimum size copper bonding jumpers must be installed to bond properly the electrical service for the residence discussed in this text? _____

  _____

  _____

  _____

  _____

28. a. What is a Type S fuse? _____

_____

   b. Where must Type S fuses be installed? _____

_____

29. a. What is the maximum voltage permitted between conductors when using plug fuses?

_____

   b. May plug fuses (Type S) be installed in a switch that disconnects a 120/240-volt clothes dryer? _____

   c. Give a reason for the answer to (b). _____

_____

_____

30. Will a 20-ampere, Type S fuse fit properly into a 15-ampere adapter? _____

31. What part of a circuit breaker causes the breaker to trip

   a. on an overload? _____   b. on a short circuit? _____

32. What is meant by an ambient-compensated circuit breaker? _____

_____

_____

_____

_____

_____

_____

_____

_____

_____

_____

_____

33. List the standard sizes of fuses and circuit breakers up to and including 100 amperes.

_____

_____

34. Using the method shown in this Unit, what is the approximate short-circuit current available at the terminals of a 50-kVA single-phase transformer rated 120/240 volts? The transformer impedance is 1%.

   a. line-to-line?

   b. line-to-neutral?

35. Where is the main panel for this residence located? _____

36. a. On what type of wall is Panel A fastened? _____

   b. On what type of wall is Panel B fastened? _____

37. State four possible combinations of service equipment which meet the requirements of *Sections 110-9* and *110-10* of the Code.

   a. _____

   _____

   _____

   b. _____

   _____

   _____

   c. _____

   _____

   _____

   d. _____

   _____

   _____

38. When conduits pass through the wall from outside to inside, the conduit must be _____ to prevent air circulation through the conduit.

39. Briefly explain why electrical systems and equipment are grounded. _____

   _____

   _____

   _____

   _____

   _____

40. What Code section states that all overcurrent devices must have adequate interrupting ratings for the available fault current to be interrupted? _____

   _____

41. All electrical components have some sort of "withstand rating." The *NEC*® and U.L. standards refer to this as the components short-circuit current rating. This rating indicates the ability of the component to withstand a specified amount of fault current for the time it takes the overcurrent device upstream of the component to open the circuit. *Section* _____ of the *NEC*® refers to short-circuit current rating.

42. Arcing fault damage is closely related to the value of _____

43. In general, systems are grounded so that the maximum voltage to ground does not exceed (Circle one)
   a. 120 volts.   b. 150 volts.   c. 300 volts.

44. To ensure a complete grounding electrode system (Circle one)
   a. everything must be bonded together.
   b. all metal pipes and conduits must be isolated from one another.
   c. the service neutral is grounded to the water pipe only.

45. An electric clothes dryer is rated at 5,700 watts. The electric rate is 10.091 cents per kilowatt-hour. The dryer is used continuously for 3 hours. Find the cost of operation, assuming the heating element is on continuously.

46. A heating cable rated at 750 watts is used continuously for 72 hours to prevent snow from freezing in the gutters of the house. The electric rate is 8.907 cents per kilowatt-hour. Find the cost of operation.

47. When used as service equipment, a load center must be _____ that it is being used as a service disconnect, *Section 230-70*. The panel must also be marked to identify it as being _____ for use as service equipment, *Section 230-66*.

48. The utility company has provided a letter to the contractor stating that the available fault current at the line side of the main service-entrance equipment in a residence is 17,000 amperes RMS symmetrical, line-to-line. In the space following each statement, write in "Meets Code" or "Violation" of *Sections 110-9* and *110-10* of the Code.

    a. Main breaker 10,000-ampere interrupting rating; branch breakers 10,000-ampere interrupting rating. _____

    b. Main current-limiting fuse having a 200,000-ampere interrupting rating; branch breakers having 10,000-ampere interrupting rating. The panel is "Series-Rated."

    _____

    c. Main breaker has 22,000-ampere interrupting rating; branch breakers have 10,000-ampere interrupting rating. The panel is marked "Series-Connected."

    _____

49. While working on the main panel with the panel energized, the electrician inadvertently causes a direct short circuit (line-to-ground) on one of the branch-circuit breakers. The available fault current at the main service equipment is rather high. The panel is labeled "Series-Connected." The main breaker is rated 100 amperes. What will happen?

	True	False
a. Only the branch breaker will trip off.	_____	_____
b. The branch breaker will not trip off.	_____	_____
c. The main breaker and the branch breaker will probably both trip off, resulting in a total power outage.	_____	_____
d. All of the 120-volt connected loads in the panel will lose power.	_____	_____
e. All of the 240-volt connected loads and one-half of the 120-volt connected loads will lose power. There will not be a total power outage.	_____	_____
f. One fuse will open; therefore, all 240-volt connected loads and one-half of the 120-volt connected loads will lose power. There will not be a total power outage.	_____	_____

50. Repeat question 49 on a main service panel that consists of 100-ampere main current-limiting fuses and breakers for the branch-circuits. (Mark True or False.)

    a. _____

    b. _____

    c. _____

    d. _____

    e. _____

    f. _____

51. Here are five commonly used terms in the electrical industry. Enter the letter of the term before its definition.

    a. Grounding electrode conductor

    b. Main bonding jumper

    c. Grounded circuit conductor

    d. Equipment grounding conductor

    e. Lateral service conductors

    _____ This is the neutral conductor.

    _____ This is the term used to define underground service-entrance conductors that run between the meter and the utilities connection.

    _____ This is the conductor (sometimes a large threaded screw) that connects the neutral bar in the service equipment to the service-entrance enclosure.

    _____ This is the conductor that runs between the neutral bar in the main service equipment to the grounding electrode (water pipe, ground rod, etc.).

    _____ This is the bare copper conductor found in nonmetallic-sheathed cable.

52. The electrician mounted the disconnect switch for a central air-conditioning unit 8 feet (2.4 m) above the ground. This was easy to do because all he had to do was run the conduit across the ceiling joists in the basement, through the outside wall, and directly into the back of the disconnect switch. The electrical inspector turned the job down, citing *Section* _____ of the *NEC*® that requires that the disconnecting operating handle must not be higher than _____ feet above the ground.

53. *Section* _____ of the *NEC*® prohibits the connecting of fuses and/or circuit breakers into the grounded circuit conductor.

54. In general, overcurrent protective devices are inserted where the conductor receives its

    _____ .

55. According to the *NEC*® definition of a wet location, a basement cement wall that is in direct contact with the earth is considered to be a wet location. A panelboard mounted on this wall must have at least (¼ inch) (½ inch) (1 inch) space between the wall and the panel. Underline the correct answer.

56. When using re-bars as the concrete encased electrode, the re-bars must be (a) insulated with plastic material so they will not rust, or (b) electrically conductive so as to provide excellent conductivity. Underline the correct answer.

57. The circuit directory in a panelboard (may) (shall) be filled out according to *Section* _____ of the *NEC.* Circle the correct answer, and enter the correct *Section* number.

58. A main service panel is located in a dark corner of a basement, far from the basement light. In your opinion, does this installation meet the requirements of *Section 110-26(d)*?

_____ .

59. a. Does the *NEC*® permit panelboards to be installed in clothes closets? _____

b. Does the *NEC*® permit panelboards to be installed in bathrooms? _____

60. Does the electric utility in your area allow the location of the meter to be on the back side of a residence? _____

61. The term used when the utility charges different rates during different periods during the day is _____

62. A ground rod is driven below the meter outside of the house. An armored grounding electrode conductor connects between the meter base and the ground rod. The ground clamp is buried under the surface of the soil. This is permitted if the ground clamp is _____ for direct burial according to *Section* _____ .

63. Copper ground rods (grounding electrode) are the most commonly used. These rods must:

a. be at least _____ inch in diameter.

b. be driven to a depth of at least _____ feet unless solid rock is encountered, in which case the rod may be driven at a _____ degree angle, or it may be laid in a trench that is at least _____ feet deep.

c. be separated by at least _____ feet when more than one rod is driven.

d. have a ground resistance of not over _____ ohms for one rod.

64. What section of the Code prohibits the use of sheet metal screws as a means of attaching grounding conductors to enclosures? *Section* _____

65. What section of the *NEC*® prohibits using the space below and in front of an electrical panel to be used as storage space? *Section* _____ .

Service-Entrance Calculations

OBJECTIVES

After studying this Unit, the student will be able to

- determine the total calculated load of the residence.
- calculate the size of the service entrance, including the size of the neutral conductors.
- understand the Code requirements for services, *Article 230*.
- understand how to read a watt-hour meter.
- fully understand special Code rules for single-family dwelling service-entrance conductor sizing.
- derate service-entrance conductor if installation is located in extremely hot climate.
- do an optional calculation for computing the required size of service-entrance conductors for residence.

The various load values determined in earlier Units of this text are now used to illustrate the proper method of determining the size of the service-entrance conductors for the residence. The calculations are based on *NEC®* requirements. The student must check local and state electrical codes for any variations in requirements that may take precedence over the *NEC*.

SIZE OF SERVICE-ENTRANCE CONDUCTORS AND SERVICE DISCONNECTING MEANS

Section 230-42 tells us that service-entrance conductors must be sufficient to carry the load as computed according to *Article 220*, and shall have an ampacity not less than the rating of the service disconnect.

▶ For a one-family residence, the minimum size main service disconnecting means is 100 amperes, three-wire, *Section 230-79(c)*. ◀

The receptacle for the refrigerator is permitted to be connected to one of the required 20-ampere small-appliance branch-circuits, or it can be connected to a separate 15- or 20-ampere branch-circuit, *Section 210-52(b)(1), Exception No. 2.*

The choice is yours as to which branch-circuit you want to feed the receptacle for the refrigerator.

▶ If you choose to run a separate 15- or 20-ampere branch-circuit for the refrigerator receptacle, you do not have to add another 1,500 volt-amperes for the service-entrance calculations, *Section 220-16(a), Exception.* ◀ You really have not added any load to the service. You have merely transferred the refrigerator load from one of the required 20-ampere small-appliance branch-circuits to a branch-circuit of its own.

Two methods for calculating services on homes are permitted by the *NEC®*. Method 1 is outlined in *Article 220, Parts A* and *B*. Method 2 is given in *Article 220, Part C*.

Method 1—*Article 220, Parts A and B*

All of the volt-ampere values in the following calculations are taken from previous Units of this text.

Single-Family Dwelling Service-Entrance Calculations

1. General Lighting Load, *Section 220-3(a)*

    3,232 sq. ft. @ 3 VA per sq. ft. = 9,696 VA

2. Minimum Number of Lighting Branch-Circuits

    9,696 VA ÷ 120 volts = 80.8 amperes

    then, $\dfrac{\text{amperes}}{15} = \dfrac{80.8}{15}$ = 5.387 (round up to 6) 15-ampere branch-circuits

3. Small Appliance Load, *Sections 210-11(c)(1)* and *220-16(a)*
    (Minimum of two 20-ampere branch-circuits)

Kitchen	3
Clothes Washer	1*
Workshop	2**

    6   branch-circuits @ 1,500 VA each = 9,000 VA

4. Laundry Branch-Circuit, *Sections 210-11(c)(2)* and *220-16(b)*
    (Minimum of one 20-ampere branch-circuit)

    1   branch-circuit(s) @ 1,500 VA = 1,500 VA

5. Total General Lighting, Small Appliance, and Laundry Load

    Lines 1 + 3 + 4 = 20,196 VA

6. Net computed General Lighting, Small Appliance, and Laundry Loads (less ranges, ovens, and "fastened-in-place" appliances). Apply Demand Factors from *Table 220-11*.

    a. First 3,000 VA @ 100%   = 3,000 VA

    b. Line 5 20,196 − 3,000 = 17,196 @ 35% = 6,019 VA

    Total a + b = 9,019 VA

7. Electric Range, Wall-Mounted Ovens, Counter-Mounted Cooking Units *(Table 220-19)*

Wall-mounted oven	7,450 VA
Counter-mounted cooking unit	6,600 VA
Total	14,050 VA (14 kW)

    14 kW exceeds 12 kW by 2 kW

    2 kW × 5% = 10% increase, therefore: 8 kW + 0.8 kW = 8.8 kW = 8,800 VA

---

* This is in addition to the required minimum of one laundry branch-circuit. Thus, the laundry is served by a 20-ampere branch-circuit B18 for the automatic washer and another 20-ampere branch-circuit B20 supplying two receptacles for plugging in an electric iron, steamer, or clothes press.

** Although not true small appliance circuits as defined by the *NEC*,® it is better to add these circuits in as 1,500 VA small appliance circuits rather than including them in the general lighting load calculations. This results in a more adequate load calculation.

8. Electric Dryer, *Table 220-18* = 5,700 VA

9. Electric Furnace, *Section 220-15*, Air Conditioner, Heat Pump, *Article 440*

> Air conditioner: $26.4 \times 240 = 6,336$ VA
> Electric Furnace: 13,000 VA
> (Enter largest value, *Section 220-21*) = 13,000 VA

10. Net Computed General Lighting, Small Appliance, Laundry, Ranges, Ovens, Cooktop Units, HVAC.

> Lines $6 + 7 + 8 + 9 = 36,519$ VA

11. List "Fastened-in-Place" Appliances *in addition* to Electric Ranges, Air Conditioners, Clothes Dryers, Space Heaters.

Appliance	VA Load
Water Heater	3,000 VA
Dishwasher*	1,000 VA

> * Motor: $7.2 \times 120 = 864$ VA
>   Heater: 1,000 VA (watts)
>   Enter largest value, *Section 220-21*.

Garage Door Opener $5.8 \times 120 =$		696 VA
Food Waste Disposer $7.2 \times 120 =$		864 VA
Water Pump 8×240	$=$	1,920 VA
Hydromassage Tub 10×120	$=$	1,200 VA
Attic exhaust fan 5.8×120	$=$	696 VA
Heat/Vent/Lights $1,500 \times 2$	$=$	3,000 VA
Freezer 5.8×120	$=$	696 VA
	Total	13,072 VA

12. Apply 75% Demand Factor, *Section 220-17,* if Four or More "Fastened-in-Place" Appliances. If Less than Four, Figure at 100%. Do not include electric ranges, clothes dryers, space heating, or air conditioning equipment.

> Line 11 $13,072 \times 0.75 = 9,804$ VA

13. Total Computed Load (Lighting, Small Appliance, Ranges, Dryers, HVAC, "Fastened-in-Place" Appliances)

> Line 10 $36,519$ + Line 12 $9,804 = 46,323$ VA

14. Add 25% of Largest Motor *(Section 220-14)*

> This is the water pump motor:
> $8 \times 240 \times 0.25 = 480$ VA

15. Grand Total Line 13 + Line 14 = 46,803 VA

16. Minimum Ampacity for Ungrounded Service-Entrance Conductors

$$\text{Amperes} = \frac{\text{Line 15}}{240} = \frac{46,803}{240} = 195 \text{ amperes}$$

17. Minimum Ampacity for Neutral Service-Entrance Conductor, *Sections 220-22* and *310-15(b)(6)*. Do Not Include Straight 240-Volt Loads.

    a. Line 6   9,019 VA

    b. Line 7 $8,800 \times 0.70$ =   6,160 VA

    c. Line 8 $5,700 \times 0.70$ =   3,990 VA

    d. Line 11 (include only 120-volt loads)

Freezer	696 VA
Food Waste Disposer	864 VA
Garage Door Opener	696 VA
Heat/Vent/Light	3,000 VA
Attic Fan	696 VA
Hydromassage tub	1,200 VA
Dishwasher heater	1,000 VA
Total	8,152 VA

    e. Line d total @ 75% Demand Factor: $8,152 \times 0.75 = 6,114$ VA

    f. Add 25% of largest 120-volt motor
        This is food waste disposer:
        $7.2 \times 120 \times 0.25$ =   216 VA

    g. Total a + b + c + e + f   = 25,499 VA

$$\text{Amperes} = \frac{\text{volt} - \text{amperes}}{\text{volts}} = \frac{25,499}{240} = 106.245 \text{ amperes}$$

18. Ungrounded Conductor Size (copper)   2/0THWN

    Note: This could be a No. 3/0 THW, THHW, THWN, XHHW, or THHN per *Table 310-16*, or a No. 2/0 (same types) using *Table 310-15(b)(6)*. *Table 310-15(b)(6)* may only be used for 120/240-volt, 3-wire, residential single-phase service-entrance conductors, service lateral conductors, and feeder conductors that serve as the main power feeder to a dwelling unit. This table *cannot* be applied to the feeder to Panel B because that feeder carries only part of the load in the residence.

19. Neutral Conductor Size (copper), *Section 220-22*   No. 1 AWG

    Note: The calculations indicate that a No. 2 AWG copper neutral conductor would be adequate based upon the unbalanced load calculations. However, *Table 310-15(b)(6)* states that the neutral conductor is permitted to be smaller than the ungrounded "hot" conductors if the requirements of *Sections 215-2, 220-22,* and *230-42* are met. The Specifications for this residence specify that a No. 1 neutral conductor be installed. The neutral conductor must not be smaller than the grounding electrode conductor, *Section 250-24(b)(1)*.

20. Grounding Electrode Conductor Size (copper), *Table 250-66*   No. 4 AWG

21. Conduit Size (See Figure 28-15 for calculations)   1¼ inch

Table 29-1 replicates *Table 310-15(b)(6)* in the *NEC.*® The current ratings for a given size wire are higher than the allowable ampacities listed in *Table 310-16* because of the tremendous diversity of electricity use in homes.

Be careful. Locations exposed to the weather are considered to be wet locations, per the definition in the *NEC.*® Therefore, most electrical inspectors will require that service-entrance conductors, from the line side terminals of the service disconnect to the service head, be suitable for wet locations. This would mean that the conductors must have a "W" designation in their type lettering.

High Temperatures

In certain parts of the country, such as the southwestern desert climates where extremely hot temperatures are common, the authority having jurisdiction may require that service-entrance conductors and any other conductors exposed to direct sunlight be corrected (derated) according to the "Correction Factors" found below *Table 310-16*, See Figure 29-1.

EXAMPLE: If the U.S. weather bureau lists the average summer temperature as 113°F (45°C), then a correction factor of 0.82 must be applied. For instance, a 3/0 XHHW copper conductor per *Table 310-16* is 200 amperes at 86°F (30°C). At 113°F (45°C) the conductor's ampacity is

$$200 \times 0.82 = 164 \text{ amperes}$$

A properly sized conductor capable of carrying 200 amperes safely requires

$$\frac{200}{0.82} = 243.9 \text{ amperes}$$

Therefore, according to *Table 310-16* and the applied correction factor, a 250 kcmil Type XHHW copper conductor is required.

Check this with the electrical inspection department where this situation might be encountered.

Neutral Conductor

Why is the neutral conductor permitted to be smaller than the "hot" ungrounded conductors?

The neutral conductor must be able to carry the computed maximum neutral current, *Section 220-22*. Straight 240-volt loads are not included in the calculation of neutral conductors. The calcula-

▶ **SPECIAL AMPACITY RATINGS FOR RESIDENTIAL 120/240-VOLT, 3-WIRE, SINGLE-PHASE SERVICE-ENTRANCE CONDUCTORS, SERVICE LATERAL CONDUCTORS, FEEDER CONDUCTORS, AND CABLES THAT SERVE AS THE MAIN POWER FEEDER TO A DWELLING UNIT.** ◀

Copper Conductor Size (AWG) for RH, RHH, RHW-2, THHN, THHW, THW, THW-2, THWN, THWN-2, XHHW, XHHW-2, SE, USE, USE-2	Aluminum or Copper-Clad Aluminum Conductors (AWG)	Allowable Special Ampacity
4	2	100
3	1	110
2	1/0	125
1	2/0	150
1/0	3/0	175
2/0	4/0	200
3/0	250 kcmil	225
4/0	300 kcmil	250
250 kcmil	350 kcmil	300
350 kcmil	500 kcmil	350
400 kcmil	600 kcmil	400

Table 29-1 This table is similar to the table in *Table 310-15(b)(6)*, and is *only* for residential 120/240-volt, 3-wire, single-phase service-entrance conductors, service lateral conductors, and residential feeder conductors that carry the main power to the dwelling. This table is based on the great diversity of electrical use in homes. In the residence discussed in this text, this table cannot be used to determine the feeder size to Panel B because the feeder to Panel B does *not* carry the main power to the home. The neutral conductor may be smaller than the phase conductors when maximum unbalanced load calculations prove that the requirements of *Sections 215-2, 220-22, and 230-42* have been met. Reduced neutrals require 240-volt loads, such as electric ranges, dryers, water heaters, central air-conditioners, or heat pumps, where none of or only some of the load current passes through the neutral conductor.

tions resulted in a computed load of 195 amperes for the ungrounded conductors, and 106 amperes for the neutral conductor. Referring to *Table 310-16*, this could be two No. 3/0 THW, THHW, THWN, XHHW, or THHN copper conductors and one No. 2 neutral conductor. The neutral conductor may be insulated or bare per *Sections 230-30 and 230-41*.

Table 310-15(b)(6) provides special consideration for 120/240-volt, 3-wire, single-phase service-entrance conductors, service lateral conductors, and

TEMPERATURES ON OUTSIDE AND INSIDE OF CONDUIT CAN EASILY REACH OR EXCEED 200°F.

Figure 29-1   Example of high-temperature location.

feeder conductors that serve as the main power feeder to a dwelling unit. For a 200-ampere residential service, No. 2/0 THW, THHW, THWN, XHHW, or THHN copper conductors are adequate. See Table 29-2.

Section 310-15(b)(6) also permits the neutral conductor to be smaller than the "hot" ungrounded conductors, but only if it can be verified by calculations that the requirements of *Sections 215-2, 220-22,* and *230-42* are met. We verified this in our calculations for the residence discussed in this text.

The Specifications for this residence call for two

No. 2/0 ungrounded "hot" conductors and one No. 1 neutral conductor.

The service neutral conductor serves another purpose. In addition to carrying the unbalanced current, the neutral conductor must also be capable of safely carrying any fault current that it might be called upon to carry, such as a line-to-ground fault in the service equipment. Sizing the neutral conductor only for the normal neutral unbalance current could result in a neutral conductor too small to safely carry fault currents. The neutral conductor might burn off, causing serious voltage problems,

	SERVICE-ENTRANCE CONDUCTOR SIZE					
	COPPER			ALUMINUM		
Size (amperes)	Phase (Hot) Conductors	Neutral Conductor	Conduit Size (inches)	Phase (Hot) Conductors	Neutral Conductor	Conduit Size (inches)
100	#2 AWG	#4 AWG	1½	#1 AWG	#2 AWG	1½
125	#1 AWG	#4 AWG	1½	00 AWG	#1 AWG	1½
150	0 AWG	#4 AWG	1½	000 AWG	0 AWG	2
175	00 AWG	#2 AWG	2	0000 AWG	00 AWG	2
200	000 AWG	#2 AWG	2	250 kcmil	000 AWG	2

Table 29-2   Service-entrance conductor sizing used in some cities so that service-entrance calculations do not have to be made for each service-entrance installation.

damage to equipment, and the creation of fire and shock hazards.

In sizing neutral conductors for dwellings, this generally is not a problem. But installing reduced size neutrals on commercial and industrial services can present a major problem. The grounded conductor of *any* service must never be smaller than the required grounding electrode conductor for that particular service. Refer to *Sections 230-23, 230-31, 230-42(c), 250-2, 250-24(b), 250-90,* and *250-96.* This topic is covered in *Electrical Wiring— Commercial* (Delmar Publishers).

Subpanel B Calculations

General lighting load
(Sections 220-3(a),
220-11)
1,420 sq. ft. @ 3 volt-
   amperes per sq. ft. =   4,260 volt-amperes
   (kitchen, living room,
   laundry, rear entry
   hall, powder room,
   recreation room)

Small appliance load
(Section 220-16)
   Kitchen   3
   Laundry   1
   Automatic
      washer   <u>1</u>
     Total   5 @ 1,500 VA
              per
              circuit =   <u>7,500 volt-amperes</u>

Total general lighting and
   small appliance load = 11,760 volt-amperes

Application of demand
   factors *(Table 220-11)*
   3,000 volt-amperes
      @100% =   3,000 volt-amperes
   11,760 − 3,000 = 8,760
      @ 35% =   <u>3,066 volt-amperes</u>

A Net computed load (less
     range and "fastened-in-
     place" appliances)   6,066 volt-amperes

Wall-mounted oven and
   counter-mounted range:

Table 220-19, Note 4:
   Wall-mounted oven   7,450 volt-amperes
   Counter-mounted range   <u>6,600 volt-amperes</u>
     Total   14,050 volt-amperes

14 kW exceeds 12 kW by 2 kW:
   2 kW × 5% = 10%
     increase
     8 kW × 0.10 = 0.8 kW
     (Column A, Table 220-19)

B 8 + 0.8 = 8.8 kW =   8,800 volt-amperes

C Net computed load (with
     range) = A + B = C   14,866 volt-amperes

Dryer *(Section 220-18)*   5,700 volt-amperes
Dishwasher
   Motor: 7.2 × 120 =
     864 volt-amperes
   Heater: 1,000 watts
     (volt-amperes)
     (maximum demand
     is 1,000 VA because
     motor and heater do
     not operate at the
     same time,
     Section 220-21)   1,000 volt-amperes
Food waste disposer
   7.2 × 120 =   864 volt-amperes
Garage door opener
   5.8 × 120 =   <u>696 volt-amperes</u>
Total "fastened-in-place"
   appliance load   8,260 volt-amperes
Lighting, small appliance
   circuits, ranges, dryer = <u>14,866 volt-amperes</u>

D Add 25% of largest motor.
     7.2 × 120 × 0.25 =   216 volt-amperes

E Net computed load
     (lighting, small appli-
     ance circuits, ranges,
     dryer, dishwasher, food
     waste disposer, garage
     door opener)   23,342 volt-amperes

$$\text{Amperes} = \frac{\text{volt-amperes}}{\text{volts}} = \frac{23,342}{240} = 97.3 \text{ amperes}$$

The feeder conductors supplying Panel B could be No. 3 THHW, THW, THHN, or XHHW per *Table 310-16*. We have not shown the calculations for the neutral to Panel B. The specifications and also Figure 28-2 require that three No. 3 THHN feeder conductors supply Panel B.

Method 2 (Optional Calculations — Dwellings)—*Article 220, Part C*

A second method for determining the load for a one-family dwelling is given in *Section 220-30*. This method simplifies the calculations. But remember, the minimum size service for a one-family dwelling is 100 amperes, *Sections 230-42(b)* and *230-79(c)*.

Let's take a look at *Part C of Article 220*. This is an alternative method of computing service loads and feeder loads. It is referred to as the *optional method*.

Section 220-30(a) tells us that we are permitted to calculate service-entrance conductors and feeder conductors (both phase conductors and the neutral conductor) using an optional method. *Section 220-30* addresses single-family dwellings. *Section 220-32* addresses multifamily dwellings.

Section 220-30(b) tells us to

1. include 1500 volt-amperes for each 20-ampere small appliance circuit.

2. include 3 volt-amperes per square foot for lighting and general-use receptacles.

3. include the nameplate rating of appliances:
   - that are fastened in place, or
   - that are permanently connected, or
   - that are located to be connected to a specific circuit

   Appliances listed in *Section 220-30(b)(3)* are ranges, wall-mounted ovens, counter-mounted cooking units, clothes dryers, and water heaters.

4. include nameplate ampere rating or kVA for motors and all low-power-factor loads. The intent of the reference to low power factor is to address such loads as low-cost, low-power-factor fluorescent ballasts of the type that might be used in the recessed lay-in fluorescent fixtures in the recreation room. It is

always recommended that high-power-factor ballasts be installed.

Section 220-30(c) tells us that for heating and air-conditioning loads, select the largest load from a list of six types of loads:

1. 100% of air conditioning and cooling.

2. 100% of heat pump compressors and supplemental heating unless both cannot operate at the same time.

3. 100% electric thermal storage heating if expected to be continuous at full nameplate value.

4. 65% of central electric space heating that also has supplemental heating in heat pumps where the compressor and supplemental heating will not operate at the same time.

5. 65% of electric space heating of less than four separately controlled units.

6. 40% of electric space heating if four or more separately controlled units.

So let's begin our optional calculation for the residence discussed in this text. The residence has an air conditioner and an electric furnace.

Air conditioner

$$26.4 \times 240 = 6{,}336 \text{ volt-amperes}$$

Electric furnace

$$13{,}000 \times 0.65 = 8{,}450 \text{ volt-amperes}$$

Therefore we will select the electric furnace load for our calculations because it is the largest load. It is also a noncoincidental load, as defined in *Section 220-21*. We can omit the air-conditioner load from our calculations from here on.

We now add up all of the other loads:

General lighting load
3,232 sq. ft. @ 3 volt-
amperes per sq. ft. =   9,696 volt-amperes
Small appliance circuits (7)
@ 1,500 volt-amperes
each   10,500 volt-amperes
Wall-mounted oven
(nameplate rating)   6,600 volt-amperes
Counter-mounted cooking
unit (nameplate rating)   7,450 volt-amperes
Water heater
(nameplate rating)   3,000 volt-amperes

Clothes dryer
(nameplate rating)   5,700 volt-amperes
Dishwasher (maximum
demand, heater only)   1,000 volt-amperes
Food waste disposer
7.2 × 120   864 volt-amperes
Water pump 8 × 240   1,920 volt-amperes
Garage door opener
5.8 × 120   696 volt-amperes
Heat-vent-lights (2)
1,500 × 2   3,000 volt-amperes
Attic exhaust fan
5.8 × 120   696 volt-amperes
Hydromassage tub
10 × 120   1,200 volt-amperes

Total other loads   52,322 volt-amperes

We can now complete our optional calculation:

Enter electric furnace load
$13,000 \times 0.65 =$ 8,450 volt-amperes
Plus first 10 kVA of all
other loads at 100%   10,000 volt-amperes
Plus remainder all other
loads at 40%:
$52,322 - 10,000 = 42,322$
$42,322 \times 0.40$ 16,929 volt-amperes
Total   35,379 volt-amperes

$$\text{Amperes} = \frac{\text{volt-amperes}}{\text{volts}} = \frac{35,379}{240} = 147.4 \text{ amperes}$$

Using the optional calculation method, and referring to *Table 310-15(b)(6)*, we find that for the 147+ amperes computed load, a No. 1 copper conductor with any of the insulation types listed has a special ampacity of 150 amperes and could be installed for this service-entrance. This seems a bit small, and that is why some local electrical codes do not permit the use of this special ampacity table.

The specifications for the residence discussed in this text call for a full 200-ampere service consisting of two No. 2/0 THW, THHN, THWN, or XHHW copper phase conductors and one No. 1 bare copper neutral conductor.

Service-Entrance Conductor Size Table

Table 29-2 has been taken from one major city that prefers to show minimum service-entrance con-

ductor size requirements rather than having electrical contractors make calculations each and every time they install a service.

Note that the conductor sizes in Table 29-2 are larger than conductor sizes permitted by *Table 310-15(b)(6)* of the *NEC.*® But because the Code is considered to be a minimum standard, it is within the realm of local authorities to publish code requirements specifically for their communities.

Main Disconnect

The main service disconnecting means in this residence is rated 200 amperes, discussed in Unit 28.

Grounding Electrode Conductor

Grounding electrode conductors are sized according to *Table 250-66* of the Code.

As an example, this residence is supplied by No. 2/0 AWG copper service-entrance conductors. Checking *Table 250-66*, we find that the minimum grounding electrode conductor must be a No. 4 AWG copper conductor.

The grounding electrode conductor for the service in this residence is a No. 4 AWG copper armored ground cable, discussed in Unit 28.

TYPES OF WATT-HOUR METERS

Figure 29-2 shows typical single-phase watt-hour meters. The first (A) is a conventional five-dial meter. The second (B) is a five-dial meter where one set of dials registers total kilowatt-hours and the second set of dials registers premium time kilowatt-hours. The third (C) is a programmable electronic digital display watt-hour meter that can register kilowatt-hour consumption for up to four different rate schedules. Utilities that offer special "time-of-use" rates install meters of the second and third types.

Wiring diagrams for the metering and control of electric water heaters, air conditioners, and heat pumps were discussed in Unit 19. Electric utilities across the country have different requirements for hooking up these loads. Always consult your local utility for their latest information regarding electric service.

Table 250-66. Grounding Electrode Conductor for Alternating-Current Systems

Size of Largest Service-Entrance Conductor or Equivalent Area for Parallel Conductors[1]		Size of Grounding Electrode Conductor	
Copper	Aluminum or Copper-Clad Aluminum	Copper	Aluminum or Copper-Clad Aluminum[2]
2 or smaller	½ or smaller	8	6
1 or 1/0	2/0 or 3/0	6	4
2/0 or 3/0	4/0 or 250 kcmil	4	2
Over 3/0 through 350 kcmil	Over 250 kcmil through 500 kcmil	2	1/0
Over 350 kcmil through 600 kcmil	Over 500 kcmil through 900 kcmil	1/0	3/0
Over 600 kcmil through 1100 kcmil	Over 900 kcmil through 1750 kcmil	2/0	4/0
Over 1100 kcmil	Over 1750 kcmil	3/0	250 kcmil

Notes:

1. Where multiple sets of service-entrance conductors are used as permitted in Section 230-40, Exception No. 2, the equivalent size of the largest service-entrance conductor shall be determined by the largest sum of the areas of the corresponding conductors of each set.

2. Where there are no service-entrance conductors, the grounding electrode conductor size shall be determined by the equivalent size of the largest service-entrance conductor required for the load to be served.

[1]This table also applies to the derived conductors of separately derived ac systems.

[2]See installation restrictions in Section 250-64(a).

FPN: See Section 250-24(b) for size of ac system conductor brought to service equipment.

(Reprinted with permission from NFPA 70-1999)

Figure 29-2   Photo "A" shows a typical single-phase watt-hour meter. Photo "B" shows a single-phase watt-hour meter that has two sets of dials. One registers the total kilowatt-hours. The second registers premium time kilowatt-hours. Photo "C" shows a programmable electronic digital display watt-hour meter that is capable of registering four different "Time-Of-Use" rates. *Courtesy* of Landis & Gyr Energy Management, Inc.

READING WATT-HOUR METERS

Reading a digital watt-hour meter is easy. Just read the numbers.

For dial-type meters, starting with the first dial, record the last number the pointer has just passed. Continue doing this with each dial until all dial readings have been recorded. The reading on the five-dial watthour meter in Figure 29-3 is 18,672 kilowatt-hours.

If the meter reads 18,975 one month later, as indicated in Figure 29-4, by subtracting the previous reading of 18,672, it is found that 303 kilowatt-hours of electricity were used during the month. The 303 kilowatt-hours of electricity used is multiplied by the rate per kilowatt-hour, and the power company bills the consumer for the energy used. The utility may also add a fuel adjustment charge or other charges.

Figure 29-3   The reading of this five-dial meter is 18,672 kilowatt-hours.

Figure 29-4   One month later, the meter reads 18,975 kilowatt-hours, indicating that 303 kilowatt-hours were used during the month.

REVIEW

Note: Refer to the Code or the plans where necessary.

1. According to *Section* _____ , the minimum size service required for a one-family dwelling is _____ amperes.

2. a. What is the unit load per square foot for the general lighting load of a residence?

   _____

   b. What are the demand factors for the general lighting load in dwellings? _____

   _____

   _____

   _____

3. a. What is the ampere rating of the circuits that are provided for the small appliance loads? _____

   b. What is the minimum number of small appliance circuits permitted by the Code?

   _____

   c. How many small appliance circuits are included in this residence? _____

4. Why is the air-conditioning load for this residence omitted in the service calculations?

   _____

   _____

   _____

   _____

   _____

   _____

5. What demand factor may be applied when four or more fixed appliances are connected to a service, in addition to an electric range, air conditioner, clothes dryer, or space heating equipment? *(Section 220-17)* _____

   _____

6. What load may be used for an electric range rated at not over 12 kW? *(Table 220-19)*

   _____

7. What is the computed load for an electric range rated at 16 kW? *(Table 220-19)* Show calculations.

8. What is the computed load when fixed electric heating is used in a residence? *(Section 220-15)* _____

9. On what basis is the neutral conductor of a service entrance determined? _____

   _____

   _____

   _____

   _____

10. Why is it permissible to omit an electric space heater, water heater, and certain other 240-volt equipment when calculating the neutral service-entrance conductor for a residence? _____

    _____

    _____

11. a. What section of the Code contains an optional method for determining residential service-entrance loads? _____

    b. Is this section applicable to a two-family residence? _____

12. Calculate the minimum size of copper service-entrance conductors and grounding electrode conductor required for a residence containing the following: floor area is 24' × 38' (7.3 m × 11.6 m). The dwelling will have:

    - a 12-kW, 120/240-volt electric range

    - a 5-kW, 120/240-volt electric clothes dryer consisting of a 4-kW, 240-volt heating element, and a 120-volt motor, and 120-volt lamp that have a combined load of 1 kW

    - a 2,200-watt, 120-volt sauna heater

    - six individually controlled 2-kW, 240-volt baseboard electric heaters

    - a 12-ampere, 240-volt air conditioner

    - a 3-kW, 240-volt electric water heater

    - a 1,000-watt, 120-volt dishwasher

    - a 7.2-ampere, 120-volt food waste disposer

    - a 5.8-ampere, 120-volt attic exhaust fan

    Determine the sizes of the ungrounded "hot" conductors and the neutral conductor. Use Type THWN. Use *Table 310-15(b)(6)*. Also determine the correct size grounding electrode conductor.

    Do not forget to include the required two 20-ampere small appliance circuits and the laundry circuit.

    The panelboard is marked 75°C.

    You may use the blank *Single-Family Dwelling Service-Entrance Calculation* form in the Appendix of this text, or use the form as a guide to make your calculations in the proper steps.

    Two No. _____ THWN ungrounded "hot" conductors

    One No. _____ THWN or bare neutral conductor

    One No. _____ grounding electrode conductor

STUDENT CALCULATIONS

13. Read the meter shown. Last month's reading was 22796. How many kilowatt-hours of electricity were used for the current month?

14. After studying this Unit, we realize that electric utilities across the country have many ways to meter residential customers. They may have lower rates during certain hours of the day. A common term used to describe these programs is _____

_____

Swimming Pools, Spas, Hot Tubs

OBJECTIVES

After studying this Unit, the student will be able to

- recognize the importance of proper swimming pool wiring with regard to human safety.

- discuss the hazards of electrical shock associated with faulty wiring in, on, or near pools.

- describe the differences between permanently installed pools and pools that are portable or storable.

- understand and apply the basic Code requirements for the wiring of swimming pools, spas, hot tubs, and hydromassage bathtubs.

POOL WIRING *(ARTICLE 680)*

Because of the complexity of swimming pool wiring, a detailed drawing of the *NEC®* requirements for swimming pools appears on Sheet 10 of 10 in the Plans found in the back of this text. You will want to refer to this drawing continually as you study this Unit.

To protect people against the hazards of electric shock associated with swimming pools (water and electricity are not a good combination), extreme care is required when wiring the equipment in and around pools. The basic code requirements found in *Chapters 1* through *4* of the *NEC®* are to be followed, but, in addition, *Article 680* amends the requirements of *Chapters 1* through *4* with many special rules. These rules are unique to swimming pools, wading pools, hydromassage bathtubs, therapeutic pools, decorative pools, hot tubs, and spas.

ELECTRICAL HAZARDS

Caution: A person can suffer an immobilizing or lethal shock in a residential-type pool in either of two ways.

1. **Direct contact:** An electrical shock can be deadly to someone in a pool who touches a "live" wire or touches an object such as a metal ladder, metal fence, metal parts of a lighting fixture, metal enclosures of electrical equipment, or appliances that have become "live" by an ungrounded conductor coming in contact with the metal enclosure or exposed metal parts of an appliance. Figure 30-1 shows a person in the water touching a faulty appliance.

2. **Indirect contact:** An electrical shock can be lethal to someone in a pool by merely being in the water when voltage gradients in the water are present. This is illustrated in Figure 30-2.

As shown in Figure 30-2, "rings" of voltage radiate outward from the radio to the pool walls and floor. These rings can be likened to the rings that form when a rock is thrown into the water. The voltage rings or gradients range from 120 volts at the radio to zero volts at the pool walls and floor. The pool walls and floor are assumed to be at ground or zero potential. The gradients, in varying degrees, are found in the entire body of water.

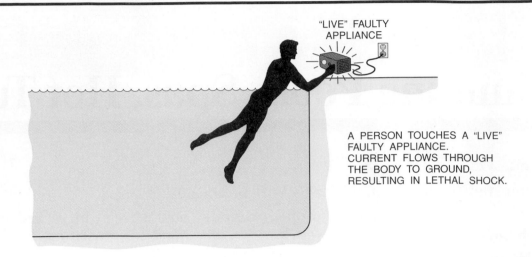

Figure 30-1   Touching a "live," faulty appliance can cause lethal shock.

Figure 30-2 shows voltage gradients in the pool of 90 volts and 60 volts. This figure is a simplification of the actual situation in which there are many voltage gradients. In this case, the voltage differential, 30 volts, is an extremely hazardous value. The person in the pool, who is surrounded by these voltage gradients, is subject to severe shock, immobilization (which can result in drowning), or actual electrocution. Tests conducted over the years have shown that a voltage gradient of 1½ volts per foot can cause paralysis.

Study the Code requirements for swimming pool wiring appearing in the plans. You will note

that, in general, underwater swimming-pool lighting fixtures must be positioned at least 18 inches (457 mm) below the normal water level. The reasoning behind this rule is that when a person is in the water close to an underwater lighting fixture, the relative position of the fixture is well below the person's heart. Fixtures that are permitted to be installed no less than 4 inches (102 mm) below the normal water level have undergone tougher, abnormal impact tests, similar to the abuse given by cleaning tools or other mechanical objects. The lenses on these fixtures can withstand these abnormal impact tests, whereas the standard underwater

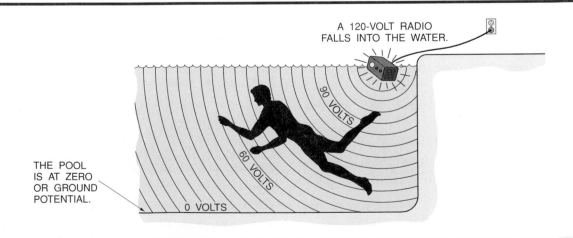

Figure 30-2 Voltage gradients surrounding a person in the pool can cause severe shock, drowning, or electrocution.

fixtures that are required to be 18 inches (457 mm) below the water level are subjected to normal impact tests that duplicate the impact of a person kicking the lens.

The shock hazard to a person in a pool is quite different from that of the normal "touch" shock hazard to a person not submersed in water. The water makes contact with the entire skin surface of the body rather than just at one "touch" point. Also, skin wounds, such as cuts and scratches, reduce the body's resistance to shock to a much lower value than that of the skin alone. Body openings such as ears, nose, and mouth further reduce the body resistance. As Ohm's law states, for a given voltage, the lower the resistance, the higher the current.

Another hazard associated with spas and hot tubs is prolonged immersion in hot water. If the water is too hot and/or the immersion too long, lethargy (drowsiness, hyperthermia) can set in. This can increase the risk of drowning. The Underwriters Laboratories (U.L.) Standard 1563 establishes maximum water temperature at 104°F. The suggested maximum time of immersion is generally 15 minutes.

Instructions furnished with spas and hot tubs specify the maximum temperature and time permitted for using the spa or hot tub.

Hot water, in combination with drugs and alcohol, presents a real hazard to life. Caution must be observed at all times.

Wiring Methods

Throughout *Article 680*, you will find that the wiring must be installed in rigid metal conduit, intermediate metal conduit, rigid nonmetallic conduit, Type MC cable, or, in some cases, electrical metallic tubing. For wet-niche fixtures, brass or other corrosion-resistant metal must be used.

Of the many Code rules covering pools, note in *Section 680-25(c)* that the circuit conductors and the equipment grounding conductor for pool-associated motors must be installed in rigid metal conduit, intermediate metal conduit, rigid nonmetallic conduit, or Type MC cable. Electrical metallic tubing (EMT) is permitted where installed on or within buildings.

Section 680-25(c), allows any type of wiring method recognized in *Chapter 3* of the Code for that

portion of the interior wiring of one-family dwelling installations that supplies pool-associated pump motors, provided the wiring method has an insulated or covered equipment grounding conductor not smaller than No. 12 AWG. The outer jacket of any wiring method recognized in *Chapter 3* of the *NEC*, such as nonmetallic-sheathed cable, is the covering referred to above.

Ground-Fault Circuit Interrupters

Many lives have been saved because circuits have been protected against ground faults with ground-fault circuit interrupters, commonly referred to as GFCIs. Throughout *Article 680* you will find the need for GFCIs, either the receptacle type or circuit breaker type. Generally, once you begin to run wires beyond (on the load side) a GFCI device, you are not permitted to have other conductors in the same raceway, box, or enclosure. The exception to this is when the GFCI is installed in a panelboard, when obviously there will be a "mix" of other circuit conductors that are not GFCI protected.

CODE-DEFINED POOLS

Section 680-4 describes a *permanently installed swimming, wading, or therapeutic pool* as one that is constructed in the ground or partially in the ground, other pools that have a water depth of greater than 42 inches (1.07 m), and all pools installed outside of a building regardless of water depth and whether it is served by electrical circuits of any nature.

Section 680-4 describes a *storable swimming pool or wading pool* as one that is constructed above the ground, is capable of holding a water depth of not over 42 inches (1.07 m), or a pool with nonmetallic, molded polymeric walls, or inflatable fabric walls regardless of dimension. The construction of this type of pool is such that it can be readily disassembled for storage, and reassembled to its original form.

Cord-connected filter pumps for storable pools are generally "double-insulated." The cord itself has an equipment grounding conductor in it. The attachment plug cap on the cord is the 3-prong type. All of the electrical equipment associated with a storable pool must be GFCI protected.

Storable pools are covered in *Article 680, Part C.*

GROUNDING AND BONDING OF SWIMMING POOLS

Grounding *(Section 680-24)*

Section 680-24 requires that all of the following items *must* be grounded, Figure 30-3:

▶• wet-niche and dry-niche lighting fixtures, unless they are low-voltage systems "Listed" for application without an equipment grounding conductor.◀

- dry-niche underwater lighting fixtures.
- all electrical equipment within 5 feet (1.52 m) of the inside wall of the pool.
- all electrical equipment associated with the recirculating system of the pool.
- junction boxes.
- transformer enclosures.
- GFCIs.

Figure 30-3 **Grounding of important metal parts of a permanently installed swimming pool as covered in** *Sections 680-24* **and** *680-25.* **Also refer to Plan 10/10.**

- panelboards not part of the service equipment that supply the electrical equipment associated with the pool.

Proper grounding and bonding ensures that all of the metal parts in and around the pool area are at the same ground potential, thus reducing the shock hazard. Proper grounding and bonding practices also facilitate the opening of the overcurrent protective device (fuse or circuit breaker) in the event of a fault in the circuit. Grounding is covered in other Units of this text.

Equipment Grounding Conductors

Equipment grounding conductors *must*:

- be run in the same conduit with the circuit conductors, or be part of an approved flexible cord assembly, as used for the connection of wet-niche underwater lighting fixtures.

- terminate on equipment grounding terminals provided in the junction box, transformer, GFCI, subpanel, or other specific equipment.

It is important to note that metal conduit by itself is *not* considered an adequate grounding means for equipment grounding in and around pools.

Bonding *(Section 680-22)*

Section 680-22 requires that all metal parts of a pool installation *must* be bonded together by connecting the parts to a common bonding grid, Figure 30-4. The No. 8 or larger solid copper bonding wire need not be connected to any service equipment, a remote panelboard, or any grounding electrode. It need only "tie everything together." This helps keep all metal parts in and around the pool at the same voltage potential, reducing the shock hazard brought about by stray voltages and voltage gradients.

The bonding grid, *Section 680-22(b)*, may be:

- structural steel reinforcing rods in the concrete where the rods are bonded together with tightly twisted steel tie wires, or

- the wall of a bolted or welded metal pool, or

- a solid copper wire no smaller than No. 8.

Double-insulated pumps that are used with permanently installed pools are permitted to be plug- and cord-connected with a three-wire cord that contains the circuit conductors plus the equipment grounding conductor where the equipment grounding conductor serves only to ground the internal, nonaccessible, noncurrent-carrying metal parts of the pump. These pumps are "listed" and clearly marked as being "double-insulated." These pumps *do not* have to be bonded to the common bonding grid as illustrated in Figure 30-4. See *Section 680-28*.

Bonding Conductors

Bonding conductors:

- need *not* be installed in conduit.

- may be connected directly to the equipment that requires bonding, by means of brass, copper, or copper-alloy clamps.

UNDERWATER LIGHTING FIXTURES

Article 680, Part B covers underwater lighting fixtures for permanently installed pools.

There are three types of underwater lighting fixtures defined in *Section 680-4*.

1. **Dry-niche:** Dry-niche lighting fixtures are mounted in the side walls of the pool with the top of the lens not less than 18 inches (457 mm) below the normal water line, unless otherwise marked. They are designed to be relamped from the rear. These fixtures are waterproof. They are designed so water does not enter the niche. They have a means for connecting equipment grounding conductor(s) and have a provision to drain out any water that might accumulate in the fixture.

2. **Wet-niche:** Wet-niche lighting fixtures are mounted in the side walls of the pool with the top of the lens not less than 18 inches (457 mm) below the normal water line, unless otherwise marked. They are designed to be relamped from the front. The supply cord is stored inside the fixture. The cord should be long enough to reach the top of the pool deck for lamp replacement. The forming shell to which the fixture attaches is intended to fill up with pool water. The fixture is completely surrounded by water.

3. **No-niche:** No-niche lighting fixtures are mounted on a bracket on the inside wall of the pool with the top of the lens not less than 18 inches (457 mm) below the normal water line, unless otherwise marked. They are designed to be relamped from the front. The sup- ply conduit terminates on this bracket. The supply cord runs through this conduit to a deck box for connection to the circuit wiring. The extra supply cord is stored in the space behind the no-niche fixture. The cord should be long enough to reach the top of the pool

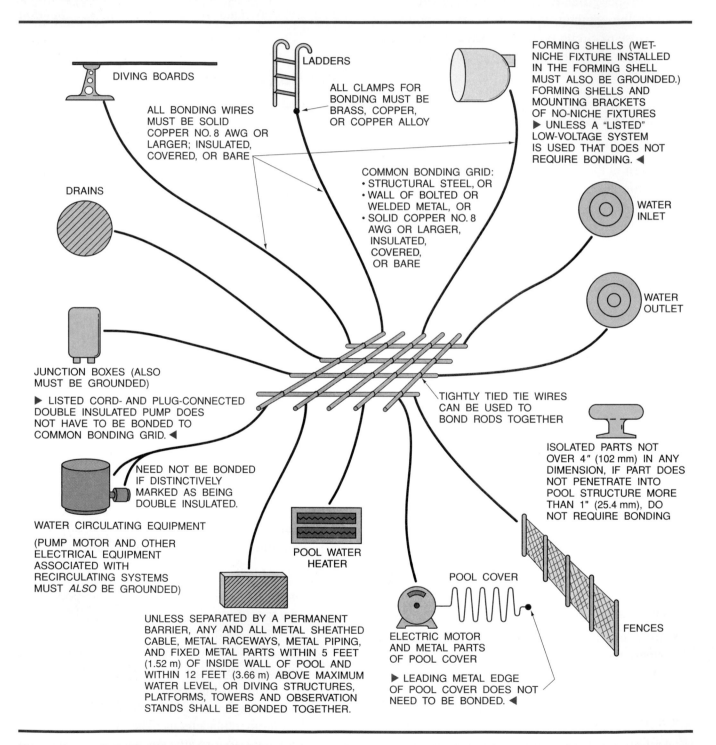

DIVING BOARDS

LADDERS

ALL CLAMPS FOR BONDING MUST BE BRASS, COPPER, OR COPPER ALLOY

FORMING SHELLS (WET-NICHE FIXTURE INSTALLED IN THE FORMING SHELL MUST ALSO BE GROUNDED.) FORMING SHELLS AND MOUNTING BRACKETS OF NO-NICHE FIXTURES ▶ UNLESS A "LISTED" LOW-VOLTAGE SYSTEM IS USED THAT DOES NOT REQUIRE BONDING. ◀

ALL BONDING WIRES MUST BE SOLID COPPER NO. 8 AWG OR LARGER; INSULATED, COVERED, OR BARE

DRAINS

COMMON BONDING GRID:
• STRUCTURAL STEEL, OR
• WALL OF BOLTED OR WELDED METAL, OR
• SOLID COPPER NO. 8 AWG OR LARGER, INSULATED, COVERED, OR BARE

WATER INLET

WATER OUTLET

JUNCTION BOXES (ALSO MUST BE GROUNDED)

▶ LISTED CORD- AND PLUG-CONNECTED DOUBLE INSULATED PUMP DOES NOT HAVE TO BE BONDED TO COMMON BONDING GRID. ◀

TIGHTLY TIED TIE WIRES CAN BE USED TO BOND RODS TOGETHER

ISOLATED PARTS NOT OVER 4" (102 mm) IN ANY DIMENSION, IF PART DOES NOT PENETRATE INTO POOL STRUCTURE MORE THAN 1" (25.4 mm), DO NOT REQUIRE BONDING

NEED NOT BE BONDED IF DISTINCTIVELY MARKED AS BEING DOUBLE INSULATED.

WATER CIRCULATING EQUIPMENT

(PUMP MOTOR AND OTHER ELECTRICAL EQUIPMENT ASSOCIATED WITH RECIRCULATING SYSTEMS MUST *ALSO* BE GROUNDED)

POOL WATER HEATER

POOL COVER

ELECTRIC MOTOR AND METAL PARTS OF POOL COVER

FENCES

UNLESS SEPARATED BY A PERMANENT BARRIER, ANY AND ALL METAL SHEATHED CABLE, METAL RACEWAYS, METAL PIPING, AND FIXED METAL PARTS WITHIN 5 FEET (1.52 m) OF INSIDE WALL OF POOL AND WITHIN 12 FEET (3.66 m) ABOVE MAXIMUM WATER LEVEL, OR DIVING STRUCTURES, PLATFORMS, TOWERS AND OBSERVATION STANDS SHALL BE BONDED TOGETHER.

▶ LEADING METAL EDGE OF POOL COVER DOES NOT NEED TO BE BONDED. ◀

Figure 30-4 Bonding the metal parts of a permanently installed swimming pool, *Section 680-22*. In addition to bonding, some of the above items must also be grounded. Refer to *Sections 680-24* and *25* for specifics. Also see Plan 10/10.

deck for lamp replacement. This type of fixture is mainly installed in retrofit installations and for aboveground pools.

Electric Swimming Pool Water Heaters

The branch-circuit conductor ampacity and the rating of the branch-circuit overcurrent devices for swimming pool water electric water heaters must not be less than 125% of the ampere rating of the electric heater, *Section 680-9*. Electric water heaters for swimming pools will be marked in volts, amperes, and watts.

ELECTRIC HEATING OF SWIMMING POOL DECKS

Section 680-27 of the *NEC*® lists the requirements for electric heating of swimming pool decks. These requirements are in addition to those in *Article 422* for electric appliances.

Electric Heating Branch-Circuit Sizing and Marking

Electric heaters are appliances. *Section 424-3(b)* requires that the branch-circuit sizing for electric

appliances is figured at 125% of the nameplate rating. These electric heaters are marked with their voltage and wattage, or with their voltage and amperage, or with voltage, amperage, and wattage.

Types of Electric Heating Permitted

Where electric heating is located within 20 feet (6.1 m) from the inside wall of the pool some Code restrictions apply. These are illustrated in Figure 30-5.

Unit heaters:

- shall be rigidly mounted on the structure.
- shall be totally enclosed or guarded.
- shall not be located over the pool.
- shall be at least 5 feet (1.52 m) back from the inside walls of the pool.

Permanently wired radiant heaters:

- shall be suitably guarded.
- shall be securely fastened.
- shall not be located over a pool.
- shall be at least 5 feet (1.52 m) back from the inside walls of the pool.

Figure 30-5   Deck-area electric heating within 20 feet (6.1 m) from inside edge of swimming pools, *Section 680-27.*

- shall be mounted at least 12 feet (3.66 m) vertically above pool deck, unless approved otherwise.

Radiant heat cables:

- are not permitted to be embedded in or below the pool because of possible breakdown of the insulation, resulting in current leakage.

SPAS AND HOT TUBS
(ARTICLE 680, PART D)

There are basically two types of spas and hot tubs.

1. **Self-contained spa or hot tub:** A factory fabricated unit consisting of a spa or hot tub vessel with all water circulating, heating, and control equipment integral to the unit. Equipment may include pumps, air blowers, heaters, lights, controls, sanitizers, or generators.

2. **Packaged spa or hot tub assembly:** A factory fabricated unit consisting of water circulating, heating, and control equipment mounted on a common base, intended to operate a spa or hot tub. Equipment may include pumps, air blowers, heaters, lights, controls, sanitizers, or generators.

The basic difference between a spa and a hot tub is that a hot tub is constructed of wood such as redwood, teak, cypress, or oak, whereas a spa is made of plastics, fiberglass, concrete, tile, or other man-made products.

Spas and hot tubs contain electrical equipment for heating and recirculating water.

Spas and hot tubs are intended to be filled and used. They are not drained after each use, as in the case of a regular bathtub.

Some spas and hot tubs are furnished with a cord-and-plug set when shipped by the manufacturer. Where these can be converted from cord-and-plug connection to "hard wiring" they are listed and identified as "convertible" spas and hot tubs. The manufacturer must furnish clear instructions as to how the conversion from cord-and-plug set to "hard wiring" is to be done.

▶ *Section 680-38* requires that for spas and hot tubs, a clearly labeled emergency shutoff switch for the control of the recirculation system and jet system shall be installed at least 5 feet (1.52 m) away, adjacent to and within sight of the spa or hot tub. This requirement does not apply to single-family dwellings, but it does apply to multifamily dwellings. ◀

This new disconnect requirement for spas and hot tubs came about because of reported drownings that have occurred when a person has been trapped under water by the tremendous suction at an unprotected drain, i.e., the grate over the drain has been removed, or is broken. It has been reported that three adult males could not pull a basketball from a drain because of the suction!

Outdoor Installations, *Section 680-40*:

Spas and hot tubs installed outdoors *must* conform to the installation requirements discussed previously for regular swimming pools, *Article 680, Parts A* and *B*.

The metal bands and hoops that secure the wood staves of a spa or hot tub do not have to be grounded.

Listed packaged units are permitted to have a cord not longer than 15 feet (4.56 m). Such units must be GFCI protected.

Where there is metal-to-metal mounting to the same metal base, that is acceptable as the required bonding.

▶ Listed packaged units that have a factory-installed remote-control panel may be connected with liquidtight flexible conduit not over 6 feet (1.83 m) long. ◀

Indoor Installation, *Section 680-41*:

Spas and hot tubs installed indoors *must* conform to the following requirements:

- must be connected using the wiring methods covered in *Chapter 3* of the Code.

- may be cord- and plug-connected if rated 20 amperes or less.

Receptacles, *Section 680-41(a)*:

For indoor installations:

- at least one 125-volt receptacle must be installed at least 5 feet (1.52 m) but not more than 10 feet (3.05 m) from the inside wall of the spa or hot tub. ▶This receptacle must be connected to a general-purpose branch-circuit. ◀

▶• any receptacles installed on the property must be installed at least 5 feet (1.52 m) measured horizontally from the inside wall of the spa or hot tub. ◀

- any 125-volt receptacle located within 10 feet (3.05 m) of the inside walls of a spa or hot tub must be GFCI protected.

- any receptacles that supply power to a spa or hot tub must be GFCI protected no matter how far they are from the spa or hot tub.

Lighting Fixtures, Lighting Outlets, and Ceiling Fans, *Section 680-41(b)*:

- all lighting outlets, fixtures, and ceiling fans located within 5 feet (1.52 m) from the inside wall of the spa or hot tub or directly over the spa or hot tub must be at least 7½ feet (2.29 m) above the maximum water level and must be GFCI protected.

  There are a few exceptions to this:

  1. GFCI protection is not required if the fixture or ceiling fan is 12 feet (3.66 m) or more above maximum water level.

  2. Recessed lighting fixtures may be installed less than 7½ feet (2.29 m) above the hot tub or spa if the fixture:
     - is GFCI protected, and
     - has a glass or plastic lens, and
     - has a nonmetallic trim suitable for use in damp locations, or
     - has an electrically isolated metal trim suitable for use in damp locations.

  3. Surface-mounted fixtures may be installed less than 7½ feet (2.29 m) above the hot tub or spa if the fixture:
     - is GFCI protected, and
     - has a plastic or glass globe, and
     - has a nonmetallic body, or
     - has a metallic body that is isolated from contact.
     - is suitable for use in damp locations.

- if any underwater lighting is to be installed, the rules discussed previously for regular swimming pools apply.

Wall Switches, *Section 680-41(c)*:

- wall switches must be located at least 5 feet (1.52 m) from the inside walls of the spa or hot tub.

Bonding, *Sections 680-41(d)* and *(e)*:

The bonding requirements for spas and hot tubs are similar to those for regular swimming pools.

Bond together:

- all metal fittings within or attached to the spa or hot tub.

- all metal parts of electrical equipment associated with the circulating system, including pump motors.

- all metal pipes, conduits, and metal surfaces within 5 feet (1.52 m) of the inside wall of the spa or hot tub. Bonding is not required if the materials are separated from the spa or hot tub by a permanent barrier, such as a wall or building.

- all electrical devices and controls *not* associated with the spa or hot tub located less than 5 feet from the inside wall of the spa or hot tub. If located 5 feet (1.52 m) or more from the hot tub or spa, bonding to the spa or hot tub system is not required.

Bonding is to be accomplished by means of threaded metal piping and fittings, by metal-to-metal mounting on a common base or frame, or by means of a No. 8 or larger, solid, insulated, covered or bare bonding jumper.

Grounding, *Section 680-41(f)* and *(g)*:

The requirements for grounding spas and hot tubs are:

- ground all electrical equipment within 5 feet (1.52 m) of the inside wall of the spa or hot tub.

- ground all electrical equipment associated with the circulating system, including the pump motor.

- grounding methods *must* conform to all of the applicable Code rules of *Article 250*.

- all equipment shall be connected using wiring methods as found in *Chapter 3* of the Code.

- if equipment is connected by means of a flexible cord, the grounding conductor must be part of the flexible cord, and must be fastened to a fixed metal part of the equipment.

Electric Water Heaters, *Section 680-41(h)*:

- must be listed for the purpose.

- must have their heating elements divided into loads not over 48 amperes, and protected at not over 60 amperes. This requirement is usually met by the manufacturer of the heater.

• must have branch-circuit conductors rated not less than 125% of the total load as indicated on the nameplate of the heater.

Ground-Fault Protection, *Section 680-42:*

An outlet that supplies a spa or hot tub (packaged or self-contained) must be GFCI protected unless the "listed" unit contains integral GFCI protection, and the unit is so marked.

An outlet that supplies a field-assembled spa or hot tub that has a heater load of 50 amperes or less must be GFCI protected.

DISCONNECTING MEANS

To provide for people's safety, *Section 680-12* requires that a disconnecting means be installed (1) within sight, (2) in an accessible location, and (3) at least five feet (1.52 m) horizontally from the inside walls of a pool, spa, or hot tub equipment.

HYDROMASSAGE BATHTUBS
(ARTICLE 680, PART G)

All 125-volt, single-phase receptacles that are located within 5 feet (1.52 m) measured horizontally from the inside walls of the hydromassage bathtub must be GFCI protected.

Hydromassage bathtubs, together with their associated electrical components, must be GFCI protected, *Section 680-70.*

Because a hydromassage bathtub does not constitute any more of a shock hazard than a regular bathtub, the *NEC®* permits all other wiring (fixtures, switches, receptacles, and other equipment) in the same room but not directly associated with the hydromassage bathtub to be installed according to all the normal Code requirements covering installation of that equipment in bathrooms.

This residence has a hydromassage tub in the master bathroom. It is a prewired unit furnished with a 3-foot length of ½-inch watertight conduit that contains a black, a white, and a green equipment grounding conductor.

The hydromassage tub is powered by a ½-hp, 115-volt, 3450 rpm, single-phase motor rated at 10 amperes.

A hydromassage tub and all of its associated equipment generally provided by the manufacturer must be connected to a circuit that has GFCI protection.

The hydromassage tub is connected to a separate 120-volt, 20-ampere Circuit A9. For more details on the circuitry for the hydromassage bath-tub see Unit 22.

A hydromassage tub is also known as a whirlpool tub or whirlpool bathtub.

Figure 22-12 illustrates one manufacturer's current production-model hydromassage bathtub.

By definition in *Article 680*, a hydromassage bathtub is intended to be filled, used, then drained after each use, whereas the spa or hot tub is filled with water and not drained after each use.

FOUNTAINS *(ARTICLE 680, PART E)*

The residence discussed in this text does not have a fountain. The installation requirements for a small self-contained portable fountain that has no dimensions over 5 feet (1.52 m) is not too stringent. Merely follow the instructions that are furnished with the fountain. For fountains having any dimensions over 5 feet (1.52 m) and/or if the fountain shares water with a regular swimming pool, the fountain wiring must conform to all of the requirements found in *Part E* of *Article 680.*

UNDERWRITERS LABORATORIES STANDARDS

The U.L. standards of interest for swimming pools, spas, and hot tubs are:

U.L. 676   Underwater Lighting Fixtures

U.L. 943   Ground-Fault Circuit Interrupters

U.L. 1081   Swimming Pumps, Filters, and Chlorinators

U.L. 1241   Junction Boxes for Underwater Lighting Fixtures

U.L. 1261   Electric Water Heaters for Pools and Spas

U.L. 1563   Electric Hot Tubs, Spas, and Associated Equipment

U.L. 1795   Hydromassage Bathtubs

SUMMARY

The Figures in this Unit and a plan-sized diagram present a detailed overview of the Code requirements for pool wiring.

REVIEW

Note: Refer to the Code or the plans where necessary.

1. The article of the *NEC*® that covers most of the requirements for wiring of swimming pools is *Article* _____ .

2. Name the two ways in which a person may sustain an electrical shock when in a pool.

   _____

   _____

   _____

   _____

3. The Code in *Article 680* discusses two types of pools. Name and describe each type.

   _____

   _____

   _____

   _____

   _____

4. Use *Section 680-24* of the Code to determine if the following items must be grounded.

	True	False
a. Wet-and-dry-niche lighting fixtures.	_____	_____
b. Electrical equipment located within 5 feet (1.52 m) of inside walls of pool.	_____	_____
c. Electrical equipment located within 10 feet (3.05 m) of inside walls of pool.	_____	_____
d. Recirculating equipment and pumps.	_____	_____
e. Lighting fixtures and ceiling fans installed more than 15 feet (4.56 m) from inside walls of pool.	_____	_____
f. Junction boxes, transformers, and GFCI enclosures.	_____	_____
g. Panelboards that supply the electrical equipment for the pool.	_____	_____
h. Panelboards 20 feet (6.1 m) from the pool that do not supply the electrical equipment for the pool.	_____	_____

5. Grounding conductors (must) (may) be run in the same conduit as the circuit conductors. Circle correct answer.

6. Grounding conductors (may) (may not) be spliced with wire-nut types of wire connectors. Circle correct answer.

7. The purpose of grounding and bonding is to _____

   _____

   _____

   _____

8. What parts of a permanently installed pool must be bonded together? _____

_____

_____

_____

_____

_____

_____

_____

_____

9. May electrical wires be run above the pool? Explain. _____

_____

_____

10. What is the closest distance that a receptacle may be installed to the inside wall of a pool? _____

_____

_____

_____

_____

_____

_____

_____

11. Receptacles located within 20 feet (6.08 m) of the inside wall of a pool must be protected by a _____ .

12. Lighting fixtures installed over a pool must be mounted at least (10 feet), (12 feet), (15 feet), above the maximum water level. Circle correct answer.

13. Grounding conductor terminations in wet-niche metal-forming shells, as well as the conduits entering junction boxes or transformer enclosures where the conduit runs directly to the wet-niche lighting fixture, must be _____ with a (an) _____ to prevent corrosion to the terminal and to prevent the passage of air through the conduit, which could result in corrosion.

14. Wet-niche lighting fixtures are accessible (from the inside of the pool) (from a tunnel) (on top of a pole). Circle correct answer.

15. Wet-niche lighting fixtures operating at above 15 volts must be protected by _____

_____

16. Dry-niche lighting fixtures are accessible (from a tunnel, or passageway, or deck) (from the inside of the pool) (on top of a pole). Circle correct answer.

17. In general, it is *not* permitted to install conduits under the pool or within 5 feet (1.52 m) measured horizontally from the inside wall of the pool. True or false? Explain.

_____

_____

_____

_____

_____

_____

_____

_____

_____

_____

18. Junction boxes, wet-niche lighting fixtures, and transformer and GFCI enclosures have one thing in common. They all (are made of bronze) (have threaded hubs) (must be mounted at least 8 inches, or 203 mm, above the deck). Circle correct answer.

19. Circle the correct answer(s). Existing lighting fixtures are permitted to be less than 5 feet (1.52 m) measured horizontally from the inside walls of a pool if:

   a. they are made entirely of plastic.

   b. they are GFCI protected.

   c. they are rigidly fastened in place.

   d. they are controlled by a wall switch.

20. The Code permits radiant heating cable to be buried in or below the deck of a pool. True or false? _____

21. For indoor spas and hot tubs, the following statements are either true or false. Check one.

	True	False
a. Receptacles may be installed within 5 feet (1.52 m) from the edge of the spa or hot tub.	_____	_____
b. All receptacles within 10 feet (3.05 m) of the spa or hot tub must be GFCI protected.	_____	_____
c. Any receptacles that supply power to pool equipment must be GFCI protected.	_____	_____
d. Wall switches must be located at least 10 feet (3.05 m) from the pool.	_____	_____
e. Lighting fixtures above the pool or within 5 feet (1.52 m) from the inside wall of the pool must be GFCI protected.	_____	_____

22. Bonding and grounding of electrical equipment in and around spas and hot tubs (are required by the Code) (are not required by the Code) (are decided by the electrician). Circle correct answer.

23. Where a spa or hot tub is installed in an existing bathroom, an existing receptacle outlet within 5 feet (1.52 m) of the tub is permitted to remain, but only if the circuit feeding the receptacle outlet has _____ protection.

24. a. Hydromassage bathtubs and their associated electric components (shall) (shall not) be protected by a ground-fault circuit interrupter. Circle the correct answer.

    b. All receptacles located within 5 feet (1.52 m) of a hydromassage tub (shall) (shall not) be protected by a ground-fault circuit interrupter. Circle the correct answer.

25. What section of the Code requires that no-niche underwater lighting fixtures be grounded? *Section* _____ .

26. Underwater pool lighting fixtures shall be mounted so that the top of the lens is not less than _____ inches below the normal water line.

27. What section of the Code tells us that all pool associated pump motors shall be grounded? *Section* _____ .

28. What section of the Code tells us that double insulated pumps for permanently installed pools do not have to be bonded to the common bonding grid? *Section* _____ .

Wiring for the Future: Advanced Home Automation Systems

OBJECTIVES

After studying this Unit, the student will be able to

- understand the "three generations" of advanced home systems.
- understand the two primary technologies for advanced home systems today.
- understand how carrier-current control systems differ from hard-wired ones.
- understand the basics of X-10 systems.
- understand some factors that interfere with the efficient operation of X-10 systems, and how these can be corrected.

INTRODUCTION

An increasing trend in upper-end residential wiring is the installation of advanced home systems. Normally these are installed in larger, more expensive homes. Previously they were called "home automation" systems, but over the last five years or so, the concept of a computerized electronic home is being replaced by the concept of home entertainment systems and rapid computer access to the World Wide Web on the Internet.

There are two different trends in advanced home systems for upper-end residences today:

1. Remote-control systems that operate by means of carrier-current signals carried on conventional power wiring.

2. Information-entertainment distribution systems called Structured Premises Wiring Systems. Some people refer to this concept as "future proofing," which means prewiring homes today, during original construction, for all the signals and applications that homeowners may want tomorrow.

To see how we got where we are today, it is necessary to take a short look at the history of advanced home systems.

Three Generations of Advanced Home Systems

There have always been custom installers who can design and install remote-control systems for expensive homes. Generally speaking, these systems involve extensive use of low-voltage "bell wire," and labor-intensive integration by the installer. Integration is programming and fine-tuning applications such as low-voltage lighting control, security, climate control, and audio-video distribution so that they all work together as a more-or-less unified system and can be controlled by the homeowner from a "user interface" such as a control panel, keypad, or telephone handset. These custom-designed home automation systems typically require a good deal of continuing maintenance from their installers.

First Generation. In the early 1980s, several different home automation systems intended for mass installation in new homes appeared. These new systems, like SMART HOUSE sponsored by the National Association of Home Builders, involved central controllers, extensive remote control features (turning things on and off at a distance, programming events to occur in the future), and complex, specialized wiring. Smart House hybrid cable combined ac power, telephone, coaxial cable, and low-voltage control conductors inside a single cable jacket. It was large, expensive, and so complicated that it required its own Code rules found in *Article 780* of the *NEC.*® Figure 31-1 shows a typical hybrid cable.

Another leading home control system was called CEBus (for Consumer Electronic Bus) and sponsored by the Electronic Industries Association; there were also a number of others including Unity Systems, Mastervoice, and LonWorks. None of this first generation of home automation systems sold very well or were installed in very many houses. Consumers did not show much interest in home automation, possibly because of a lack of compatible appliances and products. Most of the home automation devices actually installed in homes used a simpler, less sophisticated technology called X-10, which sends control signals over ac power wiring and, thus, did not require expensive, specialized cabling systems.

Second Generation. While few houses were built with home automation controllers, stand-alone home products were rapidly becoming smarter and more sophisticated. The increasing popularization of cell phones, large screen TVs, CD and laserdisc players, and ever-more powerful personal computers turned the home buying public's attention away from programming the clothes washer and more

toward home entertainment and communications technologies.

The second generation of advanced home systems was entertainment-based. They focused not on home automation and control, but on routing audio and video signals from around the home. Systems like Square D's *Elan* not only included home theater installations, but allowed homeowners to watch security camera pictures on any TV in the home, listen to a baby monitor through the intercom system, or play a movie on the basement VCR and watch it on a TV elsewhere in the home. Entertainment systems of this sort are becoming more popular in upper-end homes, but have never become widespread. Like the earliest home automation systems, each different installation is basically custom designed and requires a good deal of integration by installers.

Third Generation. Following development of the World Wide Web in 1995, a third generation of home wiring systems has been introduced by a number of different manufacturers. Structured premises wiring systems, sometimes called "home LANs" (for local area networks) are intended to allow distribution of high-speed data, audio, and video signals throughout the home. They offer such advantages as faster Internet access for home computers, and convenient networking of data devices (computers, printers, fax machines) in home offices. Structured premises wiring systems get their name from EIA/TIA standards 568 and 570, which describe "structured" or standardized cabling systems for business and homes (see box on home wiring standards).

The next two sections of this Unit describe in greater detail the two current directions in advanced home systems: X-10 and structured premises wiring.

X-10: BASIC HOME AUTOMATION

The simplest home automation technology on the market today—also the cheapest and most successful—is made by X-10 Powerhouse based in Northvale, NJ. X-10 is a powerline carrier system that transmits control signals over existing ac power wiring, instead of using dedicated conductors such as coaxial cable or telephone twisted pairs.

Powerhouse introduced the technology in 1978; and since then it has been licensed to many other companies. X-10 compatible products are sold

Figure 31-1 A typical hybrid cable consisting of AC power, telephone, coaxial cable, and low-voltage control conductors.

under many brand names including Radio Shack, Stanley Lightminder, Leviton, Enerlogic, Honeywell TotalHome, Lucent Technologies HomeStar,® and IBM Home Director.

X-10 is a "plug-n-play" system. This means the system can be bought piecemeal and expanded gradually. A basic system would start with a pushbutton controller that sends coded signals over the ac power wiring to control eight ON/OFF modules located anywhere in the house.

These modules plug into ordinary receptacle outlets, and lights and appliances are plugged into the modules. This means that from a living room controller, you can turn off the stereo in your son's room upstairs or have single point control of inside and outside lights, all without installing special wiring.

Sophistication—and cost—can be added with accessories such as a 16-module controller, timers, dimmers, thermostat interface, drapery controller, wireless remote, or a telephone responder (turn off your son's stereo from work). Also, personal computer software can be programmed to regulate dozens of functions such as turning lights and appliances on and off at certain times or arming and disarming a security system.

Leviton Decora is an upscale line of X-10 modules and controls that replaces conventional receptacles and wall switches, giving a more finished appearance.

Technical Overview

X-10 products, in general, are designed to carry out simple two-state functions, such as on/off control of lights and appliances, dimming and brightening of lights, or security contact closures. They do not offer all the features of more complete home automation systems, such as control of audio/video equipment for changing channels and controlling volume. X-10 is basically a one-way communications system. Some devices "talk" and others "listen," but no X-10 devices do both (except for PC control software).

X-10 technology, though simple, is the least costly home automation system especially for retrofit systems. It is also very flexible, since special wiring is not required for basic functions. More advanced applications (to be described in a future article) do require dedicated low-voltage wiring.

The principle behind X-10 technology is simple. Transmitting devices take a small amount of power from the 120-volt power line, modulate or change it to a higher frequency, and then superimpose this high frequency signal back on the ac circuit. At the other end of the wiring, a receiving device, such as a lamp or appliance module, senses the high frequency burst and is programmed to perform a certain task, such as turning on or off a lamp plugged into that module.

The X-10 signal is a one-millisecond burst of 120 kHz at the zero crossing point of 60 Hz electrical power. X-10 was described above as a two-state system. The presence of a 120-kilohertz (kHz) burst represents binary 1, and its absence represents binary 0.

Receiving modules react according to whether they receive a 1 or 0. An X-10 control signal consists of three separate sections: a start code, a house code, and a key number. Each receiving module can be assigned a unique address such as A2 or H25.

Because it requires no special wiring, X-10 is an excellent technology for retrofits. For example, if you decide that you would like to be able to control a security light on the back porch from a switch in the master bedroom, you can do so quickly and easily by installing an X-10 lamp or appliance module at the fixture to be controlled, and an X-10 switch in the bedroom.

X-10 Modules (Receivers)

The most common types of X-10 modules are shown in Figure 31-2 and described below:

Dimmer. An X-10 dimmer switch replaces a standard 120-volt wall switch. It fits in a standard electrical device box and connects to the same wires. When the button on the switch face is pressed manually, it turns the light to which it is connected on or off. When the button is held down, it dims or brightens the light. The X-10 dimmer switch does everything that a standard dimmer switch does, but it can also respond to 120-kHz control signals. X-10 wall switch modules come in both dimming and non-dimming versions, and in three- and four-way switch versions.

Lamp Module. A lamp module is plugged into a standard wall outlet, and a lamp or other portable appliance is plugged into a two-blade receptacle on the bottom of the unit. The lamp module turns lamps

Figure 31-2 Various types of X-10 receiver modules. *Artist rendition based on illustrations courtesy* **Leviton Mfg. Co., Inc.**

on or off by means of an internal relay, or it can dim or brighten them in response to X-10 control signals. X-10 lamp modules can only be used with standard incandescent fixtures. They should not be used with fluorescent fixtures or appliances.

Appliance Module. The appliance module is very similar to the lamp module but it only turns devices on or off. It has no dimming capability. Appliance modules can be used with lamps (including fluorescent fixtures), but lamp modules should not be used to control appliances.

Universal Interface Module. The universal interface module is similar to the lamp module, but instead of a two-blade receptacle, it has two binding screws on the bottom of the unit. The universal interface module is used for on/off control of any small, 120-volt device and does not plug into a standard receptacle.

Receptacle Module. A receptacle module installs in a wall box to replace a standard receptacle,

providing a more finished appearance. Like the plug-in appliance module, it switches lamps or appliances on or off in response to control signals received over the 120-volt power wiring; it does not have a dimming function. Receptacle modules have three-prong grounding-type receptacles for plugging in appliances, unlike appliance modules, which have two-prong receptacles.

X-10 Controllers (Transmitters)

X-10 controllers, such as illustrated in Figure 31-3, plug directly into any standard wall receptacle in order to transmit 120-kHz powerline control signals to receiver modules. An X-10 signal consists of three parts: house code, unit code, and function code. Function codes are *on*, *off*, *all lights on*, *all off*, *bright*, and *dim*.

X-10 basic controllers have a dial for the house code and push buttons that correspond to unit codes and functions. The commands *all lights on*, *bright*, and *dim* can only be received by lamp modules and dimming wall switches; sending them to any other device has no effect. The commands *on*, *off*, and *all off* can be received by any X-10 module. *All off* turns off every X-10 device in the house.

System Addressing. X-10 modules are addressed by means of two small dials that can be rotated with a small screwdriver. One of the dials is labeled "house code" and has alphabetical characters ranging from A to P. The other dial is called the "unit code" and has numbers from 1 to 16. These two dials can be adjusted to any of 256 combinations; 256 is the maximum number of separate addresses on any X-10 home automation system. Each receiving module on an X-10 system can be assigned to the same address to let them operate together.

House codes can also be used to group modules together for common operation. X-10 controllers can send messages either to a specific address (house code plus unit code, such as B13) to control a single module or to a specific house code. For example, an *off* signal sent to a house code D would turn off all modules with D in their addresses. An *on* signal sent to D4 would activate only the module(s) set to house code D and unit code 4.

Bright and *dim* commands can also be sent to either individual module addresses or to a house code. Such codes have no effect on those modules

Figure 31-3   A typical X-10 controller makes it easy to transmit preprogrammed signals to receiver modules. *Artist rendition based on illustrations courtesy* Leviton Mfg. Co., Inc.

without dimming capability (appliance modules, universal interface modules, or regular wall switches). The *all lights on* command turns on all lamp modules regardless of address; and *all off* deactivates every X-10 device in the house.

There are also more advanced computer controllers that allow more comprehensive control of home functions beyond simply turning lights and other devices on and off.

Limitations of X-10 Technology

X-10 modules are susceptible to interference on the ac power line environment. This can be a problem because the ac power line environment is noisy, or electrically "dirty." Many things can generate noise on 120-volt power lines, including fluorescent fixture ballasts, appliance motors, wireless intercoms, and electric charcoal lighters. And, because standard home wiring (most often Type NM cable) is unshielded, it can pick up unintended induced signals in the same way that an antenna does, even from devices that are not attached to the same circuit.

In fact, where two or more houses are connected to the secondary of the same utility transformer, unwanted signals from one house may be transmitted to the other. This means that the amount of data that can be transmitted via X-10, even under "normal" conditions, is limited.

It is also limited because X-10 signals are relatively slow. The maximum data rate is 60 bits per second (the CEBus home automation system, by contrast, supports up to 10,000 bps). This means there can be a noticeable delay between the time you activate a control and the time a controlled action takes place.

X-10 system functioning can be optimized through use of a "phase coupler" at the circuit breaker panel. If a transmitting device (plug-in controller) is connected to one leg of the 120/240-volt, single-phase, three-wire service, and a receiving module is on the other leg, operation may be erratic. This can be improved by installing a capacitive phase coupler at the service panel to act as a signal bridge. This is illustrated in Figure 31-4.

There are also X-10 filters that install at the service panel to block interference from outside the house; however, this is an expensive option that should only be considered as a troubleshooting option.

X-10 systems are usually subject to considerable tinkering and fine-tuning in the field to make everything work in an optimal fashion. The home automation magazine *Electronic House/H.A. Pro* runs a regular column called "Ask Mister Module," which lets users ask technical questions about troubleshooting and getting the most out of their X-10 systems.

Code Considerations

Interestingly, there are no *National Electrical Code®* requirements that apply specifically to power-line carrier control systems like X-10 and CEBus. *Section 810-1* specifically exempts "equipment and antennas used for coupling carrier current to powerline conductors." Otherwise, the low-voltage *Articles* of *Chapters 7* and *8* do not mention powerline carrier systems.

STRUCTURED PREMISES WIRING SYSTEMS

Structured premises wiring systems consist of video, telecommunications, and control cabling distributed from throughout the house over a standardized wiring network. In this view, providing an information infrastructure during original house construction will soon become the norm rather than

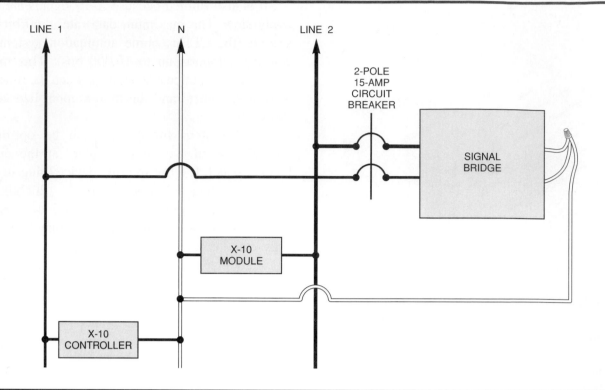

Figure 31-4 Connecting a phase coupler to a panel. This acts as a signal bridge to improve X-10 operation. *Artist rendition based on illustrations courtesy* **Leviton Mfg. Co., Inc.**

an exception, just as central air-conditioning went from being an upscale luxury to a standard feature. In short, structured premises wiring systems aim to prevent the "functional obsolescence" of residential wiring.

Combination Outlets

Telecom, coaxial, and sometimes fiberoptic connectors can be combined in the same wallplates, with multiple cables to each outlet location, Figure 31-5.

Central Distribution

Cables from room outlets homerun to the central cabinet, referred to as a "Service-Center." Except for outlets, most system components are located at this cabinet, which also functions as a patch panel for making most wiring changes at a central location, Figure 31-6.

Standards

The home automation wars of the past were caused partly by lack of standardization. SMART HOUSE was not compatible. Perhaps because of that

Figure 31-5 A combination telecom and coaxial cable wall outlet. *Courtesy* **AMP, Inc.**

Figure 31-6 A "service-center" with telecom, audio, and home automation controllers installed. *Courtesy* AMP, Inc.

history, today's structured premises wiring systems are respectful of standards. All three systems described below comply with the two primary home wiring standards, EIA/TIA-570 and CEBus (see sidebar). While they are not interchangeable, this does mean that all three systems provide essentially the same wiring infrastructure and attached product interfaces (i.e., outlets).

A Closer Look at Three Structure Premises Wiring Systems

Features of advanced home wiring and distribution systems are constantly evolving. At the time of publication, three of the leading structured premises wiring systems were the following:

IES Technologies
179 East Main Street
Sandy, UT 84070
(801) 566-8892; (800) 903-4237

Cabinet. FIN Panels (for *FutureSmart Interactive Network*) come in different models providing between 4 and 32 zones; each zone is a room with telephone, video, and fiber outlets. The IES system accommodates both standard and PBX/key system telephones. An Ethernet hub is available for installation in FIN Panels (also called Node 0).

Cables. IES Speedwrap cable contains six conductors:

> Two RG6 coaxial cables for video
>
> One CAT 5, 4-pair, unshielded twisted pair (UTP) cables for telecom
>
> Dual 62.5 fiberoptic/125 µm fiberoptic cable for future services
>
> Optional control wiring for systems such as MediaLink, NEST and Echelon

Receptacles. FIR Receptacles (for *FutureSmart Interactive Receptacles*) come in three models with different combinations of RJ45 telecom, F-type coaxial, and ST-style fiberoptic connectors. Signal receptacles can be combined with standard ac receptacles or an RF modulator in multi-gang wallplates.

AMP OnQ
939 East Park Drive
Harrisburg, PA 17105-3608
(717) 592-6233; (800) 321-2343

OnQ offers more of a mix-and-match approach with modular components and optional kits. This lets installers and customers tailor a system for particular home applications.

Cabinet. A variety of distribution head-end components can be installed in the Service Center, which serves up to 24 rooms (24 homeruns each of telecom and video cable) and comes in two models. The Series 150 service center distributes telecom and video only; the Series 200 accommodates telecom, video, and control components. The OnQ system is compatible with both standard and PBX/key system telephones.

Cables. The standard OnQ cable layout uses the following. The system also supports the future use of fiberoptic cables.

> Two Series 6 coaxial cables (this has somewhat more demanding specs than standard RG6)

One CAT 5, 4-pair UTP for telecom

CAT 3 UTP and Class 2 cables are used as control wiring for systems including SMART HOUSE, AMP Lighting Control, and Smart-ONE

Outlets. Not surprisingly, since AMP is one of the world's leading outlet and connector manufacturers, the OnQ system offers the largest selection of different telecom, video, and fiberoptic outlets, which use Decora-style wallplates. OnQ signal outlets are not intended to be combined with ac receptacles, although there is nothing to prevent this.

Greyfox Systems
P.O. Box 861
Carnegie, PA 15106-0861
(412) 429-0580; (800) 480-3380

Greyfox, which calls itself "The CEBus Compliant Home Network System," uses a central cabinet that includes CEBus electronics. The 10" × 16" FoxBox cabinet, which is physically the smallest of the three, also distributes cable TV and telephone.

Cables.

Two RG6 coaxial cables for video

One CAT 3, 4-pair UTP for telecom

Dual 62.5 fiberoptic/125 μm fiberoptic cable for future services

Outlets. Decora-style combination outlets include RJ45 telecom, F-type coaxial, and ST-style fiberoptic connectors. Coaxial outlets use CEBus coding for internal and external cables. Signal receptacles can be combined with standard AC receptacles in multi-gang wallplates.

STANDARDS FOR ADVANCED HOME WIRING SYSTEMS

A number of different organizations publish standards that affect the design of advanced home wiring systems.

EIA/TIA 570, Residential Telecommunications Cabling Standard

This standard, jointly developed by the Electronic Industries Association (EIA) and Telecommunications Industry Association (TIA), describes a premises wiring system that uses twisted four-pair telecommunications cabling, modular jack outlets, and star-series wiring to each room from the distribution device.

CEBus (Consumer Electronic Bus)

The CEBus standard was developed by the Electronic Industries Association (EIA). It establishes a uniform high-quality cabling system to handle all of the data communications needs of the home (audio, video, CEBus signaling, high-speed data); specifies how products send and receive messages; and provides a common application language for built-in interoperability and system integration. The CEBus standard requires four-pair telecommunications cabling for all home wiring.

Two pairs are used for voice and data, with CEBus signaling on the remaining two pairs.

Home Automation Association

Wiring America's Homes is a national campaign involving more than 25 leading manufacturers of advanced home products and systems. *Wiring America's Homes* promotes the benefits of prewiring homes during original installation with a combination of twisted-pair telecommunications wiring, coaxial cable for video, and optional fiberoptic lines for applications requiring a lot of bandwidth, such as multimedia or video teleconferencing.

FCC

At present, Federal Communications Commission rules allows two-pair "quad" telephone wiring for home telephone service. However, a growing chorus of voices is crying that yesterday's phone wire is inadequate for tomorrow's voice and data information highway needs. BICSI, the industry association of telecommunications installers, has proposed an FCC rule change that would require minimum Category 3 cabling for all residential construction.

REVIEW

1. There have been three "generations" of advanced home systems to date. SMART HOUSE and CEBus belong to which generation?

   a. First

   b. Second

   c. Third

2. Which of the following are *not* typical characteristics of a residential structured premises wiring system? (Pick all that apply.)

   a. Central distribution of cables

   b. System controller

   c. Combination outlets

   d. GFCI protection for telephone wiring

3. List three types of cables typically used in structured premises wiring systems:

   a. _____

   b. _____

   c. _____

4. Powerline control systems like X-10 use two coaxial cables for transmitting signals and two #24 AWG wires for transmitting 12-VDC control power to X-10 receiver modules.

   a. True

   b. False

5. Which of the following *are* typical addresses for X-10 modules? (Pick all that apply.)

   a. A8

   b. B12

   c. K32

   d. M16

6. Briefly describe the use of a "phase coupler" to improve the performance of X-10 systems.

   _____

   _____

   _____

   _____

7. An X-10 system transmits data at which of the following speeds?

   a. 60 bits per second (bps)

   b. 10,000 bps

   c. 28,800 kbps

8. Which of the following organizations write standards that apply to residential wiring systems? (Pick all that apply.)

   a. Electronic Industries Association (EIA)

   b. National Electrical Manufacturers Association (NEMA)

   c. Telecommunications Industry Association (TIA)

   d. Underwriters Laboratories Inc. (UL)

Specifications for Electrical Work— Single-Family Dwelling

1. **GENERAL:** The "General Clause and Conditions" shall be and are hereby made a part of this division.

2. **SCOPE:** The electrical contractor shall furnish and install a complete electrical system as shown on the drawings and/or in the specifications. Where there is no mention of the responsible party to furnish, install, or wire for a specific item on the electrical drawings, the electrical contractor will be responsible completely for all purchases and labor for a complete operating system for this item.

3. **WORKMANSHIP:** All work shall be executed in a neat and workmanlike manner. All exposed conduits shall be routed parallel or perpendicular to walls and structural members. Junction boxes shall be securely fastened, set true and plumb, and flush with finished surface when wiring method is concealed.

4. **LOCATION OF OUTLETS:** The electrical contractor shall verify location, heights, outlet and switch arrangements, and equipment prior to rough-in. No additions to the contract sum will be permitted for outlets in wrong locations, in conflict with other work, and so on. The owner reserves the right to relocate any device up to 10 feet (3.05 m) prior to rough-in, without any charge by the electrical contractor.

5. **CODES:** The electrical installation is to be in accordance with the latest edition of the *National Electrical Code,* all local electrical codes, and the utility company's requirements.

6. **MATERIALS:** All materials shall be new and shall be listed and bear the appropriate label of Underwriters Laboratories, Inc., or another nationally recognized testing laboratory for the specific purpose. The material shall be of the size and type specified on the drawings and/or in the specifications.

7. **WIRING METHOD:** Wiring, unless otherwise specified, shall be nonmetallic-sheathed cable, armored cable, or electrical metallic tubing, adequately sized and installed according to the latest edition of the *National Electrical Code®* and local ordinances.

8. **PERMITS AND INSPECTION FEES:** The electrical contractor shall pay for all permit fees, plan review fees, license fees, inspection fees, and taxes applicable to the electrical installation and shall be included in the base bid as part of this contract.

9. **TEMPORARY WIRING:** The electrical contractor shall furnish and install all temporary wiring for handheld tools and construction lighting per latest OSHA standards and *Article 305, NEC®* and include all cost in base bid.

10. **WORKSHOP:** Workshop wiring is to be installed in electrical metallic tubing using steel compression gland fittings.

11. **NUMBER OF OUTLETS PER CIRCUIT:** In general, not more than (10) lighting and/or receptacle outlets shall be connected to any one lighting branch-circuit. Exceptions may be made in the case of low-current-consuming outlets.

12. **CONDUCTOR SIZE:** General lighting branch-circuits shall be No. 14 AWG copper protected by 15-ampere overcurrent devices. Small appliance circuits shall be No. 12 AWG copper protected by 20-ampere overcurrent devices.

   All other circuits: wire and overcurrent device as required by the Code.

13. **LOAD BALANCING:** The electrical contractor shall connect all loads, branch-circuits, and feeders per Panel Schedule, but shall verify and modify these connections as required to balance connected and computed loads to within 10% variation.

14. **SPARE CONDUITS:** Furnish and install two empty ½-inch thinwall (EMT) conduits between workshop and attic for future use.

15. **GUARANTEE OF INSTALLATION:** The electrical contractor shall guarantee all work and materials for a period of one full year after final acceptance by the architect/engineer, electrical inspector, and owner.

16. **APPLIANCE CONNECTIONS:** The electrical contractor shall furnish all wiring materials and make all final electrical connections for all permanently installed appliances such as, but not limited to, furnace, water heater, water pump, built-in ovens and ranges, food waste disposer, dishwasher, and clothes dryer.

   These appliances are to be furnished by owner.

17. **CHIMES:** Furnish and install two (2) two-tone door chimes where indicated on the plans, complete with two (2) push buttons and suitable chime transformer. Allow $150.00 for above items. Chimes and buttons to be selected by owner.

18. **DIMMERS:** Furnish and install dimmer switches where indicated.

19. **EXHAUST FANS:** Furnish, install, and provide connections for all exhaust fans indicated on the plans, including, but not limited to, ducts, louvers, trims, speed controls, and lamps. Included are recreation room, laundry, rear entry powder room, range hood, and bedroom hall ceiling fan. Allow a sum of $700.00 in base bid for this. This allowance does not include the two bathroom heat/vent/light fixtures.

20. **FIXTURES:** A fixture allowance of $1,500.00 shall be included in the electrical contractor's bid for all surface-mounted fixtures and post lamp, not including the ceiling paddle fans. Fixtures to be selected by owner. This allowance includes the three (3) medicine cabinets.

   The ceiling paddle fans, speed controls, and lamps for same shall be furnished by the owner and installed by the contractor.

   The electrical contractor shall:
   a. furnish and install all incandescent and fluorescent recessed fixtures.
   b. install all surface, recessed, track, strip, pendant, hanging fixtures, and post lamp.
   c. furnish and install two 2-lamp fluorescent fixtures above the workbench.
   d. furnish and install all lamps except as noted. Fluorescent lamps shall be Watt-Mizer F40 SP30/RS/WM or equivalent. Incandescent lamps shall be energy-efficient Watt-Mizer lamps or equivalent. Lamps for ceiling paddle fans to be furnished by owner.
   e. furnish and install all porcelain pull-chain and keyless lampholders.

   This fixture allowance does not include the two bathroom ceiling heat/vent/light fixtures. See Clause 21.

21. **HEAT/VENT/LIGHT CEILING FIXTURES:** Furnish and install two heat/vent/light units where indicated on the plans complete with switch assembly, ducts, louvers, required to perform the heating, venting, and lighting operations as recommended by the manufacturer.

22. **PLUG-IN STRIP:** Where noted in the workshop, furnish and install multioutlet assembly with six outlets.

23. **SWITCHES, RECEPTACLES, AND FACEPLATES:** All flush switches shall be of the quiet ac-rated toggle type. They shall be mounted 46 inches (1.17 m) to center above the finished floor unless otherwise noted.

    Receptacle outlets shall be mounted 12 inches (305 mm) to center above the finished floor unless otherwise noted. All convenience receptacles shall be of the grounding type. Furnish and install where indicated, ground-fault circuit interrupter receptacles to provide ground-fault circuit protection as required by the *National Electrical Code.* All wiring devices are to be provided with ivory handles or faces and shall be trimmed with ivory faceplates except in the kitchen, where chrome-plated steel faceplates shall be used.

    Receptacle outlets, where indicated, shall be of the split-circuit design.

24. **TELEVISION OUTLETS:** Furnish and install 4-inch square, 1½-inch-deep outlet boxes with single-gang raised plaster covers at each television outlet where noted on the plans. Mount at the same height as receptacle outlets. Furnish and install 75-ohm coaxial cable to each television outlet from a point in the workshop near the main service-entrance switch. Allow 6 feet (1.83 m) of cable in workshop. Furnish and install television plug-in jacks at each location. Faceplates are to match other faceplates in home. All remaining work done by others.

25. **TELEPHONES:** Furnish and install a 3-inch-deep device box or 4-inch square box, 1½ inches deep, with suitable single-gang raised plaster cover at each telephone location, as indicated on the plans.

    Furnish and install four-conductor No. 18 AWG copper telephone cables from the telephone company's point of demarcation near service-entrance panel to each designated telephone location. Terminate in proper modular jack, complete with faceplates that match the electrical device faceplates. Allow 6 feet (1.83 m) of cable to hang below ceiling joists. Telephone company to furnish, install, and connect any and all equipment (including grounding connection) up to and including their Standard Network Interface device.

    Installation shall be in accordance with any and all applicable *National Electrical Code* and local code regulations.

26. **MAIN SERVICE PANEL:** Furnish and install in the Workshop where indicated on the plans one (1) 200-ampere, 120/240-volt, single-phase, 3-wire load center with a 200-ampere main circuit breaker. Load center shall be series-rated for 22,000 amperes available fault current.

    Furnish and install all active and spare breakers as indicated on Panel Schedule.

27. **SERVICE-ENTRANCE UNDERGROUND LATERAL CONDUCTORS:** To be furnished and installed by utility. Meter equipment (pedestal type) to be furnished by utility and installed by electrical contractor where indicated on plans. Electrical contractor to furnish and install all panels, conduits, fittings, conductors, and other materials required to complete the service-entrance installation from the demarcation point of the utility's equipment to and including the Main Panel.

28. **SERVICE-ENTRANCE CONDUCTORS:** Service-entrance conductors supplied by the electrical contractor shall be two (2) No. 2/0 THW, THHW, THWN, or XHHW phase conductors and one (1) No. 1 bare neutral conductor. Install 1½-inch electrical metallic tubing from Main Panel A to meter pedestal.

29. **BONDING AND GROUNDING:** Bond and ground service-entrance equipment in accordance with latest edition of the *National Electrical Code,* local, and utility code requirements. Install No. 4 AWG copper grounding electrode conductor.

30. **SUBPANEL:** Furnish and install in Recreation Room where indicated on the plans one (1) 20-circuit, 125-ampere, 120/240-volt, single-phase, 3-wire MLO load center. Load center shall be series-rated for 22,000 amperes available fault current.

   Furnish and install all active and spare breakers as indicated on Panel Schedule.

   Feed load center with three (3) No. 3 THHN or THWN conductors from 100-ampere, 2-pole breaker in Main Panel A. Install conductors in 1-inch electrical metallic tubing.

31. **CIRCUIT IDENTIFICATION:** All panelboards shall be furnished with typed-card directories with proper designation of the branch-circuit loads, feeder loads and equipment served. The directories shall be located in the panel in a holder for clear viewing.

32. **CEILING PENETRATIONS:** The electrical contractor shall seal and weatherproof all penetrations through foundations, exterior walls, and roofs.

33. **FINAL SITE CLEAN UP:** Upon completion of the installation, the electrical contractor shall review and check the entire installation, clean equipment and devices, and remove surplus materials and rubbish from the owner's property, leaving the work in neat and clean order and in complete working condition. The electrical contractor shall be responsible for the removal of any cartons, debris, and rubbish for equipment installed by the electrical contractor, including equipment furnished by the owner or others and removed from the carton by the electrical contractor.

34. **SMOKE DETECTORS:** Furnish and install all smoke detectors and associated wiring per manufacturer's instructions and all codes. See Clause 5. Detectors to be of the AC/DC type. Interconnect to provide simultaneous signaling.

35. **ALTERNATIVE BID   Low-Voltage, Remote-Control System:** The electrical contractor shall submit an alternative bid on the following:

   Furnish and install a complete low-voltage remote-control system to accomplish the same results as would be obtained with the conventional switching arrangement as indicated on the electrical plans.

   In addition, furnish and install two 8-position master selector switches, one in the master bedroom and one in the front hall or as directed by the architect or owner. Outlets to be controlled by this switch to be selected by the owner.

   All low-voltage wiring to conform to the *National Electrical Code.*®

36. **SPECIAL-PURPOSE OUTLETS:** Install, provide, and connect all wiring for all special-purpose outlets. Upon completion of the job, all fixtures and appliances shall be operating properly. See plans and other sections of the specifications for information as to who is to furnish the fixtures and appliances.

SCHEDULE OF SPECIAL-PURPOSE OUTLETS

SYMBOL	DESCRIPTION	VOLTS	HORSE-POWER	APPLIANCE AMPERE RATING	TOTAL APPLIANCE WATTAGE RATING (OR VA)	CIRCUIT AMPERE RATING	POLES	WIRE SIZE THHN	CIRCUIT NUMBER	COMMENTS
A	Hydromassage tub, master bedroom	120	1/2	10	1200	15	1	14	A9	Connect to separate circuit Class "A" GFCI circuit breaker.
B	Water pump	240	1	8	1920	20	2	12	A(5–7)	Run circuit to disconnect switch on wall adjacent to pump. Install 10-ampere dual-element fuses as "back-up" protection. Pump controller has integral overload protection.
C	Water heater: top element 2000 W. Bottom element 3000 W.	240	–0–	12.5	3000	20	2	12	A(6–8)	Connected for limited demand. Both elements cannot operate at the same time.
D	Dryer	120/240	120 V ⅙ Motor Only	23.75	5700 Total	30	2	10	B(1–3)	Provide flush-mounted 30-A dryer receptacle.
E	Overhead garage door opener	120	¼	5.8	696	15	1	14	B14	Unit comes with 3 W cord. Provide box-cover unit (fuse/switch). Install Fustat Type S fuse, 8 amperes. Unit has integral protection. Fustat fuses are additional "back-up" protection. Connect to garage lighting circuit.
F	Wall-mounted oven	120/240	–0–	27.5	6600	30	2	10	B(6–8)	
G	Countertop range	120/240	–0–	31	7450	40	2	8	B(2–4)	
H	Food waste disposer	120	⅓	7.2	864	20	1	12	B19	Controlled by S.P. switch on wall to the right of sink.
I	Dishwasher	120	⅙ Motor Only	Motor 7.20 Htr. 8.33 Total 15.53	864 1000 W 1864	20	1	12	B5	
J	Heat/vent/light master B.R. bath	120	–0–	12.5	1500	20	1	12	A12	
K	Heat/vent/light front B.R. bath	120	–0–	12.5	1500	20	1	12	A11	
L	Attic exhaust fan	120	¼	5.8	696	15	1	14	A10	Run circuit to 4" square box. Locate near fan in attic. Provide box-cover unit. Install Fustat Type S fuse, 8 amperes. Unit has integral protection. Fustat fuses provide additional back-up protection.
M	Electric furnace	240	⅙ Motor Only	Motor 3.5 Htr. 50.7 Total 54.2	13000	70	2	4	A(1–3)	The overcurrent device and branch-circuit conductors shall not be less than 125% of the total load of the heaters and motor. 54.2 × 1.25 = 67.75 (Section 424-3(b)).
N	Air conditioner	240	–0–	26.4	6336	45	2	10	A(2–4)	
O	Freezer	120	¼	5.8	696	15	1	14	A13	Install single receptacle outlet. Do not provide GFCI protection.

USEFUL FORMULAS

TO FIND	SINGLE PHASE	THREE PHASE	DIRECT CURRENT
AMPERES when kVA is known	$\dfrac{kVA \times 1000}{E}$	$\dfrac{kVA \times 1000}{E \times 1.73}$	not applicable
AMPERES when horsepower is known	$\dfrac{HP \times 746}{E \times \% \text{ eff.} \times pf}$	$\dfrac{HP \times 746}{E \times 1.73 \times \% \text{ eff.} \times pf}$	$\dfrac{HP \times 746}{E \times \% \text{ eff.}}$
AMPERES when kilowatts are known	$\dfrac{kW \times 1000}{E \times pf}$	$\dfrac{kW \times 1000}{E \times 1.73 \times pf}$	$\dfrac{kW \times 1000}{E}$
HORSEPOWER	$\dfrac{I \times E \times \% \text{ eff.} \times pf}{746}$	$\dfrac{I \times E \; 1.73 \times \% \text{ eff.} \times pf}{746}$	$\dfrac{I \times E \times \% \text{ eff.}}{746}$
KILOVOLT AMPERES	$\dfrac{I \times E}{1000}$	$\dfrac{I \times E \times 1.73}{1000}$	not applicable
KILOWATTS	$\dfrac{I \times E \times pf}{1000}$	$\dfrac{I \times E \times 1.73 \times pf}{1000}$	$\dfrac{I \times E}{1000}$
WATTS	$E \times I \times pf$	$E \times I \times 1.73 \times pf$	$E \times I$

$$\text{ENERGY EFFICIENCY} = \frac{\text{Load Horsepower} \times 746}{\text{Load Input kVA} \times 1000}$$

$$\text{POWER FACTOR (pf)} = \frac{\text{Power Consumed}}{\text{Apparent Power}} = \frac{W}{VA} = \frac{kW}{kVA} = \cos\varnothing$$

I = Amperes	E = Volts	kW = Kilowatts	kVA = Kilovolt-amperes
HP = Horsepower	% eff. = Percent Efficiency e.g., 90% eff. is 0.90		pf = Power Factor e.g., 95% pf is 0.95

EQUATIONS BASED ON OHM'S LAW:

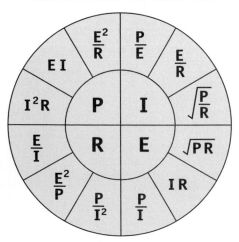

P = POWER, IN WATTS
I = CURRENT, IN AMPERES
R = RESISTANCE, IN OHMS
E = ELECTROMOTIVE FORCE,
IN VOLTS

METRIC SYSTEM OF MEASUREMENT

The following table provides useful conversions of English Customary terms to SI terms, and SI terms to English Customary terms. The metric system is known as the International System of Units (SI), taken from the French *Le Système International d'Unites*. Whenever the slant line (/) is found, say it as "per." Be practical when using the values in the following table. Present-day use of calculators and computers provide many "places" beyond the decimal point. You must decide how accurate your results must be when performing a calculation. When converting centimeters to feet, for example, one could be reasonably accurate by rounding the value of 0.03281 to 0.033. When a manufacturer describes a product's measurement using metrics, but does not physically change the product, it is called a *soft conversion*. When a manufacturer actually changes the physical measurements of the product to standard metric size or a rational whole number of metric units, it is called a *hard conversion*. When rounding off numbers, be sure to round off in such a manner that the final result does not violate a "maximum" or "minimum" value, such as might be required by the *National Electrical Code.*® The rule for discarding digits is that if the digits to be discarded begin with a 5 or more, increase the last digit retained by one. For example: If rounded to three digits, 6.3745 would become 6.37. If rounded to four digits, it would become 6.375. The following information has been developed from the latest U.S. Department of Commerce, National Institute of Standards and Technology publications covering the metric system.

Multiply This Unit(s)	By This Factor	To Obtain This Unit(s)
acre	4 046.9	square meters (m²)
acre	43 560	square feet (ft²)
ampere hour	3 600	coulombs (C)
angstrom	0.1	nanometer (nm)
atmosphere	101.325	kilopascals (kPa)
atmosphere	33.9	feet of water (at 4°C)
atmosphere	29.92	inches of mercury (at 0°C)
atmosphere	0.76	meter of mercury (at 0°C)
atmosphere	0.007 348	ton per square inch
atmosphere	1.058	tons per square foot
atmosphere	1.0333	kilograms per square centimeter
atmosphere	10 333	kilograms per square meter
atmosphere	14.7	pounds per square inch
bar	100	kilopascals (kPa)
barrel (oil, 42 U.S. gallons)	0.158 987 3	cubic meter (m³)
barrel (oil, 42 U.S. gallons)	158.987 3	liters (L)
board foot	0.002 359 737	cubic meter (m³)
bushel	0.035 239 07	cubic meter (m³)
Btu	778.16	foot-pounds
Btu	252	gram-calories
Btu	0.000 393 1	horsepower-hour
Btu	1 054.8	joules (J)
Btu	1.055 056	kilojoules (kJ)
Btu	0.000 293 1	kilowatt-hour (kWh)
Btu per hour	0.000 393 1	horsepower (hp)
Btu per hour	0.293 071 1	watt (W)
Btu per degree Fahrenheit	1.899 108	Kilojoules per kelvin (kJ/K)
Btu per pound	2.326	Kilojoules per kilogram (kJ/kg)
Btu per second	1.055 056	kilowatts (kW)
calorie	4.184	joules (J)
calorie, gram	0.003 968 3	Btu
candela per feet squared (cd/ft²)	10.763 9	candelas per meter squared (cd/m²)

Note: The former term *candlepower* has been replaced with the term candela.

candela per meter squared (cd/m²)	0.092 903	candela per feet squared (cd/ft²)
candela per meter squared (cd/m²)	0.291 864	footlambert*
candela per square inch (cd/in²)	1 550.003	candelas per meter squared (cd/m²)

Celsius = (Fahrenheit − 32) × 5/9
Celsius = (Fahrenheit − 32) × 0.555555
Celsius = (0.556 × Fahrenheit) − 17.8

Note: The term *centigrade* was officially discontinued in 1948, and was replaced by the term *Celsius*. The term *centigrade* may still be found in some publications.

Multiply This Unit(s)	By This Factor	To Obtain This Unit(s)
centimeter (cm)	0.032 81	foot (ft)
centimeter (cm)	0.393 7	inch (in)
centimeter (cm)	0.01	meter (m)
centimeter (cm)	10	millimeters (mm)
centimeter (cm)	393.7	mils
centimeter (cm)	0.010 94	yard
circular mil	0.000 005 067	square centimeter (cm^2)
circular mil	0.785 4	square mil
circular mil	0.000 000 785 4	square inch (in^2)
cubic centimeter (cm^3)	0.061 02	cubic inch (in^3)
cubic foot per second	0.028 316 85	cubic meter per second (m^3/s)
cubic foot per minute	0.000 471 947	cubic meter per second (m^3/s)
cubic foot per minute	0.471 947	liter per second (L/s)
cubic inch (in^3)	16.39	cubic centimeters (cm^3)
cubic inch (in^3)	0.000 578 7	cubic foot (ft^3)
cubic meter (m^3)	35.31	cubic feet (ft^3)
cubic yard per minute	12.742 58	liters per second (L/s)
cup (c)	0.236 56	liter (L)
decimeter	0.1	meter (m)
dekameter	10	meters (m)
Fahrenheit = (9/5 Celsius) + 32		
Fahrenheit = (Celsius × 1.8) + 32		
fathom	1.828 804	meters (m)
fathom	6.0	feet
foot	30.48	centimeters (cm)
foot	12	inches
foot	0.000 304 8	kilometer (km)
foot	0.304 8	meter (m)
foot	0.000 189 4	mile (statute)
foot	304.8	millimeters (mm)
foot	12 000	mils
foot	0.333 33	yard
foot, cubic (ft^3)	0.028 316 85	cubic meter (m^3)
foot, cubic (ft^3)	28.316 85	liters (L)
foot, board	0.002 359 737	cubic meter (m^3)
cubic feet per second (ft^3/s)	0.028 316 85	cubic meter per second (m^3/s)
cubic feet per minute (ft^3/min)	0.000 471 947	cubic meter per second (m^3/s)
cubic feet per minute (ft^3/min)	0.471 947	liter per second (L/s)
foot, square (ft^2)	0.092 903	square meter (m^2)
footcandle	10.763 91	lux (lx)
footlambert˙	3.426 259	candelas per square meter (cd/m^2)
foot of water	2.988 98	kilopascals (kPa)
foot pound	0.001 286	Btu
foot pound-force	1 055.06	joules (J)
foot pound-force per second	1.355 818	joules (J)
foot pound-force per second	1.355 818	watts (W)
foot per second	0.304 8	meter per second (m/s)
foot per second squared	0.304 8	meter per second squared (m/s^2)
gallon (U.S. liquid)	3.785 412	liters (L)
gallons per day	3.785 412	liters per day (L/d)
gallons per hour	1.051 50	milliliters per second (mL/s)
gallons per minute	0.063 090 2	liter per second (L/s)
gauss	6.452	lines per square inch
gauss	0.1	millitesla (mT)
gauss	0.000 000 064 52	weber per square inch
grain	64.798 91	milligrams (mg)
gram (g) (a little more than the weight of a paper clip)	0.035 274	ounce (avoirdupois)
gram (g)	0.002 204 6	pound (avoirdupois)
gram per meter (g/m)	3.547 99	pounds per mile (lb/mile)
grams per square meter (g/m^2)	0.003 277 06	ounces per square foot (oz/ft^2)
grams per square meter (g/m^2)	0.029 494	ounces per square yard (oz/yd^2)
gravity (standard acceleration)	9.806 65	meters per second squared (m/s^2)
quart (U.S. liquid)	0.946 352 9	liter (L)
horsepower (550 ft•lbf/s)	0.745 7	kilowatt (kW)
horsepower	745.7	watts (W)
horsepower hours	2.684 520	megajoules (MJ)
inch per second squared (in/s^2)	0.025 4	meter per second squared (m/s^2)

Multiply This Unit(s)	By This Factor	To Obtain This Unit(s)
inch	2.54	centimeters (cm)
inch	0.254	decimeter (dm)
inch	0.083 3	feet
inch	0.025 4	meter (m)
inch	25.4	millimeters (mm)
inch	1 000	mils
inch	0.027 78	yard
inch, cubic (in³)	16 387.1	cubic millimeters (mm³)
inch, cubic (in³)	16.387 06	cubic centimeters (cm³)
inch, cubic (in³)	645.16	square millimeters (mm²)
inches of mercury	3.386 38	kilopascals (kPa)
inches of mercury	0.033 42	atmosphere
inches of mercury	1.133	feet of water
inches of water	0.248 84	kilopascal (kPa)
inches of water	0.073 55	inch of mercury
joule (J)	0.737 562	foot pound-force (ft•lbf)
kilocandela per meter squared (kcd/m²)	0.314 159	lambert*
kilogram (kg)	2.204 62	pounds (avoirdupois)
kilogram (kg)	35.274	ounces (avoirdupois)
kilogram per meter (kg/m)	0.671 969	pound per foot (lb/ft)
kilogram per square meter (kg/m²)	0.204 816	pound per square foot (lb/ft²)
kilogram meter squared (kg•m²)	23.730 4	pounds foot squared (lb•ft²)
kilogram meter squared (kg•m²)	3 417.17	pounds inch squared (lb•in²)
kilogram per cubic meter (kg/m³)	0.062 428	pound per cubic foot (lb/ft³)
kilogram per cubic meter (kg/m³)	1.685 56	pound per cubic yard (lb/yd³)
kilogram per second (kg/s)	2.204 62	pounds per second (lb/s)
kilojoule (kJ)	0.947 817	Btu
kilometer (km)	1 000	meters (m)
kilometer (km)	0.621 371	mile (statute)
kilometer (km)	1 000 000	millimeters (mm)
kilometer (km)	1 093.6	yards
kilometer per hour (km/h)	0.621 371	mile per hour (mph)
kilometer squared (km²)	0.386 101	square mile (mile²)
kilopound-force per square inch	6.894 757	megapascals (MPa)
kilowatts (kW)	56.921	Btus per minute
kilowatts (kW)	1.341 02	horsepower (hp)
kilowatts (kW)	1 000	watts (W)
kilowatt-hour (kWh)	3 413	Btus
kilowatt-hour (kWh)	3.6	megajoules (MJ)
knots	1.852	kilometers per hour (km/h)
lamberts*	3 183.099	candelas per square meter (cd/m²)
lamberts*	3.183 01	kilocandelas per square meter (kcd/m²)
liter (L)	0.035 314 7	cubic foot (ft³)
liter (L)	0.264 172	gallon (U.S. liquid)
liter (L)	2.113	pints (U.S. liquid)
liter (L)	1.056 69	quarts (U.S. liquid)
liter per second (L/s)	2.118 88	cubic feet per minute (ft³/min)
liter per second (L/s)	15.850 3	gallons per minute (gal/min)
liter per second (L/s)	951.022	gallons per hour (gal/hr)
lumen per square foot (lm/ft²)	10.763 9	lux (lx)
lumen per square foot (lm/ft²)	1.0	footcandles
lumen per square foot (lm/ft²)	10.763	lumens per square meter (lm/m²)
lumen per square meter (lm/m²)	1.0	lux (lx)
lux (lx)	0.092 903	lumen per square foot (foot-candle)
maxwell	10	nanowebers (nWb)
megajoule (MJ)	0.277 778	kilowatt hours (kWh)
meter (m)	100	centimeters (cm)
meter (m)	0.546 81	fathom
meter (m)	3.2809	feet
meter (m)	39.37	inches
meter (m)	0.001	kilometer (km)
meter (m)	0.000 621 4	mile (statute)
meter (m)	1 000	millimeters (mm)
meter (m)	1.093 61	yards
meter, cubic (m³)	1.307 95	cubic yards (yd³)
meter, cubic (m³)	35.314 7	cubic feet (ft³)

Multiply This Unit(s)	By This Factor	To Obtain This Unit(s)
meter, cubic (m³)	423.776	board feet
meter per second (m/s)	3.280 84	feet per second (ft/s)
meter per second (m/s)	2.236 94	miles per hour (mph)
meter, squared (m²)	1.195 99	square yards (yd²)
meter, squared (m²)	10.763 9	square feet (ft²)
mho per centimeter (mho/cm)	100	siemens per meter (S/m)

Note: The older term *mho* has been replaced with *siemens*. The term *mho* may still be found in some publications.

micro inch	0.025 4	micrometer (μm)
mil	0.002 54	centimeters
mil	0.000 083 33	feet
mil	0.001	inches
mil	0.000 000 025 40	kilometers
mil	0.000 027 78	yards
mil	25.4	micrometers (μm)
mil	0.025 4	millimeter (mm)
miles per hour	1.609 344	kilometers per hour (km/h)
miles per hour	0.447 04	meter per second (m/s)
miles per gallon	0.425 143 7	kilometer per liter (km/L)
miles	1.609 344	kilometers (km)
miles	5 280	feet
miles	1 609	meters (m)
miles	1 760	yards
miles (nautical)	1.852	kilometers (km)
miles squared	2.590 000	kilometers squared (km²)
millibar	0.1	kilopascal (kPa)
milliliter (mL)	0.061 023 7	cubic inch (in³)
milliliter (mL)	0.033 814	fluid ounce (U.S.)
millimeter (about the thickness of a dime)	0.1	centimeter (cm)
millimeter (mm)	0.003 280 8	foot
millimeter (mm)	0.039 370 1	inch
millimeter (mm)	0.001	meter (m)
millimeter (mm)	39.37	mils
millimeter (mm)	0.001 094	yard
millimeter squared (mm²)	0.001 550	square inch (in²)
millimeter cubed (mm³)	0.000 061 023 7	cubic inches (in³)
millimeter of mercury	0.133 322 4	kilopascal (kPa)
ohm	0.000 001	megohm
ohm	1 000 000	micro ohms
ohm circular mil per foot	1.662 426	nano ohms meter (nΩ•m)
oersted	79.577 47	amperes per meter (A/m)
ounce (avoirdupois)	28.349 52	grams (g)
ounce (avoirdupois)	0.062 5	pound (avoirdupois)
ounce, fluid	29.573 53	milliliters (mL)
ounce (troy)	31.103 48	grams (g)
ounce per foot squared (oz/ft²)	305.152	grams per meter squared (g/m²)
ounces per gallon (U.S. liquid)	7.489 152	grams per liter (g/L)
ounce per yard squared (oz/yd²)	33.905 7	grams per meter squared (g/m²)
pica	4.217 5	millimeters (mm)
pint (U.S. liquid)	0.473 176 5	liter (L)
pint (U.S. liquid)	473.177	milliliters (mL)
pound (avoirdupois)	453.592	grams (g)
pound (avoirdupois)	0.453 592	kilograms (kg)
pound (avoirdupois)	16	ounces (avoirdupois)
poundal	0.138 255	newton (N)
pound foot (lb•ft)	0.138 255	kilogram meter (kg•m)
pound foot per second	0.138 255	kilogram meter per second (kg•m/s)
pound foot squared (lb•ft²)	0.042 140 1	kilogram meter squared (kg•m²)
pound-force	4.448 222	newtons (N)
pound-force foot	1.355 818	newton meters (N•m)
pound-force inch	0.112 984 8	newton meter (N•m)
pound-force per square inch	6.894 757	kilopascals (kPA)
pound-force per square foot	0.047 880 26	kilopascal (kPa)
pound per cubic foot (lb/ft³)	16.018 46	kilograms per cubic meter (kg/m³)
pound per foot (lb/ft)	1.488 16	kilograms per meter (kg/m)
pound per foot squared (lb/ft²)	4.882 43	kilograms per meter squared (kg/m²)

Multiply This Unit(s)	By This Factor	To Obtain This Unit(s)
pound per foot cubed (lb/ft³)	16.018 5	kilograms per meter cubed (kg/m³)
pound per gallon (U.S. liquid)	119.826 4	grams per liter (g/L)
pound per second (lb/s)	0.453 592	kilogram per second (kg/s)
pound inch squared (lb•in²)	292.640	kilograms millimeter squared (kg•mm²)
pound per mile	0.281 849	gram per meter (g/m)
pound square foot (lb•ft²)	0.042 140 11	kilogram square meter (kg/m²)
pound per cubic yard (lb/yd³)	0.593 276	kilogram per cubic meter (kg/m³)
quart (U.S. liquid)	946.353	milliliters (mL)
square centimeter (cm²)	197 300	circular mils
square centimeter (cm²)	0.001 076	square foot (ft²)
square centimeter (cm²)	0.155	square inch (in²)
square centimeter (cm²)	0.000 1	square meter (m²)
square centimeter (cm²)	0.000 119 6	square yard (yd²)
square foot (ft²)	144	square inches (in²)
square foot (ft²)	0.092 903 04	square meter (m²)
square foot (ft²)	0.111 1	square yard (yd²)
square inch (in²)	1 273 000	circular mils
square inch (in²)	6.4516	square centimeters (cm²)
square inch (in²)	0.006 944	square foot (ft²)
square inch (in²)	645.16	square millimeters (mm²)
square inch (in²)	1 000 000	square mils (m²)
square meter (m²)	10.764	square feet (ft²)
square meter (m²)	1 550	square inches (in²)
square meter (m²)	0.000 000 386 1	square mile
square meter (m²)	1.196	square yards (yd²)
square mil	1.273	circular mils
square mil	0.000 001	square inch (in²)
square mile	2.589 988	square kilometers (km²)
square millimeter (mm²)	1 973	circular mils
square yard (yd²)	0.836 127 4	square meter (m²)
tablespoon (tbsp)	14.786 75	milliliters (mL)
teaspoon (tsp)	4.928 916 7	milliliters (mL)
therm	105.480 4	megajoules (MJ)
ton (long) (2 240 lb)	1 016.047	kilograms (kg)
ton (long) (2 240 lb)	1.016 047	metric tons (t)
ton, metric	2 204.62	pounds (avoirdupois)
ton, metric	1.102 31	tons, short (2 000 lb)
ton, refrigeration	12 000	Btus per hour
ton, refrigeration	4.716 095 9	horsepower-hours
ton, refrigeration	3.516 85	kilowatts (kW)
ton (short) (2 000 lb)	907.185	kilograms (kg)
ton (short) (2 000 lb)	0.907 185	metric ton (t)
ton per cubic meter (t/m³)	0.842 778	ton per cubic yard (ton/yd³)
ton per cubic yard (ton•yd³)	1.186 55	tons per cubic meter (ton/m³)
torr	133.322 4	pascals (Pa)
watt (W)	3.412 14	Btus per hour (Btu/hr)
watt (W)	0.001 341	horsepower
watt (W)	0.001	kilowatt (kW)
watt-hour (Wh)	3.413	Btus
watt-hour (Wh)	0.001 341	horsepower-hour
watt-hour (Wh)	0.001	kilowatt-hour (kW/h)
yard	91.44	centimeters (cm)
yard	3	feet
yard	36	inches
yard	0.000 914 4	kilometer (km)
yard	0.914 4	meter (m)
yard	914.4	millimeters
yard, cubic (yd³)	0.764 55	cubic meter (m³)
yard, squared (yd²)	0.836 127	meter squared (m²)

* These terms are no longer used, but may still be found in some publications.

SINGLE-FAMILY DWELLING SERVICE-ENTRANCE CALCULATIONS

1. **General Lighting Load** *(Section 220-3(a))*.

   _____ sq. ft. @ 3 VA per sq. ft.   = _____ VA

2. **Minimum Number of Lighting Branch-Circuits**

   $$\frac{Line\ 1}{120} = \frac{\rule{3cm}{0.4pt}}{120} \qquad\qquad = \rule{3cm}{0.4pt}\ amperes$$

   then, $\dfrac{amperes}{15} = \dfrac{\rule{2cm}{0.4pt}}{15} = \rule{3cm}{0.4pt}$ 15-ampere branch-circuits

3. **Small Appliance Load** *(Sections 210-11(c)(1) and 220-16(a))*.
   (Minimum of two 20-ampere branch-circuits)

   _____ branch-circuits @ 1,500 VA each   = _____ VA

4. **Laundry Branch-Circuit** *(Sections 210-11(c)(2) and 220-16(b))*.
   (Minimum of one 20-ampere branch-circuit)

   _____ branch-circuit(s) @ 1,500 VA each   = _____ VA

5. **Total General Lighting, Small Appliance, and Laundry Load.**

   Lines 1 + 3 + 4   = _____ VA

6. **Net Computed General Lighting, Small Appliance, and Laundry Loads (less ranges, ovens, and "fastened-in-place" appliances). Apply Demand Factors from** *Table 220-11.*

   a. First 3,000 VA @ 100%   = ____3,000____ VA

   b. Line 5 _____ − 3,000 = _____ @ 35% = _____ VA

   Total a + b = _____ VA

7. **Electric Range, Wall-Mounted Ovens, Counter-Mounted Cooking Units** *(Table 220-19).*   = _____ VA

8. **Electric Clothes Dryer** *(Table 220-18).*   = _____ VA

9. **Electric Furnace** *(Section 220-15).*
   Air Conditioner, Heat Pump *(Article 440).*
   (Enter largest value, *Section 220-21*)   = _____ VA

10. **Net Computed General Lighting, Small Appliance, Laundry, Ranges, Ovens, Cooktop Units, HVAC.**

    Lines 6 + 7 + 8 + 9   = _____ VA

11. **List "Fastened-in-Place" Appliances** *in addition* **to Electric Ranges, Electric Clothes Dryers, Electric Space Heating, and Air-Conditioning Equipment.**

Appliance		VA Load	
Water heater:	=	_____	VA
Dishwasher:	=	_____	VA

    Motor: _____ VA

    Heater: _____ VA (watts)

    Enter largest value, *Section 220-21.*

Garage door opener:	=	_____	VA
Food waste disposer:	=	_____	VA
Water pump:	=	_____	VA
Gas-fired furnace:	=	_____	VA
Sump pump:	=	_____	VA
Other: _____	=	_____	VA
_____	=	_____	VA
_____	=	_____	VA
_____	=	_____	VA
Total	=	_____	VA

12. **Apply 75% Demand Factor** *(Section 220-17)* **if Four or More "Fastened-in-Place" Appliances. If Less Than Four, Figure @ 100%.** Do not include electric ranges, electric clothes dryers, electric space heating, or air-conditioning equipment.

    Line 11 Total: _____ × 0.75   = _____ VA

13. **Total Computed Load (Lighting, Small Appliance, Ranges, Dryer, HVAC, "Fastened-in-Place" Appliances).**

    Line 10 _____ + Line 12 _____   = _____ VA

14. **Add 25% of Largest Motor** *(Section 220-14).*

    _____ × 0.25   = _____ VA

15. **Grand Total Line 13 + Line 14.**   = _____ VA

16. **Minimum Ampacity for Ungrounded Service-Entrance Conductors.**

$$\text{Amperes} = \frac{\text{Line 15}}{240} = \frac{\rule{2cm}{0.4pt}}{240}$$ = _____ amperes

17. **Minimum Ampacity for Neutral Service-Entrance Conductor,** *Sections 220-22* **and** *310-15(b)(6).*
    Do Not Include Straight 240-Volt Loads.

    a. Line 6:   = _____ VA

    b. Line 7: _____ @ 0.70   = _____ VA

    c. Line 8: _____ @ 0.70   = _____ VA

    d. Line 11: (Include only 120-volt loads.)

    _____   _____ VA

    _____   _____ VA

    _____   _____ VA

    _____   _____ VA

    _____   _____ VA

    Total   _____ VA

    e. Line d total @ 75% Demand Factor if four or more, otherwise use 100%.

    _____ × 0.75   = _____ VA

    f. Add 25% of largest 120-volt motor.

    _____ × 0.25   = _____ VA

    Total = _____ VA

    g. Total a + b + c + e + f.   = _____ VA

    $$\text{Amperes} = \frac{\text{Line g}}{240} = \frac{\quad\quad\quad}{240}$$ = _____ amperes

18. **Ungrounded Conductor Size (copper).**   No. _____

    Note: *Table 310-15(b)(6)* may be used only for 120/240-volt, 3-wire, residential
    single-phase service-entrance conductors, service lateral conductors, and feeder
    conductors that serve as the main power feeder to a dwelling unit.

19. **Neutral Conductor Size (copper)** *(Section 220-22).*   No. _____

    Note: *Section 310-15(b)(6)* permits the neutral conductor to be smaller than the
    ungrounded "hot" conductors if the requirement of *Sections 215-2, 220-22,* and
    230-42 are met. When bare conductors are used with insulated conductors, the
    conductors' ampacity is based on the ampacity of the other insulated conductors
    in the raceway, *Section 310-15(b)(3).* The neutral conductor must not be smaller
    than the grounding electrode conductor, *Section 250-24(b)(1).*

20. **Grounding Electrode Conductor Size (copper)** *(Table 250-66).* No. _____

21. **Raceway Size.**   _____ inch

    Obtain dimensional data from *Tables 1, 4, 5,* and *8, Chapter 9, NEC.*

MEMBERSHIP APPLICATION

for
Inspector Member
Associate Member
Student Member

Annual Membership Dues $40
Students $30

Services from the IAEI for you and your organization include all this and more:

- Subscription to *IAEI News* - Bimonthly magazine with latest news on Code changes
- Local Chapter or Division meetings - New products, programs, national experts answer questions on local problems
- Section meetings to promote education and cooperation
- Promote advancement of electrical industry
- Save TIME and MONEY by having the latest information

International Association of Electrical Inspectors' Major Objectives

- Formulation of standards for safe installation and use of electrical materials, devices and appliances.
- Promotion of uniform understanding and application of the National Electrical Code®, other codes, and any adopted electrical codes in other countries.
- Promotion of uniform administrative ordinances and inspection methods.
- Collection and dissemination of information relative to the safe use of electricity.
- Representation of electrical inspectors in the electrical industry, nationally and internationally.
- Cooperation with national and international organizations in the further development of the electrical industry.
- Promotion of cooperation among inspectors, inspection departments of city, county and state at national and international levels, the electrical industry and the public.

Contact IAEI customer service department for information on our special membership categories - Section, National and International Member; Sustaining Member (Bronze, Silver, Gold or Platinum); and Inspection Agency Member.

PLEASE PRINT

| Name - Last | First | M.I. | Chapter, where you live or work, if known |

| Title | | (Division, where appropriate) |

| Employer | | For Office Use Only | Section No. | Chapter No. | Division No. |

| Address of Applicant | | If previous member, give last membership number |

| City | State or Province | Zip or Postal Code | and last year of membership 19____ |

| Student applicants give school attending ** | | Graduation Date / / |

Endorsed by: R A Y C M U L L I N

Endorser's Membership Number: 5 3 3 8 0 0

Applicant's Signature

(Area Code) Telephone Number - -

☐ Check ☐ Money Order ☐ Visa Name on Card Charge Card Number Exp. Date
☐ MasterCard ☐ Amex ☐ Discover

☐ Inspector $40 ☐ Chief Electrical Inspector * $40 ☐ Associate $40 ☐ Student ** $30

Inspector Members and Chief Electrical Inspector Members MUST sign below:

☐ I, _____ , meet the qualifications for inspector member type as described below.

Inspector members must regularly make electrical inspections for preventing injury to persons or damage to property on behalf of a governmental agency, insurance agency, rating bureau, recognized testing laboratory or electric light and power company.

*Chief electrical inspector members must regularly provide technical supervision of electrical inspectors.

**Student member must be currently enrolled in an approved college, university, vocational technical school or trade school specializing in electrical training or approved electrical apprenticeship school.

Mail to: IAEI, P. O. Box 830848 Richardson, TX 75083-0848   For information call: (972) 235-1455 (8-5 CT)

The following is a list of terms used extensively in the electrical trades. These terms are unique to the electrical industry. Words marked with an * are defined in the *National Electrical Code*.®

Accessible (wiring methods)*:   Capable of being removed or exposed without damaging the building structure or finish, or not permanently closed in by the structure or finish of the building.

Accessible (equipment)*:   Admitting close approach: not guarded by locked doors, elevation, or other effective means.

Accessible, Readily*:   Capable of being reached quickly for operation, renewal, or inspections, without requiring those to whom ready access is requisite to climb over or remove obstacles or to resort to portable ladders, chairs, etc.

Addendum: Modification (change) made to the construction documents (plans and specifications) during the bidding period.

Adjustment Factor: A multiplier (penalty) that is applied to conductors when there are more than three current-carrying conductors in a raceway or cable. These multipliers are found in *Table 310-15(b)(2)*. An *adjustment factor* is also referred to as a *derating factor*.

AL/CU: Terminal marking on switches and receptacles rated 30 amperes and greater suitable for use with aluminum, copper, and copper-clad aluminum conductors. If not marked, suitable for copper conductors only.

Ambient Temperature: The environmental temperature surrounding the object under consideration.

American National Standards Institute (ANSI):   An organization that identifies industrial and public requirements for national consensus standards and coordinates and manages their development, resolves national standards problems, and ensures effective participation in international standardization. ANSI does not itself develop standards. Rather, it facilitates development by establishing consensus among qualified groups. ANSI ensures that the guiding principles—consensus, due process, and openness are followed.

Ampacity*:   The current in amperes that a conductor can carry continuously under the conditions of use without exceeding its temperature rating.

Ampere: The measurement of intensity of rate of flow of electrons in an electric circuit. An ampere is the amount of current that will flow through a resistance of one ohm under a pressure of one volt.

Appliance*:   Utilization equipment, generally other than industrial, normally built in standardized sizes or types, that is installed or connected as a unit to perform one or more functions such as clothes washing, air conditioning, food mixing, deep frying, etc.

Approved*:   Acceptable to the authority having jurisdiction.

Arc-Fault Circuit Interrupter*:   A device intended to provide protection from the effects of arc faults by recognizing characteristics unique to arcing and by functioning to de-energize a circuit when an arcing fault is detected.

As-Built Plans: When required, a modified set of working drawings that are prepared for a construction project that includes all variances from the original working drawings that occurred during the project construction.

Authority Having Jurisdiction: A governmental body that has legal jurisdiction over electrical installations. The Authority Having Jurisdiction has the responsibility for making interpretations of the rules, for deciding upon the approval of equipment and materials, and for granting the special permission where required. See *Section 90-4* of the *NEC*.®

AWG: The American Wire Gage. Previously known as the Brown & Sharpe Gage. The smaller the AWG number, the larger the conductor. There are 40 electrical conductor sizes from No. 36 through 0000. Starting from the smallest AWG size (No. 36), each successive size is approximately 1.26 times larger than the previous AWG size. *Table 8* of *Chapter 9* in the *NEC*® lists conductor sizes commonly found in the electrical industry. Conductor sizes through 0000 are shown using the AWG designation. Conductors larger than 0000 are shown in circular mil area. The 1.26 relationship between sizes is easily confirmed in *Table 8* by checking conductor sizes No. 4 through 0000.

Ballast: Used for energizing fluorescent lamps. It is constructed of a laminated core and coil windings or solid-state electronic components.

Bonding*: The permanent joining of metallic parts to form an electrically conductive path that will ensure electrical continuity and the capacity to conduct safely any current likely to be imposed.

Bonding Jumper*: A reliable conductor to ensure the required electrical conductivity between metal parts required to be electrically connected.

Bonding Jumper, Circuit*: The connection between portions of a conductor in a circuit to maintain required ampacity of the circuit.

Bonding Jumper, Equipment*: The connection between two or more portions of the equipment grounding conductor.

Bonding Jumper, Main*: The connection between the grounded circuit conductor and the equipment grounding conductor at the service.

Branch-Circuit*: The circuit conductors between the final overcurrent device protecting the circuit and the outlet(s).

Branch-Circuit, Appliance*: A branch-circuit supplying energy to one or more outlets to which appliances are to be connected; such circuits are to have no permanently connected lighting fixtures not a part of an appliance.

Branch-Circuit, General Purpose*: A branch-circuit that supplies a number of outlets for lighting and appliances.

Branch-Circuit, Individual*: A branch-circuit that supplies only one utilization equipment.

Branch-Circuit, Multiwire*: A branch-circuit consisting of two or more ungrounded conductors having a potential difference between them, and a grounded conductor having equal potential difference between it and each ungrounded conductor of the circuit and that is connected to the neutral or grounded conductor of the system.

Branch-Circuit Selection Current*: Branch-circuit selection current is the value in amperes to be used instead of the rated-load current in determining the ratings of motor branch-circuit conductors, disconnecting means, controllers, and branch-circuit short-circuit and ground-fault protective devices wherever the running overload protective device permits a sustained current greater than the specified percentage of the rated-load current. The value of branch-circuit selection current will always be equal to or greater than the marked rated-load current.

Circuit Breaker*: A device designed to open and close a circuit by nonautomatic means and to open the circuit automatically on a predetermined overcurrent without damage to itself when properly applied within its rating.

CC: A marking on wire connectors and soldering lugs indicating they are suitable for use with copper-clad aluminum conductors only.

CC/CU: A marking on wire connectors and soldering lugs indicating they are suitable for use with copper or copper-clad aluminum conductors only.

CU: A marking on wire connectors and soldering lugs indicating they are suitable for use with copper conductors only.

CO/ALR: Terminal marking on switches and receptacles rated 15 and 20 amperes that are suitable for use with aluminum, copper, and copper-clad aluminum conductors. If not marked, they are suitable for copper or copper-clad conductors only.

Conductor*:

Bare: A conductor having no covering or electrical insulation whatsoever. (See "Conductor, Covered.")

Covered: A conductor encased within material of composition or thickness that is not recognized by this Code as electrical insulation. (See "Conductor, Bare.")

Insulated: A conductor encased within material of composition and thickness that is recognized by this Code as electrical insulation.

Continuous Load*: A load where the maximum current is expected to continue for three hours or more.

Contract Modifications: After the agreements has been signed, any additions, deletions, or modifications of the work to be done are accomplished by change order, supplemental instruction, and field order. They can be issued at any time during the contract period.

Correction Factor: A multiplier (penalty) that is applied to conductors when the ambient temperature is greater than 30°C (86°F) . These multipliers are found at the bottom of *Table 310-16* of the *NEC.*®

CSA: This is the Canadian Standards Association that develops safety and performance standards in Canada for electrical products similar to but not always identical to those of UL (Underwriters Laboratories) in the United States.

Current: The flow of electrons through an electrical circuit, measured in amperes.

Derating Factor: A multiplier (penalty) that is applied to conductors when there are more than three current-carrying conductors in a raceway or cable. These multipliers are found in *Table 310-15(b)(2)*. A *derating factor* is also referred to as an *adjustment factor*.

Details: Plans, elevations, or sections that provide more specific information about a portion of a project component or element than smaller scale drawings.

Diagrams: Nonscaled views showing arrangements of special system components and connections not possible to clearly show in scaled views. A *schematic diagram* shows circuit components and their electrical connections without regard to actual physical locations. A *wiring diagram* shows circuit components and the actual electrical connections.

Dimmer: A switch with components that permits variable control of lighting intensity. Some dimmers have electronic components; others have core and coil (transformer) components.

Disconnecting Means*: A device, or group of devices, or other means by which the conductors of a circuit can be disconnected from their source of supply.

Drawings: A graphic representation of the work to be done. They show the relationship of the materials to each other, including sizes, shapes, locations, and connections. The drawings may include schematic diagrams showing such things as mechanical and electrical systems. They may also include schedules of structural elements, equipment, finishes, and other similar items.

Dry Niche Lighting Fixture*: A lighting fixture intended for installation in the wall of a pool or fountain in a niche that is sealed against the entry of pool water.

Dwelling Unit*: One or more rooms for the use of one or more persons as a housekeeping unit with space for eating, living, and sleeping, and permanent provisions for cooking and sanitation.

Elevations: Views of vertical planes, showing components in their vertical relationship, viewed perpendicularly from a selected vertical plane.

Feeder*: All circuit conductors between the service equipment or the source of a separately derived system and the final branch-circuit overcurrent device.

Fire Rating: The classification indicating in time (hours) the ability of a structure or component to withstand fire conditions.

Fine Print Note (FPN)*: Explanatory material is in the form of Fine Print Notes (FPN). Fine Print Notes are informational only, and are not enforceable as a requirement by the Code.

Flame Detector: A radiant energy-sensing fire detector that detects the radiant energy emitted by a flame.

Fully Rated System: Panelboards are marked with their short circuit rating in RMS symmetrical amperes. In a fully-rated system, the panelboard short circuit current rating will be equal to the lowest interrupting rating of any branch-circuit breaker or fuse installed. All devices installed shall have an interrupting rating greater than or equal to the specified available fault current. (See "Series-Rated System").

Fuse: An overcurrent protective device with a fusible link that operates and opens the circuit on an overcurrent condition.

Ground-Fault Protection of Equipment (GFPE)*: A system intended to provide protection of equipment from damaging line-to-ground fault currents by operating to cause a disconnecting means to open all ungrounded conductors of the faulted circuit. This protection is provided at current levels less than those required to protect conductors from damage through the operation of a supply circuit overcurrent device.

Ground*: A conducting connection, whether intentional or accidental, between an electrical circuit or equipment and the earth, or to some conducting body that serves in place of the earth.

Ground-Fault: A condition where an ungrounded conductor comes in contact with a grounded object such as but not limited to a grounded metal water pipe. A ground-fault is often referred to as a "line-to-ground" fault. See *Section 230-95* of the *NEC.*®

Grounded*: Connected to earth or to some conducting body that serves in place of the earth.

Grounded Conductor*: A system or circuit conductor that is intentionally grounded.

Grounding Conductor*: A conductor used to connect equipment or the grounded circuit of a wiring system to a grounding electrode or electrodes.

Grounded, Effectively*: Intentionally connected to earth through a ground connection or connections of sufficiently low impedance and having sufficient current-carrying capacity to prevent the buildup of voltages that may result in undue hazards to connected equipment or to persons.

Ground-Fault Circuit-Interrupter (GFCI)*: A device intended for the protection of personnel that functions to de-energize a circuit or portion thereof within an established period of time when a current to ground exceeds some predetermined value that is less than that required to operate the overcurrent protective device of the supply circuit.

Grounding Conductor, Equipment*: The conductor used to connect the noncurrent-carrying metal parts of equipment, raceways, and other enclosures to the system grounded conductor, the grounding electrode conductor, or both, at the service equipment or at the source of a separately derived system.

Grounding Electrode Conductor*: The conductor used to connect the grounding electrode to the equipment grounding conductor, to the grounded conductor, or to both, of the circuit at the service equipment or at the source of a separately derived system.

HACR: A circuit breaker that has been tested and found suitable for use on heating, air-conditioning, and refrigeration equipment. The circuit breaker is marked with the letters HACR. Equipment that is suitable for use with HACR circuit breakers will be marked for use with HACR-type circuit breakers.

Heat Alarm: A single or multiple station alarm responsive to heat.

Hermetic Refrigerant Motor-Compressor: A combination consisting of a compressor and motor, both of which are enclosed in the same housing, with no external shaft or shaft seals; the motor operating in the refrigerant.

Household Fire Alarm System: A system of devices that produces an alarm signal in the household for the purpose of notifying the occupants of the presence of a fire so that they will evacuate the premises.

Identified*: (as applied to equipment) Recognizable as suitable for the specific purpose, function, use, environment, application, etc., where described in a particular Code requirement.

Identified* (conductor): The identified conductor is the insulated grounded conductor.

- For sizes No. 6 or smaller, the insulated grounded conductor shall be identified by a continuous white or natural gray outer finish or by three continuous white stripes on other than green insulation along its entire length.

- For sizes larger than No. 6, the insulated grounded conductor shall be identified either by a continuous white or natural gray outer finish or by three continuous white stripes on other than green insulation along its entire length, or at the time of installation by a distinctive white marking at its terminations.

Identified* (terminal): The identification of terminals to which a grounded conductor is to be connected shall be substantially white in color. The identification of other terminals shall be of a readily distinguishable different color.

IEC: The International Electrotechnical Commission is a worldwide standards organization. These standards differ from those of Underwriters Laboratories. Some electrical equipment might conform to a specific IEC Standard, but may not conform to the U.L. Standard for the same item.

Immersion Detection Circuit Interrupter: A device integral with grooming appliances that will shut off the appliance when the appliance is dropped in water.

In Sight: Where this Code specifies that one equipment shall be "in sight from," "within sight from," or "within sight," etc., of another equipment, the specified equipment is to be visible and not more than 50 ft (15.24 m) from the other.

Inductive Load: A load that is made up of coiled or wound wire that creates a magnetic field when energized. Transformers, core and coil ballasts, motors, and solenoids are examples of inductive loads.

International Association of Electrical Inspectors: A not-for-profit and educational organization cooperating in the formulation and uniform application of standards for the safe installation and use of electricity, and collecting and disseminating information relative thereto. The IAEI is made up of electrical inspectors, electrical contractors, electrical apprentices, manufacturers, and governmental agencies.

Interrupting Rating*: The highest current at rated voltage that a device is intended to interrupt under standard test conditions.

Isolated Ground Receptacle: A grounding type device in which the equipment ground contact and terminal are electrically isolated from the receptacle mounting means.

Kilowatt: One thousand watts equals one kilowatt.

Kilowatt hour: One thousand watts of power consumed in one hour. Ten 100-watt lamps burning for ten hours is one kilowatt hour. Two 500-watt electric heaters operated for two hours is one kilowatt hour.

Labeled*: Equipment or materials to which has been attached a label, symbol, or other identifying mark of an organization that is acceptable to the authority having jurisdiction and concerned with product evaluation that maintains periodic inspection of production of labeled equipment or materials and by whose labeling the manufacturer indicates compliance with appropriate standards or performance in a specified manner.

Lighting Outlet*:   An outlet intended for the direct connection of a lampholder, a lighting fixture, or a pendant cord terminating in a lampholder.

Listed*:   Equipment, materials, or services included in a list published by an organization that is acceptable to the authority having jurisdiction and concerned with evaluation of products or services, that maintains periodic inspection of production of listed equipment or materials or periodic evaluation of services, and whose listing states that either the equipment, material, or service meets appropriate designated standards or has been tested and found suitable for use in a specified purpose.

Load: The electric power used by devices connected to an electrical system. Loads can be figured in amperes, volt-amperes, kilovolt-amperes, or kilowatts. Loads can be intermittent, continuous intermittent, periodic, short-time, or varying. See the definition of "Duty" in the *NEC*.

Load Center: A common name for residential panelboards. A load center may not be as deep as a panelboard, and generally does not contain relays or other accessories as are available for panelboards. Circuit breakers "plug-in" as opposed to "bolt-in" types used in panelboards. Manufacturers' catalogs will show both load centers and panelboards. The UL standards do not differentiate.

Locked Rotor Current: The steady-state current taken from the line with the rotor locked and with rated voltage and frequency applied to the motor.

Luminaire: A complete lighting unit consisting of lamp or lamps together with the parts designed to distribute the light, to position and protect the lamps, and to connect the lamps to the power supply. This is a lighting fixture.

Mandatory Rules: Required. The terms *shall* or *shall not* are used when a Code statement is mandatory.

Maximum Continuous Current (MCC):   A value that is determined by the manufacturer of hermetic refrigerant motor compressors under high load (high refrigerant pressure) conditions. This value is established in the *NEC*® as being no greater than 156% of the marked rated amperes or the branch-circuit selection current.

Maximum Overcurrent Protection (MOP):   A term used with equipment that has a hermetic motor compressor(s). MOP is the maximum ampere rating for the equipment's branch-circuit overcurrent protective device. The ampere rating is determined by the manufacturer, and is marked on the nameplate of the equipment. The nameplate will also indicate if the overcurrent device can be a HACR-type circuit breaker, a fuse, or either.

Minimum Circuit Ampacity (MCA):   A term used with equipment that has a hermetic motor compressor(s). MCA is the minimum ampere rating value used to determine the proper size conductors, disconnecting means, and controllers. MCA is determined by the manufacturer, and is marked on the nameplate of the equipment.

National Electrical Code (*NEC*®): The electrical code published by the National Fire Protection Association. This Code provides for practical safeguarding of persons and property from hazards arising from the use of electricity. It does not become law until adopted by federal, state, and local laws and regulations. The *NEC*® is not intended as a design specification nor an instruction manual for untrained persons.

National Electrical Manufacturers Association (NEMA):   NEMA is a trade organization made up of many manufacturers of electrical equipment. They develop and promote standards for electrical equipment.

National Fire Protection Association (NFPA):   Located in Quincy, MA. The NFPA is an international Standards Making Organization dedicated to the protection of people from the ravages of fire and electric shock. The NFPA is responsible for developing and writing the *National Electrical Code*, The Sprinkler Code, The Life Safety Code, The National Fire Alarm Code, and over 295 other codes, standards, and recommended practices. The NFPA may be contacted at (800) 344-3555 or on their web site at www.nfpa.org.

Nationally Recognized Testing Laboratory (NRTL): The term used to define a testing laboratory that has been recognized by OSHA. For example, Underwriters Laboratories.

Neutral: In residential wiring, the neutral conductor is the grounded conductor in a circuit consisting of three or more conductors. There is no neutral conductor in a two-wire circuit, although electricians many times refer to the white grounded conductor as the neutral.

The voltages from the ungrounded conductors to the grounded conductor are of equal magnitude.

See the definition of "Multiwire Branch-Circuit" in the *NEC*.®

No Niche Lighting Fixture*: A lighting fixture intended for installation above or below the water without a niche.

Noncoincidental Loads: Loads that are not likely to be on at the same time. Heating and cooling loads would not operate at the same time. See *Section 220-21* of the *NEC*.®

Notations: Words found on plans to describe something.

Occupational Safety and Health Act (OSHA): This is the Code of Federal Regulations, developed by the Occupational Safety and Health Administration, U.S. Department of Labor. The electrical regulations are covered in *Part 1910, Subpart S*. So as not to shorten and simplify the regulations, *Part 1920, Subpart S* contains only the most common performance requirements of the *NEC*.® The *NEC*® must still be referred to, in conjunction with OSHA regulations.

Ohm: A unit of measure for electric resistance. An ohm is the amount of resistance that will allow one ampere to flow under a pressure of one volt.

Outlet*: A point on the wiring system at which current is taken to supply utilization equipment.

Overcurrent*: Any current in excess of the rated current of equipment or the ampacity of a conductor. It may result from overload, short circuit, or ground fault.

Overload*: Operation of equipment in excess of normal, full-load rating, or of a conductor in excess of rated ampacity that, when it persists for a sufficient length of time, would cause damage or dangerous overheating. The current flow is contained in its normal intended path. A fault, such as a short circuit or ground fault, is not an overload.

Overcurrent Device: Also referred to as an overcurrent protective device. A form of protection that operates when current exceeds a predetermined value. Common forms of overcurrent devices are circuit breakers, fuses, and thermal overload elements found in motor controllers.

Panelboard*: A single panel or group of panel units designed for assembly in the form of a single panel; including buses, automatic overcurrent devices, and equipped with or without switches for the control of light, heat, or power circuits; designed to be placed in a cabinet or cutout box placed in or against a wall or partition and accessible only from the front.

Permissive Rules: Allowed but not required. Terms such as *shall be permitted* or *shall not be required* are used when a Code statement is permissive.

Plans: Views of horizontal planes, showing components in their horizontal relationship. A set of construction drawings.

Power Supply: A source of electrical operating power including the circuits and terminations connecting it to the dependent system components.

Qualified Person*: One familiar with the construction and operation of the equipment and the hazards involved.

Raceway*: An enclosed channel of metal or nonmetallic materials designed expressly for holding wires, cables, or busbars, with additional functions as permitted in this Code. Raceways include, but are not limited to, rigid metal conduit, rigid nonmetallic conduit, intermediate metal conduit, liquidtight flexible conduit, flexible metallic tubing, flexible metal conduit, electrical nonmetallic tubing, electrical metallic tubing, underfloor raceways, cellular concrete floor raceways, cellular metal floor raceways, surface raceways, wireways, and busways.

Rated-Load Current: The rated-load current for a hermetic refrigerant motor-compressor is the current resulting when the motor-compressor is operated at the rated load, rated voltage, and rated frequency of the equipment it serves.

Rate of Rise Detector: A device that responds when the temperature rises at a rate exceeding a predetermined value.

Receptacle*:   A receptacle is a contact device installed at the outlet for the connection of a single contact device. A single receptacle is a single contact device with no other contact device on the same yoke. ►A multiple receptacle is two or more contact devices on the same yoke.◄

Receptacle Outlet*: An outlet where one or more receptacles are installed.

Resistive Load: An electric load that opposes the flow of current. A resistive load does not contain cores or coils of wire. Some examples of resistive loads are the electric heating elements in an electric range, ceiling heat cables, and electric baseboard heaters.

Root-Mean-Square (RMS): The square root of the average of the square of the instantaneous values of current or voltage. For example, the RMS value of voltage line to neutral in a home is 120 volts. During each electrical cycle, the voltage rises from zero to a peak value ($120 \times 1.4142 = 169.7$ volts), back through zero to a negative peak value, then back to zero. The RMS value is 0.707 of the peak value ($169.7 \times 0.707 = 120$ volts). The root-mean-square value is what an electrician reads on an ammeter or voltmeter.

Schedules: Tables or charts that include data about materials, products, and equipment.

Sections: Views of vertical cuts through and perpendicular to components, showing their detailed arrangement.

Series-Rated System: Panelboards are marked with their short-circuit rating in RMS symmetrical amperes. A series-rated panelboard will be determined by the main circuit breaker or fuse, and branch-circuit breaker combination tested in accordance to UL *Standard 489*. The series-rating will be less than or equal to the interrupting rating of the main overcurrent device, and greater than the interrupting rating of the branch-circuit overcurrent devices. (See "Fully-Rated System.")

Service*:   ►The conductors and equipment for delivering energy from the serving utility to the wiring system of the premises served.◄

Service Conductors*:   ►The conductors from the service point to the service disconnecting means.◄

Service Drop*:   The overhead service conductors from the last pole or other aerial support to and including the splices, if any, connecting to the service-entrance conductors at the building or other structure.

Service Equipment*:   The necessary equipment, usually consisting of a circuit breaker or switch and fuses, and their accessories, located near the point of entrance of supply conductors to a building or other structure, or an otherwise defined area, and intended to constitute the main control and means of cutoff of the supply.

Service Lateral*:   The underground service conductors between the street main, including any risers at a pole or other structure or from transformers, and the first point of connection to the service-entrance conductors in a terminal box or meter or other enclosure with adequate space, inside or outside the building wall. Where there is no terminal box, meter, or other enclosure with adequate space, the point of connection shall be considered to be the point of entrance of the service conductors into the building.

Service Point*:   Service point is the point of connection between the facilities of the serving utility and the premises wiring.

Shall: Indicates a mandatory requirement.

Short Circuit: A connection between any two or more conductors of an electrical system in such a way as to significantly reduce the impedance of the circuit. The current flow is outside of its intended path, thus the term "short circuit." A short circuit is also referred to as a fault.

Should: Indicates a recommendation or that which is advised but not required.

Smoke Alarm: A single or multiple station alarm responsive to smoke.

Smoke Detector: A device that detects visible or invisible particles of combustion.

Specifications: Text setting forth details such as description, size, quality, performance, and workmanship, etc. Specifications that pertain to all of the construction trades involved might be subdivided into "General Conditions" and "Supplemental General Conditions." Further subdividing of the Specifications might be specific requirements for the various contractors such as electrical, plumbing, heating, and masonry, etc. Typically, the electrical specifications are found in Division 16.

Split Circuit Receptacle: A receptacle that can be connected to two branch-circuits. These receptacles may also be used so that one receptacle is live at all times, and the other receptacle is controlled by a switch. The terminals on these receptacles usually have breakaway tabs so the receptacle can be used either as a split circuit receptacle or as a standard receptacle.

Standard Network Interface (SNI):   A device usually installed by the telephone company at the demarcation point where their service leaves off and the customers service takes over. This is similar to the service point for electrical systems.

Surge Suppressor: A device that limits peak voltage to a predetermined value when voltage spikes or surges appear on the connected line.

Switches*:

> **General-Use Snap Switch:**   A form of general-use switch constructed so that it can be installed in device boxes or on box covers, or otherwise used in conjunction with wiring systems recognized by this Code.

> **General-Use Switch:**   A switch intended for use in general distribution and branch-circuits. It is rated in amperes, and it is capable of interrupting its rated current at its rated voltage.

> **Motor-Circuit Switch:**   A switch, rated in horsepower, capable of interrupting the maximum operating overload current of a motor of the same horsepower rating as the switch at the rated voltage.

Symbols: A graphic representation that stands for or represents another thing. A symbol is a simple way to show such things as lighting outlets, switches and receptacles on an electrical Plan. The American Institute of Architects has developed a very comprehensive set of symbols that represent just about everything used by all building trades. When an item cannot be shown using a symbol, then a more detailed explanation using a notation or inclusion in the Specifications is necessary.

Terminal: A screw or a quick-connect device where a conductor(s) is intended to be connected.

Thermocouple: A pair of dissimilar conductors so joined at two points that an electromotive force is developed by the thermoelectric effects when the junctions are at different temperatures.

Thermopile: More than one thermocouple connected together. The connections may be series, parallel, or both.

Transient Voltage Surge Suppressors (TVSS):   A device that clamps transient voltages by absorbing the major portion of the energy (joules) created by the surge, allowing only a small, safe amount of energy to enter the actual connected load.

UL: Underwriters Laboratories is an independent not-for-profit organization that develops standards, and tests electrical equipment to these standards.

UL Listed: Indicates that an item has been tested and approved to the standards established by UL for that particular item. The UL Listing Mark may appear in various forms, such as the letters UL in a circle. If the product is too small for the marking to be applied to the product, the marking must appear on the smallest unit container in which the product is packaged.

UL Recognized: Refers to a product that is incomplete in construction features or limited in performance capabilities. A "Recognized" product is intended to be used as a component part of equipment that has been "listed." A "Recognized" product must not be used by itself. A UL product may contain a number of components that have been "Recognized."

Ungrounded Conductor: The conductor of an electrical system that is not intentionally connected to ground. This conductor is referred to as the "hot" or "live" conductor.

Volt: The difference of electric potential between two points of a conductor carrying a constant current of one ampere, when the power dissipated between these points is equal to one watt. A voltage of one volt can push one ampere through a resistance of one ohm.

Voltage (of a circuit)*: The greatest root-mean-square (effective) difference of potential between any two conductors of the circuit concerned.

Voltage (nominal)*: A nominal value assigned to a circuit or system for the purpose of conveniently designating its voltage class (e.g., 120/240 volts, 480Y/277 volts, 600 volts).

The actual voltage at which a circuit operates can vary from the nominal within a range that permits satisfactory operation of equipment.

Voltage to Ground*: For grounded circuits, the voltage between the given conductor and that point or conductor of the circuit that is grounded; for ungrounded circuits, the greatest voltage between the given conductor and any other conductor of the circuit.

Voltage Drop: Also referred to as IR drop. Voltage drop is most commonly associated with conductors. A conductor has resistance. When current is flowing through the conductor, a voltage drop will be experienced across the conductor. Voltage drop across a conductor can be calculated using Ohm's Law: $E = IR$.

Volt-ampere: A unit of power determined by multiplying the voltage and current in a circuit. A 120-volt circuit carrying 1 ampere is 120 volt-amperes.

Watt: A measure of true power. A watt is the power required to do work at the rate of 1 joule per second. Wattage is determined by multiplying voltage times amperes times the power factor of the circuit: $W = E \times I \times PF$.

Watertight: Constructed so that moisture will not enter the enclosure under specified test conditions.

Weatherproof*: Constructed or protected so that exposure to the weather will not interfere with successful operation.

(FPN): Rainproof, raintight, or watertight equipment can fulfill the requirements for weatherproof where varying weather conditions other than wetness, such as snow, ice, dust, or temperature extremes, are not a factor.

Wet Niche Lighting Fixture*: A lighting fixture intended for installation in a forming shell mounted in a pool.

The following list of Internet World Wide Web sites has been included for your convenience. These are current as of time of printing. Web sites are a ìmoving targetî with continual changes. We will do our utmost to keep these web sites current. If you are aware of Web sites that should be added, deleted, or that are different than shown, please let us know. Most manufacturers will provide catalogs and technical literature upon request. You will be amazed at the amount of electrical equipment information that is available on these web sites.

Electrical Industry *(This search engine links to over 800 sites related to the electrical industry.)*

www.electricalsearch.com

ACCUBID	www.accubid.com
Advance Transformer Co.	www.advancetransformer.com
AFC Cable Systems	www.afcweb.com
Alflex	www.alflex.com
Allen-Bradley	www.ab.com
Allied Moulded Products, Inc.	www.alliedmoulded.com
Allied Tube & Conduit	www.industry.net/allied.tube.conduit
Alpha Wire	www.alphawire.com
American Lighting Association	www.americanlightingassoc.com
American Insulated Wire Corp.	www.aiwc.com
American National Standards Institute	www.ansi.org
American Society of Heating, Refrigerating, and Air-Conditioning Engineers	www.ashrae.org
American National Standards Institute	www.ansi.org
AMP	www.amp.com
Amprobe	www.amprobe.com
Angelo	www.angelobrothers.com
Appleton Electric Co.	www.appletonelec.com
Arrow Hart	www.crouse-hinds.com
Associated Builders & Contractors	www.abc.org
Automatic Switch Co. (ASCO)	www.asco.com
Baldor Motors & Drives	www.baldor.com
Belden	www.belden.com
Best Power	www.bestpower.com
BOCA	www.bocai.org
Brady	www.whbrady.com
Bridgeport	www.bpfittings.com
Broan	www.broan.com
Burndy	www.burndy.com
Bussmann	www.bussmann.com
CABO	www.cabo.org/
Canadian Standards Association	www.csa.ca
CEBus Industry Council	www.cebus.org
CEE News magazine	www.ceenews.com/
Certified Ballast Manufacturers	www.certbal.org

Chromalox	www.chromalox.com
Construction Specifications Institute	www.csinet.org
Continental Automated Building Association	www.caba.org
Cooper Hand Tools	www.coopertools.com
Cooper Power Systems	www.cooperpower.com
Copper Development Association Inc.	www.copper.org
Copper information	www.copper.org
Crouse-Hinds	www.crouse-hinds.com
Crouse-Hinds (Hazardous locations)	www.zonequest.com
Daniel Woodhead Company	www.danielwoodhead.com
Delmar Publishers	www.thomson.com/delmar/index.htm
Dual-Lite	www.dual-lite.com
Eagle Electric	www.eagle-electric.com
Eaton/Cutler-Hammer	www.cutlerhammer.eaton.com
EC&M magazine	www.ecmweb.com
Edwards	www.edwards-signals.com
Electrical Contractor magazine	www.ecmag.com
Erico, Inc.	www.erico.com
Estimation Inc.	www.estimation.com
Factory Mutual Insurance Co.	www.factorymutual.com
Federal Pacific Co.	www.electro-mechanical.com/fpc.html
Ferraz	www.ferraz.com
First Alert	www.FirstAlert.com
Fluke Corporation	www.fluke.com
F. W. Dodge	www.fwdodge.com
Gardner Bender	gardnerbender.com
GE Electrical Distribution & Control	www.ge.com/edc
GE Lighting	www.ge.com/lighting/business
Generac Power Systems	www.generac.com
GE Total Lighting Control	www.ge.com/tlc
Gould Shawmut	www.gouldshawmut.com
Greenlee Textron Inc.	www.greenlee.textron.com
Halo (Cooper Lighting)	www.cooperlighting.com
Heyco Molded Products, Inc.	www.heyco.com
Hilti	www.hilti.com
Home Automation Association	www.creator.hometeam.com/haa
HomeStar Wiring Systems	www.lucent/netsys/homestar
Home Toys (Home automation)	www.hometoys.com
Honeywell	www.Honeywell.com
Housing Urban Development	www.HUD.gov
Houston Wire & Cable Company	www.houwire.com
Hubbell Incorporated	www.hubbell.com
Hubbell Incorporated (Lighting)	www.hubbell-ltg.com
Hubbell Incorporated (Premise)	www.hubbell-premise.com
Hubbell Incorporated (Wiring)	www.hubbell-wiring.com
Hunt Dimmers	www.huntdimming.com
Husky	www.mphusky.com
ICBO	www.icbo.org
Ideal Industries, Inc.	www.idealindustries.com

ILSCO	www.ilsco.com
Independent Elcctrical Contractors	www.ieci.org
Institute of Electrical and Electronic Engineers, Inc.	www.ieee.org
inter.Light, Inc.	www.lightsearch.com
Intermatic	www.intermatic.com
International Association of Electrical Inspectors (IAEI)	www.iaei.com
Juno Lighting, Inc.	www.junolighting.com
Klein Tools	www.klein-tools.com
Kohler Generators	www.kohlergenerators.com
Leviton Manufacturing Co.	www.leviton.com
Lightolier	www.lightolier.com
Lithonia Lighting	www.lithonia.com
Littelfuse, Inc.	www.littelfuse.com
Lutron Electronics Co., Inc.	www.lutron.com
MagneTek Lighting Products	www.magnetek.com/ballast
Manhattan/CDT	www.manhattancdt.com
McGill	www.mcgillelectrical.com
Megger	www.avointl.com
Midwest Electric Products, Inc.	www.midwestelectric.com
Milbank Manufacturing Co.	www.milbankkmfg.com
Minerallac	www.minerallac.com
Motorola Lighting	www.motorola.com/aegg/mli
MYTECH	www.lightswitch.com
National Association of Electrical Distributors	www.naed.org
National Electrical Contractors Association (NECA)	www.necanet.org
National Electrical Manufacturers Association (NEMA)	www.nema.org
National Fire Protection Association	www.nfpa.org
National Joint Apprenticeship and Training Committee	www.njatc.org
NuTone	www.nutone.com
Occupational Safety & Health Act	www.osha.gov
Onan Corporation	www.onan.com
Optical Cable Corporation	www.occfiber.com
Osram Sylvania, Inc.	www.sylvania.com
Panasonic Lighting	www.panasonic.com/lighting
Panduit Corporation	www.panduit.com
Paragon Electric Company, Inc.	www.paragonelectric.com
Penn-Union	www.penn-union.com
Philips Lamps	www.ALTOlamp.com
Philips Lighting Company	www.lighting.philips.com/nam
Progress Lighting	www.progresslighting.com
RACO	www.racoinc.com
Russelectric	www.russelectric.com
SBCCI	www.sbcci.org
S&C Electric Company	www.sandc.com

Siemens-Furnas Controls	www.furnas.com
Simplex	www.simplexnet.com
Sola/Hevi Duty	www.sola-hevi-duty.com
Southwire Co.	www.southwire.com
Square D Company	www.squared.com
Sylvania	www.sylvania.com
Thomas & Betts Corporation	www.tnb.com
Trade Service	www.tra-ser.com
Underwriters Laboratories	www.ul.com
Underwriters Laboratories (Code Authority)	www.ul.com/auth/index.html
Underwriters Laboratories (Contact)	www.ul.com/contact.html
Waukesha Electric Systems	www.waukeshaelectric.com
Wheatland Tube Company	www.wheatland.com
Wiremold	www.wiremold.com
X-10 Pro	www.x10pro.com

Note: Page numbers in **bold type** reference non-text material.